2022 IEEE 7th Optoelectronics Global Conference (OGC 2022)

Shenzhen, China
6 – 11 December 2022

IEEE Catalog Number: CFP22D33-POD
ISBN: 978-1-6654-8699-6

**Copyright © 2022 by the Institute of Electrical and Electronics Engineers, Inc.
All Rights Reserved**

Copyright and Reprint Permissions: Abstracting is permitted with credit to the source. Libraries are permitted to photocopy beyond the limit of U.S. copyright law for private use of patrons those articles in this volume that carry a code at the bottom of the first page, provided the per-copy fee indicated in the code is paid through Copyright Clearance Center, 222 Rosewood Drive, Danvers, MA 01923.

For other copying, reprint or republication permission, write to IEEE Copyrights Manager, IEEE Service Center, 445 Hoes Lane, Piscataway, NJ 08854. All rights reserved.

*** This is a print representation of what appears in the IEEE Digital Library. Some format issues inherent in the e-media version may also appear in this print version.

IEEE Catalog Number: CFP22D33-POD
ISBN (Print-On-Demand):   978-1-6654-8699-6
ISBN (Online):   978-1-6654-8698-9

Additional Copies of This Publication Are Available From:

Curran Associates, Inc
57 Morehouse Lane
Red Hook, NY 12571 USA
Phone: (845) 758-0400
Fax: (845) 758-2633
E-mail: curran@proceedings.com
Web: www.proceedings.com

Proceedings of the 7th Optoelectronics Global Conference (OGC 2022)

Table of Contents

Preface...ix
Conference Committees...x

✧ Laser Technology

Comparison of Quasi-four-frequency and Four-frequency Zeeman Laser Gyros with Different Types of Biasing
....1

Evgenii Kuznetsov, Yury Golyaev, Yury Kolbas, Igor Savelev, Tatiana Soloveva

7-Core Erbium-Ytterbium Co-dopped Microstructured Fiber Amplifier.....................6

Yifan Zhang, Yifei Zhao, Weichao Ma, Guiyao Zhou

Study on the Influence of Deposition Temperature on the Properties of Lanthanum Titanate Films.....................10

Yang Li, Junqi Xu, Junhong Su, Zheng Liu

ErYb Co-doped Double-clad Fiber Amplifiers with Average Gain of 29dB by High Concentration Doping........15

Yifei Zhao, Yifan Zhang, Shizhuo Xi, Gui Yao Zhou

Ultra Narrow Linewidth Distributed Feedback Fiber Laser Based on Self-injection Locking......................19

Meng Zou, Kai Shen, Qizhen Sun, Zhijun Yan

Er^{3+}-Pr^{3+}-Yb^{3+} Tri-doped La_2O_3-Al_2O_3-SiO_2 Glass Double Clad Fiber for C+L Amplification.....................22

Zhuoyuan Huang, Weichao Ma, Tong Wu, Jia ao Lu, Jiantao Liu, Changming Xia, Zhiyun Hou, Guiyao Zhou

✧ Optical Communication and Networks

Evaluation Method of Polarization State Characteristic in Forward Transmission 25

Zeng Xiangwei, Chen Xueye, Zhang Quanzhong, Li Xiaoyu, Li Yahong

High-Resolution Microwave Frequency Measurement Based on Optical Frequency Comb and Image Rejection Photonics Channelized Receiver.....................30

Ximin Wang, Yingxi Miao, Jialiang Chen, Caili Gong, Yongfeng Wei, Yuqing Yang

A SDN Enabled PON Controller based on Hierarchical Models.....................36

Zhenming Liang, Ziyao Yang, Jian Tang

Study of Filter-based Neuromorphic Photonic Reservoir Computing for Signal Equalization in 224Gbps Sub-carrier Modulation IM-DD Short Reach Optical Fiber Communication System 41

Penghao Luo, An Yan, Aolong Sun, Guoqiang Li, Sizhe Xing, Jianyang Shi, Ziwei Li, Chao Shen, Junwen Zhang, Nan Chi

Coupling Efficiency Analysis for Optical Fiber with Different Core Diameters 45

Yuzhong Ma, Zijing Huang, Lin Sun, Gordon Ning Liu

Underwater Wireless Optical Communication Channel Characterization Using Machine Learning Techniques 50

Abdulaziz Al-Amodi, Mudassir Masood, M. Z. M. Khan

Joint Optimization of Multidimensional Resources Allocation in Cloud Networking 55

Jialong Li, Kangqi Zhu, Nan Hua, Chen Zhao, Yanhe Li, Xiaoping Zheng, Bingkun Zhou

Optical Labels Enabled Optical Network Performance Monitoring 60

Tao Yang

Equalization for Optical PAM Data Center Interconnects 63

Gordon Ning Liu, Lin Sun, Caoyang Liu, Jiawang Xiao, Yi Cai, Gangxiang Shen

✧ Near-infrared, Mid-infrared and Far-infrared Technologies and Applications

Research on 5.5 μm Infrared Filter Applied to Infrared Thermometer 66

Suotao Dong, Xiuhua Fu

Dark Current Analysis in Type-II InAs/GaSb Superlattice LWIR Detector with M-structure Barrier 70

Dongpei Shen, Tong Sun, Pengfei Zhu, Xiaoning Guan, Baonan Jia, Haizhi Song, Pengfei Lu

✧ Quantum Optics and Information

A Single-laser System for Mobile Cold Atom Gravimeter 75

Pei Dongliang, Kong Delong, Wang Jieying, Chen Weiting, Lu Xiangxiang, Wei Junxin, Liu Weiren

✧ Fiber-Based Technologies and Applications

Intensity Compensation of Echo Pulses for Fiber Interferometers Based on uwFBG Reflectors 81

Yandong Pang, Junbin Huang, Hongcan Gu, Su Wu, Zhiqiang Zhang

Strain and Temperature Discrimination by Fourier Analyzing Transmission Spectrum of an In-fiber Mach-Zehnder Interferometer 86

Shiying Xiao, Beilei Wu, Zixiao Wang, Youchao Jiang, Chunran Sun

High Brightness Ultra Wideband Fiber Source 91

    Tongle Yuan, Yu Cheng, Ming Chen, Libo Yuan, Sumei Huang, Jing Li

BOTDR Denoising by Sparse Representation Algorithm with Preformed Dictionary.........................96

    Yuting Liu, Zhijie Sun, Ning Cui, Qing Bai, Yu Wang, Baoquan Jin

Characterization of Various Bound State Solitons Using Linear Optical Sampling Technique 101

    Jingwen Li, Zhichao Wu, Zhe Yu, Chaoyu Xu, Tianye Huang, Songnian Fu

Highly Sensitive Multi-coating Photonic Crystal Fiber Biosensor at Near-Infrared Waveband.................105

    Duanming Li, Wei Zhang, Jiangfei Hu, Minxue Gu

A Sensitive Material for Optical Fiber Sensor--$Dy_8Fe_{16-x}Co_x$ (x=0,2,3): First-principles Calculations.........110

    Yue Yuan, Tao Shen, Chi Liu, Tianyu Yang, Ai Na Gong

Isopropanol-sealed Cascaded-Peanut Taper fiber Structure for Temperature Sensing Incorporated Fiber Laser..115

    Weihao Lin, Jie Hu, Siming Sun, Perry Ping Shum, Fang Zhao, Changyuan Yu, Liyang Shao

The Micro-control Refractive Index Sensor of Dual-metal Antiresonance Optical Fiber 119

    Boyao Li, Tianrong Huang

Non-Invasive Optical Fiber Sensing Vital Signs Monitoring Based on Envelope Extraction BCG Data Processing...124

    Hanyu Zhao, Guo Zhu, Fei Liu, Xiaojun Liu, Jinhui Yuan, Xian Zhou

Accurate Measurement of Large Strain under High-temperature Environment Based on Fiber Bragg Grating 128

    Zhiyuan Wang, Jindong Wang, Tao Zhu

Design of a Hollow-core Microstructured Optical Fiber with Low Loss and High Polarization-maintaining...133

    Aoyan Zhang, Zhipeng Deng, Jialong Li, Guiyao Zhou, Perry Ping Shum

Interference Fading Suppression for Multi-frequency Φ-OTDR 137

    Yu Wang, Junhong Wang, Bin Liang, Yan Li, Qing Bai, Baoquan Jin

Taper Optical Fiber for Distributed Light-driven Soft Robots...142

    Minghui Niu, Ziyan Zhao, Jiayuan Min, Jie Hu, Huanhuan Liu, Dan Luo, Liyang Shao, Perry Ping Shum

Comparison of the Simulation Algorithms for Nonlinear Pulse Propagation in Multimode Fibers.........145

    Jiayu Lu, Lili Kong, Xiaosheng Xiao

Hollow Core Bragg Fiber-based Gas Pressure Sensor Using Parallel Fabry-Perot Interferometers..........149

    Zongru Yang, Weihao Yuan, Changyuan Yu

✧ Optoelectronic Devices and Applications

Self-adjusting Light Source Based on a Dual-Function GaN Light-Emitting Diode 154

    Yumeng Luo, Jiahao Yin, Kwai Hei Li

Modeling and Analysis of Zinc Diffusion Effect within InP-Based Mach-Zehnder Modulators..................158

    Ruoyun Yao, Wanshu Xiong, Zhangwan Peng, Yiti Xiong, Chaodan Chi, Chen Ji

Advanced Getter Solutions for Gas Contaminants Absorption in Optoelectronic Devices..........................163

    Giovanni Zafarana, Enea Rizzi, Luca Mauri, Alessio Corazza, Marco Moraja

Modeling and Analysis of High-Speed Modified Uni-Travelling-Carrier Photodiodes Under High Optical Power Injection..........................167

    Zhangwan Peng, Wanshu Xiong, Ruoyun Yao, Chaodan Chi, Yiti Xiong, Chen Ji

Research and Application System Design of Intelligent Inspection of Multispectral Segment Optoelectronic Devices Based on 5G..........................171

    Tielin Lu, Lei Yue, Bowen Lu, Xiawei Feng, Hongqi Han

Portable Microscopic Phase Retrieval System Using the Transport of Intensity Equation on Android Platform..........................177

    Yu Chen, Hong Cheng, Zhengguang Tian, Xunting Yang, Fen Zhang, Wei Li

Visible and Infrared Luminescence and Applications of Er-doped AlN Thin Films..........................183

    Zhiyuan Wang, Feihong Zhang, Sergii Golovynskyi, Zhenhua Sun, Baikui Li, Honglei Wu

Wave-front Coding Technology to Extend Depth of Field in Remote Sensing Optical System.................187

    Qixiang Gao, Yuanhang Wang, Xing Zhong, Yu Li

Influence of Target Layout on the Accuracy of Monocular Vision 6DOF Spatial Pose Measurement......191

    Liu Xingtan, Hua Baocheng, Liu Qihai, Deng Loulou, Liu jing, Tao Liqing, Wu Xiuyu, Lei Kaiyu

Design of LED Array Control Module for Optical Camera Communication..........................195

    Han Liu, JianPing Wang, HuiMin Lu

✧ Biophotonics and Optical Biomedicine

Self-Supervised Denoising of single OCT image with Self2Self-OCT Network 200

    Chenkun Ge, Xiaojun Yu, Mingshuai Li, Jianhua Mo

Adaptive Dynamic Analysis-based Optical Coherence Tomography Angiography for Blood Vessel Tail Artifacts Suppression 205

Junxiong Zhou, Yuntao Li, Jianbo Tang

✧ Data Center Optical Interconnects and Networks

Ultra-stable and Low-complexity Retiming Technique for Bandwidth-limited 112-Gbps PAM-4 Systems 210

Lin Sun, Luxiao Zhang, Yi Cai, Gangxiang Shen, Gordon Ning Liu, Bin Chen

Application-aware Configuration of All-optical Interconnects in Hyper-FleX-LION 214

Hao Yang, Zuqing Zhu

✧ Silicon Photonics

Flexible Dispersion Engineering in Thin GaP-OI Frequency Comb Resonator Design 218

Zhaoting Geng, Houling Ji, Zhuoyu Yu, Weiren Cheng, Yi Li, Qiancheng Zhao

A Solid-state FMCW Lidar System Based on Lens-assisted Beam Steering 222

Xianyi Cao, Kan Wu, Chao Li, Tianyi Li, Jiaxuan Long, Jianping Chen

Silicon Photonic Integrated Reservoir Computing Processor with Ultra-high Tunability for High-speed IM/DD Equalization 227

Aolong Sun, An Yan, Penghao Luo, Junwen Zhang, Nan Chi

Double-tip Scandium Aluminum Nitride Edge Couplers at 1550 nm Wavelength 231

Hengyu Wang, Xingyan Zhao, Shaonan Zheng, Zhengji Xu, Yuan Dong, Ting Hu

Optomechanical Cavity for Electrical Voltage Sensing 235

Qiong Yao, Xia Ji, Fuyin Wang, Chunyan Cao, Shuidong Xiong

✧ Emerging Technologies for Wide Bandgap Semiconductors and Information Displays

A Solver for Devices of Subwavelength Lamellar Gratings 239

Zhuang Wang, Chuan Shen, Liu Wang, Bin Wang, Daofeng He, Sui Wei

✧ Translational Photomedicine and Biophotonics

Optically Levitated Conveyor Belt Based on Specular-Reflection Photonic Nanojet 245

Feng Xu, Song Zhou, Jiahui Zhang, Guanghui Wang, Fei Xu

Optimal Raman Spectral Classifcation Model Based on Differentiable Architecture Search of Hybrid Structure Network for Disease Diagnosis......248

Jiaqi Hu, Jinna Chen, Chenlong Xue, Yanqun Xiang, Guoying Liu, Hong Dang, Dan Lu, Huanhuan Liu, Longqing Cong, Zhen Gao, Haibin Su, Perry Ping Shum

✦ THz Metamaterials and Device Applications

Chiral-Selective Transmission of Edge States in Terahertz Valley Topological Photonic Crystals......253

Hongyi Li, Jiajun Ma, Shilei Liu, Yi Liu, Chunmei Ouyang

Tuning Performance and Mechanism of Gate-tuned Graphene Grating for Dynamically Controlling Terahertz Wavefront 257

Qianqian Wang, Xiaotong Li, Jie Liang, Runze Li

Switchable Multifunctional Metasurfaces Based on Vanadium Dioxide in the Terahertz Region......261

Shilei Liu, JiaJun Ma, Hongyi Li, Yi Liu, Chunmei Ouyang

✦ Optical Fiber Upgrade

Numerical Simulation of C+L Broadband Single-mode Fiber 265

G. H. Zhang, W. Sun, F. Lei, W. Chen, L. Wang, Y. L. Wang, Y. T. Li, H. F. Guan, J. C. Yuan, Z. Jiang, Q. Y. Liu

✦ Computational Imaging

Coherence Retrieval and Multi-contrast Microscopy Imaging by Transport of Intensity Stack......270

Runnan Zhang, Zewei Cai, Chao Zuo

GS Iterative Phase Retrieval Algorithm Based on Fusion of Spatial Phase Gradient Descent and Frequency Domain Amplitude Linear Weighting......274

Hong Cheng, Haonan Zheng, Siwei Sun

Laser Weld Seam Tracking Sensing Technology Based on Swing Mirror 278

Tian Changyong, Song Xuhao, Yin Tie, Zhang Yi

Author Index

Preface

This issue of Proceedings gathers the papers presented at 2022 Optoelectronics Global Conference (OGC 2022) held virtually from December 6 to 11, 2022. The conference is sponsored by IEEE Photonics Society Guangdong Chapter, hosted by Department of Electrical and Electronic Engineering, Southern University of Science and Technology. Due to the long-term impact of COVID-19, the conference is held online. There are 9 symposias, 3 special sessions and 2 workshops arranged at the conference with the topics covering precision optics, optical communications, lasers, infrared applications, fiber sensors and so on.

Starting in 2015, OGC has achieved brilliant results, which successfully brought researchers from around the world to meet and exchange research ideas. As an annual conference, it is not only the integration and interpenetration of multiple sciences, but also widely applied in various scientific fields, such as infrared applications, military optoelectronics technology, etc. It offers a significant platform to present the latest research results among scholars and researchers from universities, research institutions, military enterprises, and optoelectronic companies. The aim of the conference is to promote interaction and exchange of various disciplines among professionals in academia and industry at home and abroad. Meanwhile, it serves to turn technologies into industrial applications.

Two plenary speeches are delivered at the conference: "Optoelectronics for Artificial Intelligence" given by Min Gu, who is Executive Chancellor and Distinguished Professor of University of Shanghai for Science and Technology; "Imaging Macromolecules with X-ray Laser Pulses" addressed by Henry Chapman, a director of the Center for Free-Electron Laser Science at the Deutsches Elektronen-Synchrotron and the University of Hamburg in Germany.

The proceedings of OGC 2022 include 63 full papers that were presented at the conference. Those papers are grouped into 14 chapters: Laser Technology, Optical Communication and Networks, .Near-infrared, Mid-infrared and Far-infrared Technologies and Applications, Quantum Optics and Information, Fiber-Based Technologies and Applications, Optoelectronic Devices and Applications, Biophotonics and Optical Biomedicine, Data Center Optical Interconnects and Networks, Silicon Photonics, Emerging Technologies for Wide Bandgap Semiconductors and Information Displays, Translational Photomedicine and Biophotonics, THz Metamaterials and Device Applications, Optical Fiber Upgrade and Computational Imaging. At the conference, 2 plenary speakers, 154 invited speakers from all over the world were invited to give speeches at the conference.

We are very grateful for the participation of all authors who share the highest level of international research to grand challenges in various fields. In addition, it is acknowledged that the honorary chairs, general chairs, technical program chairs, local organizing committee chairs, publicity chairs, and technical committees have put in a lot of efforts in organizing this conference. Without their active action, the conference couldn't be held successfully. Our sincere gratitude also goes to the symposium chairs, special session chairs, workshop chairs, special event chairs and international reviewers, conference secretariat who are dedicated to making the conference run smoothly and properly and ensuring the quality of the proceedings.

Optoelectronics Global Conference welcomes participation and valuable suggestions from all over the world to help us improve the conference. Sincerely look forward to your participation again next year face to face!

Best regards,

OGC Organizing Committee

Conference Committees

Honorary Chairs

Qikun Xue, Southern University of Science and Technology, China
Xiancheng Yang, Chairman of China International Optoelectronic Exposition Organizing Committee Office, China

General Chairs

Perry Shum, Southern University of Science and Technology, China
Qihuang Gong, Peking University, China
Chennupati Jagadish, Australian National University, Australia
John Dudley, Université de Franche-Comté, France
David Neil Payne, University of Southampton, UK

Program Chairs

Dan Luo, Southern University of Science and Technology, China
Sze Y. Set, The University of Tokyo, Japan
Anna Peacock, University of Southampton, UK
Ken Oh, Yonsei University, South Korea
George Humbert, CNRS, France
Neil Broderick, Auckland University, New Zealand
Xiang Zhou, Google, USA
Yiyang Luo, Chongqing University, China

Local Organizing Committee Chairs

Huanhuan Liu, Southern University of Science and Technology, China
Hai Yuan, Shenzhen Institute of Advanced Technology, Chinese Academy of Sciences, China
Zhaohui Li, Sun Yat-Sen University, China
Yunxu Sun, Harbin Institute of Technology, Shenzhen, China
Songnian Fu, Guangdong University of Technology, China

Publicity Chair

Nan Zhang, JPT, China

Treasurer

Gina Chen, Southern University of Science and Technology, China

International Advisory Committee

Songhao Liu, South China Normal University, China
Yunjie Liu, China Unicom Co. Ltd., China
Xun Hou, Xi'an Institute of Optics and Precision Mechanics, Chinese Academy of Sciences, China
Jianquan Yao, Tianjin University, China
Huilin Jiang, Changchun University of Science and Technology, China
Ziseng Zhao, Wuhan Research Institute of Posts and Telecommunications, China
Zhizhan Xu, Shanghai Institute of Optics and Precision Mechanics, Chinese Academy of Sciences, China
Shuisheng Jian, Beijing Jiaotong University, China
Dianyuan Fan, Shenzhen University, China

Lijun Wang, Changchun Institute of Optics and Fine Mechanics and Physics, Chinese Academy of Sciences, China
Wenqing Liu, Anhui Institute of Optics and Fine Mechanics, Chinese Academy of Sciences, China
Shaohua Yu, China Information Communication Technologies Group Corporation, China
Ying Gu, The General Hospital of the People's Liberation Army, China

Proceedings of the 7th Optoelectronics Global Conference (OGC 2022)

Comparison of Quasi-four-frequency and Four-frequency Zeeman Laser Gyros with Different Types of Biasing

Evgenii Kuznetsov[1,2], Yury Golyaev[1], Yury Kolbas[1], Igor Savelev[1], Tatiana Soloveva[1*]

[1]"POLYUS" Research Institute of M.F. Stelmakh, Moscow, Russia
[2]Peoples' Friendship University of Russia, Moscow, Russia
*e-mail: momentmail@mail.ru

Abstract—**Among laser gyros, being the most precise inertial sensors nowadays, multi-oscillator laser gyros of a new generation are especially promising. Instead of mechanical biasing for eliminating frequency lock-in, they use physical optical methods, increasing their accuracy and resistance to external mechanical impacts. Here we consider Zeeman laser gyros, whose important advantage is the "pure optical path" (without intra-cavity elements). Two modifications of such laser gyros are considered – quasi-four-frequency and four-frequency. The results of experimental measurements of the input-output characteristics with different types of rectangular alternating frequency biasing are presented: with a single biasing (meander) and with a combined biasing (two different meanders). Comparative results confirm the advantages of the combined frequency biasing, improving the linearity of the input-output characteristic by 4 times.**

Keywords-laser gyro; nonplanar resonator; Zeeman biasing; four-frequency mode; combined frequency biasing

I. INTRODUCTION

The laser gyros (LG) have become the mainstream sensors for inertial navigation and control systems. Due to the advantages of high sensitivity and resistance to severe environment, they are widely used in aerospace, marine transportation, oil exploration, and other fields. LG are especially demanded in the range of 0.005 – 0.1 °/h bias stability and extra high scale factor stability, unachievable for other types of gyros, such as fiber, hemispheric, MEMS gyros, and others [1-6].

LG operation is based on Sagnac effect, which is described by fundamental formula [7] $f_\Omega = 4S\Omega/\lambda L$, where f_Ω – frequency difference between two counter-propagating light waves in the ring laser; S and L – square and length of the ring laser resonator; Ω – angular rate; λ – wavelength of the laser.

So, the ideal input-output characteristic of a LG is described by simple expression $f_\Omega = k\Omega$, where $k = 4S/\lambda L$ is called a scale factor. For a more compact form of this expression a scale factor k is often included in Ω [7], thereby making the direct relation between the LG output signal and measured angular rate more obvious.

Key problem in LG is synchronization of counter-propagating wave frequencies at the very low rotation rates yielding zero rotation information at the LG output. It is general phenomena in oscillation physics, when waves with

closed frequencies are synchronized. As for LG, implemented in the concrete technical design, the mutual coupling between the counter-propagating waves occurs due to backscattering from the ring laser intra-cavity components. So, the real LG characteristic differs from ideal linear (dotted line), as shown in Fig. 1, where characteristic nonlinearity appears due to the lock-in region.

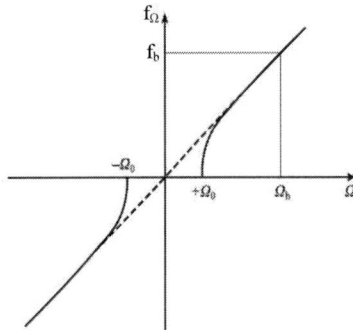

Figure 1. Static input-output characteristic of the laser gyro: Ω_0 – lock-in threshold, Ω_b and f_b – amplitude of the biasing and corresponding frequency difference at the laser gyro output.

The equation which describes input-output characteristic Fig. 1 (outside lock-in region) $f_\Omega = [\Omega^2 - \Omega_0^2]^{1/2}$, where Ω_0 is lock-in threshold is the basic formula for LG [7].

To solve lock-in problem, various types of biasing are used in order to move the working point far from the dead zone (Fig.1). Evident and simple way is permanent rotation of the ring laser – constant unidirectional or alternating (with self-compensation in the output signal). In practical usage, mechanical rapid alternating rotation (dither) is still dominated, as it was a first simple concept realized in mass-produced devices. Although the mechanically dithered gyros reached size and precision requirements for guidance and navigation applications, the moving parts they use result in many disadvantages such as mechanical noise, acoustic noise, dither cross-coupling, coning error. That is why there are many fields in which dither approach is not applicable [5]. Lastly, the usage of a mechanical dither motion on this highly elegant LG goes against the esthetic values of the designers [7]. More science-based solution to avoid lock-in problem is magneto-optical biasing which makes LG a pure optical device without any moving parts. This allows such LG to be used in many additional applications requiring

978-1-6654-8699-6/22 $31.00 © 2022 IEEE

very high precision, low noise and resistance to severe environmental conditions. Since this LG ideally solves the lock-in problem with purely optical method, it has attracted great interests from many companies and universities, [3-6, 8-10], but most of them were not successful in overcoming the technical problems [7].

LGs with magneto-optical biasing are called "laser gyro of second generation" [10]. Important feature of such biasing is the possibility of four-frequency operation, which gives means to radical improving the LG accuracy. In this gyro, two counter-propagating right-circularly polarized (RCP) modes and two counter-propagating left-circularly polarized (LCP) modes are travelling. A reciprocal biasing is used to split the frequency of co-propagating modes having different circular polarizations whereas a non-reciprocal biasing splits the frequencies of counter-propagating modes having the same polarization. General view of four-frequency LG spectrum is presented in Fig. 2.

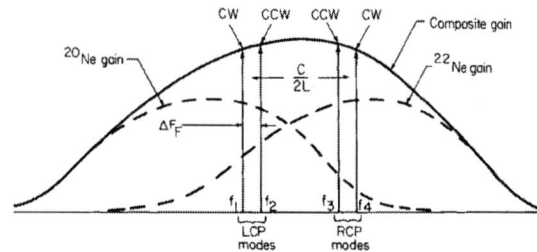

Figure 2. Spectrum of laser gyro with magneto-optical biasing, which provides a splitting of the resonator frequencies into four f_1, f_2, f_3, f_4.

The non-reciprocal biasing can be achieved by Faraday effect in a cell located inside the resonator (ΔF in Fig. 2 [11]) or by Zeeman effect in the gain medium itself [8].

Of Western firms, only Northrop Grumman (after acquired Litton) [5,10] realized an approach with a magneto-optic biasing, but its disadvantage is the introduction of a Faraday cell inside the resonator, which complicates the design and creates a thermo-sensitive, scattering-added, optically inhomogeneous zone in the path of the laser beam.

Various papers were published on development of four-frequency LG with nonplanar resonator in China, and some of them [4,6,12].

"POLYUS" was the first to develop in 1967 an original approach with nonplanar resonator and direct applying magnetic field to the laser medium causing Zeeman effect, and thus creating nonreciprocal biasing of the frequencies of counter-propagating waves [13]. This approach does not require any optical components within the laser resonator, so it is called "clear-path resonator" in later American publications [9]. In Russia, we call it "empty resonator".

To imagine a four-frequency LG operation the device may be interpreted as a hybrid of two "sub-gyros" sharing a common resonator. One "sub-gyro" has RCP waves and the other has LCP waves. The "sub-gyros" are split in frequency using reciprocal biasing due to nonplanar

resonator. Two modes of each "sub-gyro" are split again in frequency using nonreciprocal biasing due to Zeeman effect in the gain medium, which accomplished by the electrical coils wound around the resonator internal channels with He-Ne active medium. Summarizing the output signals of both "sub-gyros" in a definite way enables self-compensation of correlated errors and doubling of the scale factor, thus being the key reason for improving accuracy of the four-frequency LG.

Any kind of alternating biasing, both mechanical and magneto-optical, cannot completely eliminate the nonlinearity of input-output characteristic. The shape of characteristic after introducing the alternating biasing is presented in Fig. 3. In contrast to the characteristic without the biasing, called "static", it is called "dynamic", because the biasing causes a continuous real or virtual rotation of the LG. Instead of a big lock-in region near zero ("static lock-in zone") numerous "dynamic lock-in zones" appear along the characteristic (see Fig. 3).

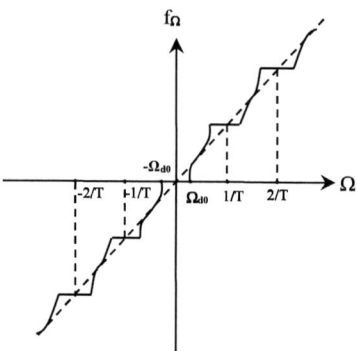

Figure 3. Dynamic input-output characteristic of the laser gyro with alternating biasing, which has switching period T.

As it is seen in Fig. 3, the static lock-in zone is significantly decreased to a much smaller width with threshold Ω_{d0} due to the biasing. It is called the "dynamic zero lock-in zone". For further improvement of the linearity of input-output characteristic, the combined biasing is used, i.e. a sum of two types of biasing. Dithered LG combines alternating rapid rotation with introduction of noise onto the dither amplitude [7]. The dither concept permits only combination of the periodic and noise components and inherently restricts the shapes of the biasing functions, while the magneto-optic biasing technique allows a variety of their shapes to be easily controlled, thus opens effective ways for linearization of characteristic, offering excellent scale factor stability [5,8,10].

The development of Zeeman LG is carried out with all possibilities of improvement in approaches, theory, designs, algorithms, modeling [1,2,14,15-19].

II. ZEEMAN FOUR-FREQUENCY AND QUASI-FOUR-FREQUENCY LASER GYROS IMPLEMENTATION

"POLYUS" Zeeman LG are realized in 2 modifications: two-frequency (later transformed to quasi-four-frequency) and four-frequency. Both modifications are based on the same concept of nonplanar resonators (Fig. 4), but with different angles of nonplanarity, which results different spectrums of resonator frequencies (Fig. 5).

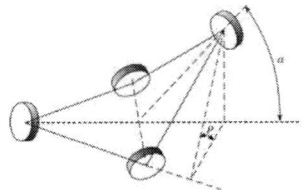

Figure 4. Basic scheme of nonplanar resonator with fold angle α; ρ is the angle between the planes of light beam incidence on the neighboring mirrors.

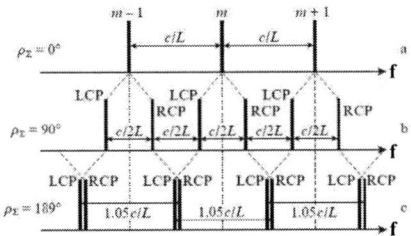

Figure 5. Spectrum of resonators with different angle ρ: a – planar resonator with linear polarization, b – nonplanar two-frequency resonator, c – nonplanar four-frequency resonator.

Each of the modifications of Zeeman LG has a specific design that provides appropriate conditions for the stable existence of two-frequency or four-frequency generation.

Design of two-frequency Zeeman LG is simpler, but lost the advantage of multi-oscillating concept leads to absence of self-compensation of correlated errors, in particular, significant sensitivity to the external magnetic field. To overcome this disadvantage and to increase the Zeeman two-frequency LG accuracy the approach of "quasi-four-frequency" was developed. This approach is provided by electronic switching (reversing) of two pairs of frequencies (modes) with orthogonal circular polarizations in pairs periodically every 2 min. Thus instead of complicating the design the realization of multi-oscillating concept is passed to electronic and software-mathematical methods of the mode switching with further processing of the gyros output signals, that is easily decided at the modern level of microcomputer.

The drawback of quasi-four-frequency approach is short-time loss of information at the moment of mode switching. Basing on state-of-the art level of digital technique this drawback is compensated with algorithmic digital correction by restoration of the lost information using output signal values before and after mode switching. We developed complex algorithms and software for quasi-four-frequency modification, which perform a comprehensive step-by-step correction of LG errors: temperature correction of the nonmagnetic drift component, full compensation of the magnetic drift component and replacing wrong outputs of the LG in the moment of RCP-LCP mode switching with proper data by approximating method. Thus the accuracy of

quasi-four-frequency modification increased up to 15-fold in comparison with two-frequency modification.

In spite of the perfect results for the quasi-four-frequency modification, we continued the development of the true four-frequency modification, since the first modification has a larger sampling error and the signal noise component than the second one. In this regard, according to our estimates, the theoretical accuracy limit for the quasi-four-frequency LG with a perimeter of 20 cm is 0.0035 °/h, whereas for a true four-frequency LG it may be better 0.001 °/h. This is important for high-precision applications, thus the use of four-frequency mode in the Zeeman LG opens up new areas for them.

Let's consider the implementation of two kinds of rectangular alternating biasing (meander) in Zeeman LGs: single and combined meanders. In the case of a single meander the widths of dynamic lock-in zones at the input-output characteristic are determined by the following equations

$$\Omega_{di} = \Omega_d^m |\sin(\pi TA/2)|, i = 0,2,4,\dots$$

$$\Omega_{di} = \Omega_d^m |\cos(\pi TA/2)|, i = 1,3,5,\dots$$

$$\Omega_d^m = (2\Omega_0/\pi TA)\sqrt{1 + \pi^2 \tau A},$$

where Ω_{di} – half-widths of zero, even and odd dynamic zones; T – switching period of the biasing; A – amplitude of the biasing; Ω_d^m – RMS value of the half-widths of neighboring even and odd dynamic lock-in zones; Ω_0 – half-width of the static lock-in zone for a specified mode; τ – duration of the edge of the biasing switching.

The variability of the magneto-optic biasing allows us to reduce even many more times the lock-in zones at longer times compared to switching period of the biasing.

If we introduce by means of a special electrical coil an additional meander, which has a much smaller amplitude and frequency in comparison with the main meander (we add a "slow meander" to the "fast meander"), only small fractional dynamic lock-in zones remain on the characteristic. RMS values of their half-widths Ω_{dj}^s reduce to

$$\Omega_{dj}^s = 2\Omega_{di}/\pi \, T_s \, A_s,$$

where T_s – switching period of the "slow meander"; A_s – amplitude of the "slow meander"; j – number of fractional dynamic lock-in zone between dynamic lock-in zones i and i+1.

III. INPUT-OUTPUT CHARACTERISTICS OF ZEEMAN FOUR-FREQUENCY AND QUASI-FOUR-FREQUENCY LASER GYROS

In the literature, there are no properly derived expressions for the input-output characteristic of the four-frequency Zeeman LG with an alternating biasing, which takes into account the coupling of counter-propagating waves due to backscattering. Here we introduce our results considering this coupling.

In case of different scale factors and bias (null shift) of input-output characteristic for every mode the frequency differences of the counter-propagating waves

$$f_{rcp}^{+} = k_{rcp}(\Omega + \Delta\Omega_{rcp}),$$

$$f_{lcp}^{+} = k_{lcp}(\Omega + \Delta\Omega_{lcp}),$$
$$f_{rcp}^{-} = k_{rcp}(\Omega + \Delta\Omega_{rcp}),$$
$$f_{lcp}^{-} = k_{lcp}(\Omega + \Delta\Omega_{lcp}),$$

where f_{rcp}^{+} and f_{rcp}^{-} – frequency differences of the counter-propagating waves for RCP mode in a positive and a negative half-periods of biasing; f_{lcp}^{+} and f_{lcp}^{-} – frequency differences of the counter-propagating waves for LCP mode in a positive and a negative half-periods of biasing; k_{rcp} and k_{lcp} – scale factors for RCP and LCP modes; $\Delta\Omega_{rcp}$ and $\Delta\Omega_{lcp}$ – biases for RCP and LCP modes.

We defined the half-period of the Zeeman biasing when the sign of the magnetic field is positive as the "positive half-period" and the half-period of the Zeeman biasing when the sign of the magnetic field is negative as the "negative half-period".

The output signal of the four-frequency LG is the algebraic sum of the frequency differences, with its average value for the period f_Ω determined by the equations

$$f_\Omega = \left(f_{rcp}^{+} + f_{lcp}^{+} + f_{rcp}^{-} + f_{lcp}^{-}\right)/2 =$$
$$= k_{rcp}\left(\Omega + \Delta\Omega_{rcp}\right) + k_{lcp}\left(\Omega + \Delta\Omega_{lcp}\right) =$$
$$= \Omega\left(k_{rcp} + k_{lcp}\right) + k_{rcp}\Delta\Omega_{rcp} + k_{lcp}\Delta\Omega_{lcp}.$$

The above equations show that output signal of the four-frequency LG is proportional to the angular rate multiplied by the sum of the scale factors for RCP and LCP modes. These results are also valid for a quasi-four-frequency LG with optimal periods of mode switching and output signal processing.

Experiments for plotting of the Zeeman LG input-output characteristics were carried out on a turntable with angular rate stability not worse than 0.0005%. The Zeeman four-frequency LG with resonator perimeter 20 cm was rigidly fixed on the turntable with the sensitivity axis perpendicular to the Earth's surface.

During the experiment the angular rate of the turntable was changed in the range from - 8.5°/s to + 8.5°/s with a small angular acceleration ($0.01°/s^2$) and the LG output signal f_Ω, corresponding to angular rate, was measured.

At the first stage of experiment we used "fast meander" only with amplitude A = 5400 Hz, switching period T = 1 ms, and switching edge τ = 5 μs at 0.9 level. At the second stage we used combination of "fast meander" and "slow meander" with a large switching period T_s = 1s and small amplitude 200 Hz.

The polarization separation of modes in the ring laser gyro beam combining optics enabled the selection each of the orthogonal mode signals. This allowed us to plot the input-output characteristics as the virtual two-frequency "sub-gyros", as the four-frequency modification. We could

also simulate the quasi-four-frequency modification by processing the outputs of the "RCP sub-gyro" and the "LCP sub-gyro" with reversing. The results are presented as the graphs of f_Ω versus the angular rate Ω for two types of biasing: "fast meander" only and combination of "fast meander" + "slow meander" (Fig. 6 and Fig. 7).

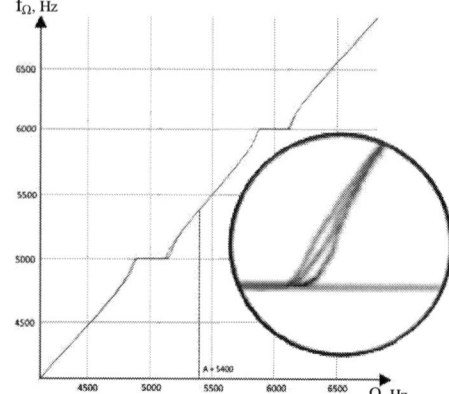

Figure 6. Input-output characteristic of the Zeeman four-frequency laser gyro with the single rectangular alternating biasing ("fast meander"): — RCP mode, — LCP mode, — four-frequency mode.

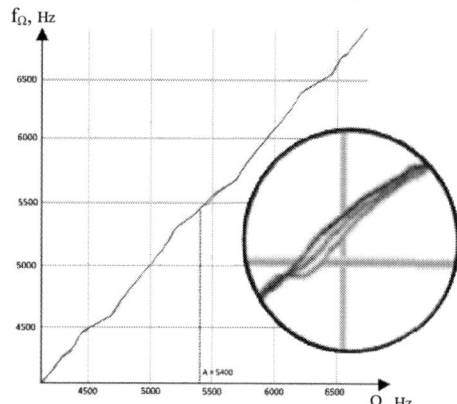

Figure 7. Input-output characteristic of the Zeeman four-frequency laser gyro with combination the "fast meander" + "slow meander": — RCP mode, — LCP mode, — four-frequency mode.

As noted in part I, often the scale factor is included in the angular rate, so we made this, recalculated the value of Ω and specified it in Hz.

The area near the amplitude A of the Zeeman biasing is zoomed and inset in Fig. 6 and Fig. 7.

As it is clearly seen in Fig. 6 and Fig. 7, the input-output characteristic of the Zeeman laser gyro in four-frequency operation is the average between those for RCP and LCP modes, that proves our theoretical calculations.

Naturally, the Zeeman LG in quasi-four-frequency operation has the same characteristic, which is the average between those for the RCP and LCP modes.

The values of the static zones are different for RCP and LCP modes: $\Omega_0^{rcp} \approx 290$ Hz and $\Omega_0^{lcp} \approx 310$ Hz. The static lock-in zone values depend on the size and type of scattering sources on the laser mirrors, as well as on the polarization of the mode. For this reason, it can be significantly different for modes with orthogonal polarizations, which appears in the difference in their input-output characteristics, as it is clear seen in Fig. 6, 7, particularly in the area of dynamic zones.

RMS values of half-widths of even and odd dynamic lock-in zones, calculated by equations from part II, gives for RCP and LCP modes $\Omega_d^{mrcp} \approx 38$ Hz, $\Omega_d^{mlcp} \approx 41$ Hz, and their ratio 0.93; the ratio of even and odd dynamic lock-in zones $\Omega_{d0}/\Omega_{d1} \approx 1.33$.

Experimental data: $\Omega_d^{mrcp} \approx 27$ Hz, $\Omega_d^{mlcp} \approx 32$ Hz, their ratio 0.84. For the four-frequency mode $\Omega_d^m \approx 27$ Hz, i.e. the smaller of the values for the PCP and LCP modes (see Fig. 6), which is the result of algebraic summing of their signals $f_{rcp}^+, f_{rcp}^-, f_{lcp}^+, f_{lcp}^-$ at the LG output. The widths of the even and odd dynamic lock-in zones for the four-frequency mode also correspond to the smaller values from those obtained for the PCP and LCP modes (Fig. 7). Their ratio in the experiment is $\Omega_{d0}/\Omega_{d1} \approx 1.36$. The noticeably smaller value of the dynamic lock-in zones observed in the experiment is apparently due to the turntable rotation non-uniformity, which causes a desynchronizing effect.

RMS values of the fractional dynamic zone half-widths Ω_{dj}^s for four-frequency mode are: by calculation 0.04 Hz, by experiment 0.03 Hz. Thus, by adding a "slow meander", the dynamic lock-in zones are reduced by a factor of 1000, and, in addition, the characteristic nonlinearity for angular rates close to the main biasing amplitude is reduced four-fold.

Note that the theoretical and experimental results coincide well.

IV. CONCLUSION

Theoretical calculations and experiments with Zeeman laser gyros gave us an ability to evaluate the input-output characteristics of the four-frequency and quasi-four-frequency modifications with the different types of alternating biasing.

The input-output characteristics of both modifications are average between those of the "sub-gyros" with orthogonal polarizations of the travelling waves. The four-frequency laser gyro is experimentally verified as providing an improved input-output characteristic compared to those of two-frequency "sub-gyros".

Our research confirmed the theoretically predicted effectiveness of using the combined rectangular alternating biasing (meander). Implementation in Zeeman laser gyro the combination of "fast meander" (switching period T = 1 ms) together with "slow meander" (switching period T_s = 1s) reduces the dynamic lock-in zones by a factor of 1000 and improves the linearity of the characteristic fourfold. The experimental comparative graphs clearly illustrate this improvement.

REFERENCES

[1] Lukyanov, D., Filatov, Yu., Golyaev, Yu., Kuryatov, V., Solovieva, T., Vasiliev, V., Buzanov, V., Spectorenko, V., Klochko, O., Vinogradov, V., Schreiber, K.-U., Perlmutter, M., "50th anniversary of the laser gyro," 20th Saint-Petersburg International Conference on Integrated Navigation Systems ICINS 2013 - Proceedings, 36-49 (2013).

[2] Kuznetsov, E., Golyaev, Yu., Kolbas, Yu., Kofanov, Yu., Kuznetsov, N., Vinokurov, Yu., Soloveva, T., "Thermal computer modeling of laser gyros at the design stage: a promising way to improve their quality and increase the economic efficiency of their development and production," Optical and Quantum Electronics 53(10), 596, 1-15 (2021).

[3] Passaro, V. M. N., Cuccovillo, A., Vaiani, L., De Carlo, M., Campanella, C.E., "Gyroscope technology and applications: a review in the industrial perspective," Sensors 17, 2284, 1-22 (2017).

[4] Shaodi Wang, Zhili Zhang, "Research on principle, application and development trend of laser gyro," Journal of Physics: Conference Series 1549(2), 022118 (2020).

[5] Volk, C.H., Gillespie, S.C., Mark, J.G., Tazartes, D.A., "Multioscillator ring laser gyroscopes and their applications," in Optical gyros and their application, RTO AGARDograph 339, 4-1-4-26 (1999).

[6] Zhiguo Wang, Xingwu Long, Fei Wang, "Bias characteristics of a multioscillator ring laser gyro with consideration of differential losses," Optics &LaserTechnology 48, 285–293, (2013).

[7] Aronowitz, F., "Fundamentals of the ring laser gyro," in Optical Gyros and their Application, RTO AGARDograph 339, 3-1-3-45 (1999).

[8] Azarova, V.V., Golyaev, Yu.D., Dmitriev, V.G., Drozdov, M.S., Kazakov, A.A., Melnikov, A.V., Nazarenko, M.M., Svirin, V.N., Soloviova, T.I., Tikhmenev, N.V., "Zeeman Laser Gyroscopes," in Optical Gyros and their Application, RTO AGARDograph 339, 5-1-5-29 (1999).

[9] Dorschner, T.A., Smith, I.W., "Clear-Path Four-Frequency Resonators for Ring Laser Gyros," Proc. of the Annual Optical Society of America Meeting, 24-27 (1978).

[10] Volk, C.Y., Longstaff, I., Canfield, J.M., Gillespie, S.C. "Litton's second generation ring laser gyroscope", Proceedings of the 15th Biennial Guidance Test Symposium. Holloman Air Force Base, New Mexico, USA, 493-502 (1991).

[11] Statz, H., Dorschner, T.A., Holtz, M., Smith, I.W., "The multioscillator ring laser gyroscope" in Laser Handbook, 229–332 (1985).

[12] Weibin Zhu, Gang Zheng, Linfeng Chen, Ke Liang, Yao Huang, and Zi Xue, "Modified adaptive filter for digital subdivision of a four-frequency differential laser gyro," Appl. Opt. 60, 342-350 (2021).

[13] Ribakov, B.V., Melnikov, A.V., Skulachenko, S.S., Khromikh, A.M., Prosvetov, V.K., "Ring lasers with nonplanar resonator," USSR Patent N46006 (1967).

[14] Azarova, V.V., Golyaev, Y.D., Savelyev, I.I., "Zeeman laser gyroscopes," Quantum Electron. 45(2), 171–179 (2015).

[15] Golyaev,Yu.D., Kolbas, Yu.Yu., "Application of ring lasers to determine the directions to the poles of Earth's rotation," Quant. Electronics 42(10), 949-952 (2012).

[16] Golyaev, Yu.D., Dronov, I.V., Ivanov, M.A., Kolbas, Y.Y., Soloveva, T.I., "Zeeman laser gyro in quasifourmode regime," Innovative Information Technologies: Materials of the Int. Scientific-Practical Conf., Part 3, 158-159, Moscow (2014).

[17] Kuznetsov, E., Kolbas, Y., Kofanov, Y., Kuznetsov, N., Soloveva, T., "Method of computer simulation of thermal processes to ensure the laser gyros stable operation," Mechanisms and Machine Science 75, 295-300 (2020).

[18] Kuznetsov, E., Golyaev, Y., Kolbas, Y., Kofanov, Y., Kuznetsov, N., Soloveva, T., Kurdibanskaya, A., "The method of intelligent computer simulation of laser gyros behavior under vibrations to ensure their reliability and cost-effective development and production," Proc. of SPIE 2020, 11523, 115230B (2020).

[19] Varenik, A.I., Gorshkov, V.N., Grushin, M.E., Ivanov, M.A. Kolbas, Yu.Yu., Savelyev, I.I. "Digital system for frequency regulation and stabilisation of a four-frequency Zeeman laser gyroscope", Quantum Electron. 51 (3), 276–282 (2021).

7-Core Erbium-Ytterbium Co-dopped Microstructured Fiber Amplifier

Yifan Zhang
School of Information and Optoelectronic Science and Engineering
South China Normal University
Guangdong, China
e-mail: wai450658444@163.com

Yifei Zhao
School of Information and Optoelectronic Science and Engineering
South China Normal University
Guangdong, China
e-mail: zhaoyifei8028@163.com

Weichao Ma
School of Information and Optoelectronic Science and Engineering
South China Normal University
Guangdong, China
e-mail: mwch9797@foxmail.com

Guiyao Zhou*
School of Information and Optoelectronic Science and Engineering
South China Normal University
Guangdong, China
e-mail: gyzhou@scnu.edu.cn

Abstract—Recently, multi-core fibers have performed prominently in the field of space division multiplexing (SDM) due to their multi-channel characteristics. As an important component of the SDM system, multi-core fiber amplifiers face huge challenges compared to single-channel amplifiers in front-end and back-end coupling devices, pumping methods, and amplification performance. Based on this, we propose and develop a 7-core erbium-ytterbium co-doped microstructured fiber amplifier (7CEYDFA) for efficient multi-core fiber amplification. By using air holes to surround the inner cladding, the double cladding structure can effectively isolate the pump light. At the same time, the specially prepared high concentration doped core rare earth material can ensure that 7CEYDFA can achieve efficient amplification within only 90cm. Through experimental tests, 7CEYDFA achieved a gain of 20.9dB at 1532nm.

Keywords-muticore fiber;microstructured fiber;fiber amplifier

I. INTRODUCTION

With the development of modulation technology based on wavelength division multiplexing and polarization multiplexing, the transmission rate of a single optical fiber has reached 100Tbit/s, which is close to the Shannon limit due to the effects of nonlinear effects [1]. Using a single core fiber for data transmission can not meet the growing demand for communication capacity. Therefore, space division multiplexing technology has been proposed as an effective method for channel capacity expansion. The common implementation of optical fiber SDM technology is to combine multiple single-mode fibers to form multiple channels. Accordingly, it is necessary to provide pumps and optical receivers for each channel in the amplifier, which undoubtedly increases the complexity of the system to a

great extent. The multi-core fiber can realize multi-channel transmission in one fiber by introducing multiple cores into the same fiber, thus overcoming the limitation of Shannon limit. The matched multi-core fiber amplifier needs to optimize the core material and fiber structure, and amplify the multi-channel signal under the action of a small amount or even a single pump, so as to show better integration and practical value.

During the last decade, SDM transmission and its components have made great progress [2-4]. From the perspective of coupling strength, multi-core amplifier can be divided into randomly coupled multi-core fiber amplifier [5] weakly coupled multi-core fiber amplifier [6,7]. Randomly coupled fiber forms supermodes through inter core coupling, and its transmission delay is lower. However, the MIMO decoupling system of multi-core fiber is complex and difficult to realize in application. Weakly coupled multi-core fiber amplifier means that the signal light of multiple channels is transmitted independently and amplified in their respective channels. Since the designed distance between cores is generally greater than 25um, the coupling degree between cores is low, and each core has strong independence, which is convenient for the decoupling of the output [8]. In the weakly coupled multi-core fiber amplifier, the signal light of multiple channels is transmitted independently and amplified in their respective channels. Since the designed distance between cores is generally greater than 25um, the coupling degree between cores is low, and each core has strong independence, which is convenient for self decoupling at the output [8]. Although the decoupling technology of multi-core single-mode amplifier is relatively mature, the information capacity supported by its channel is still limited. The multi-core few mode fiber amplifier [9] combines SDM and mode division multiplexing (MDM) technology to expand the number of channels and the support capacity of each channel to achieve higher dimensional signal amplification. In addition, the gain medium determines the amplification performance and spectral width of SDM fiber.

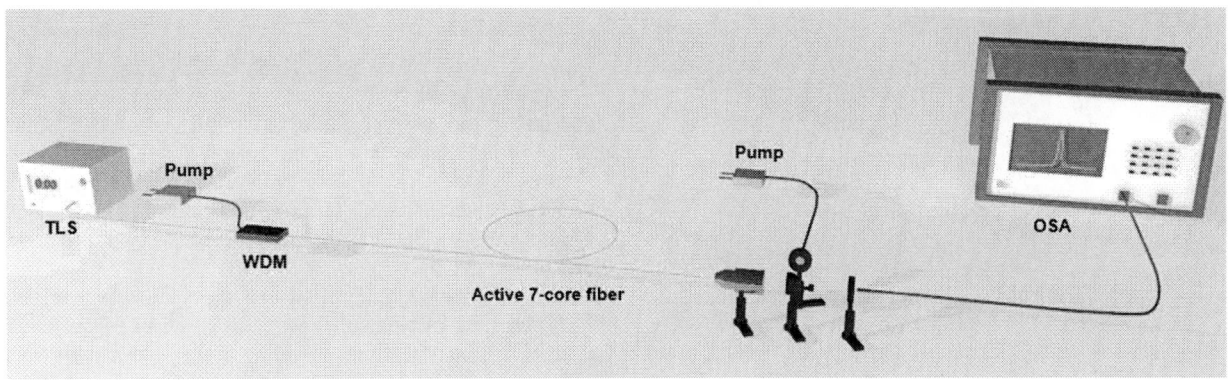

Figure 1. Experimental setup for testing the amplification perfomence of the proposed 7CEYDFA.

The common multi-core erbium-doped fiber amplifier (MC-EDFA) [10] has been suitable for C+L-band signal amplification. However, due to the limited absorption of pump energy by erbium-doped fiber, in order to improve the utilization efficiency of pump light and achieve better amplification effect, erbium-ytterbium co-doped rare earth materials are introduced into the multi-core fiber amplifier system [11].

In this paper, we propose and fabricate an erbium-ytterbium co doped seven core few mode fiber amplifier. By optimizing the structure and material parameters, it shows good optical and amplification characteristics. First of all, seven core microstructure fiber we designed needs to consider the core size to support few mode transmission. At the same time, the distance between cores needs to ensure sufficient distance to form weak coupling transmission and amplification. Through the iteration of simulation and experiment, the core location of multi-core fiber is determined. Then, in order to reduce the system complexity and make the pumping of each core more uniform, cladding pumping has greater advantages than core pumping [9]. Therefore, we designed a multi-core fiber double cladding structure surrounded by air holes, which can highly isolate the energy of the inner cladding and the outer cladding, so that the pump light can efficiently transmit in the inner cladding and act on each fiber core. In addition, for the gain medium of the amplifier, erbium-ytterbium co-doped core material rods are prepared by Non-modified Chemical Vapor Deposition (NMCVD) method. Compared with the concentration limitation of the traditional Modified Chemical Vapor Deposition (MCVD) method and the problem that the gas phase method cannot accurately control the doping ratio, NMCVD method can achieve high concentration doping and accurately regulate the components. Finally, according to the optimized design results, 7CEYDF with end faces as shown in Figure 3 was prepared, and one of the cores was amplified and tested by cladding pumping. Due to its excellent structure and unique material preparation method, the developed multi-core fiber has the potential to show better performance in high energy efficiency, high power amplification, low nonlinearity and so on.

II. THEORETICAL DESIGN OF 7CEYDF AND PLATFORM CONSTRUCTION

In order to explore the influence of the structural parameters of the seven core microstructure fiber on the signal transmission and amplification, the finite element vector analysis method is used to analyze and simulate the seven core fiber. By adjusting the core diameter d_0 and the core distance Λ, seven symmetrical cores are uniformly arranged in the structure of Figure 2a with the inner cladding diameter d_1 and the outer cladding diameter d_2. Figure 2b shows the 7CEYDF mode field distribution of simulation calculation. Meanwhile, in Figure 2b, each core supports 4 modes, and a total of 28 channels can be formed. Under high concentration of doping ratio, although NMCVD method can support more accurate ratio control, adding Al_2O_3 can also increase the insoluble phase region of quartz. Therefore, the ratio of each component is very important in material control. Through several iterative experiments, the doped region of the core was prepared by $Al_2O_3-Yb_2O_3-Er_2O_3$ system combined with SiO_2 glass substrate, and was evenly divided into seven parts as the core rod. In addition to structural design and material control, the stability of core size and inter core distance also plays an important role in the connection with front-end input devices and inter core gain difference. Therefore, a stable drawing process plays a decisive role in the performance of the amplifier. The arranged precast rods are prepared by stack-and-draw method, and 7CEYDF as shown in Figure 3 is obtained, with d_0=11.9um, Λ =28.5um, d_1=98um, d_2=160um.

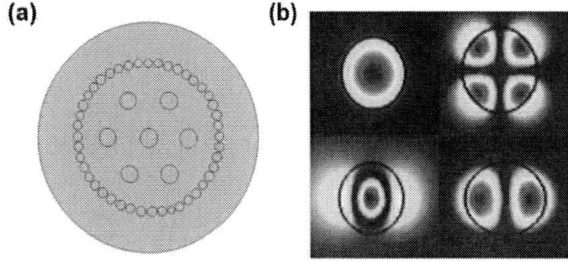

Figure 2. (a) Theoretical design of 4CEYDF. (b) Mode distribution in one of the seven cores.

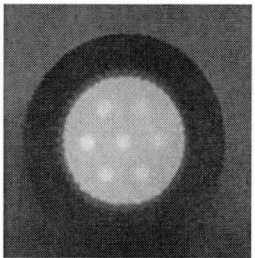

Figure 3. Cross section of the prepared 7CEYDF

To characterize the optical properties of 7CEYDF, pump light absorption is measured by the method shown in Figure 4. The 976nm multimode pump light is fused with the active fiber through a wavelength division multiplexer (WDM) coupling pigtail. The pigtail of WDM is a low refractive index coated double clad fiber, so the pump light can enter 7CEYDF and transmit in its inner cladding. At the backward end, an optical power meter is connected at the 60cm, 80cm, 100cm, 120cm and 140cm of the optical fiber through the method of truncation, and the output power is tested and recorded. At the same time, the signal light with the wavelength of 1532nm is emitted by the tunable laser source (TLS) and enters one of the seven cores of 7CEYDF through the core of WDM pigtail. After truncation, the power of the gain fiber at 60cm,80cm,100cm,120cm,140cm is recorded with a power meter, and its absorption loss is calculated.

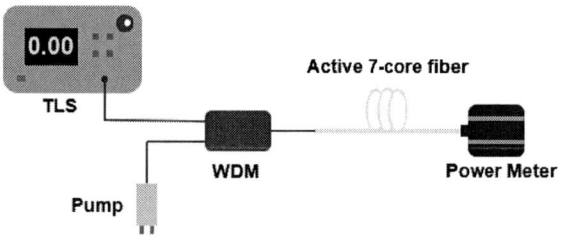

Figure 4. Setup used to measure the cladding absorption of the pump power(976nm).

The amplification test scheme for one of the cores of 7CEYDF is shown in Figure 1. The signal light is transmitted by TLS and enters the signal and pump combiner after being attenuated by the attenuator. At the same time, the pump pigtail with a wavelength of 976nm is connected to the other WDM connection port and is emitted from the double clad pigtail after beam combining. Then, the double-clad pigtail was spliced with 7CEYDF by using the LZM-100 multifunctional fiber fusion splicer. During the welding process, it is necessary to ensure sufficient strength and avoid the collapse of air holes in microstructure optical fibers. At the backward end of the gain fiber, a spatial optical coupling system is built for backward-end pumping and signal acquisition after amplification. The backward end pump light is expanded by the lens and reflected into the objective lens by the incident dichroic mirror (reflection below 1000nm and projection above 1000nm)，then focused by the objective lens and incident into 7CEYDF. After the amplified signal passes through the objective lens and dichroic lens, it is focused through the collimating lens into a multimode fiber connected to the spectrometer.

III. EXPERIMENTAL RESULT

Amplification spontaneous emission (ASE) of gain fiber can reflect the spectral range and amplification performance of amplifier to a certain extent. Using the scheme shown in Figure 1, the ASE of 7CEYDF is tested first under the condition that no signal is input and only pump is injected. Figure 5 shows the distribution of ASE when the length of 7CEYDF is 90cm.

In the characterization of the absorption properties of 7CEYDF, the pump with power of 12w,14w,16w,18w,20w is injected into the inner cladding layer respectively in the absence of signal input. The pump light acts on the seven cores of the inner cladding and is absorbed. A power meter with a maximum power of 40W is used as the receiver. Figure 6 shows the pump light intensity after absorption at different fiber lengths. In addition, the output of 7CEYDF at 90cm under different pump light power is shown in figure 7. After calculation, the absorption of pump light by 7CEYDF

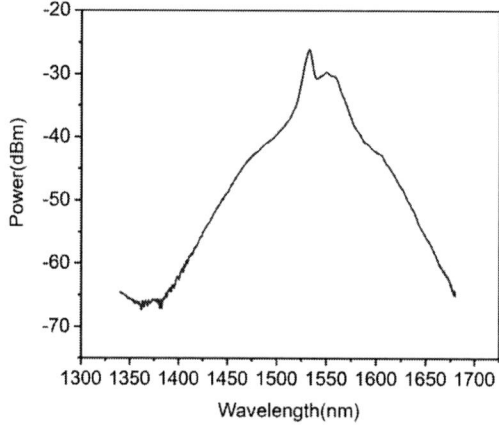

Figure 5. ASE spectrum of 7CEYDFA when signal is not injected.

inner cladding is 12.5dB/m. At the same time, in order to test the signal absorption of the weakly coupled seven core fiber, a signal light of 1532nm is selected to inject into one of the seven cores as the tested wavelength according to the ASE distribution. Using the same truncation test method, the signal absorption at 1532nm is 30.4dB/m.

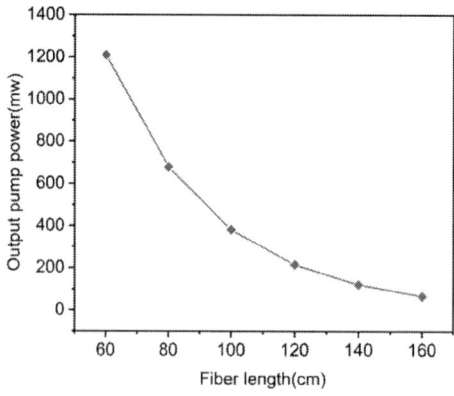

Figure 6. Measured output pump power with different fiber length.

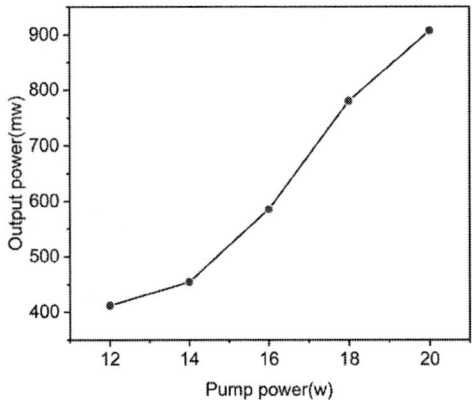

Figure 7. Measured output power with different forward-end injected pump power.

In order to explore the amplification performance of 7CEYDFA, one of the seven cores was amplified in the way shown in Figure 1. When the forward pump is 20W and the backward pump is 15W, the signal input wavelength is injected into the intermediate core. The amplification spectrum range of the optical fiber is affected by the length of the optical fiber [8], and the microstructure optical fiber amplifier can achieve efficient amplification with only a very short length because of its good isolation characteristics to the pump light. Through the optimization of the fiber length, when the length of 7CEYDF is 90cm, combined with the ASE gain characteristics, the gain reaches 20.9dB at the input wavelength of 1532nm. It is worth mentioning that, due to the pump light coupling efficiency, the actual pump power entering the core is far less than the output value of the pump pigtail, so the corresponding energy conversion efficiency will be higher.

Compared with single-mode multi-core fiber amplification and transmission, introducing more modes into each fiber core can improve the signal capacity of SDM fiber in a higher dimension, that is, the combination of SDM and MDM. In 7CEYDF, the refractive index of each core is tested as 1.45, its core diameter is 11.9um, and the inner cladding material is quartz, which supports the transmission and amplification of few modes. Since the fiber is a weakly coupled fiber, the transmission mode of each core is the same. In Figure 2b, the transmission mode in the fiber core is studied by full vector finite element analysis. It can be seen that each core supports 4 transmission modes, which are LP_{01}, LP_{11}, LP_{12} and LP_{21} modes. Therefore, 7CEYDF supports up to 28 channels.

IV. CONCLUSION

In this paper, a seven core erbium-ytterbium co-doped microstructural fiber amplifier is designed and fabricated. Optical properties, amplification characteristics and transmission modes of gain fiber are studied theoretically and analyzed experimentally. Due to the special structure of 7CEYDF and the high concentration doped core materials prepared by NMCVD methods, only 90cm gain fiber is required to achieve a gain of more than 20dB at 1532nm. Therefore, the proposed new multi-core microstructure fiber amplifier will have great potential in high energy utilization, low nonlinear amplification at high power and so on.

ACKNOWLEDGMENT

This work was supported by the National Key Research and Development Program of China（Grant No. 2019YFB2204002).

REFERENCES

[1] Essiambre R J , Mecozzi A . Capacity limits in single-mode fiber and scaling for spatial multiplexing. IEEE, 2012.

[2] Tsuchida Y , Maeda K , Yu M , et al. Amplification characteristics of a multi-core erbium-doped fiber amplifier. IEEE, 2012.

[3] Yu M , Tsuchida Y , Maeda K , et al. Batch Multicore Amplification with Cladding-Pumped Multicore EDF[J]. Technical Report of Ieice Ocs, 2012.

[4] Mukasa, K. , K. Imamura , and R. Sugizaki . "Multi-core Few-mode optical fibers with large Aeff." European Conference & Exhibition on Optical Communications IEEE, 2014.

[5] Jung, Y. , et al. "High Spatial Density 6-Mode 7-Core Multicore L-Band Fiber Amplifier." Optical Fiber Communication Conference 2019.

[6] Sakaguchi, J. , et al. "19-core MCF transmission system using EDFA with shared core pumping coupled via free-space optics." Optics express 22.1(2014):90.

[7] Abedin, K. S. , et al. "Seven-core erbium-doped double-clad fiber amplifier pumped simultaneously by side-coupled multimode fiber." Optics Letters (2014).

[8] Charles, C. , et al. "Multicore cladding-pumped fiber amplifier with annular erbium doping for low gain compression." Journal of Lightwave Technology PP.103 (2022).

[9] Jung, Y. , et al. "High Spatial Density 6-Mode 7-Core Fiber Amplifier for L-band Operation." Journal of Lightwave Technology PP.99(2020).

[10] Chen, H. , et al. "Integrated cladding-pumped multicore few-mode erbium-doped fibre amplifier for space-division-multiplexed communications." Nature Photonics (2016).

[11] Jain, S. , et al. "32-core erbium/ytterbium-doped multicore fiber amplifier for next generation space-division multiplexed transmission system." Optics Express 25.26(2017):32887.

Study on the Influence of Deposition Temperature on the Properties of Lanthanum Titanate Films

Yang Li
Xi'an Technological University
Shaanxi Province Thin Films Technology and Optical
Test Open Key Laboratory
Xi'an, China
e-mail: liyang01@st.xatu.edu.cn

Junqi Xu*
Xi'an Technological University
Shaanxi Province Thin Films Technology and Optical
Test Open Key Laboratory
Xi'an, China
e-mail: jqxu2210@163.com

Junhong Su
Xi'an Technological University
Shaanxi Province Thin Films Technology and Optical
Test Open Key Laboratory
Xi'an, China
e-mail: sujunhong@mail.xait.edu.cn

Zheng Liu
Chinese Academy of Sciences
Xi'an Institute of Optics and precision machinery Joint
Laboratory of advanced optical manufacturing
technology
Xi'an, China
e-mail: 18222671133@163.com

Abstract—The work aims to study the effect of deposition temperature on optical properties and residual stresses in Lanthanum titanate (H4) films. The LaTiO$_3$ films were deposited by electron-beam thermal evaporation technique. The residual stress of LaTiO$_3$ films on fused silica was characterized macroscopically and microscopically, using laser interferometry and AFM. The residual stresses and surface profile shape change were simulated using finite element analysis methods. It was confirmed that the deposition temperature did not affect the optical properties of the films, but did for residual stresses. The residual stress of LaTiO$_3$ films changes from decreasing tensile stress to compressive stress as the deposition temperature increases. The deposition temperature is used to modulate the magnitude and transition of the residual stress in the films. There is a strong dependence between the residual stresses and the densities of surface columnar structures in LaTiO$_3$ films. The effect of density of surface columnar structures is found as follows: the film with the lower density of surface columnar structures generally shows a tensile and high density easily transform into compress stress. This conclusion is also verified by the increase of the corresponding refractive index. The simulated surface profiles are basically overlapping with the measured data. The proposed model is validated for the simulation of residual stresses in monolayers.

Keywords-component; Lanthanum titanate films; residual stress; surface morphology; finite element analysis(FEA)

I. INTRODUCTION

The coated optical components are required not only to have good optical properties but also to have precise surface shapes. Such as mirrors in various laser systems and astronomical telescopes [1][2]. However, it caused large stress to deposit film solid materials on optical surfaces by electron-beam thermal evaporation technique. It leads to the bending deformation of the mirror surface and spectral drift and increasing scatter of the coated components. At the same time, its mechanical properties are drastically reduced and even lead to the invalid of coated components due to the rupture of films [3]. The generation of residual stresses in multilayer films is complex and no defined model is available. Therefore, it is essential to clarify the residual stress mechanisms of the monolayer and to match the stresses between the layers for reducing residual stresses and surface shape changes of the coated components.

The mechanisms of stress generation are complex. It is usually acted upon by a combination of mechanisms. Meanwhile, some experimental studies have been made on monolayer stresses, where the residual stresses in monolayers are regulated by controlling the deposition temperature [4], deposition angle [5], ion source assistance [6], annealing treatment [7], etc. Numerous studies [8-10] have shown that the transformation of the film crystal structure corresponds to the change of the residual stress. The kinetic energy of evaporating particles is different with the change of deposition conditions. The changes in the stacking density, aggregation density, microscopic crystalline state [11], crystal growth direction [12], and grain size [13] of the films were found to be the root reasons for the changes in film stress. In order to clarify the residual stress distribution in films and guide experiments, simulation methods [14][15] had emerged. Specifically, residual stresses were loaded as external forces [16] or equivalent substitution [17] was used to simulate the residual stresses and surface shape changes in film-substrate. However, it also faces problems such as over-idealized models, complicated interface treatment and large computational efforts.

Due to the complexity and variety of theoretical models of residual stresses in monolayers, this paper combines experimental studies and simulation analysis to make some studies on the optical constants and stresses in LaTiO$_3$ films at different deposition temperatures. It is significant to clarify

978-1-6654-8699-6/22 $31.00 © 2022 IEEE

the effect of deposition temperature on the residual stresses of LaTiO₃ films for regulating the residual stresses of film-substrate and reducing the surface shape change of the coated components.

II. EXPERIMENTS

A. Sample Preparation

The LaTiO₃ material was selected from the Beijing Research Institute of Nonferrous Metals. The substrate is fused silica (JGS1, ϕ 30 × 2mm). In the vacuum chamber coater (ZZS500-2/G), samples were deposited at different deposition temperatures (40℃, 80℃, 120℃, 160℃, 200℃) by electron-beam thermal evaporation technique. The film thickness was monitored by the photoelectric polarimetry method (λ_0 = 520 nm). The optical thickness, deposition rate, the oxygen charge rate, the working vacuum and the electron gun beam current were $3\lambda_0/4$, 0.38-0.42 nm/s, 5 sccm, 2.2×10^{-2} Pa, and 100 mA (8 KV), respectively.

B. Young's Modulus of LaTiO₃ Films

There are large differences between the film and bulk material in terms of characteristic parameters [18]. Therefore, the experiments were conducted using Agilent nanomechanical test system (G200), and the continuous stiffness measurement module was chosen to test Young's modulus of LaTiO₃ films. In order to eliminate errors and obtain Young's modulus accurately, the samples at different temperatures and points were tested. Fig. 1 shows Young's modulus in relation to the press-in depth for LaTiO₃ films with several measurements, where the black curve and the orange areas are the average value and the error band, respectively.

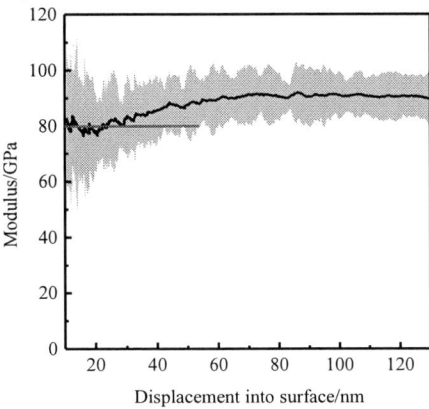

Figure 1. Young's modulus of LaTiO₃ film in differentt pressure depths

According to Fig. 1, Young's modulus remains constant after reaching 60 nm. It indicates that the properties of fused silica substrate have been expressed at this point, and then the peak can be considered as the maximum estimate of Young's modulus of the film material. It is also pointed out in the literature [19] that it is not to ignore the interference of mechanical properties of the substrate on the measurement results when the pressed depth of the indenter is bigger than 10%-25% of the film thickness. Therefore, the value here

can be approximated as Young's modulus of the film. The Young's modulus of LaTiO₃ film can be determined to be 80.70 GPa by the excellent agreement of the several multi-point test curves in Fig. 1.

C. Samples Testing

The optical constants were tested by a broad-spectrum variable-angle ellipsometer (J.A. WOOLLAM, M-2000UI). The surface shapes of substrates before and after coated were measured by Zygo laser interferometer (Model: 4″ Verifire PE, Standard Mirror: 121031) in the shading environment and air-floating optical vibration isolation platform. Among them, the monolayer thermal stress and residual stress were calculated by Equation (1) and Equation (2), respectively.

$$\alpha_{th} = \frac{E_f}{1-\mu_f}\left(\alpha_f - \alpha_s\right)\Delta T \qquad (1)$$

$$\sigma = \frac{4E_s}{3(1-\mu_s)}\frac{t_s^2}{t_f}\frac{\Delta PV}{D_s^2} \qquad (2)$$

E_f, α_f, t_f and μ_f are Young's modulus, thermal expansion coefficient, thickness and Poisson's ratio of the film. E_s, α_s, t_s, μ_s and D_s are Young's modulus, thermal expansion coefficient, thickness, Poisson's ratio and diameter of the substrate. ΔT is the temperature difference values between deposition temperature and room temperature. ΔPV is the difference value between the peak and valley of the substrate surface before and after being coated. The parameters of substrate and film materials are shown in Table I. The surface morphology of LaTiO₃ films in a square area (2 μm × 2 μm) was tested using an atomic force microscope (Bruke, multimode 8).

TABLE I. MATERIAL PARAMETERS

Material	Elasticity modulus(GPa)	Poisson's ratio	Coefficient of thermal expansion(10^{-6}/℃)
Fused silica[17]	72.00	0.17	0.58
H4	88.70	0.27	3.60

III. RESULTS

A. Effect of Deposition Temperature on Optical Constants of LaTiO₃ Films

The refractive index curves of LaTiO₃ films at different deposition temperatures are shown in Fig. 2(a). The refractive index decreases with increasing wavelength and satisfies the normal dispersion relationship. With the higher deposition temperature, the refractive indices tend to increase in a small range. When the deposition temperature was increased, film material particles easily gained extra energy, and the mobility on the substrate became greater, thus improving the cohesion coefficient and aggregation density of films during film-forming, which resulted in higher refractive indices of films. As the deposition temperature increased, the higher refractive index reflected the gradual improvement of the packing density of the film.

As shown in Fig. 2(b), the extinction coefficients of LaTiO₃ films were analyzed. There is no obvious law for the change of the extinction coefficient of LaTiO₃ films at different deposition temperatures. At 1064 nm, the extinction coefficient of films is maximum at 80°C (1.41×10^{-3}) and minimum at 200°C (3.24×10^{-5}). Although there are some differences in the extinction coefficients of the films prepared at different deposition temperatures, the values are small and the influence on absorption can be neglected. It demonstrates that the LaTiO₃ is a high refractive index material with low absorption.

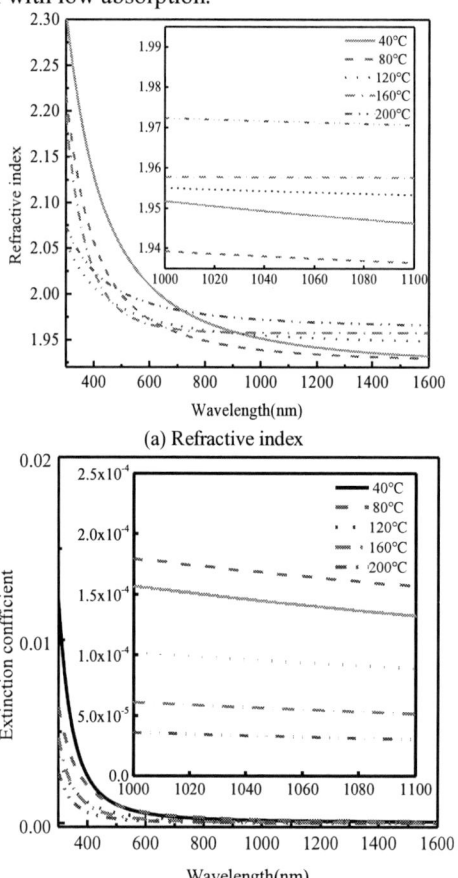

Figure 2. Optical constants of LaTiO₃ films

B. Effect of Deposition Temperature on the Stress of LaTiO₃ Films

The residual stresses in optical films are strongly related to the film deposition parameters, where the deposition temperature is one of the important factors. The thermal and residual stresses of the films were calculated according to the material parameters (Table 1), the face shape changes measured by the interferometer, and equations (1) and (2). The thermal and residual stresses of the LaTiO₃ films on fused silica substrates deposited at different temperatures are shown in Fig. 3.

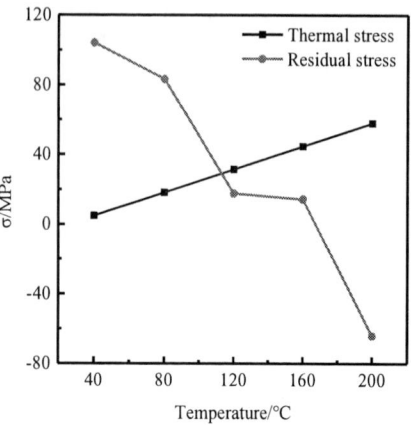

Figure 3. Stress of LaTiO₃ films at different deposition temperatures

The thermal stress of LaTiO₃ films on fused silica substrates is tensile stress that becomes larger as deposition temperature increases. It is due to the fact that the thermal expansion coefficient of the LaTiO₃ film is larger than the fused silica substrate and the testing temperature is lower than the deposition temperature. The increasing substrate temperature and film aggregation density lead to the reduction of film tensile stress or even conversion to compressive stress. Therefore, the residual stresses of LaTiO₃ films are the result of the interaction of tensile stresses caused by thermal and compressive stresses generated by film densification. As the substrate temperature is raised, the increasing compressive stress from film densification compensates for the growing tensile stress caused by thermal, which results in a decrease and zero stress in the residual stress of the film-substrate. As the deposition temperature continually increases, the residual stresses transform into compressive stresses dominated by internal stresses. Equation (1) shows that the thermal stress of the film is a linearly increasing function related to the change of temperature. Fig. 3 shows that the thermal and residual stresses of the LaTiO₃ films have opposite trends. Then the internal stresses play a major role in residual stresses along with the increase in temperature. In summary, the deposition temperature has a large effect on the residual stress of LaTiO₃ films. It is possible to control the residual stress of the film with zero stress or the transition from tensile to compressive stress by appropriately increasing the deposition temperature. A similar conclusion on the adjustment of residual stresses of ZrO₂ films by deposition temperature was also obtained by Shu-ying Shao [20].

C. Effect of deposition temperature on the surface morphology of LaTiO₃ films

From the above, it is clear that internal stresses have a strong influence on the residual stresses of LaTiO₃ films with increasing deposition temperatures. The surface morphology of LaTiO₃ films at different deposition temperatures is shown in Fig. 4.

978-1-6654-8699-6/22 $31.00 © 2022 IEEE

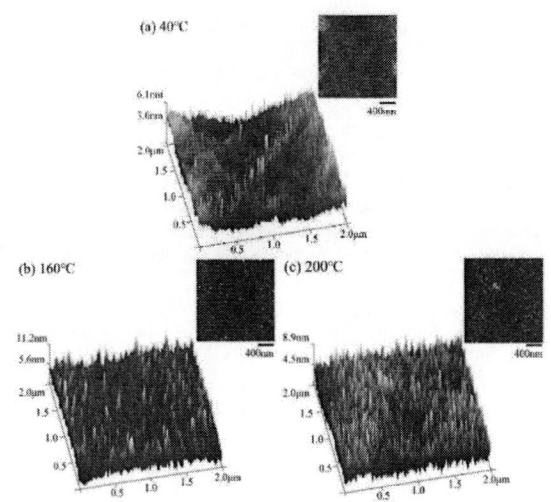

Figure 4. Surface morphology of LaTiO₃ films at different deposition temperatures

Based on the tested samples, the surface morphology of the LaTiO$_3$ films deposited by thermal evaporation was columnar at different deposition temperatures. As shown in Fig. 4(a), when the deposition temperature is low, the surface undulation height of the film is low and the grown film becomes uneven and accompanied by some small grooves. The film material particles acquire less energy to transport on the substrate surface and the columnar structure is loosely distributed. As shown in Fig. 4(b), when the deposition temperature gradually increases, the surface undulation height of the film grows higher and voids on the surface become obvious. The film material particles obtain more energy for lateral transport and growth as a result of which the distribution of columnar structures becomes dense. When the deposition temperature is 200°C, as shown in Fig. 4(c). The surface undulation height of the film is slightly reduced. Then the surface grooves disappear and the voids are reduced. The film material particles are provided with sufficient energy to transport enough on the growth surface and the columnar structure of the film becomes more closely distributed. In correspondence to the rising deposition temperature, the residual stress of the LaTiO$_3$ film changes from reduced tensile stress to compressive stress. Therefore, the residual stresses of the LaTiO$_3$ films are strongly dependent on the density of the surface columnar structure. Equally, low densities tend to exhibit tensile stresses and high densities easily transform into compressive stresses. Similarly, Pei-fu Gu[21] proposed that the film aggregation density is an important factor contributing to the internal stresses of TiO$_2$ and SiO$_2$ films.

D. Simulation of Residual Stresses and Surface Shape Change of LaTiO₃ Films

The internal stress effect was replaced by the equivalent reference temperature (ERT). The residual stress distribution and surface shape change of LaTiO$_3$ films were simulated using the finite element analysis method. The model was meshed by sweeping and refining the interface between the film layers and the film-substrate to improve accuracy. In the coordinates with reference to Fig. 5, the boundary conditions were defined to fix the displacements of the x=0 and the y=0 plane in their own direction, then the center point of the circle was also fixed. Finally, the model was loaded with temperature and solved.

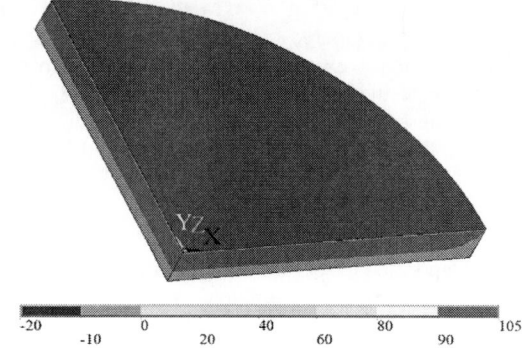

Figure 5. The Z axis residual stress cloud diagram of LaTiO₃ film-substrate system (Mpa)

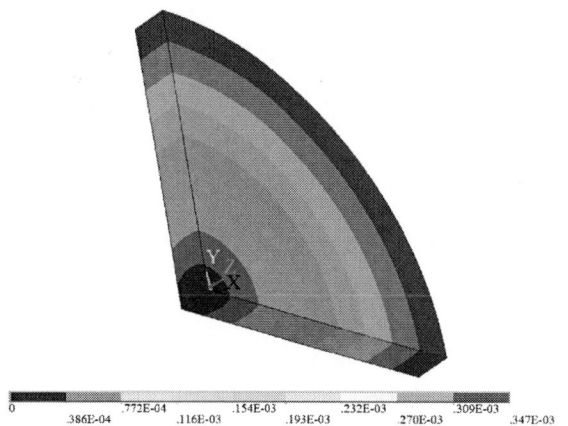

(a) The Z-axis displacement diagram of LaTiO₃ monolayer(mm)

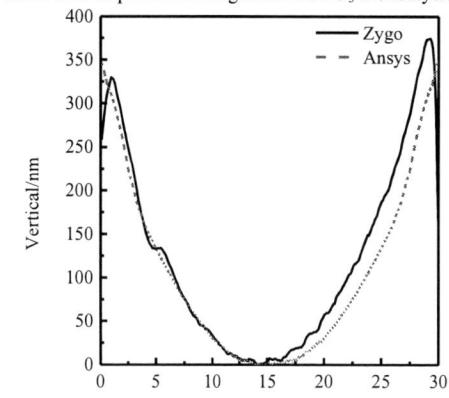

(b) Comparison of the deformation from Zygo and Ansys

Figure 6. Surface shape change of LaTiO₃ film-substrate system

Fig. 5 shows the Z-axis residual stress cloud diagram of the LaTiO$_3$ film-substrate system. The overall residual stresses show a layered distribution, changing from tensile to compressive and then from compressive to tensile stresses from the substrate to the films. The neutral surface of the system is located at a distance of one-third from the bottom of the substrate. The residual stresses in the substrate are small, unevenly distributed, and slightly larger at the edges. The residual stresses in the film are larger. The Z-axis deformation caused by residual stresses is circularly distributed, as shown in Fig. 6(a). Measurements and Ansys simulations were compared by extracting data on the nodal deformation in the Z-axis and the results are shown in Fig. 6(b). Because of the idealized processing in the simulation and not being strictly symmetrical in the actual coated components, the two curves basically match within a certain error.

IV. CONCLUSION

In this paper, the properties of $LaTiO_3$ films deposited by electron beam thermal evaporation at different deposition temperatures have been investigated, and the conclusions are as follows.

(1) The deposition temperature has little effect on the optical constants of $LaTiO_3$ films, which proves that it is a high refractive index material with small absorption and good optical properties. It can be used for depositing laser multilayer protection films.

(2) On the fused silica substrates, the residual stresses of $LaTiO_3$ films are heavily influenced by the deposition temperature and the internal stresses contribute mainly to the residual stresses with increasing temperature. It is possible to control the residual stress of the film with zero stress or the transition from tensile to compressive stress by appropriately increasing the deposition temperature.

(3) The residual stress of $LaTiO_3$ films by electron beam-evaporated has a strong dependence on the density of its surface columnar structure. A low degree of density tends to show tensile stress, and a high degree of density tends to transform into compressive stress.

(4) The simulations and the tested surface profiles basically agree with each other, which verifies that the equivalent reference temperature and the birth-death element are applied to the simulation of residual stresses in monolayers.

REFERENCES

[1] BAI Jin-lin, LIU Hua-song, JIANG Yu-gang, et al. "Ultra-low stress high reflective film by dual-ion-beam sputtering deposition," Optical engineering, vol. 61, Mar. 2021, doi: 10.1117/1.OE.61.3.031203.

[2] YAO You-wei, Brandon D. Chalifoux, Ralf K. Heilmann,et al. "Progress of coating stress compensation of silicon mirrors for Lynx x-ray telescope mission concept using thermal oxide patterning method," Journal of Astronomical Telescopes, Instruments, and Systems, vol. 5, Apr. 2019, doi: 10.1117/1.JATIS.5.2.021011.

[3] LIU Zhi-wei, The research on interface crack and buckling problems in film/substrate system, Harbin: Harbin Engineering University, 2008.

[4] Paloma López-Reyes, Belén Perea-Abarca, Carlos Honrado-Benítez, et al. "Optimization of the deposition parameters of MgF2/LaF3 narrowband reflective FUV multilayers", Opt. Mater. Express, vol. 11, Jun. 2021, pp. 1678-1691, doi: 10.1364/OME.424742.

[5] Cheng-Chung Jaing, "Thermal expansion coefficients of obliquely deposited MgF2 thin films and their intrinsic stress", Applied Optics, vol. 50, Mar. 2011, pp. C159-C163, doi: 10.1364/AO.50.00C159.

[6] Aaron Davenport, Emmett Randel, Carmen S. Menoni, "Ultra-low stress SiO_2 coatings by ion beam sputtering deposition", Applied Optics, vol. 59, Mar. 2020, pp. 1871-1875, doi: 10.1364/AO.380844.

[7] SHEN Jie-nan, ZENG Yi-bo, XU Ma-hui,et al. "Effects of annealing parameters on residual stress and piezoelectric performance of ZnO thin films studied by X-ray diffraction and atomic force microscopy", Journal of applied crystallography, vol. 52, Oct. 2019, pp. 951-959, doi: 10.1107/S1600576719010124.

[8] XIAO Qi-ling, HE Hong-bo, SHAO Shu-ying, et al. "Influence of deposition temperature on residual stress of yttria-stabilized zirconia thin films", Acta Optica Sinica, vol. 28, May. 2008, pp. 1007-1011.

[9] J. Chakraborty, T. Oellers, R. Raghavan, et al. "Microstructure and residual stress evolution in nanocrystalline Cu-Zr thin films", Journal of Alloys and Compounds, vol. 896, Mar. 2022, doi:10.1016/j.jallcom.2021.162799.

[10] LI Hai-tao, SUN Peng-fei, CHENG Dong-hai, et al. "Effects of deposition temperature on structure, residual stress and corrosion behavior of Cr/TiN/Ti/TiN films", Ceramics International,. vol. 47, Dec. 2021, pp. 34909-34917, doi: 10.1016/j.ceramint.2021.09.032.

[11] Borroto, S.Bruyere, S.Migot, et al. "Growth kinetics and origin of residual stress of two-phase crystalline–amorphous nanostructured films", Journal of Applied Physics, vol. 129, Apr. 2021, doi: 10.1063/5.0044029.

[12] I-Hsin Tseng, Yun-Ting Hsu, Jihperng Leu, et al. "Effect of thermal stress on anisotropic grain growth in nano-twinned and un-twinned copper films", Acta Materialia, vol. 206, Mar. 2021, doi: 10.1016/j.actamat.2021.116637.

[13] Philipp Klose, V Roddatis,Astrid Pundt, "Tuning the stress state in Nb-thin films by lateral size confinement", Acta Materialia, vol. 222, Jan. 2022, doi: 10.1016/j.actamat.2021.117454.

[14] Erman Citirik, Taha Demirkan, Tansel Karabacak, "Residual stress modeling of density modulated silicon thin films using finite element analysis", Journal of Vacuum Science & Technology A, vol. 33, Mar. 2015, doi: 10.1116/1.4902953.

[15] Mohammad Shishehsaz, HamedRaissi, Shapour Moradi, "Stress distribution in a five-layer circular sandwich composite plate based on the third and hyperbolic shear deformation theories", Mechanics of Advanced Materials and Structures, vol. 27, Jun. 2020, pp. 927-940, doi: 10.1080/15376494.2018.1502379.

[16] ZHANG Li-sha, XU Hong. Influence of oxygen partial pressure on HfO_2 residual stresses and its finite element analysis [J].High power laser and particle beams,2008(06):894-898.

[17] GAO Chun-xue. Study of the distribution and control of stress in optical thin films[D]. Nanjing :Southeast University,2015.

[18] BAI Jin-lin,JIANG Yu-gang, WANG Li-shuan, et al.Study on the design and preparation technology of ultra-low profile wideband high reflection thin films[J].Infrared and Laser Engineering, vol. 50, Feb. 2021, pp. 128-133, doi: 10.3788/IRLA20200413.

[19] TIAN Xiao-xi.Study on thermodynamic matching between substrate and films in optical thin film technology[D].Chengdu: Institute of optics, Chinese Academy of Sciences,2020.

[20] SHAO Shu-ying, FAN Zheng-xiu, FAN Rui-ying, et al.Influence of deposition temperature on the properties of ZrO_2 films prepared by electron beam evaporation [J].Chinese journal of lasers,2004,31(06):701-704.

[21] GU Pei-fu, ZHENG Zhen-rong, ZHAO Yong-jiang, et al. Study on the mechanism and measurement of stress of TiO_2 and SiO_2 thin-films[J].Acta Physica Sinica,2006(12):6459-6463.

Proceedings of the 7th Optoelectronics Global Conference (OGC 2022)

ErYb Co-doped Double-clad Fiber Amplifiers with Average Gain of 29dB by High Concentration Doping

Yifei Zhao
School of Information and Optoelectronic Science and Engineering
South China Normal University
Guangzhou, China
e-mail: zhaoyifei8028@163.com

Yifan Zhang
School of Information and Optoelectronic Science and Engineering
South China Normal University
Guangzhou, China
e-mail: wai450658444@163.com

Shizhuo Xi
School of Information and Optoelectronic Science and Engineering
South China Normal University
Guangzhou, China
e-mail: xsz1997jensen@163.com

GuiYao Zhou*
School of Information and Optoelectronic Science and Engineering
South China Normal University
Guangzhou, China
e-mail: gyzhou@scnu.edu.cn

Abstract—We proposed an ErYb co-doped double-clad fiber amplifier. By adjusting the ion doping concentration of the gain medium in the core, the gain characteristics of the EYDFA are theoretically analyzed. In the experiment, through the setup of the cladding pumped optical amplifier system, the absorption value of the fiber to 976nm pump light reaches 2.06db/m, and the average gain of 29dB is obtained in the 1531-1567nm band. The corresponding noise figure varies from 4.9 to 11.6 dB.

Keywords-ErYb co-doped; double cladding fiber; fiber optic amplifiers; multimode pumping

I. INTRODUCTION

With the increase of communication capacity year by year, the traditional single-mode optical fiber (SMF) used to transmit signals has gradually failed to meet the need of communication because of its few-mode channels and nonlinear effects [1]. The amplification scheme with high gain characteristics and wide gain wave band is very important to broaden the bandwidth of the wavelength division multiplexing (WDM) [2]. With the introduction of WDM and Erbium-Ytterbium co-doped fiber amplifier (EYDFA), the research and development of fiber amplifier have further moved towards high gain bandwidth, which has pushed the optical fiber communication system to the direction of high-speed, large capacity and long-distance [3]. The use of ErYb Co-doped gain medium effectively reduces the concentration quenching effect of Er^{3+} ions through the action of Yb^{3+} ions. At the same time, combined with the cladding pumping method, it can greatly improve the absorption of pump light by the amplification system [4].

At the same time, the gain bandwidth can be broadened by adjusting the type of doped ions [5].

In this paper, we propose an ErYb co-doped double-clad fiber amplifier to be used in the C-band. By adjusting the ion concentration of the gain medium and adopting the structure design of the fiber amplifier system with multi-mode cladding pumping, at the input of -20dBm signal light per wave (the wavelength range is 1531-1567nm), we obtain a 25dB high gain of 1531-1567nm in C-band, with an average gain of 29dB, and the corresponding noise figure varying from 4.9 to 11.6 dB.

II. NUMERICAL SIMULATION AND ANALYSIS

Starting from the luminescence process of ErYb co-doped materials, after obtaining the rate equation of the corresponding doped materials, we bring in the ion concentration and emission absorption cross section calculated by J-O theory, and further combine the theoretical model of the corresponding pump light, signal light and ASE light transmission along the optical fiber, establish the transfer equation about the number of energy-level particles through MATLAB, and solve the gain characteristics of EYDFA at different doping concentration ratios of multiple wavelengths [6].

TABLE I.   MOLAR PERCENTAGE OF ERBIUM-YTTERBIUM ION CONCENTRATION

	Doped rare earth ions(c%)	
	Yb^{3+}	Er^{3+}
Low doping concentration	0.15	0.075
High doping concentration	0.3	0.075

978-1-6654-8699-6/22 $31.00 © 2022 IEEE

For the gain media with different ion doping concentrations, see Table I. Bring them into EYDFA model to calculate the correlation between their gain and wavelength. By comparing the gain spectra of low doping and high doping theoretically calculated in Fig. 1 and Fig. 2, it can be seen that the gain of 1528-1545nm is low at low doping ion concentration. When the concentration of Yb^{3+} ion is doubled, the gain of the short-wave part is greatly improved, and the overall high gain band covers 1528-1563nm. This is because the increase of Yb^{3+} ion concentration enhances the absorption of pump light by the gain medium.

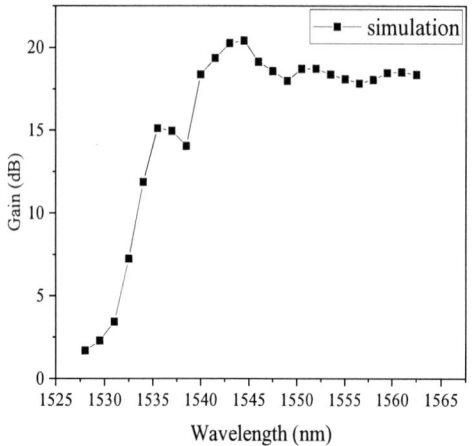

Figure 1. The theoretical calculation of the gain of the optical fiber amplifier with low ion doping concentration

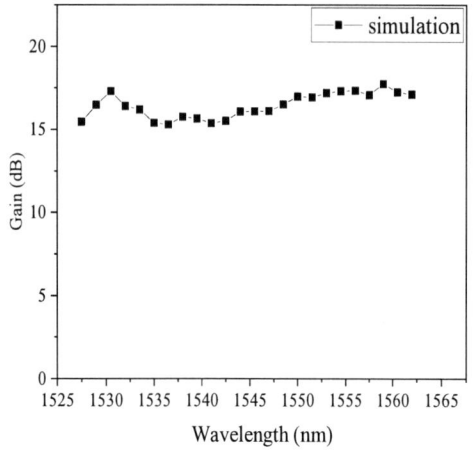

Figure 2. The theoretical calculation of the gain of the optical fiber amplifier with high ion doping concentration

III. EASE OF USE EXPERIMENTAL RESULTS AND DISCUSSION

As shown in Fig. 3, -2.9dBm of input signal per wave is obtained by single-mode TLS (wavelength range 1528-1563nm) and coupled into the EYDF amplifier. Through the design of the double cladding structure, the 976nm multimode pump power is coupled into the inner cladding of the active fiber via WDM. The gain was then calculated by using the obtained input and output signal.

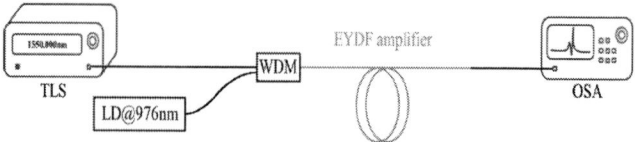

Figure 3. Diagram of an EYDFA with a LD@976nm. TLS, tunable laser source; WDM, wavelength division multiplexing; OAS, optical spectrum analyzer.

When the core is made of low doped ErYb co-doped quartz glass material, the cladding absorption at a wavelength of 975nm was found to be 1.5dB/m. The gain curve of the amplifier system is shown in Fig. 4. It can be seen that the 15dB high gain band is only 1542-1563nm, and the average gain is 16.52dB.

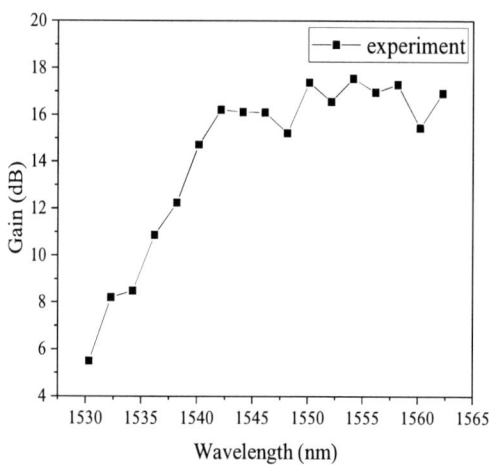

Figure 4. Experimental values for the gain curve of fiber amplifiers with low ion doping concentrations

Under the same input of signal and pump power, the quartz glass material with high concentration doping is used. The absorption coefficient of 976nm pump power by the optical fiber cladding is 2.06dB/m. The gain curve of the amplifier system is shown in Fig. 5. The 15dB high gain band is 1528-1563nm, and the average gain is 16.44dB.

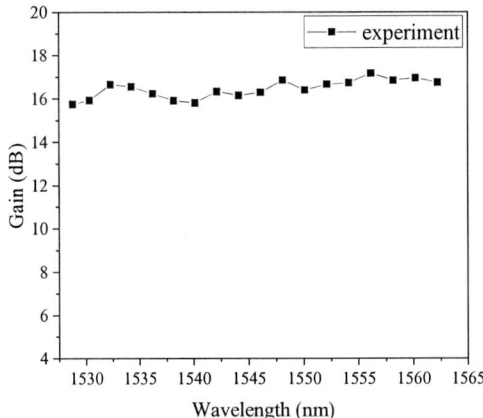

Figure 5. Experimental values for the gain curve of fiber amplifiers with high ion doping concentrations

See Fig. 6 for comparison of the gain curve of the low and high doping concentration fiber amplifier. It is found that the gain in the low gain region increases with high concentration doping, and the gain curve is basically consistent with the theoretical results. This is because increasing the Yb^{3+} ion doping concentration can improve the pump power absorption efficiency of the doped material, thus saturating the system gain and greatly broadening the gain bandwidth.

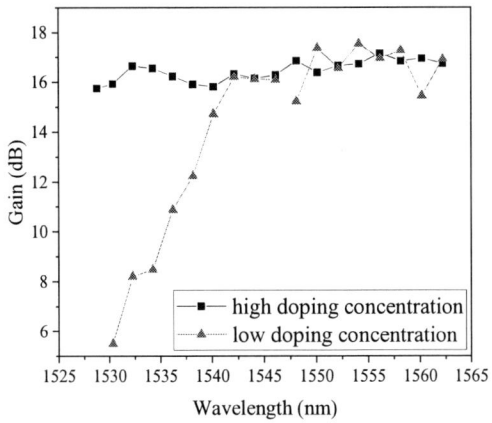

Figure 6. The black line is high doping concentration, and the red line is the low dopant concentration fiber amplifier gain curve

For EYDFA with high Yb^{3+} ion doping, as shown in Fig. 7, when the incident signal power is -20dBm per wave, the pump light power is regulated. With the increase of pump power, the gain increases gradually. When the pump power is 5W, the gain tends to be saturated. As shown in Fig. 8, the 25dB high gain band is 1531-1567nm, and the average gain is 29.6dB. Within the same wavelength range, the corresponding noise figure varies from 4.9 to 11.6 dB. This is due to the double clad fiber structure and the system setting of multi-mode pump. When the multi-mode pump

power is increased, the system can couple the high-power pump light without obvious thermal effect.

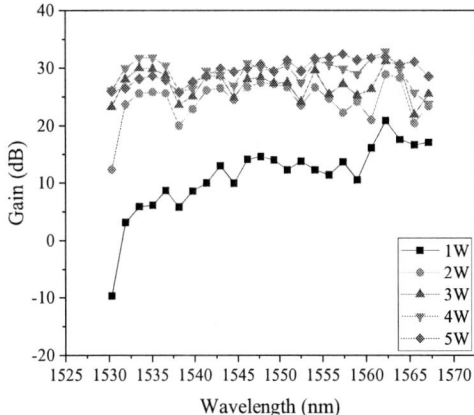

Figure 7. Gain of high concentration doped fiber amplifier at different pumping power

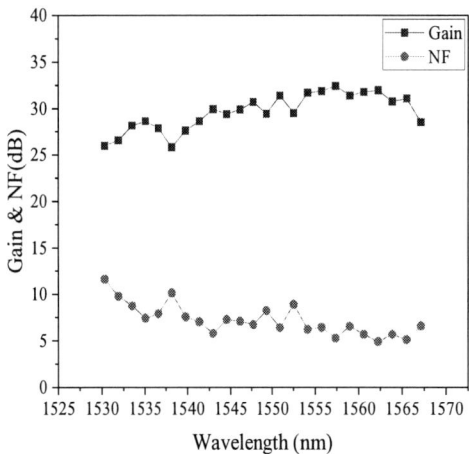

Figure 8. Gain and NF of high concentration doped fiber amplifier at 5w pump power

IV. CONCLUSION

In summary, an efficient wideband and flat gain EYADF amplifier is experimentally investigated and developed, utilizing cladding-pump scheme. The proposed amplifier comprises a high doping concentration EYDF as active fiber to fulfill amplification in C-bands. The performance of the proposed EYDF amplifier is examined in both low and high doping concentration schemes. It can be concluded that the high doping concentration amplifier performed better than the low doping concentration amplifier. At the optimum laser diodes power, the cladding-pumping amplifier attained the average gain of 29dB, throughout a wide bandwidth of 36nm, that is from 1531nm to 1567nm. Using double-cladding technique, the proposed amplifier demonstrates not only an efficient performance, but also reduces the cost because of the use of inexpensive multimode pump source. At the same time, the structure provides technical support for the development of multi-core fiber amplifier in the future. Provides a new solution for multi-core fiber cladding pumping solutions [7][8].

ACKNOWLEDGMENT

Funding：This work was supported by National Key Research and Development Program of China (Grant No. 2019YFB2204002).

REFERENCES

[1] J. Richardson, J. M. Fini, and L. E. Nelson, "Space-division multiplexing in optical fibres," Nature Photon., vol. 7, no. 5, pp. 354–362, Apr. 2013

[2] A. K. Srivastava and Y. Sun, "Advances in erbium-doped fiber amplifiers," Optical Fiber Telecommunications IVA: Components, I. P. Kamikov and T. Li, Eds. New York, NY, USA: Academic, pp. 174–212, 2002

[3] S. Jain, N. K. Thipparapu, P. Barua and J. K. Sahu, "Cladding-Pumped Er/Yb-Doped Multi-Element Fiber Amplifier for Wideband Applications," in IEEE Photonics Technology Letters, vol. 27, no. 4, pp. 356-358, 2015

[4] S. Jain, T. C. May-Smith, and J. K. Sahu, "Cladding-pumped Er/Yb-doped multi-element fiber amplifier," Proc. Workshop Specialty Opt. Fibers, paper W5.4, 2013

[5] Mario Christian Falconi, Dario Laneve, Vincenza Portosi, Stefano Taccheo, and Francesco Prudenzano, "Design of a Multi-Wavelength Fiber Laser Based on Tm:Er:Yb:Ho Co-Doped Germanate Glass," Journal of Lightwave Technology. vol. 38, no. 8, pp. 2406-2413, 2020

[6] C. Matte-Breton et al., "Modeling and Characterization of Cladding-Pumped Erbium-Ytterbium Co-Doped Fibers for Amplification in Communication Systems," Journal of Lightwave Technology, vol. 38, no. 7, pp. 1936-1944, 2020

[7] K. S. Abedin et al., "Seven-core erbium-doped double-clad fiber amplifier pumped simultaneously by side-coupled multimode fiber," Opt. Lett, vol. 39, no. 4, pp. 993–996, 2014

[8] B. J. Puttnam et al., "High Data-Rate and Long Distance MCF Transmission With 19-Core C+L band Cladding-Pumped EDFA," Journal of Lightwave Technology, vol. 38, no. 1, pp. 123-130, 2020

Proceedings of the 7th Optoelectronics Global Conference (OGC 2022)

Ultra Narrow Linewidth Distributed Feedback Fiber Laser Based on Self-injection Locking

Meng Zou

School of Optical and Electronic Information
Huazhong University of Science and Technology
Wuhan, China
Wuxi Research Institute，Huazhong University of
Science and Technology, Wuxi, China
e-mail: D202180853@hust.edu.cn

Kai Shen

Wuxi Research Institute，Huazhong University of
Science and Technology, Wuxi, China
e-mail: shenkai@hust-wuxi.com

Qizhen Sun

School of Optical and Electronic Information
Huazhong University of Science and Technology
Wuhan, China
Wuxi Research Institute，Huazhong University of
Science and Technology, Wuxi, China
e-mail: qzsun@hust.edu.cn

Zhijun Yan

School of Optical and Electronic Information
Huazhong University of Science and Technology
Wuhan, China
Wuxi Research Institute，Huazhong University of
Science and Technology, Wuxi, China
e-mail: yanzhijun@hust.edu.cn (corresponding author)

Abstract—We have reported an ultra narrow linewidth fiber laser based on π phase-shift fiber Bragg grating (π-FBG) and self-injection locking, in which the π-FBG is inscribed on Erbium-ytterbium co-doped fiber with scanning phase mask method. Using self-injection locking, the relaxation oscillation frequency (ROF) peak was reduced about 25 dB from -103 dB/Hz to -128 dB/Hz. The 20-dB linewidth of the laser was suppressed to around 500 Hz.

Keywords-fiber laser; narrow linewidth; low noise; π-FBG; self-injection locking

I. INTRODUCTION

Single frequency, narrow linewidth fiber lasers have attracted considerable interest for a variety of applications due to their promising performances of low noise and compact structure, such as optical sensing, coherent optical communications, LIDAR and spectroscopy[1-3]. However, in the high precision optical measurement field, such as gravitational wave detection[4], optical atomic clock[5,6] and optical frequency comb[7], lower noise, narrower linewidth and stable laser is desirable, so further suppressing the noise and compressing the linewidth of single frequency fiber laser have become the focus of current research.

Various techniques have been reported to obtain narrow linewidth laser at 1550 nm. Ring cavity fiber laser associated with narrow band filters can realize narrow linewidth lasers[8,9], but the filters introduce large loss and the ring cavity is susceptible to the environmental fluctuations. Stimulated Brillouin and Rayleigh fiber lasers can compress the laser linewidth to a few hertz[10,11], but these lasers usually need very long fiber to generate stimulated Brillouin or Raleigh scattering laser, which make them difficult to engineer. Short line cavity fiber lasers include distributed

Bragg reflector fiber laser (DBR-FL) and distributed feedback fiber laser (DFB-FL), due to the short cavity length, DBR-FL and DFB-FL can avoid longitude hopping and realize single longitude mode laser, DBR-FL consist of a pair of fiber Bragg gratings (FBGs) and a length of active fiber, and DFB-FL is composed of a π-FBG inscribed on the active fiber, the π-FBG is formed by introducing a π phase shift in the middle of the FBG, the cavity length of DBR-FL and DFB-FL are on the order of centimeters, which are easy to be integrated.

Generally, techniques of suppressing laser intensity noise or compressing laser linewidth are as follows: photoelectric feedback, saturable absorption effect and optical injection locking. Based on photoelectric feedback, low frequency intensity noise of the laser can be effectively inhibited but the suppression bandwidth is limited by the circuit[12]. Saturable absorber such as semiconductor optical amplifier (SOA)[13] and unpumped doped fiber[14] can suppress the intensity noise of the laser due to the saturation absorption effect, however, the disadvantages are that the saturable absorber will reduce the optical power and introduce spontaneous radiated noise. Compared with external cavity injection locking, self-injection locking does not need a high quality factor external cavity or another ultra stable laser, researchers have theoretically studied the self-injection locking of fiber lasers[15].

In this paper, we reported a ultra narrow linewidth DFB-FL based on a homemade π-FBG and self-injection locking. 10-m delay fiber was employed to extend the feedback loop of self-injection locking. The π-FBG, delay fiber and other passive optical devices were encapsulated in a sealed box with vibration isolation and insulation materials. After self-injection locking, the ROF peak of the laser decreased to -128 dB/Hz which was better than commercial narrow

978-1-6654-8699-6/22 $31.00 © 2022 IEEE

linewidth fiber laser, and the linewidth was narrowed to 500 Hz.

II. EXPERIMENTAL SETUP

The configuration of the ultra narrow linewidth DFB-FL based on self-injection locking is shown in Fig. 1. A 980 nm laser diode (LD) pump source was launched into the homemade π-FBG through a 980/1550 wavelength division multiplexer (WDM). The π-FBG of 5 cm length was fabricated in 6-cm long Erbium-ytterbium co-doped fiber (Coractive EY305) spliced with pigtails of passive fiber. The transmission spectrum of the π-FBG was tested by a broadband source and an optical spectral analyzer (OSA). The test result is shown in Fig. 2, from which we can see that a transmission peak is located in the middle of reflection bandwidth of the π-FBG, and the reflectivity of the FBG is over 99.9%. The absorption coefficient of the EY305 active fiber is 2.4 dB/m at 980 nm and 85 dB/m at 1535 nm respectively. An isolator (ISO) was adopted to ensure the unidirectional light propagation. The 10-m length delay fiber was employed to extend photon lifetime in the cavity. The delay fiber was spliced to the 1 port of a circulator (CIR) and the 2 port of the CIR was spliced to the 1550 nm port of the WDM, and the 3 port was used as the output port. The π-FBG was encapsulated with vibroisolating material, the temperature near the π-FBG was monitored by a thermistor, a thermoelectric cooler (TEC) was used to cooling the π-FBG and keep temperature constant inside. The entire optical path was packaged in a sealed box and fixed with flexible glue.

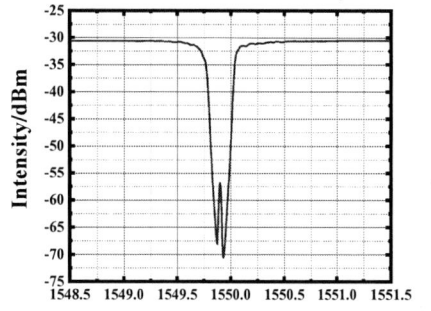

Figure 1. Schematic of DFB-FL based on self-injection locking.

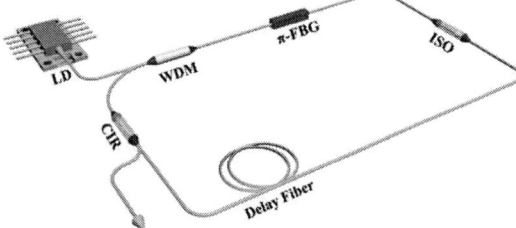

Figure 2. The transmission spectrum of the π-FBG.

The laser fundamental thermal noise is inversely proportional to the efficient cavity length, and the relation between the ROF and the cavity photon lifetime is [16,17]:

$$f_R = \sqrt{\frac{AP_0}{\tau_c} - \frac{1}{2}\left(\frac{1}{\tau_g} + AP_0\right)^2} \quad (1)$$

Therefore, one can increase the laser efficient cavity length to compress the laser linewidth and reduce the intensity noise. By self-injection locking, the laser efficient cavity length would be extended while the laser physical length remains unchanged, we adopted 10-m delay fiber in the experiment, considering environmental disturbances, longer delay fiber was inappropriate.

III. EXPERIMENTAL RESULTS

Firstly, we measured the laser output power varies with the pump power. As shown in Fig. 3(a), the threshold of the laser was approximately 62 mW, corresponding to the slope efficiency of about 0.15%. The measured optical spectrum of the DFB-FL is shown in Fig. 3(b), the center wavelength of the laser is 1549.8 nm, and the optical signal to noise ratio (SNR) is nearly 70 dB. The single longitudinal mode output of the DFB-FL was confirmed by using a scanning Fabry-Perot interferometer (FPI) with free spectral range of 4 GHz and fineness of 2000 as illustrated in Fig. 4(a), the absence of any peaks between the main resonances of the FPI indicates the operation on only one longitudinal mode. The relative intensity noise (RIN) was measured with the frequency analyzer (ESA), the background noise of the ESA and the photoelectric detector (PD) were also measured, as a contrast, we also tested the commercial NKT E15 fiber laser, the results were shown in Fig. 4(b), from which the ROF shifted from 500 kHz to 400 kHz and the ROF peak was reduced by about 25 dB from -103 dB/Hz to -128 dB/Hz after self-injection locking, and the noise level was better than that of E15.

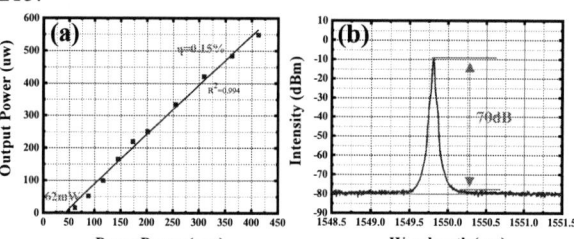

Figure 3. (a) Output power versus pump power, (b) measured optical spectrum of the DFB-FL.

We evaluated the linewidth of the DFB-FL with the self-heterodyne method involving a 50-km fiber delayed Mach-Zehnder interferometer (MZI) and a 200-MHz fiber-coupled acousto-optical modulator (AOM), the result is shown in Fig. 5(a), from which we can see that the 20-dB Lorentz linewidth of the free running DFB-FL is about 5 kHz, and decreased to 500 Hz after self-injection locking. The phase noise of the DFB-FL was measured with the 50-m fiber delayed MZI and AOM, the result is shown in Fig. 5(b), the phase noise decreased from -30 dBc/Hz to -40 dBc/Hz at 3 kHz, while the NKT E15 fiber laser has lower phase noise of -50 dBc/Hz at 3kHz.

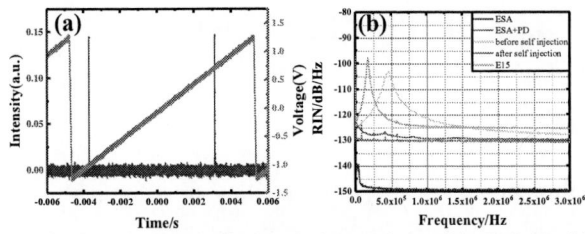

Figure 4. (a) Longitudinal mode characteristics of the DFB-FL, (b) Measured RIN of the DFB-FL.

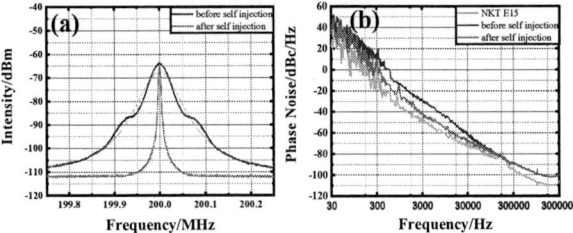

Figure. 5. (a) Measured self-heterodyne spectra of the DFB-FL, (b) measured phase noise of the DFB-FL.

IV. CONCLUSION

In this paper, we demonstrated an ultra narrow linewidth DFB-FL based on a homemade π-FBG and self-injection locking. After self-injection locking, the ROF peak of the laser decreased to -128 dB/Hz. The laser linewidth was narrowed to 500 Hz, and the phase noise decreased to -40 dBc/Hz at 3 kHz.

ACKNOWLEDGMENT

This work was supported in part by the National Key Research and Development Program of China (2018YFB2100902); in part by the National Natural Science Foundation of China (No. 62075071).

REFERENCES

[1] Y. L. Liu, W. T. Zhang and T. W. Xu, "Fiber laser sensing system and its applications," Photonic Sensors, 1(1), 43-53(2011)

[2] E. Ronnekleiv, S. W. Lovseth and J. T. Kringlebotn, "Er-doped fiber distributed feedback lasers-properties, applications and design considerations," Photonics Fabrication, 4943, 69-80(2002)

[3] H. F. Qi, Z. Q. Song, J. S. Ni, et al., "An amplified distributed feedback fiber laser for distributed and interference sensing," Optoelectronics and Microelectronics Technology and Application, 10244, 1-5(2016)

[4] B. P. Abbott, "Observation of gravitational waves from a binary black hole merger," Physical Review Letters, 116(061102), 1-16(2016)

[5] S. L. Campbell, R. B. Hutson, "A fermi-degenerate three-dimensional optical lattice clock," Science, 358(6359), 90–94(2017)

[6] A. D. Ludlow, M. M. Boyd, "Optical atomic clocks," Reviews Of Modern Physics, 87(2), 637-701(2014)

[7] G. M. Brodnik, M. W. Harrington, J. H. Dallyn, et al, "Optically synchronized fibre links using spectrally pure chip-scale lasers," Nature Photonics, 15, 588-593(2021)

[8] S. Junjie, W. Zefeng, et al., "Watt-level tunable 1.5 μm narrow linewidth fiber ring laser based on a temperature tuning π-phase-shifted fiber Bragg grating," Applied Optics. 56(32), 9114-9118(2017)

[9] X. P. Cheng, P. Shum, et al., "Single-Longitudinal-Mode Erbium-Doped Fiber Ring Laser Based on High Finesse Fiber Bragg Grating Fabry–Pérot Etalon," Photonics Technology Letters, 20(12), 976-978(2008)

[10] T. Zhu, X. Y. Bao, et al., "A Single Longitudinal-Mode Tunable Fiber Ring Laser Based on Stimulated Rayleigh Scattering in a Nonuniform Optical Fiber," Journal of Lightwave Technology. 29(12),1802-1807(2011)

[11] S. H. Huang, T. Zhu, et al., "Tens of hertz narrow-linewidth laser based on stimulated Brillouin and Rayleigh scattering," Optics Letters, 42(24), 5286-5289(2017)

[12] Y. Xiao, C. Li, S. H. Xu, et al., "Simultaneously Suppressing Low-Frequency and Relaxation Oscillation Intensity Noise in a DBR Single-Frequency Phosphate Fiber Laser,"Chinese Physics Letters, 32(6), 064205(2015)

[13] D. Gwennael, B. Francois, "GHz bandwidth noise eater hybrid optical amplifier: design guidelines," Optics Letters, 39(14), 4239-4242(2014)

[14] Z. Q. Pan, J. Zhou, "Low-frequency noise suppression of a fiber laser based on a round-trip EDFA power stabilizer," Laser Physics, 23, 035105(2013)

[15] H. Xiang, Z. Qilai, L. Wei, et al., "Linewidth suppression mechanism of self-injection locked single-frequency fiber laser," Optics Express, 24(17), 18907-18916(2016)

[16] F. Scott, "Fundamental limits on 1õf frequency noise in rare-earth-metal-doped fiber lasers due to spontaneous emission," Phys. Rev. 78, 013820(2008)

[17] M. Shupei, H. Xiang, et al., "Compact slow-light single-frequency fiber laser at 1550 nm," Applied Physics Express, 8,082703(2015)

Proceedings of the 7th Optoelectronics Global Conference (OGC 2022)

Er^{3+}-Pr^{3+}-Yb^{3+} Tri-doped La$_2$O$_3$-Al$_2$O$_3$-SiO$_2$ Glass Double Clad Fiber for C+L Amplification

Zhuoyuan Huang[a,b], Weichao Ma[a,b], Tong Wu[a,b], Jia ao Lu[a,b], Jiantao Liu[a,b], Changming Xia[a,b,*], Zhiyun Hou[a,b], Guiyao Zhou[a,b]

[a] Guangzhou Key Laboratory for Special Fiber Photonic Devices and Applications, South China Normal University, Guangzhou, 510006, China
[b] Specially functional fiber Engineering Technology Research Center of Guangdong Higher Education Institutes, South China Normal University, Guangzhou, 510006, China
* Corresponding author. E-mail address: xiacmm@126.com (C. Xia).

Abstract—**Er^{3+}-Pr^{3+}-Yb^{3+} tri-doped double clad fiber was fabricated by the stack-and-draw technology. The fiber core Er^{3+}-Pr^{3+}-Yb^{3+} tri-doped La$_2$O$_3$-Al$_2$O$_3$-SiO$_2$ glass was prepared using the conventional melting method. The fluorescence properties of fiber were experimentally investigated. The result suggested Er^{3+}-Pr^{3+}-Yb^{3+} tri-doped double clad fibers are potential material for broadband light source and C+L amplification.**

Keywords- Double clad fiber; Er^{3+}-Pr^{3+}-Yb^{3+} tri-doped La$_2$O$_3$-Al$_2$O$_3$-SiO$_2$ glass; rare earth doped; C+L band; broadband emission

I. INTRODUCTION

Recently few years, the C+L (1530-1625nm) fiber amplifiers and light sources have been attracting more attention. Broadband fiber light source has broad applications in the areas of optical measurements [1-3]. And C+L amplifiers could be used in the areas of optical communications [4-12].

In most cases, the gain medium is a glass fiber doped with rare earth ions such as erbium-doped fiber amplifier [5], Bi/Er co-doped Optical fiber [3], Bi/Er/La co-doped silica optical fiber [1]. In addition, some rare earth glasses have been used for light sources with C+L such as Yb/Er/Tm co-doped glass [2] and Ce/Yb/Er triply doped bismuth borosilicate glass [12].

In this paper, Er^{3+}-Pr^{3+}-Yb^{3+} tri-doped La$_2$O$_3$-Al$_2$O$_3$-SiO$_2$ glass was prepared. With the prepared glass as fiber core, Er^{3+}-Pr^{3+}-Yb^{3+} tri-doped double clad fiber was firstly fabricated. And the optical property of the photonic crystal fiber was experiment investigated.

II. EXPERIMENT

A. Glass Preparation

The glass with the composition of 0.1Er$_2$O$_3$-0.2Pr$_6$O$_{11}$-0.3Yb$_2$O$_3$-69.4SiO$_2$-21Al$_2$O$_3$-9La$_2$O$_3$ (mol%) was prepared. Firstly, the raw materials were mixed thoroughly, after heating at 1720 °C for 3 h, then further annealed at 950 °C for 2 h was carried out. Finally, they were cooled to room temperature. Finally, Er^{3+}-Pr^{3+}-Yb^{3+} tri-doped glasses were successfully prepared [13].

B. Er^{3+}-Pr^{3+}-Yb^{3+} Tri-doped Double Clad Fibers Fabrication

Double clad fiber preform was fabricated using the stack-and-draw technology. The glass was polished with a diameter of 3mm for fiber core. Afterwards, the preform was dried and sealed. It was then placed in the feeder of the drawing tower. Finally, the preform was drawn into fiber under the suitable temperature. The cross section of fiber is shown in Fig. 1.

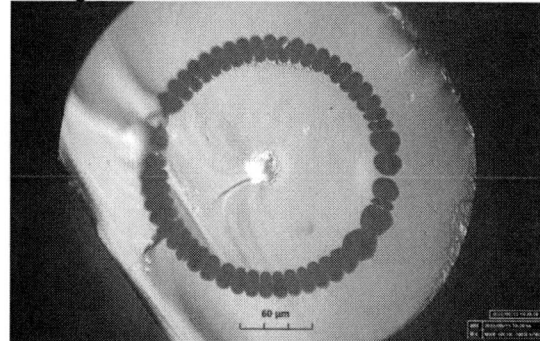

Figure 1.   Cross section of double clad fiber

C. Optical Properties Test

In order to investigate the optical properties, the glass samples were polished with great care to the ideal optical quality. The thickness was 3.54mm. The optical properties of the glass samples and fibers were measured by using a spectrometer (Maya 2000 Pro, Ocean Optics) and another spectrometer (NIRQUEST 256, Ocean Optics) in the range of 1200 nm to 1900 nm under the excitation of 976nm laser. All optical measurements were performed at ambient temperature.

III. RESULT AND DISCUSSION

Fig. 1 shows the cross section of double clad fiber, Er^{3+}-Pr^{3+}-Yb^{3+} tri-doped La$_2$O$_3$-Al$_2$O$_3$-SiO$_2$ glass is used as the core. The diameter of the fiber is ~240 μm and the doped core region diameter is ~24 μm with high refractive index. The inner cladding refractive index is lower than core refractive index. The air hole outer cladding created a low refractive index region that gradually limits the light

978-1-6654-8699-6/22 $31.00 © 2022 IEEE

propagation. This structure is helpful to reduce the leakage of light and improve the utilization efficiency of pump light.

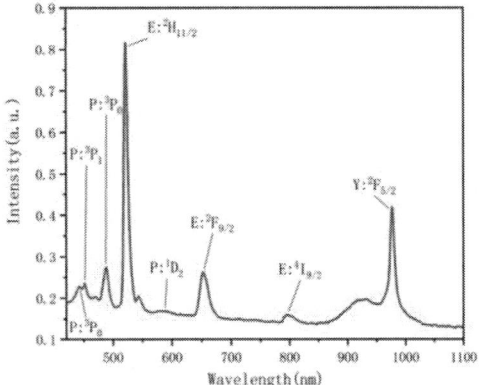

Figure 2. Absorption spectrum of Er^{3+}-Pr^{3+}-Yb^{3+} tri-doped La_2O_3-Al_2O_3-SiO_2 glass

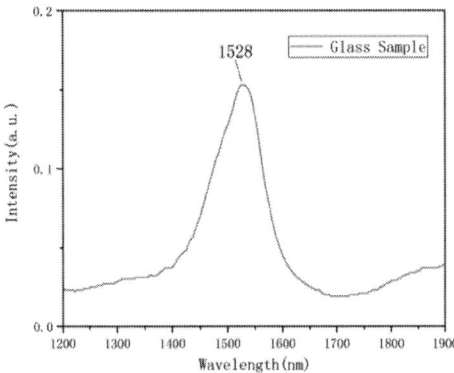

Figure 3. Near-infrared absorption spectrum of Er^{3+}-Pr^{3+}-Yb^{3+} tri-doped La_2O_3-Al_2O_3-SiO_2 glass

Figure 2 and Figure 3 shows the absorption spectra of the prepared glass samples. And the transition is discussed. Two Yb^{3+} absorption peaks with centers at 930 and 976 nm, which assigned to the transition from the ground state $^2F_{7/2}$ to the excited state $^2F_{5/2}$. For the absorption spectrum of Pr^{3+}, we have four peaks at 440, 470, 487, 588 and 1528 nm due to the transition from the ground state $^3H_4$ to the excited states $^3P_2$, $^3P_1$, $^3P_0$, $^1D_2$ and $^3F_3$, respectively. Furthermore, the absorption peaks located at 520, 650, 800 and 1528 nm, which correspond to the transition of Er^{3+} from the ground state $^4I_{15/2}$ to the excited states $^2H_{11/2}$, $^2F_{9/2}$, $^4I_{9/2}$ and $^4I_{13/2}$, respectively. And 976 nm excitation was chosen because of the intense absorption.

Fig. 4 shows the near-infrared emission of the Er^{3+}-Pr^{3+}-Yb^{3+} tri-doped glass under a 976 nm excitation. There is an intense emission peak centered at 1542nm which can cover the whole C+L band. And the FWHM is up to 100nm. which provides a way to achieve C+L bands amplification.

To investigate the near-infrared emission property of the prepared double clad fiber, the optical measurement was carried out. Figure 5 shows the experiment setup of double clad fiber. And Figure 6 shows the emission of the Er^{3+}-Pr^{3+}-Yb^{3+} tri-doped double clad fiber. The emission centered at

1598nm with a 110nm FWHM covering C+L band is observed. There is great potential for C+L bands amplification.

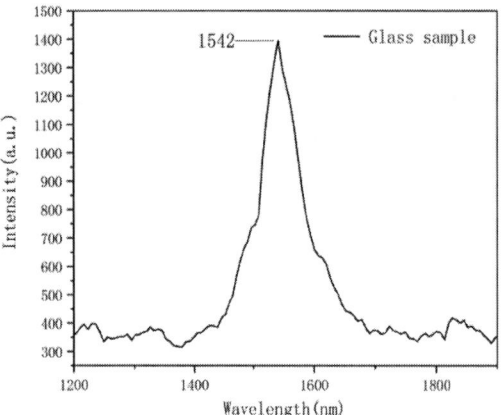

Figure 4. Near-infrared emission spectrum of the Er^{3+}-Pr^{3+}-Yb^{3+}tri-doped glass samples under a 976 nm excitation.

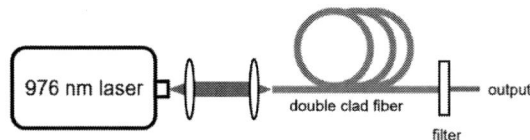

Figure 5. The diagram for measuring the fluorescence of Er^{3+}-Pr^{3+}-Yb^{3+} tri-doped double clad fiber

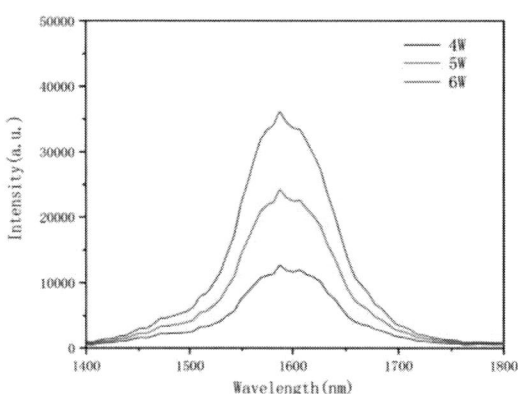

Figure 6. Near-infrared emission spectrum of the Er^{3+}-Pr^{3+}-Yb^{3+}tri-doped double clad fiber under a 976 nm excitation.

IV. CONCLUSION

Er^{3+}-Pr^{3+}-Yb^{3+} tri-doped La_2O_3-Al_2O_3-SiO_2 glass was prepared for fiber core glass. And Er^{3+}-Pr^{3+}-Yb^{3+} tri-doped double-clad fiber was fabricated. The near-infrared emission spectrum of the Er^{3+}-Pr^{3+}-Yb^{3+}tri-doped glass under a 976 nm excitation can cover the whole C+L wave band and the broadband emission FWHM centered at 1598nm is up to 110nm. It suggests that Er^{3+}-Pr^{3+}-Yb^{3+} triply doped fiber is a promising host material for C+L amplification.

ACKNOWLEDGMENT

National Key Research and Development Program of China (Grant No. 2019YFB2204002)

REFERENCES

[1] L Yang, J Wen, L Zeng, Y Wu, S Huang, F Pang, et al., "Light sources in L-band based on a Bi/Er/La co-doped silica optical fiber," 2021 19th International Conference on Optical Communications and Networks (ICOCN), 2021, pp. 01-03, doi: 10.1109/ICOCN53177.2021.9563835.

[2] Pisarski WA, Pisarska J, Lisiecki R, Ryba-Romanowski W. Broadband Near-Infrared Luminescence in Lead Germanate Glass Triply Doped with Yb3+/Er3+/Tm3+. Materials. 2021; 14(11):2901. https://doi.org/10.3390/ma14112901

[3] D. Song, J. Zhang, S. Fang, W. Sun, Z. M. Sathi, Y. Luo and G. Peng, "Bismuth and Erbium Co-doped Optical Fiber for a White Light Fiber Source," Optics and Photonics Journal, Vol. 3 No. 2B, 2013, pp. 175-178. doi: 10.4236/opj.2013.32B042.

[4] Ravi Shankar Mishra and Aditya Goel, "Analysis of C+L Band Erbium Doped Fiber Amplifier for Optical Network using Dual Forward Pumping", BSSS Journal of Computer, ISSN : 0975-7228, Vol.1, Issue 1 (2009) pp. 20-29

[5] J Li, L Wang, J Du, Z He , C Cai , L Zhu, et al., "C + L Band Distributed Few-mode Raman Amplification with Flattened Gain for Mode-division-multiplexed Optical Transmission over 75-km Few-mode Fiber," 2018 Optical Fiber Communications Conference and Exposition (OFC), 2018, pp. 1-3.

[6] Supe A, Olonkins S, Udalcovs A, Senkans U, Mūrnieks R, Gegere L, et al. Cladding-Pumped Erbium/Ytterbium Co-Doped Fiber Amplifier for C-Band Operation in Optical Networks. Applied Sciences. 2021; 11(4):1702. https://doi.org/10.3390/app11041702

[7] D. Aguiar, G. Grasso, A. Righetti and F. Meli, "EDFA with continuous amplification of C and L bands for submarine applications," 2015 SBMO/IEEE MTT-S International Microwave and Optoelectronics Conference (IMOC), 2015, pp. 1-4, doi: 10.1109/IMOC.2015.7369198.

[8] Lipatov DS, Lobanov AS, Guryanov AN, Umnikov AA, Abramov AN, Khudyakov MM, Likhachev ME, Morozov OG. Fabrication and Characterization of Er/Yb Co-Doped Fluorophosphosilicate Glass Core Optical Fibers. Fibers. 2021; 9(3):15. https://doi.org/10.3390/fib9030015

[9] Zakis K, Olonkins S, Udalcovs A, Lukosevics I, Prigunovs D, Grube J, et al. Cladding-Pumped Er/Yb-Co-Doped Fiber Amplifier for Multi-Channel Operation. Photonics. 2022; 9(7):457. https://doi.org/10.3390/photonics9070457

[10] L Wang, D He, S Feng, C Yu, L Hu, J Qiu et al. Yb/Er co-doped phosphate all-solid single-mode photonic crystal fiber. Sciencetific Report, Vol. 4, 6139 (2014). https://doi.org/10.1038/srep06139

[11] T Sakamoto, A Mori, H Masuda, H Ono. "Wideband Rare-earth-doped Fiber Amplification Tech nologies—Gain Bandwidth Expansion in the C and L bands Special Feature." Ntt Technical Review, Vol. 2.12(2004).

[12] Y Chu, J Ren, J Zhang, G Peng, J Yang, P Wang, et al., $Ce^{3+}/Yb^{3+}/Er^{3+}$ triply doped bismuth borosilicate glass: a potential fiber material for broadband near-infrared fiber amplifiers. Scientific Reports, (2016) ,6(1), 33865. doi:10.1038/srep33865

[13] Z Huang, J Yang, Z Mo, J Lu, C Xia, Z Hou, G Zhou, Visible and near-infrared emission and energy transition of Er^{3+}-Pr^{3+}-Yb^{3+} tri-doped lanthanum aluminium silicate glasses, Journal of Non-Crystalline Solids, Volume 591,2022,121718, ISSN 0022-3093, https://doi.org/10.1016/j.jnoncrysol.2022.121718.

Proceedings of the 7th Optoelectronics Global Conference (OGC 2022)

Evaluation Method of Polarization State Characteristic in Forward Transmission

Zeng Xiangwei, Chen Xueye, Zhang Quanzhong,
Li Xiaoyu
Ludong University
College of Transportation
Yantai, China
e-mail: zengxw163@163.com

Li Yahong
Dalian Polytechnic University
Research Institute of Photonics
Dalian, China
e-mail: lrene1129@163.com

Abstract—**An evaluation method to evaluate polarization state characteristic in forward transmission is introduced in this paper. To remedy the evaluation insufficient of Stokes parameters, we defined a new dimensionless parameter: *RoPS* (retention rate of polarization state). Its meaning is to describe the retention rate of polarization state of forward scattered light. Compared with Stokes parameters, the new parameter can avoid the effect introduced by calculation of orthogonal component intensity difference. In short, *RoPS* is a transformation of Stokes-Mueller calculation form. *RoPS* can avoid the effect introduced by calculation of orthogonal component intensity difference. This research offers potential application values in communication, detection, and so on.**

Keywords-Polarization; scattering; optical testing

I. INTRODUCTION

When light transmits in scattering environments such as clouds, fog, haze, and turbid waters, the energy distribution of scattered light usually peaks in forward direction. This means that the mostly scattered light transmits forward [1]. So, it is an important research work to explore the forward propagation characteristics of light. Among the forward propagation characteristics, the polarization state persistence (also referred to as "polarization memory" [2] acts an important role in detection [3-5], communications [6], medicine [7], biology [8.9], navigation [10] and so on [11-13].

Stokes parameters [14] and their transformation forms (such *DoP* [15], *DoP*diff [16,17]) have been widely used to distinguish the change of polarization state. Stokes parameters are related to the intensity difference between the defined six specific polarization states. This leads to Stokes parameters are incompletely related to the change of polarization state. It not only affects the reliability of Stokes parameters to distinguish the change of polarization state, but also restricts the in-depth development of the polarization transmission law and mechanism. Therefore, how to reasonably evaluate the polarization state characteristic of light in forward transmission is an urgent problem.

To remedy the evaluation insufficient of Stokes parameters, we defined a new parameter: *RoPS* (retention rate of polarization state) in section 2. Its core is to describe the retention rate of polarization state of forward scattered light. In section 3 and section 4, we explored the relationship and difference of *RoPS* and Stokes parameters. Compared with Stokes parameters, *RoPS* has higher calculation accuracy.

II. DEFINITION

To describe the retention rate of polarization state of forward scattered light, we defined a new parameter: *RoPS* (retention rate of polarization state). *RoPS* is express

$$RoPS = \frac{\pm P_{\tau\text{-forward}}}{P_{0\text{-forward}}} \tag{1}$$

where, $P_{0\text{-forward}}$ is the total intensity of forward transmitted light, $P_{\tau\text{-forward}}$ is the intensity of forward transmitted τ-polarized light. τ-polarized light has the same polarization state as the initial incident light. $\pm$ represents the type of τ-polarized light as shown in figure1. When τ-polarized light is + type, the range of *RoPS* is at [0,1]; when τ-polarized light is - type, the range of *RoPS* is at [-1,0].

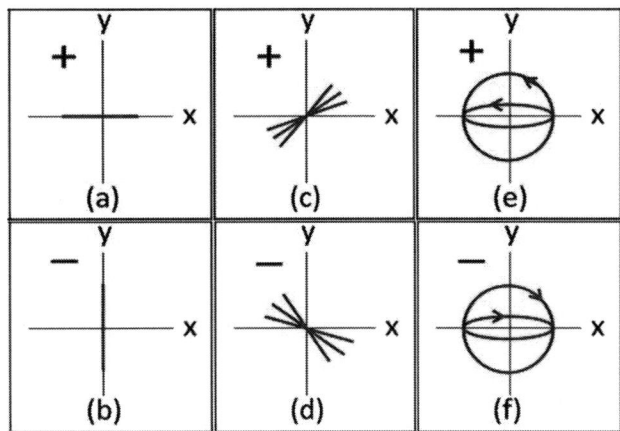

Figure 1. Express of τ-polarized light. The τ-polarized light has the same polarization state as the initial light. $\pm$ represents the type of τ-polarized light, such as (a) Horizontal, (b) Vertically, (c) (0°,90°), (d) (-90°,0°), (e) Left-handed or left-elliptic, (f) Right-handed or right-elliptic.

III. EXPERIMENT

We designed a set of comparative experiments to compare Stokes parameter and *RoPS*. Experimental environments were polystyrene suspensions composed of polystyrene microspheres and water. Their densities were 0,

978-1-6654-8699-6/22 $31.00 © 2022 IEEE

1.04, 2.08, 4.17, 8.33, 12.5, 16.6, and 20.8$\mu g/\mu m^3$, respectively. The particle diameter of polystyrene suspensions was 1±0.05μm. Initial incident light is 45° linearly polarized light.

Experiment setup is showed in Fig. 2. Firstly, a 532nm laser passed though polarizer P1 to generate fully 45° linearly polarized light. It was recorded as S_{in}. Then, 45° linearly polarized light passed through polystyrene suspensions. It was recorded as S_{out}. Finally, light was received by detector after passing through objective lens and polarizer P2. It was recorded as S_{mea}.

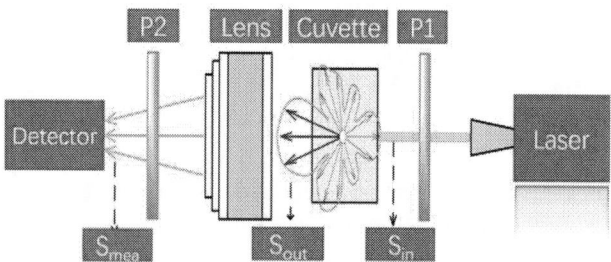

Figure 2. Schematic diagram of polystyrene suspension experiment

Fig 3 shows the experimental set up. The detector is an AvanSpec-ULS2048LTEC spectrometer made by AVANTES. We debugged and calibrated the setup by letting 45°linearly polarized light transmit through pure water solution. $s_{2\text{-forward}}$ and *RoPS* was 0.9879 and 0.9903, respectively. It suits verification experiments.

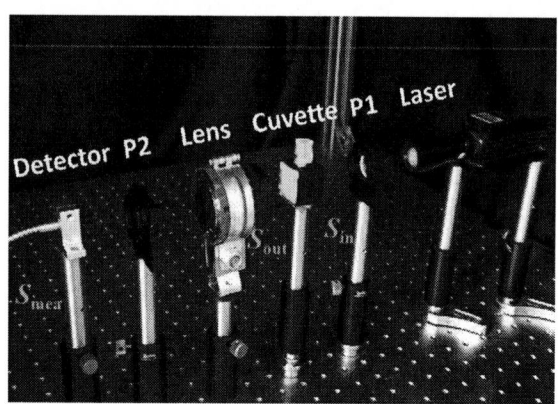

Figure 3. Experimental setup

We measured *RoPS* and normalized Stokes parameters of output light in the experiment. They can be obtained by measuring the Mueller Matrix of scattering media. To solve the Mueller matrix of scattering media, we chose 4 different kinds of polarized light as the input light and measured their corresponding output light. All input Stokes vectors were assembled together to establish the input matrix whose rank is 4. The same way was used to establish the output matrix. All Stokes vectors were normalized to dismiss the absorption effect in our experiments. Then, the relationship between input matrix and output matrix can be described as

$$[\mathbf{S}_{out-1}, \mathbf{S}_{out-2}, \mathbf{S}_{out-3}, \mathbf{S}_{out-4}] = \mathbf{M}_p \cdot [\mathbf{S}_{in-1}, \mathbf{S}_{in-2}, \mathbf{S}_{in-3}, \mathbf{S}_{in-4}] \quad (2)$$

where $\mathbf{M}_p$ is Muller matrix of polystyrene suspensions. S_{out} and S_{mea} is calculated as

$$S_{out} = \mathbf{M}_P S_{in} \quad (3)$$

$$S_{mea} = P_2 \mathbf{M}_P S_{in} \quad (4)$$

Muller matrix of polystyrene suspensions is set as

$$\mathbf{M_P} = \begin{bmatrix} M_{00} & M_{01} & M_{02} & M_{03} \\ M_{10} & M_{11} & M_{12} & M_{13} \\ M_{20} & M_{21} & M_{22} & M_{23} \\ M_{30} & M_{31} & M_{32} & M_{33} \end{bmatrix}_P \quad (5)$$

S_{out} can be solved by P2 and S_{mea}. Then, S_{out} can be expanded as

$$S_{out} = \mathbf{M}_P S_{in} = \begin{bmatrix} M_{00} + M_{02} \\ M_{10} + M_{12} \\ M_{20} + M_{22} \\ M_{30} + M_{32} \end{bmatrix}_P \quad (6)$$

All input Stokes vectors were assembled together to establish the input matrix which rank is 4. The relationship between input matrix and output matrix can be described as

$$\begin{aligned} &[S_{out-1}, S_{out-2}, S_{out-3}, S_{out-4}] \\ &= \mathbf{M}_p [S_{in-1}, S_{in-2}, S_{in-3}, S_{in-4}] \end{aligned} \quad (7)$$

After solving the Mueller matrix $\mathbf{M}_p$, the normalized Stokes parameter $s_{2\text{-forward}}$ can be calculated as

$$s_{2\text{-forward}} = \frac{M_{20} + M_{22}}{M_{00} + M_{02}} \quad (8)$$

RoPS is calculated as

$$RoPS = \frac{1}{2}\left(\frac{M_{20} + M_{22}}{M_{00} + M_{02}} + 1\right) \quad (9)$$

Experimental results are shown in figure 4. *RoPS* values become higher than $s_{2\text{-forward}}$ values with the increase of polystyrene suspension's density. $s_{2\text{-forward}}$ is related to the intensity difference between 45° and -45° linearly polarized light. With the increase of scattering events, the effect caused by the intensity difference is continuously expanded. Finally, it causes $s_{2\text{-forward}}$ values became lower than *RoPS* values.

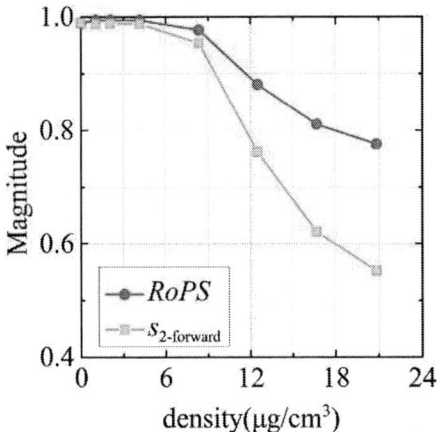

Figure 4. Experimental results of *RoPS* and $s_{2\text{-forward}}$ in polystyrene suspensions

IV. DISCUSSION

This section explores the relationship and difference of *RoPS* and Stokes parameters.

+ 45° linearly polarized light passing through a droplet was taken as an example as shown in Fig. 5. Spherical droplet radius is 5μm. Its index of refraction is 1.597. The wavelength of incident light is 670nm.

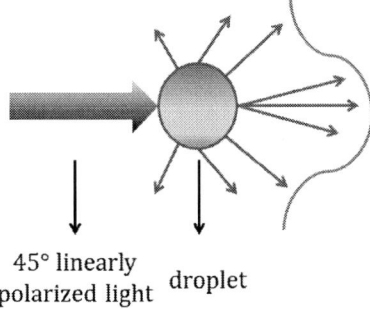

45° linearly polarized light   droplet

Figure 5. Simulation scenario of 45°-linearly polarized light passes through a droplet

The normalized Stokes parameter $s_{2\text{-forward}}$ can be calculated as

$$s_{2\text{-forward}} = \frac{P_{45\text{-forward}} - P_{-45\text{-forward}}}{P_{45\text{-forward}} + P_{-45\text{-forward}}} \quad (10)$$

where, $P_{45\text{-forward}}$ and $P_{-45\text{-forward}}$ represent the intensity of +45° and -45° linearly polarized light in forward scattered light, respectively.

Whether $s_{2\text{-forward}}$ can distinguish the polarization state change of forward scattered light? To answer this question, we calculated $s_{2\text{-forward}}$ in the range of -90° to 90° scattering angle. As shown in Fig. 6, $s_{2\text{-forward}}$ has some negative values in -80° to -90° and 80° to 90°. When $s_{2\text{-forward}}$ value is negative, the proportion of -45° linearly polarized light is higher than +45° linearly polarized light. It is difficult to perceive the intensity proportion of +45° linearly polarized light to output light. This means that the normalized Stokes

parameter $s_{2\text{-forward}}$ cannot distinguish the change of polarization state of forward scattered light in this situation.

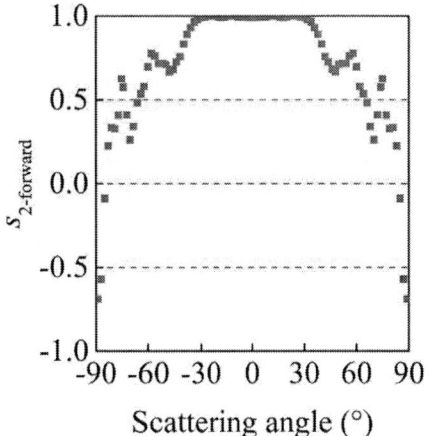

Figure 6. $s_{2\text{-forward}}$ values versus scattering angle from -90° to 90°

Then, we analyzed the evaluation performance of *RoPS*. According to formula (1), *RoPS* is transformed into

$$RoPS = \frac{P_{45\text{-forward}}}{P_{45\text{-forward}} + P_{-45\text{-forward}}} \quad (11)$$

We calculated *RoPS* in the range of -90° to 90° scattering angle. As shown in Fig. 7, *RoPS* values are always positive. Compare with the normalized Stokes parameter $s_{2\text{-forward}}$, *RoPS* exactly represent the intensity proportion of +45° linearly polarized light to forward-scattered light. Therefore, *RoPS* has superior performance to evaluate the change of polarization state in forward transmission.

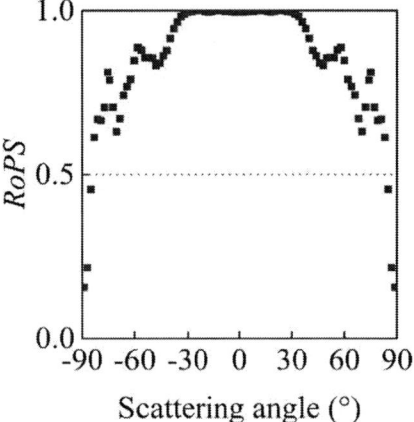

Figure 7. *RoPS* values versus scattering angle from -90° to 90°

In Fig. 6 and Fig. 7, the trend of $s_{2\text{-forward}}$ and *RoPS* seems to be consistent. Next, we analyzed the connection between s2-forward and *RoPS*.

To illustrate the relationship, consider a generic Mueller matrix $\mathbf{M}_{\text{angle}}$ in any angle of -90° to 90° is

$$\mathbf{M}_{angle} = \begin{bmatrix} M_{00} & M_{01} & M_{02} & M_{03} \\ M_{10} & M_{11} & M_{12} & M_{13} \\ M_{20} & M_{21} & M_{22} & M_{23} \\ M_{30} & M_{31} & M_{32} & M_{33} \end{bmatrix}_{angle} \quad (12)$$

$s_{2\text{-forward}}$ in any angle of -90° to 90° is calculated as

$$
\begin{aligned}
s_{2\text{-angle}} &= \frac{P_{45\text{-angle}} - P_{-45\text{-angle}}}{P_{45\text{-angle}} + P_{-45\text{-angle}}} \\
&= \frac{M_{20\text{-angle}} + M_{22\text{-angle}}}{M_{00\text{-angle}} + M_{02\text{-angle}}}
\end{aligned} \quad (13)
$$

Then, equation (11) is transformed into

$$
\begin{aligned}
2 &\frac{P_{45\text{-angle}}}{P_{45\text{-angle}} + P_{-45\text{-angle}}} \\
&= \frac{M_{20\text{-angle}} + M_{22\text{-angle}}}{M_{00\text{-angle}} + M_{02\text{-angle}}} + 1
\end{aligned} \quad (14)
$$

So, *RoPS* in any angle of -90° to 90° can be calculated as

$$RoPS_{angle} = \frac{1}{2}\left(\frac{M_{20\text{-angle}} + M_{22\text{-angle}}}{M_{00\text{-angle}} + M_{02\text{-angle}}} + 1\right) \quad (15)$$

Equations (13) and (15) are the general solutions for this example. It indicates that there is a conversion relationship between $s_{2\text{-forward}}$ and *RoPS*. Although there is a conversion relationship between $s_{2\text{-forward}}$ and *RoPS*, they have essential difference. Stokes parameters are related to the intensity difference between the defined six specific polarization states. It leads to Stokes parameters and their transformation forms are incompletely related to the change of polarization state. Compare with Stokes parameters, *RoPS* merely reflects the intensity change of light with specific polarization state.

In short, *RoPS* is Stokes-Mueller matrix transformation form, but it avoids the effect introduced by calculation of orthogonal component intensity difference. When evaluating the polarization state characteristic of light waves during forward transmission, *RoPS* has higher calculation accuracy than Stokes parameters.

V. CONCLUSION

This paper describes a new metric *RoPS* to evaluate polarization state characteristic of light during forward transmission. Its core is to describe the retention rate of polarization state of forward scattered light. *RoPS* is a transformation of Stokes-Mueller calculation form. However, *RoPS* can avoid the effect introduced by calculation of orthogonal component intensity difference. When evaluating the polarization state characteristic of light during forward transmission, *RoPS* has higher calculation accuracy than Stokes parameters. In short, *RoPS* is a transformation of Stokes-Mueller calculation form. *RoPS* can avoid the effect introduced by calculation of orthogonal component intensity difference.

This work is carried out in forward transmission of light. To expand the evaluation method, future work will be carried out in comparison with Mueller matrix, photon transmission, and backward transmission. The evaluation method has potential application values on atmospheric optics, detection, communications, and so on.

Funding. National Natural Science Foundation of China (CN) (62105136)

REFERENCES

[1] Bohren C F, Huffman D R. Absorption and scattering of light by small particles[M], John Wiley & Sons ,2008.

[2] MacKintosh F C, Zhu J X, Pine D J, et.al. Polarization memory of multiply scattered light[J] Physical Review B, 1989,40(13): 9342–9345.

[3] Guo Z Y, Wang XY, Li D K, et al. Advances on theory and application of polarization information propagation [J]. Infrared and Laser Engineering, 2020, 49(6): 20201013-1-20201013-19.

[4] Van der Laan J D, Wright J B, Kemme S A, Scrymgeour D A. Superior signal persistence of circularly polarized light in polydisperse, real-world fog environments [J] Applied Optics, 2018, 57(19): 5464-5473.

[5] Tyo J S, Goldstein D L, Chenault D B, et.al. Review of passive imaging polarimetry for remote sensing applications[J]. Applied Optics, 2006, 45(22): 5453-5469.

[6] Han Y, and Li G F. Coherent optical communication using polarization multiple-input-multiple-output [J]. Optics Express, 2005, 13(19): 7527-7534.

[7] Tuchin V V. Polarized light interaction with tissues [J]. Journal of Biomedical Optics, 2016, 21(7): 071114.

[8] Lerner A, Shashar N. Polarized light and polarization vision in animal sciences [M]. Berlin, Springer, 2014.

[9] Bok M J, Porter M L, Place A R. Biological sunscreens tune polychromatic ultraviolet vision in mantis shrimp [J]. Current Biology, 2014, 24(14): 1636-1642.

[10] Chu J K, Zhao K C, Zhang Q, et. Al. Construction and performance test of a novel polarization sensor for navigation [J]. Sensors and actuators a-physical, 2008, 148(1): 75-82.

[11] Chu J K, Wu Q M, Zeng X W, et,al. Forward transmission characteristics in polystyrene solution with different concentrations by use of circularly and linearly polarized light [J]. Appied Optics, 2019, 58(25): 6750-6754.

[12] Shen F, Zhang B M, Guo K, et.al. The depolarization performances of the polarized light in different scattering media systems [J]. IEEE Photonics Journal, 2018, 10(2): 1-12.

[13] Williams M W. Depolarization and cross polarization in ellipsometry of rough surfaces [J]. Appled Optics, 1986, 25(20): 3616-3622.

[14] Liao R W, Zeng N, Zeng M M, et.al. Estimation and extraction of the aerosol complex refractive index based on Stokes vector measurements [J]. Optics Letters, 2019, 44(19): 4877-4880.

978-1-6654-8699-6/22 $31.00 © 2022 IEEE

[15] Sankaran V, Walsh J T, and Maitland D J. Polarized light propagation through tissue phantoms containing densely packed scatterers [J]. Optics Letters, 2000, 25(4): 239-241.

[16] Van der Laan J D, Wright J B, Scrymgeour D A, et.al. Effects of collection geometry variations on linear and circular polarization persistence in both isotropic-scattering and forward-scattering environments [J]. Appled Optics, 2016, 55(32): 9042-9048.

[17] Zeng X W, Chu J K, Cao W D, et.al. Visible–IR transmission enhancement through fog using circularly polarized light [J]. Appled Optics, 2018, 57(23): 6817-6822 ,

High-Resolution Microwave Frequency Measurement Based on Optical Frequency Comb and Image Rejection Photonics Channelized Receiver

Ximin Wang, Yingxi Miao, Jialiang Chen, Caili
Gong, Yongfeng Wei*
Institute of Electronic Information Engineering
Inner Mongolia University
Hohhot, China
e-mail: wxm@mail.imu.edu.cn,
*weiyongfeng@imu.edu.cn

Yuqing Yang
China Mobile Communication Group Qinghai CO., Ltd
Xining, China
e-mail: yangyuqing@qh.chinamobile.com

Abstract—A high-resolution microwave signal frequency measurement scheme based on optical frequency comb (OFC) and an image rejection microwave photonics channelized receiver is proposed. The scheme consists of two branches. The OFC is generated by cascaded Mach-Zehnder modulators (MZMs) in the upper branch. The optical carrier is frequency shifted by the optical frequency shifter (OFS) in the lower branch. The shifted optical carrier is sent to polarization modulator (PolM) to be modulated by the RF signal to be measured. The signal from the upper and lower branches are injected into 90-degree optical hybrid and divided into four outputs. The optical signal of each output is divided into channels by the wavelength division multiplexer (WDM) and beat by the balance photodetector (BPD). The back-end of the scheme adopts image rejection down-conversion method to prevent spectral aliasing in the measurement process. Simulation verifies the effectiveness of this scheme. The results show that the scheme can accurately measure microwave signals within the frequency range of 1-79 GHz.

Keywords-microwave photonics; frequency measurement; channelized receiver; frequency analysis; optical frequency comb

I. INTRODUCTION

Frequency analysis of unknown signals plays a crucial role in wireless communication systems such as radar and electronic warfare [1]. With the rapid development of wireless communication systems, the frequency range of signals in modern communications is large, for example, receivers performing different functions usually operate in different frequency bands, so the frequency of the received signal needs to be measured accurately [2]. The conventional electrical frequency measurement schemes can achieve high resolution, but unable to perform instantaneous frequency measurements and limited by disadvantages such as slow speed, limited bandwidth and vulnerable to electromagnetic interference (EMI) cannot meet the needs of the system[3]. In this context, microwave photonics has attracted great research interest from researchers because of its advantages such as large bandwidth, low loss, anti-electromagnetic interference, and small size for transmitting and processing microwave signals in wireless communication systems [4, 5]. Many photonics-assisted frequency measurement schemes

have been proposed, and these schemes can be divided into three categories. The first category is based on photonics-assisted scanning receiver. The narrow-band filtering effect in stimulated Brillouin scattering (SBS) and fiber Bragg gratings in the photonic link used to achieve scanning frequency receiver. The conventional scanning receiver can achieve multi-band coverage, but it is difficult to receive instantaneous frequency [6]. The second category is microwave photonics instantaneous frequency measurement (IFM) scheme, the frequency of the signal to be measured is associated with easily measured parameters such as signal power, time, phase, and amplitude information. The scheme is simple structure, large measurement range, high measurement accuracy. However, IFM scheme can only realize single frequency point measurement, and cannot measure multi-frequency signal [7]. The third category is the microwave photonics channelized receiver, which divides the broadband signal into multiple independent narrowband channels for parallel processing. The power of the signal to be measured is analyzed by observing the output power of each sub-channel [8].

In [9], a microwave photonics channelized scheme based on a Fabry-Perot cavity was proposed. The frequency measurement range of this scheme is 0.5-11.5 GHz, and the measurement accuracy is ±0.5 GHz. However, the instability of the Fabry-Perot cavity and the inaccuracy of wavelength alignment severely limit the performance. The power and carrier rejection ratio of the signal to be measured greatly affect the measurement accuracy. To avoid the use of unstable optical filters, a multi-frequency measurement scheme based on optical frequency comb (OFC) and microwave photonics channelized receiver was proposed in [7], which can accurately derive multiple frequency components by analyzing the beating frequency signals. The measurement system has achievability, stability, and flexibility. However, this scheme does not consider the possible spectral aliasing during the measurement process.

In this paper, a new high-resolution frequency measurement scheme based on OFC and microwave photonics channelized receiver is proposed. The optical carrier is divided into upper and lower branches by the optical coupler (OC), and the upper branch optical carrier is

978-1-6654-8699-6/22 $31.00 © 2022 IEEE

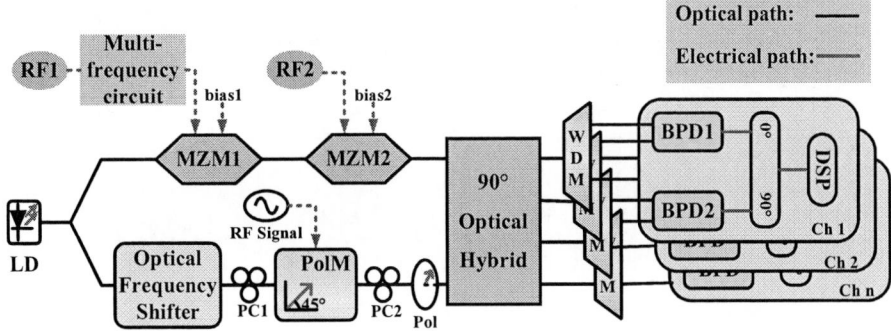

Figure 1. Schematic diagram of the proposed frequency measurement scheme (LD: laser diode; MZM: Mach-Zehnder modulator; PolM: polarization modulator; PC: polarization controller; WDM: wavelength division multiplexing; BPD: balanced photodetector; Pol: polarizer).

modulated by Mach-Zehnder modulator1 (MZM1) driven by frequency multiplication signal cascaded with MZM2 to generate 8-line ultra-flat OFC with 20GHz spacing. In the lower branch, the optical carrier is firstly injected into optical frequency shifter (OFS) for frequency shifting, and the microwave signal received by the RF antenna is modulated by the polarization modulator (PolM). After the polarizer (Pol), the ±1st order harmonic carrying the microwave signal is generated. The optical signals of the upper and lower branches are injected into the S port and L port of the 90-degree optical hybrid respectively and output in four channels. Each optical signal is divided into 8 channels by the wavelength division multiplexer (WDM) and beat by the balance photodetector (BPD). If the photoelectric conversion is directly carried out by the photoelectric detector (PD), it will cause serious spectrum aliasing and affect the receiving accuracy. The beating signal is injected into the 90-degree electrical hybrid, the frequency of the microwave signal to be measured can be calculated by detecting the electric power of each channel. The simulation results show that the proposed scheme can measure the microwave signal with a frequency of 1-79 GHz, and the measurement error is less than 0.05GHz. The spectrum aliasing in the measurement process is avoided by adopting the image rejection structure.

II. PRINCIPLE AND MODEL ANALYSIS

The structure of the high-resolution frequency measurement scheme based on OFC and microwave photonics channelized receiver is shown in Fig. 1. The scheme consists of a laser diode (LD), an optical frequency shifter, a PloM, an OFC generation module, four WDMs, a 90-degree optical hybrid, two BPDs, a 90-degree electrical hybrid. The 90-degree optical hybrid combined with BPD and 90-degree electrical hybrid is structured to achieve image rejection down-conversion to prevent spectral aliasing and improve the in-band rejection ratio of the channel.

The continuous optical carrier wave from LD is divided into two branches by OC, and it can be expressed as $E_{in}(t) = E_0 \exp(j\omega_c t)$, where E_0 and ω_c are the amplitude and angular frequency, respectively. The upper branch optical carrier wave is injected into MZM1 for modulation, and the RF signal of MZM1 is modulated by the multi-

frequency circuit. The structure of the multi-frequency circuit is shown in Fig. 2, which consists of two electrical phase shifters and two electrical multipliers, where the phase shifts of the electrical phase shifters are $\pi/4$ and $-\pi/4$. The RF signal of the multi-frequency circuit is $V_1(t) = V_{RF1}\sin(\omega_{RF1}t)$, where V_{RF1} and ω_{RF1} are the amplitude and angular frequency of the RF signal respectively. After the modulation of the multi-frequency circuit, the output RF signal is $(V_{RF1}^3/4)[\sin(\omega_{RF1}t) - \sin(3\omega_{RF1}t)]$, compared with the original RF signal, the modulated RF signal has a new RF frequency component. The optical carrier modulated by MZM1 can be expressed as:

$$E_{MZM1}(t) = \frac{E_0}{4}e^{j\omega_c t}\left[\begin{array}{l}\sum_{n=-\infty}^{\infty}J_n(m_{RF1})e^{jn\omega_{RF1}t} - \sum_{n=-\infty}^{\infty}(-1)^n J_n(m_{RF1})e^{jn\omega_{RF1}t} \\ +\sum_{n=-\infty}^{\infty}J_n(m_{RF1})e^{jn3\omega_{RF1}t} - \sum_{n=-\infty}^{\infty}(-1)^n J_n(m_{RF1})e^{jn3\omega_{RF1}t}\end{array}\right] (1)$$

$$= \frac{E_0}{4}\left[\begin{array}{l}J_1(m_{RF1})e^{j(\omega_c+\omega_{RF1})t} + J_1(m_{RF1})e^{j(\omega_c-\omega_{RF1})t} \\ +J_1(m_{RF1})e^{j(\omega_c+3\omega_{RF1})t} + J_1(m_{RF1})e^{j(\omega_c-3\omega_{RF1})t}\end{array}\right]$$

where $J_n(\bullet)$ is the Bessel function of first kind of order n, while $m_1 = \pi V_{RF1}/V_\pi$ denotes the modulation depth of the MZM1 and V_π is the half-wave voltage of the modulator.

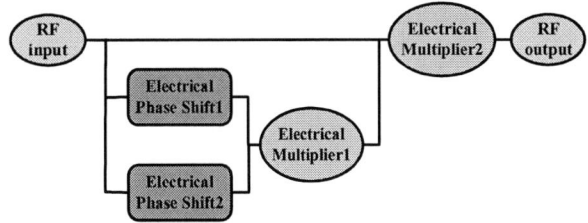

Figure 2. The structure of the multi-frequency circuit.

Then, each comb line generated by MZM1 is injected into the MZM2 as a light source for broadening. The frequency of the RF driving signal of MZM2 is $f_{RF1}/2$. The 8-line OFC at the output of MZM2 is:

$$E_{MZM2}(t) = E_{MZM1}(t)[\exp(-j\omega_{RF2}t) + \exp(j\omega_{RF2}t)] \quad (2)$$

978-1-6654-8699-6/22 $31.00 © 2022 IEEE   31

In the lower branch, the optical carrier is injected into OFS for frequency shifting, and the frequency shifted optical carrier can be expressed as $E_{in2}(t) = E_0 \exp[j(\omega_c + \omega_s)t]$. The

microwave signal to be measured is $V_3(t) = V_{RF3}\sin(\omega_{RF3}t)$, which is injected into the PolM for carrier suppressed double-sideband (CS-DSB) modulation, and the modulation

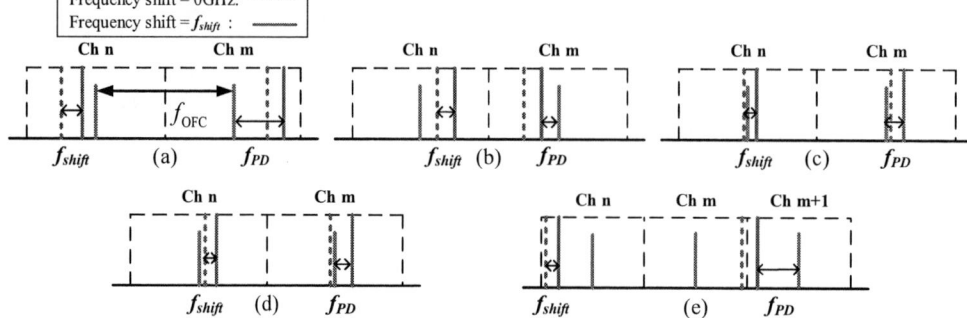

Figure 3. Schematic diagrams of different frequency distributions in the channel.

index of the PolM is $m_3 = \pi V_3 / V_\pi$. After Pol, the output signal can be expressed as:

$$E_{PolM}(t) = \frac{E_0}{2} e^{j(\omega_c + \omega_s)t} J_1(m_3)[e^{j\omega_{RF3}t} + e^{-j\omega_{RF3}t}] \quad (3)$$

The optical signals from the upper and lower branches are injected into 90-degree optical hybrid and the optical signals from the four outputs can be expressed as:

$$\begin{bmatrix} E_{out1} \\ E_{out2} \\ E_{out3} \\ E_{out4} \end{bmatrix} = \begin{bmatrix} 1 & 1 \\ 1 & -1 \\ 1 & j \\ 1 & -j \end{bmatrix} \begin{bmatrix} E_{MZM2}(t) \\ E_{PolM}(t) \end{bmatrix} = \begin{bmatrix} E_{MZM2}(t) + E_{PolM}(t) \\ E_{MZM2}(t) - E_{PolM}(t) \\ E_{MZM2}(t) + jE_{PolM}(t) \\ E_{MZM2}(t) - jE_{PolM}(t) \end{bmatrix} \quad (4)$$

After channelization of each output by WDM, E_{out1} and E_{out2} are input into BPD1, E_{out3} and E_{out4} are input into BPD2. The electrical signals obtained after beating can be expressed as:

$$E_{BPD1} = |E_{out1}|^2 - |E_{out2}|^2 = |E_{MZM2} + E_{PloM}|^2 - |E_{MZM2} - E_{PloM}|^2 \quad (5)$$

$$E_{BPD2} = |E_{out3}|^2 - |E_{out4}|^2 = |E_{MZM2} + jE_{PloM}|^2 - |E_{MZM2} - jE_{PloM}|^2 \quad (6)$$

If the outputs of the upper and lower branches are directly injected into PD for photoelectric conversion, it will cause serious spectrum aliasing and affect the correct reception of the signal. The signal after two BPDs beating are injected into the 90-degree electric hybrid to complete image rejection down-conversion and get the beating value. By calculating the specific frequency relationship and substituting the beating value, the frequency of the signal to be measured can be measured. When the optical comb generated by the signal to be measured is in different positions of the channel, the required calculation formulas are different, which are discussed in the following cases:

It is assumed that the number of channels is k and the signal to be measured is detected in channel n and channel m ($n < m$). The frequency shift of OFS is set to 0 GHz and observing the beating frequency of each channel. As shown in Fig. 3(a), when increasing the frequency shift of OFS to

f_{shift}, if the beating frequency of channel n decreases and the beating frequency in channel m increases, the frequency of the microwave signal to be measured is calculated as:

$$f = [m - (k+1)/2]f_{OFC} + f_{PD} - f_{shift} \quad (7)$$

As shown in Fig. 3(b), after changing the frequency shift of OFS, the frequency of the beat signal in channel n increases and the frequency of the beat signal in channel m decreases. The calculation of the microwave signal to be measured in this case can be expressed as:

$$f = [m - (k+1)/2]f_{OFC} - f_{PD} - f_{shift} \quad (8)$$

As shown in Fig. 3(c, d), when the frequency shift of OFS is 0GHz, the frequencies of the beat signals in channel n and channel m are all less than 1GHz. When the frequency shift of OFS is increased to f_{shift}, if the frequency of the beat signal in channel m is greater than 1GHz, the calculation of the microwave signal to be measured can be expressed respectively as:

$$f = [m - (k+1)/2]f_{OFC} + f_{PD} - f_{shift} \quad (9)$$

$$f = [m - (k+1)/2]f_{OFC} - f_{PD} + f_{shift} \quad (10)$$

As shown in Fig. 3(e), when the frequency shift of OFS is 0 GHz, the beat signal can be observed in both channel n and channel m. When the frequency shift of OFS is increased to f_{shift}, if the beat signal can be detected in channel $m+1$, or the optical signal after the frequency shift coincides with the channel boundary, the signal can be detected in both channel m and channel $m+1$ at this time. The frequency of the microwave signal to be measured in these two cases can be expressed as:

$$f = [m - (k+1)/2]f_{OFC} - f_{PD} - f_{shift} \quad (11)$$

In the process of frequency measurement, by observing the position of the OFC generated by the signal to be

978-1-6654-8699-6/22 $31.00 © 2022 IEEE

measured in the channel before and after the optical carrier frequency shift. The frequency of the signal to be measured can be obtained by using the frequency calculation formula in the corresponding case.

III. SIMULATION RESULTS AND ANALYSIS

In order to verify the feasibility, this system is simulated by "Optisystem 15.0". The parameter settings in the simulation process are as follows: the frequency of the LD is 193.1THz and the power is 28dBm.

The upper branch optical carrier enters MZM1 for modulation, and the MZM1 is based at the minimum transmission point (MITP) by adjusting the DC bias voltage. The RF signal of MZM1 is modulated by the muti-frequency circuit and be injected into MZM1. The frequency of the input RF signal is set to 20GHz and the amplitude is set to 1V. A new frequency component of 60GHz is added to the original frequency after muti-frequency circuit. Now, two different frequencies of RF signals to drive MZM1. The optical carrier is modulated by MZM1 to generate 4-line ultra-flat OFC.

Then, each comb of OFC is injected into MZM2 as a new light source for spectrum broadening. The DC bias voltage is set to 4V so that the MZM2 is biased at MITP. The amplitude of RF2 is set to 1V and the frequency is set to 10 GHz. The 8-line flat OFC generated with comb-line spacing of 20 GHz, flatness <0.5 dB and side mode suppression ratio (SMSR) of 35 dB as shown in Fig. 4(a). In the lower branch, the microwave signal to be measured is modulated by PolM and is set to 33 GHz. After Pol, the ±1st order harmonic sidebands can be generated. The spectrogram after output is shown in Fig. 4(b).

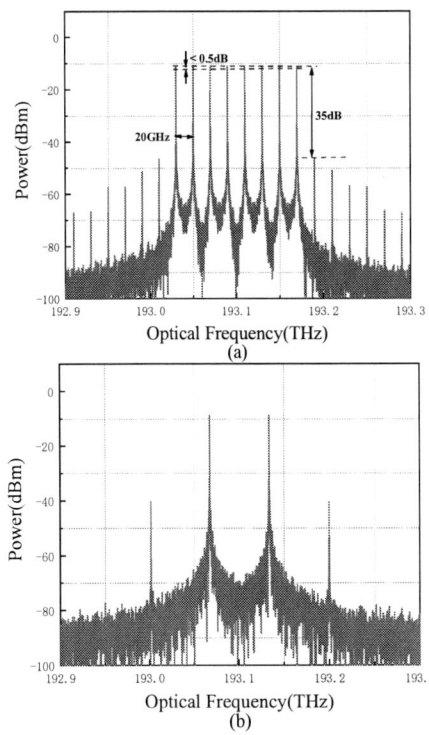

Figure 4. The output spectrum of upper branch (a) and lower branch (b).

The upper and lower branch optical signals are coupled by 90-degree optical hybrid as shown in Fig. 5. The coupled signal is divided into 8 channels by WDM, and the bandwidth of each channel is 20GHz. Each channel adopts image rejection down-conversion method to prevent spectral aliasing. By analyzing the beating signal and selecting the appropriate calculation formula, the frequency of microwave signal can be measured.

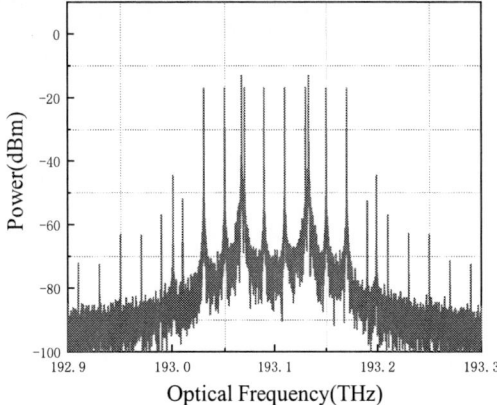

Figure 5. The output spectrum after 90-degree optical hybrid.

As shown in Fig. 5, the OFC generated by the signal to be measured is in channel 3 and channel 6. The electrical signal can be detected in channel 3 and channel 6, as shown in Fig. 6 (a, b), and the signal cannot be detected in other channels as shown in Fig. 6 (c).

After changing the frequency shift of OFS to 2GHz, the coupled signal as shown in Fig. 7(a). Compared to the signal when the frequency shift is 0GHz. The optical combs in both channel 3 and channel 6 move to the right. The beating signal detected in channel 3 and channel 6 as shown in Fig. 7(b, c).

After changing the frequency shift of OFS from 0GHz to 2GHz, the frequency of the beat signal in channel 3 decreases, while the frequency of the beating signal in channel 6 increases. The microwave signal frequency is:

$$f = [m-(k+1)/2]f_{OFC} + f_{PD} - f_{shift}$$
$$= (6-9/2)*20GHz + 5GHz - 2GHz \qquad (12)$$
$$= 33GHz$$

After calculation, the frequency of the microwave signal is 33GHz, which is consistent with the frequency of the RF signal to be measured. When the frequency of the signal to be measured is changed, there will exist some errors. After simulation, it is verified that the error between the measured value and the theoretical value in 1-79 GHz is shown in the Fig. 8. The measurement error is less than 0.05GHz.

IV. CONCLUSION

A high-resolution frequency measurement scheme based on OFC and image rejection microwave photonics channelized receiver is proposed and verified. In this scheme, the upper branch generates an 8-line ultra-flat high-quality OFC with a comb spacing of 20GHz by cascaded MZMs. In the lower branch the microwave signal to be measured is modulated by PolM to generate ±1st order harmonic sidebands. The signal from the upper and lower branches is injected into 90-degree optical hybrid and divided into four outputs. The optical signal of each output is channelized by WDM and beat by BPD. By analyzing the beating signal and selecting the appropriate calculation formula, the frequency of microwave signal can be measured. This scheme overcomes the shortcomings of the traditional frequency measurement scheme that only supports single frequency point measurement and the instability of optical filter. This scheme measures the frequency of microwave signal in 1-79 GHz and the measurement error is less than 0.05GHz. The image rejection down-conversion structure is adopted to avoid the spectrum aliasing in the measurement process. This scheme can be used in wireless communication systems such as radar and electronic warfare.

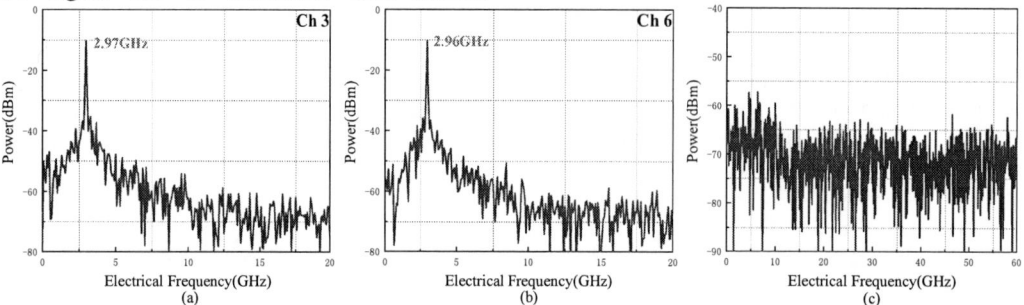

Figure 6. The beating signal detected in channel 3 (a)，channel 6 (b) and other channels (c) when the frequency shift of OFS is 0GHz.

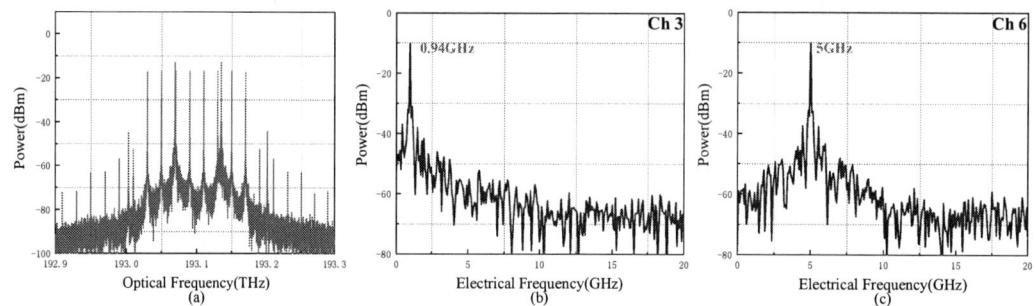

Figure 7. The output spectrum after 90-degree optical hybrid (a), the beating signal detected in channel 3 (b) ，channel 6 (c) when the frequency shift of OFS is 2GHz. Conclusion

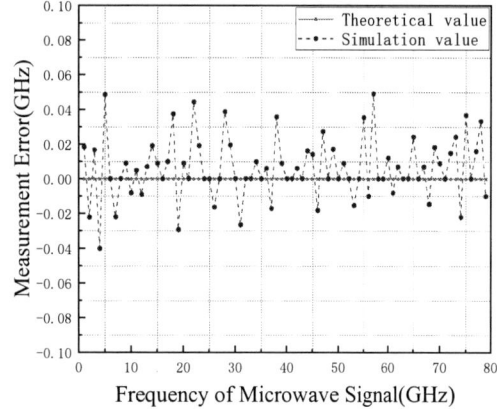

Figure 8. The measurement error between the measured value and the theoretical value in 1-79 GHz of this scheme.

ACKNOWLEDGMENT

The work was supported by the National Natural Science Foundation of China (No.62161034, No.61561037, No.61861034, No.61763034.)

REFERENCES

[1] J. Shi, F. Zhang, D. Ben, and S. Pan, "Photonics-Based Broadband Microwave Instantaneous Frequency Measurement by Frequency-to-Phase-Slope Mapping," IEEE Trans. Microw. Theory Tech. 67, 544-552 (2019).

[2] H. Zhang and S. Pan, "High Resolution Microwave Frequency Measurement Using a Dual-Parallel Mach–Zehnder Modulator," IEEE Microw. Wireless Compon. Lett. 23, 623-625 (2013).

[3] L. B. Pan, L. K. Jiang, Y. Wang, W. Dong, X. D. Zhang, and S. P. Ruan, "Instantaneous Microwave Frequency Measurement with Ultra-wide Range and High-resolution Based on Stimulated Brillouin Scattering," Acta Photonica Sinica. 46, 108-113 (2017).

[4] J. P. Yao, "Microwave Photonics," J. Lightwave Technol. 27, 314-335 (2009).

[5] T. Kawanishi, H. Hayashi, K. Inagaki, A. Kanno, and N. Yamamoto, "Instantaneous freuqnecy measurement for broadband radio signals using optical single sideband modulation," 2017 International Symposium on Antennas And Propagation (Isap 2017).

[6] X. Lu, W. Pan, X. Zou, W. Bai, P. Li, L. Yan, et al., "Wideband and Ambiguous-Free RF Channelizer Assisted Jointly by Spacing and Profile of Optical Frequency Comb," IEEE Photo. J. 12, 1-11 (2020).

[7] J. Shen, S. Wu, D. Li, and J. Liu, "Microwave multi-frequency measurement based on an optical frequency comb and a photonic channelized receiver," Appl Opt. 58, 8101-8107 (2019).

[8] L. A. Bui, "Recent advances in microwave photonics instantaneous frequency measurements," Progress in Quantum Electronics. 69, 100237 (2020).

[9] Z. Li, X. M. Zhang, H. Chi, S. L. Zheng, X. F. Jin, and J. P. Yao, "A Reconfigurable Microwave Photonic Channelized Receiver Based on Dense Wavelength Division Multiplexing Using an Optical Comb," Opt. Commun. 285, 2311-2315 (2012).

Proceedings of the 7th Optoelectronics Global Conference (OGC 2022)

A SDN Enabled PON Controller based on Hierarchical Models

1st Zhenming Liang
Institute of Network Technology
China Telecom Research Institute
Shanghai, China
liangzhm@chinatelecom.cn

2nd Ziyao Yang
Institute of Network Technology
China Telecom Research Institute
Shanghai, China
yangziyao@chinatelecom.cn

3rd Jian Tang
Institute of Network Technology
China Telecom Research Institute
Shanghai, China
tangj10@chinatelecom.cn

Abstract—With the evolution of network architecture and changes in service requirements, operators are posing new challenges to Passive Optical Network (PON) operation and maintenance. In order to solve the lack of intelligent operation and maintenance capabilities of PON, this paper proposes a software defined network (SDN) enabled PON controller based on hierarchical models. PON controller is a codeless frameworks based on model-driven and graphical programming. The standardized YANG model enables the flexible and quick modelling and services handling of different professional levels' services and networks in PON system. Based on Telemetry technology, a real-time and high-precision collection solution for PON network can be built, realizing all-round and fined real-time visibility of network services. Besides, by introducing the artificial intelligence (AI) technology represented by machine learning and deep learning, automatic closed-loop operating status monitoring, analysis and decision-making of industrial PON networks can be realized, the intelligent level of network operation and maintenance can be improved, which aims to build an automated, visualized and intelligent PON.

Keywords- PON controller; NETCONF; YANG; Telemetry; Model driven; AI

I. INTRODUCTION

At present, the architecture presented by the operator's broadband access network is vertical, and most of them are closed. Traditional closed network equipment is used for service development. Due to the understanding of service processes, development of control protocols, provision of private interfaces to integration verification, the entire process of developing a new service is completed by equipment manufacturers, and due to the differences in the technical implementation methods of equipment manufacturers, the development of new services affects the whole body, resulting in a long development cycle. The traditional network architecture and operation system are increasingly difficult to adapt to the needs of Internet service development. Network transformation and architecture

reconstruction have become the common pursuit of mainstream operators. With the deepening of SDN research, the adoption of the SDN architecture concept to achieve centralized network control and capability opening has become the direction of network transformation unanimously supported by major telecom operators and cloud service providers around the world [1].

Academia and industry continue to pay attention to and study SDN technology, and put forward solutions to the problems encountered in the evolution from non SDN network to SDN network [2,3]. Some possible security problems caused by SDN using RESTful interface [4] and performance problems caused by network collection [5] are researched. Standards organizations such as BBF have also formulated TR-383 [6] and TR-385[7] management model standards for the application of SDN technology in access network. In this paper, we investigate a PON controller based on SDN. A PON controller architecture based on hierarchical models is proposed for this purpose. The rest of this paper is organized as follows: Section II provides details of the PON controller architecture; Section III proposed the application methods of key technologies of hierarchical models in PON and achieves practical results; conclusions are drawn in section IV.

II. PON CONTROLLER ARCHITECTURE

The characteristics of the large number of devices in the PON access network and the development direction of the full-service carrying have put forward the requirements for unified management and flexible opening of the access network. The characteristics of PON access network close to users and service granularity strengthen the requirements for service quality assurance and refined operation and maintenance. In order to improve network agility, simplify network operation and maintenance, and improve user experience, it is necessary to build an automatic, visual and intelligent PON network based on PON controller. The control module shields the network complexity and simplifies the configuration to realize the automation of the

978-1-6654-8699-6/22 $31.00 © 2022 IEEE

access network; Realize access network visualization through more real-time, more comprehensive and more detailed network data collection and big data analysis of the collection module; based on the open application programming interface (API) and visual big data of automatic access network, and then realize the intelligence of the access network by introducing the analysis module of AI technology.

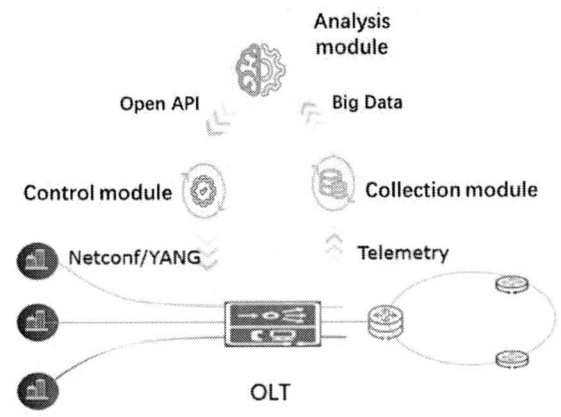

Figure 1.   Automated, visualized, intelligent construction.

The PON controller adopts a layered design: it consists of a basic configuration layer and a service function layer, and supports independent evolution and expansion of software at different layers. The southbound interface of the basic configuration layer uses network configuration protocol (NETCONF) to uniformly model the multi-vendor PON equipment through the defined standard PON equipment YANG model. And it has the collection capability based on Telemetry technology to achieve diversified, second-level, real-time network data collection. The basic configuration layer completes the orchestrator of the atomic interface defined by the YANG model of the PON equipment to the atomic operation capability, and provides the basic configuration layer interface that is standard for the device capability at the basic service level. Based on model driven and graphical programming, the codeless framework is realized to enable the agile development of PON network control function and the visual management of the whole life cycle of service process. The service function layer completes the transformation from the network real-time visual function and the basic configuration layer interface to the service north interface. The application scenario and customer differentiated requirements can be customized, and various interfaces such as RESTful can be used as needed. At the same time, the service function layer introduces AI technologies such as deep learning and machine learning, analyzes user services and network characteristics, applies AI technology to build fault analysis, dynamic warning prediction models, generates SDN collaborative control optimization strategies, and drives the intelligent management and control of PON network.

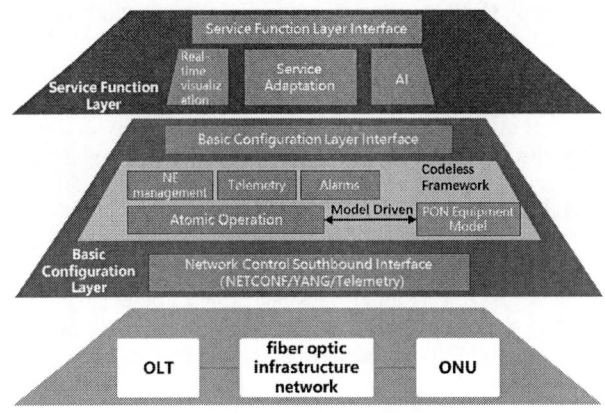

Figure 2.   PON Controller Architecture.

III.   KEY TECHNOLOGIES OF PON CONTROLLER

A.   Standardization of southbound interfaces based on the YANG model

Most of the traditional access network devices use complex proprietary interfaces such as TL1, simple network management protocol (SNMP)/management information base (MIB), extensible markup language (XML) for configuration and management. The format of message interface is complex and the degree of privatization is high. Based on the functions and services of the PON network, various southbound interfaces of the controller are analyzed, such as NETCONF/YANG, Openflow, SNMP/MIB, Open vSwitch Database (OVSDB) management protocol, etc. NETCONF/YANG is more suitable for the PON controller Southbound interface protocol type.

The IETF launched the network configuration management protocol NETCONF in 2006 [8]. The previous protocol adopted the structure of the message specified in the protocol, and read and parsed it according to the byte stream. In NETCONF, the message structure is defined by the model, and its message parsing is also completed by the peer model. In this way, the protocol does not need to define the message structure explicitly in bytes, so it has scalability.

Facing the complex and closed interfaces of traditional access network devices, it is crucial to open up device capabilities by establishing a standard and unified device capability model and API [9]. China Telecom has formulated the standard YANG modules for PON equipment, which unifies the northbound interface of the access network optical line terminal (OLT), making the OLT equipment capabilities truly open, and providing a capability foundation for the rapid launch of new services.

For the control contents of system, hardware, software, virtual local area network (VLAN), quality of service (QoS), alarm, multicast, gigabit capable passive optical network (GPON) (including 10 gigabit PON (XG(S)-PON)), based on the actual requirements of PON access network, based on the IETF model, BBF TR-383 [6], BBF TR-385 [7] and other standard models, supplemented according to existing

978-1-6654-8699-6/22 $31.00 © 2022 IEEE

network requirements, and independently proposed the Yang model of slicing, ethernet passive optical network (EPON) (including asymmetric 10G-EPON and symmetric 10G-EPON) and virtual extensible local area network (VxLAN). More importantly, the data structure requirements of the basic configuration are proposed, which not only unifies the capabilities of the access network equipment in the model and interface, but also proposes a unified method for the specific configuration of the model.

The hardware model defined in RFC 8348 [10] and BBF TR-383 is used for hardware management. The hardware objects of OLT (frame, backplane, fan, board slot, board, module slot, optical module, etc.) are instantiated through the component in the model. Specific hardware attributes of access network (such as optical module type, alarm threshold of dual-mode optical module, etc.) supplement IETF and BBF models through customized Yang. An example of instantiation relationship is shown in Fig. 3.

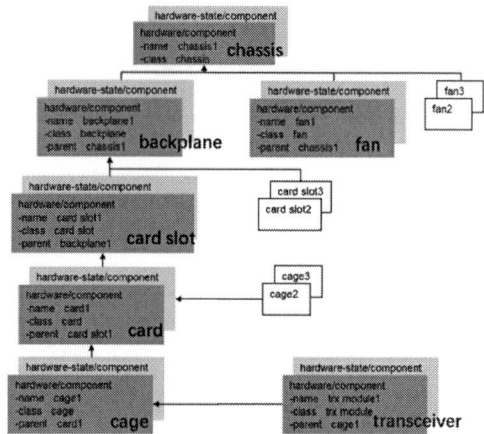

Figure 3. Schematic representation of the instantiation relationship of the hardware model.

The GPON model defined by BBF TR-385 is used for GPON, XG-PON, and XGS-PON management, including transmission convergence (TC) layer operations, physical media dependent (PMD) layer parameters, PON resources (PON link, Alloc-ID, G-PON encapsulation mode (GEM), etc.) configuration, fault management and performance management, etc. The correlation between GPON model and hardware model is completed by augmenting a *hw-port-ref* node in channel-termination to index the PON port node in the corresponding hardware model. BBF TR-383 defines VLAN, QoS and other related service models. The core is to carry related service configurations through VLAN sub-interfaces. Thus, a unified model of three core management and control functions in the PON network of hardware management, PON configuration, and service delivery is formed. The schematic diagram is as shown in Fig. 4.

For EPON Yang model, in order to make the configuration of PON port protection consistent with GPON and EPON, the possible scenarios of 10G-EPON and 50G-PON OLT in the future are also considered. The EPON model tries to maintain a data structure similar to GPON

model. The differences between EPON and GPON are redefined, such as logical link identifier (LLID), dynamic bandwidth allocation (DBA) parameters, EPON/asymmetric 10G-EPON dual-mode and EPON /symmetric 10G-EPON dual-mode scenarios, Queue Set configuration, etc. The relationship with hardware and service model is the same as that of GPON model.

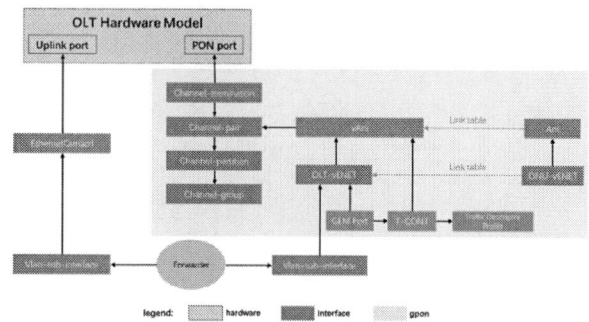

Figure 4. Relationship between hardware model, GPON configuration model, and service model.

B. Fine-grained collection of PON network based on Telemetry

Fiber to the home (FTTH) based on PON/10G-PON technology has been deployed and applied on a large scale. With the deployment and application of SDN network equipment and high-speed services such as 5G, virtual reality (VR), and edge computing, the access network is increasing the bandwidth to meet higher forwarding requirements. At the same time, it is necessary to provide better monitoring methods to realize intelligent operation and maintenance guarantee. The traditional method of collecting information based on SNMP and command-line interface (CLI) on the existing network adopts the pull mode, which has low execution efficiency. Considering the equipment capabilities in large-scale applications Insufficient, affecting network problem analysis and planning and tuning. Telemetry technology adopts active push mode, supports structured data, has higher execution efficiency and more real-time second-level collection accuracy, and has little impact on the function and performance of the device itself. Combined with SDN applications, it can quickly locate network problems, network Quality optimization and adjustment provides the most important big data analysis foundation to meet the operation and maintenance requirements of refined, visualized, and intelligent monitoring.

The mainstream Telemetry uses the Google remote procedure call (gRPC) protocol [12] to subscribe and collect device information. gRPC is a high-performance open source software framework released by Google and based on the hyper text transfer protocol (HTTP) 2.0 transport layer protocol. According to the characteristics of the PON network, the Telemetry protocol has been supplemented. The NETCONF/YANG configuration management channel of the reusable PON controller is added. In view of the large number of PON network devices and the excessive

978-1-6654-8699-6/22 $31.00 © 2022 IEEE

consumption of transmission control protocol (TCP) long connections to keep gRPC message reporting, the user datagram protocol (UDP) reporting mode is added.

The collected data received by the PON controller, whether using gRPC or UDP, is serialized by GPB coding, and the data structure of the collected data is described by the proto file. Through the standard data collection model, the data collection standardization of multi-level traffic collection, optical link information collection and ONU information collection in PON network is realized.

At present, China Telecom has completed ONU traffic, service flow traffic (VLAN granularity) and other traffic collection, optical link information collection, ONU information collection laboratory and live network test verification. Various traffic data collection intervals can reach 1 second or 3 seconds, which can provide a big data foundation for SDN applications.

C. A codeless framework based on model-driven and graphical programming

The unified modeling of the southbound interface of the PON network controller is completed through the YANG model of the PON equipment, which solves the problem of the manufacturer's private network management protocol information base and realizes the decoupling of the PON controller and the PON equipment. The YANG model of PON equipment in Chapter III.A is an abstract modeling of the internal underlying configuration object of PON equipment. Compared with the original control surface, the granularity of the South interface is finer, and the number and complexity of interfaces are also greatly increased. In addition, the coding and implementation workload of the interface and function implementation of PON controller is greatly increased. Based on the YANG tools component in OpenDaylight [13], the PON controller translates the YANG model into Java code, provides the deserialization and serialization methods of XML and javascript object notation (JSON), and generates the API interface based on the YANG model. Based on the model drive, the automatic coding of the southern interface from the YANG model of PON equipment to the controller is realized.

The southbound interfaces generated by the YANG model of the PON equipment are numerous and have strong professional attributes, which are still not friendly to development and operation and maintenance personnel. Although Chapter III.A proposes the data structure requirements of the basic configuration, there are so many use cases that the coding implementation is a very challenging task. The northbound API must hide the complexity of the YANG model of the PON network, and at the same time, the corresponding YANG model data structure and northbound interface need to be quickly constructed for new service functions. Therefore, the agile development of the mapping from the southbound interface of the PON controller to the northbound interface is realized through the graphical coding framework. The PON controller provides a visual programming environment built on the underlying model-driven and codeless framework. Operation and maintenance personnel can graphically draw

service components, interactions between components, and the entire workflow, and then use the PON controller to provide an external PON network control interface.

The automated workflow design center of the PON controller, shown in Fig. 5, provides developers with atomic capabilities-based visual orchestration capabilities to generate RESTful interfaces for northbound applications. Hand-coding and architectural decisions are prone to error, and codeless frameworks can improve functional code consistency by model-driven. The model will serve as the basis for the entire controller and become a long-term core asset. The application of the codeless framework can save the cost of enterprises, reduce the threshold for use and developers, improve the development speed, and help the digital transformation of PON network operation and maintenance.

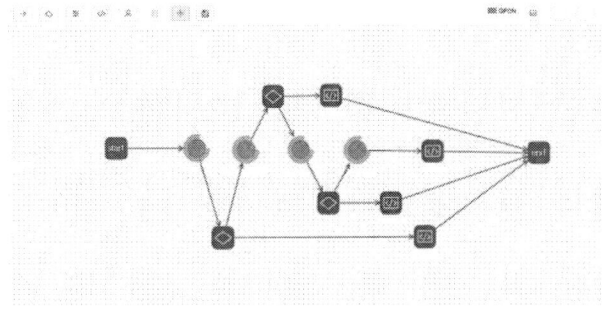

Figure 5.   Model-Driven Graphical Programming.

D. AI-based intelligent operation and maintenance of PON network

The blowout growth of applications and equipment, the innovation of network traffic mode and connection mode put forward higher requirements for operation and maintenance efficiency, operation cost and service level agreement (SLA) guarantee. AI technology has natural advantages in deep data analysis and dynamic generation of strategies. It will play an important role in reducing simple and repetitive network operations, forward-looking prevention and prediction based on historical data, high complexity multi-dimensional analysis and seeking the optimal solution of resources and service needs.

AI technology can be applied to PON network to realize intelligent network fault diagnosis, network quality analysis and network strategy optimization, realize automatic closed-loop monitoring, analysis and decision-making of PON network operation status, and improve the intelligent level of network operation and maintenance.

With the demand for PON to carry new services, the bandwidth pressure is prominent, and the demand for accurate operation and maintenance of PON port bandwidth is urgent. In the intelligent evolution of China Telecom's PON network, the first focus is on PON traffic analysis and prediction. Because the PON network is close to users, the flow granularity is fine and fluctuates greatly. In addition, the short-term characteristics of the PON port traffic are affected by the user's work and life rules, and also show a certain periodicity (such as daily, weekly, etc.). Therefore, neither a single short-correlation model nor a long-correlation model can effectively fit the PON port traffic, and the nonlinear neural network model is more conducive to characterizing the traffic characteristics of the PON port under different time spans.

The linear prediction model of the autoregressive integrated moving average (ARIMA) model has a simple algorithm and is less time-consuming to calculate, but the actual PON port traffic has random and time-varying characteristics, resulting in large prediction errors. For the prediction model of Long Short-Term Memory (LSTM) [14], in the case of a small amount of data, a short *step_time* cannot reflect the periodicity of the PON port traffic, and the use of a long *step_time* LSTM greatly increases the calculation time. In this paper, the highly correlated traffic values are independently extracted as feature data, and the periodicity of the PON port traffic is introduced into the LSTM prediction model. As shown in Table I, compared with the ARIMA and LSTM models, the prediction accuracy is the highest, and the calculation time required by the LSTM prediction model is greatly reduced, achieving a good balance between prediction accuracy and calculation time.

TABLE I.   COMPARISON OF THREE PREDICTION MODELS

model	ARIMA	LSTM	LSTM with periodic features
prediction bias	23.5%	12.3%	7.3%
time	3.1min	43.6min	7.2min

IV. CONCLUSIONS

This paper reviews SDN enabled PON controller technology. The architecture design and application of PON controller based on hierarchical models are studied. First, the main functions and operations of each layer in the hierarchical architecture of PON controllers are presented. Secondly, the application mode and the effect of model technology in hierarchical architecture are analyzed and practiced. Through PON device model, atomic operation capability, network service model, AI and self-service capability, PON controller continuously adapts to the requirements of large volume, complexity-sensitive and smooth evolution of access network, and finally builds an intelligent operation and maintenance management system for the entire life cycle of PON network.

REFERENCES

[1] Jie Chen. Traditional access network and cloud access network will coexist for a long time. People's Posts and Telecommunications, 2020-09-08(006)

[2] L. Csikor, M. Szalay, G. Rétvári, G. Pongrácz, D. P. Pezaros and L. Toka, "Transition to SDN is HARMLESS: Hybrid Architecture for Migrating Legacy Ethernet Switches to SDN," in IEEE/ACM Transactions on Networking, vol. 28, no. 1, pp. 275-288, Feb. 2020, doi: 10.1109/TNET.2019.2958762.

[3] O. Salman, I. H. Elhajj, A. Chehab and A. Kayssi, "QoS guarantee over hybrid SDN/non-SDN networks," 2017 8th International Conference on the Network of the Future (NOF), 2017, pp. 141-143, doi: 10.1109/NOF.2017.8251237.

[4] S. Woo, S. Lee, J. Kim and S. Shin, "RE-CHECKER: Towards Secure RESTful Service in Software-Defined Networking," 2018 IEEE Conference on Network Function Virtualization and Software Defined Networks (NFV-SDN), 2018, pp. 1-5, doi: 10.1109/NFV-SDN.2018.8725649.

[5] M. Aslan and A. Matrawy, "On the Impact of Network State Collection on the Performance of SDN Applications," in IEEE Communications Letters, vol. 20, no. 1, pp. 5-8, Jan. 2016, doi: 10.1109/LCOMM.2015.2496955.

[6] BBF TR-383, "Common YANG Modules for Access Networks, " https://wiki.broadband-forum.org/display/BBF/TR-383%3A+Common+YANG+Modules+for+Access+Networks

[7] BBF TR-385, "ITU-T PON YANG Modules, " https://wiki.broadband-forum.org/display/BBF/TR-385%3A+ITU-T+PON+YANG+Modules

[8] IETF RFC 4741 "NETCONF Configuration Protocol, " Dec. 2006, https://www.rfc-editor.org/info/rfc4741

[9] Leping Wei.Strategic thinking on SDN.Telecommunications Science, 2015, 31 (1) :7-12.

[10] IETF RFC 8348 "A YANG Data Model for Hardware Management, " Mar. 2018, https://www.rfc-editor.org/info/rfc8348

[11] Google, Protocol Buffers, https://developers.google.com/protocol-buffers

[12] Google, gRPC, https://grpc.io/docs/

[13] OpenDaylight, https://www.opendaylight.org/#more

[14] Shao H,Soong B H. "Traffic flow prediction with Long Short-Term Memory Networks (LSTMs)." Region 10 Conference . 2017

978-1-6654-8699-6/22 $31.00 © 2022 IEEE

Proceedings of the 7th Optoelectronics Global Conference (OGC 2022)

Study of Filter-based Neuromorphic Photonic Reservoir Computing for Signal Equalization in 224Gbps Sub-carrier Modulation IM-DD Short Reach Optical Fiber Communication System

Penghao Luo[1], An Yan[1], Aolong Sun[1], Guoqiang Li[1], Sizhe Xing[1], Jianyang Shi[1], Ziwei Li[1,2], Chao Shen[1,2], Junwen Zhang[1,2]* and Nan Chi[1,2]

[1]Key Laboratory for Information Science of Electromagnetic Waves (MoE),Department of Communication Science and Engineering, Fudan University, No. 220, Handan Road. Shanghai 200433, China
[2]Peng Cheng Laboratory, Shenzhen, 518055, China
junwenzhang@fudan.edu.cn

Abstract—The ever-increasing requirements for bandwidth in edge places higher demands on the transmission capacity and data rate of short-reach intensity-modulation and direct-detection (IM/DD) optical fiber communication systems. Advanced digital signal processing (DSP), such as neural network (NN), is verified to be a good way to improve system performance, but the complicated DSP process always means high power consumption and slow processing speed. Reservoir Computing (RC) is a machine learning algorithm suitable for time-series-based problem, which has a faster computing speed than recurrent NN (RNN). The inherent randomness of RC makes us find its potential of signal equalization in all-optical domain. In this paper, we numerically studied a neuromorphic photonic RC signal processing scheme in IM/DD system with low hardware complexity, and realize the all-optical RC through two sets of optical filter nodes. Subcarrier modulation (SCM) signal is applied to study the filter-based neuromorphic photonic RC scheme, in comparison to traditional equalization methods. Simulation results show that the photonic RC equalization can bring orders of magnitude improvement in BER over traditional schemes, and the performances of different Quadrature Amplitude Modulation (QAM) formats are also studied. Finally, the architecture implementation of photonics RC for 224Gbps SCM signal over 80km standard single-mode fiber (SSMF) transmission in C-band is numerically demonstrated.

Keywords-photonic reservoir computing; signal equalization; subcarrier modulation

I. INTRODUCTION

The upcoming of 6G, XR and edge computing services makes data traffic increase at an alarming rate. This places higher requirements on bandwidth of optical communication system, especially in the short-reach, in terms of transmission capacity or speed [1]. Many digital signal processing (DSP) methods have been proposed and applied for the signal equalization to improve the overall performance, including but not limited to LMS, Volterra and MLSE, and they have been proved to be effective for the equalization of linear and nonlinear impairments [2-4].

With the rapid development of Machine Learning (ML), Neural Network (NN) has been proved to have great potential in signal nonlinear equalization, as it theoretically can fit any nonlinear functions. Compared with traditional artificial Neural Network (ANN), recurrent Neural Network (RNN) has better performances in dealing with time-series-based problem in signal processing as it considers the sequential relation [5]. However, RNN has the disadvantages of complex training, large amount of calculation and slow convergence speed. To solve this problem, reservoir computing (RC) is proposed as a more efficient neuromorphic paradigm [6,9,12].

RC reduces the complexity by sparsifying and randomizing the connections between the nodes in Reservoir (the hidden layer), with an output layer in which all training is taking place in a linear manner, and meanwhile retains the performance. Randomness is an inherent feature of RC because the Reservoir layer is randomly generated, which limits the impact of manufacturing deviation on the performance of RC [7,8]. In this case, integrated photonics is well suited for RC. Meanwhile, considering its the high-power consumption and slow processing speed in digital domain, photonic RC is more attractive in future high-dense, high-speed and large capacity optical communication networks. Photonic RC solve this problem by equalizing signal in the optical domain and reduce the overall complexity of signal processing [9,10].

In this paper we studied the optoelectronic signal processing scheme based on photonic RC in SCM IM/DD short reach optical fiber communication system. The demonstrated scheme achieves good signal equalization performance with low hardware complexity, and reduces the complexity of DSP, providing faster processing speed and lower power consumption. In our work, only a simple Feed Forward Equalization (FFE) is used after the RC signal processing, and low bit error rate (BER) can be achieved. SCM signal is applied to study the filter-based neuromorphic photonic RC scheme, in comparison to traditional equalization methods, i.e., FFE and Volterra algorithms. Simulation results show that the photonic RC equalization can bring orders of magnitude improvement in BER over traditional schemes, and the performances of both SCM-16QAM and SCM-32 QAM modulation formats were studied. Finally, the architecture implementation of photonics RC for

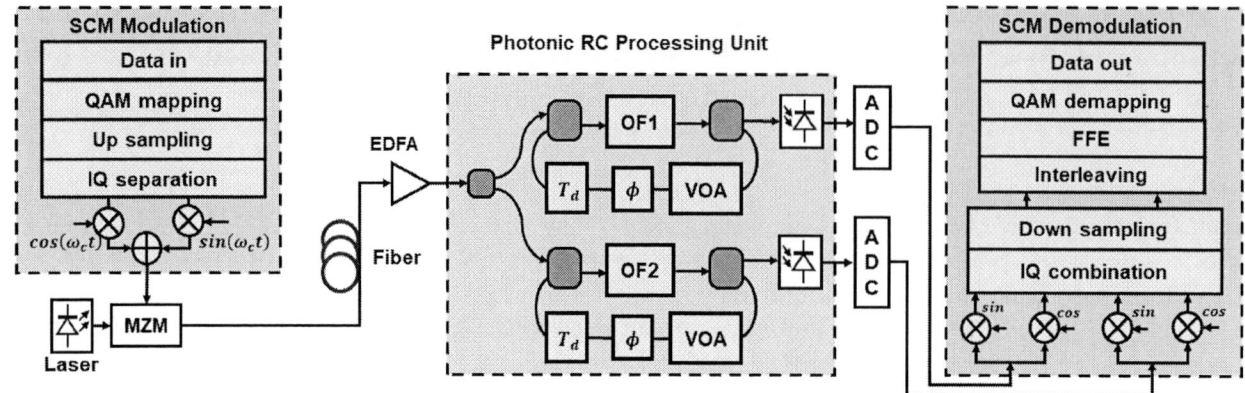

Figure 1. The system setup of data tranmssion of SCM IM/DD signal and the optoeectronic signal processing based on photonic RC

224Gbps SCM signal over 80km standard single-mode fiber (SSMF) transmission in C-band is numerically demonstrated.

II. SYSTEM DESCRIPTION

Fig. 1 shows the numerical setup of data transmission and processing of photonic RC. The core component of the photonic RC unit is an optical filter (OF), which acts as a recurrent node in the Reservoir. Apart from the filter, a 3dB coupler and a feedback loop of delay T_d are employed, together with a phase shift ϕ and a variable optical attenuator (VOA). T_d and ϕ can control the time delay and the phase shift of the feedback loop, and VOA is employed to adjust the feedback strength. Considering the optical filter can be integrated on Mach-zehnder delay interferometers (MZDI) or micro-ring resonators (MRR) [10], the complexity of the system is relatively low, and it is hardware implementable. In such a filter, the transfer function is known as that of the classical band-pass filter.

A group of such optical filter nodes form a photonic Reservoir, which recursively slice the signal in different spectral areas [11]. These optical filters selectively extract different frequency components by weighting it with different values, and thus solving time series based problems such as signal equalization. Here, we select only two groups of such nodes, that is, the spectrum is decomposed into 2 slices, and 2 PDs are used to receive the signal, similar to the setup in [11]. Slicing the spectrum with two filters means that there is a center-frequency detuning between them. Detuning is considered to be as important a variable as feedback strength, delay, etc. Choosing a suitable detuning for the filter can make the system achieve a lower BER. Increasing the number of nodes is expected to improve the performance of the system, but it will lead to more filters and PDs, and will increase the complexity of the system. Certainly, the input layer of the RC scheme is the signal transmitted after the fiber into the receiver, and the output layer is composed of PDs and ADCs. Thus, the whole RC is attained in optical domain.

As shown in Fig. 1, signals are first mapped into a QAM constellation, and then the real and imaginary parts of the signal are multiplied by the orthogonal subcarrier and added together. The Nyquist subcarrier is filtered by root-rising cosine and the roll-off factor is set to be 0.3. Thus, SCM-QAM

signal is generated. The signal will be sent into a standard single-mode fiber by a launched power of $0dBm$ for transmission. At the receivers-side, the signal first passes through the proposed photonic RC unit after an EDFA with 12dB-gain, and is then received by two pairs of PDs and sampled by analog-to-digital convertors (ADCs). For offline DSP of signal recovery, SCM demodulation is applied with down-conversion and IQ channels combinations. Before the QAM de-mapping, we employ a simple linear regression feed-forward equalizer (FFE) for signal combining, interleaving and post equalization. In such a system we numerically tested the performance of 56Gbaud SCM-16QAM signal and 37Gbaud SCM-32QAM signal, and examined the influence of launched power, received optical power (ROP), detuning and the delay T_d on the photonic RC system.

III. RESULTS

The two optical filters are the core components of the RC scheme, therefore, we studied the influence of the bandwidth and detuning of the two filters first. This verification was carried out in the scenery that, SCM-16QAM signal is transmitted 60km in an SSMF at the rate of 56Gbaud. Later discussion about other parameters were also carried out in this scenery. Fig. 2 demonstrates the BER performance versus

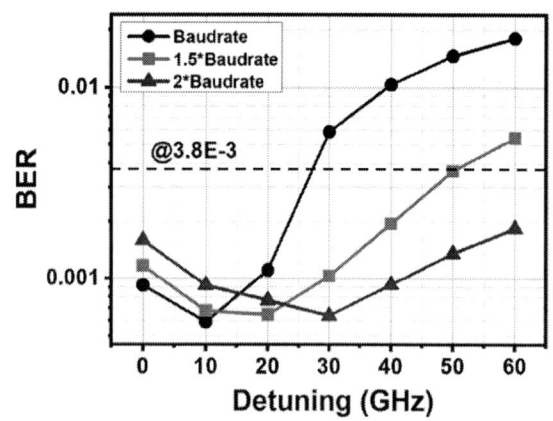

Figure 2. BER versus detuning for 56G SCM-16QAM signal

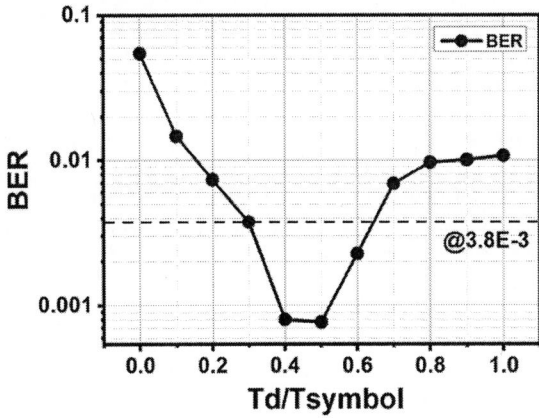

Figure 3. The influence of T_d on BER

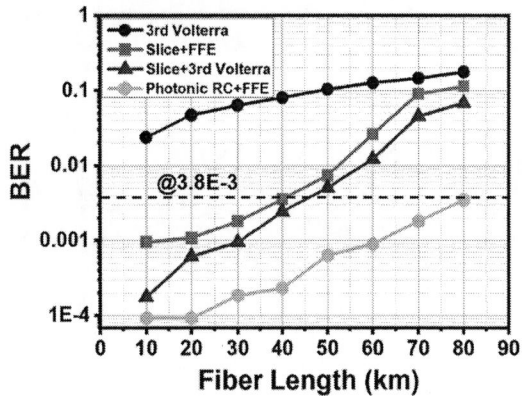

Figure 4. BER versus transmission distance for SCM-16QAM

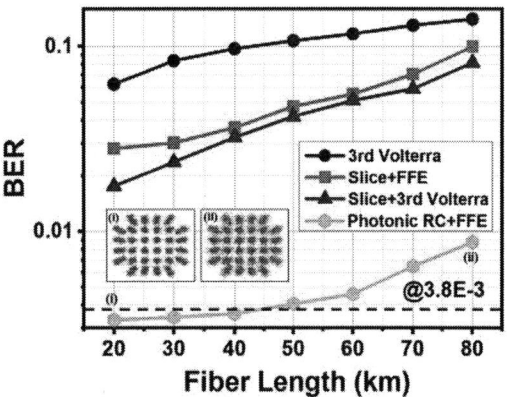

Figure 5. BER versus transmission distance for SCM-32QAM

detuning under different filter bandwidths. The black line employed a filter bandwidth of $1 \times Baudrate$, and the red line and the blue line employed a filter bandwidth of $1.5 \times Baudrate$ and $2 \times Baudrate$ respectively. As can be seen, under each filter bandwidth we can find a best detuning for our system. Although the minimum BER being almost the same, when the bandwidth is larger, the overall BER performance is better. In other words, when the bandwidth is small, the system is more sensitive to detuning, and unsuitable detuning will significantly decrease the performance. It can be seen from the Fig. 2 that when $1 \times baudrate$ is taken as the bandwidth, detuning larger than 30GHz will lead to BER exceeding the threshold; when baudrate is $2 \times Baudrate$, all the detuning within 60GHz can reach the threshold.

The influence of delay and feedback strength are also analyzed. Fig. 3 shows the effect T_d brings to the system. Delays approaching $T_{symbol}/2$ can achieve a BER below 1E-3, when the transmission distance is 60km. Both smaller and larger T_d will make the performance worse. As for feedback strength, we set the value to 50% so that it not only ensures enough feedback that enhances the memory of RC system, but also bring limited impact on ROP.

The proposed scheme employs two optical filters as the nodes of RC, and each filter using a loop to provide a feedback, thus realizing selective processing of different frequency components of the signal. As we know, even without the loop, the two optical filters can slice the signal spectrum and process different frequency components separately. In addition, the two filters and PD are equivalent to one more sampling, which also improves the BER performance. Therefore, in order to prove the effect of the proposed photonic RC scheme, we compared it with not only the effect of traditional Volterra algorithm, but also with the effect of spectrum slicing only (without the loop). In our work we considered four options. The first option is to receive the SCM signal directly without filtering, and equalize the signal with third-order Volterra. The second and third options are spectrally sliced by two filters and equalized by FFE and Volterra respectively. The last one employs the photonic RC scheme and uses FFE for DSP.

Fig. 4 shows the BER performance of 56Gbaud SCM-16QAM signal versus different transmission distance in an SSMF. With 7% FEC of 3.8E-3 as a BER threshold, the proposed RC scheme can transfer 80km, compared with the method of Volterra, which cannot reach the threshold at all. That's because Volterra has little effects on dispersion distortion. The red and the blue line are the performance of spectrum slicing only, with the red using FFE for DSP and the blue using Volterra. It's evident that spectrum slicing can bring a certain degree of improvement (a transmission of over 40km), but our RC scheme with a feedback loop is more efficient. It has been verified that both the two filters and the feedback loops can improve the performance of the system. Compared with the other three methods, the RC scheme can bring orders of magnitude improvement, and with the transmission distance becomes longer, the improvement becomes larger. For the reason, the nonlinearity becomes stronger, and the RC scheme has a higher potential for equalizing nonlinearity. The same trend can be seen in 37Gbaud SCM-32QAM signal. As the bandwidth limits the modulation order, the overall performance of 37Gbaud SCM-32QAM signal is not so good as 56Gbaud SCM-16QAM signal, but we can demonstrate the improvement RC scheme brings to the system.

Additionally, the impact of optical power is examined in Fig. 6, including the launched power (a) and ROP (b). When the launched power is below 9dBm, with the increase of launched power, the signal-to-noise ratio (SNR) will increase

978-1-6654-8699-6/22 $31.00 © 2022 IEEE 43

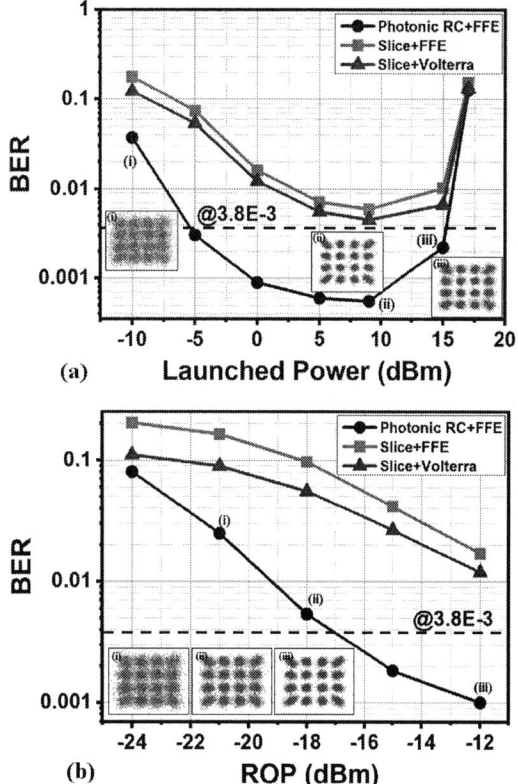

Figure 6. BER performance of SCM-16QAM signal with different (a) launched power and (b) ROP

SCM signal is applied to study the filter-based neuromorphic photonic RC scheme, in comparison to traditional equalization methods, i.e., FFE and Volterra algorithms. Simulation results show that the photonic RC equalization can bring orders of magnitude improvement in BER over traditional schemes, and the performances of both SCM-16QAM and SCM-32 QAM modulation formats were studied. Finally, the architecture implementation of photonics RC for 224Gbps SCM signal over 80km standard single-mode fiber (SSMF) transmission in C-band is numerically demonstrated.

ACKNOWLEDGMENT

This work is partially supported by National Natural Science Foundation of China (61925104, 62031011, 62171137), Natural Science Foundation of Shanghai (21ZR1408700), National Key Research and Development Program of China (2021YFB2801804) and The Major Key Project of PCL.

REFERENCES

[1] Tsirigotis A, Tsilikas I, Sozos K, et al. Filter-based photonic reservoir computing as a key-enabling platform for all-optical, high-speed processing of time-stretched images and telecomm data[C]//AI and Optical Data Sciences III. SPIE, 2022, 12019: 111-118.

[2] Least-mean-square adaptive filters[M]. John Wiley & Sons, 2003.

[3] Nowak R D, Van Veen B D. Volterra filter equalization: A fixed point approach[J]. IEEE Transactions on signal processing, 1997, 45(2): 377-388.

[4] Forney G. Maximum-likelihood sequence estimation of digital sequences in the presence of intersymbol interference[J]. IEEE Transactions on Information theory, 1972, 18(3): 363-378.

[5] Zaremba W, Sutskever I, Vinyals O. Recurrent neural network regularization[J]. arXiv preprint arXiv:1409.2329, 2014.

[6] Schrauwen B, Verstraeten D, Van Campenhout J. An overview of reservoir computing: theory, applications and implementations[C]//Proceedings of the 15th european symposium on artificial neural networks. p. 471-482 2007. 2007: 471-482.

[7] Sozos K, Bogris A, Bienstman P, et al. High Speed Photonic Neuromorphic Computing Using Recurrent Optical Spectrum Slicing Neural Networks[J]. arXiv preprint arXiv:2203.15807, 2022.

[8] Du C, Cai F, Zidan M A, et al. Reservoir computing using dynamic memristors for temporal information processing[J]. Nature communications, 2017, 8(1): 1-10.

[9] Ranzini S M, Dischler R, Da Ros F, et al. Experimental investigation of optoelectronic receiver with reservoir computing in short reach optical fiber communications[J]. Journal of Lightwave Technology, 2021, 39(8): 2460-2467.

[10] Sozos K, Bogris A, Bienstman P, et al. Photonic Reservoir Computing based on Optical Filters in a Loop as a High Performance and Low-Power Consumption Equalizer for 100 Gbaud Direct Detection Systems[C]//2021 European Conference on Optical Communication (ECOC). IEEE, 2021: 1-4.

[11] Ranzini S M. Optoelectronic receiver in short reach optical fiber communications[J]. 2021.

[12] Jaeger H. The "echo state" approach to analysing and training recurrent neural networks-with an erratum note[J]. Bonn, Germany: German National Research Center for Information Technology GMD Technical Report, 2001, 148(34): 13.

and the BER performance will be improved. However, when it reaches 9dBm, nonlinearity, which is now the main factor to limit the system, becomes stronger and stronger, resulting in the increase of BER. In Fig. 6(a), (i) with a small launched power shows the constellation diagram of recovered signal with low SNR, and (iii) with a large launched power shows the effect of nonlinearity on the recovered signal. The launched power of 0 to 10dBm performs well in our system, the BER of which can reach 6E-4. With respect to ROP, it is demonstrated in Fig. 6(b) that with ROP increases, BER becomes lower and lower, as we haven't set any bandwidth limitation for PDs. When $ROP > -17dBm$, the BER of 56Gbaud SCM-16QAM signal reaches the threshold.

IV. CONCLUSION

We studied the optoelectronic signal processing scheme based on photonic RC in SCM IM/DD short reach optical fiber communication system. The RC scheme utilized two optical filter nodes, each node having a feedback loop with T_d, ϕ and VOA to control the recurrent parameters of the Reservoir. We analyzed the influence of these parameters on the BER of the system. In our work, only a simple FFE is used after the RC signal processing, and low BER can be achieved.

978-1-6654-8699-6/22 $31.00 © 2022 IEEE

Proceedings of the 7th Optoelectronics Global Conference (OGC 2022)

Coupling Efficiency Analysis for Optical Fiber with Different Core Diameters

Yuzhong Ma, Zijing Huang, Lin Sun, Gordon Ning Liu*

Suzhou Key Laboratory of Advanced Optical Communication Network Technology, School of Electronic and Information
Engineering, Soochow University, Suzhou, Jiangsu 215006, China
e-mail: gordonnliu@suda.edu.cn

Abstract—**The loss of optical fiber link has a significant impact on the performance of optical fiber communication. In the short-distance optical interconnection, the quality of optical fiber connection is one of the main factors affecting the link loss, especially the influence of optical fiber coupling offset on the optical fiber coupling efficiency. In this paper, we build a model to quantitatively analyze the relationship between coupling offset and coupling efficiency. The simulation results show that a large core fiber has a higher tolerance to coupling offset compared with a traditional fiber.**

Keywords-fiber optics; geometrical optics; coupling loss; ray tracing

I. INTRODUCTION

Driven by the emergence and development of Internet of Things, mobile broadband, cloud computing, autonomous driving and other services, the demand for data center interconnection, mobile fronthaul, broadband access and in-vehicle network is growing rapidly [1]-[3]. The multi-mode fiber (MMF) suitable for large capacity and short distance transmission has attracted the attention of many researchers. In the optical link, the connection between optical fibers is essential. With the inevitable mechanical tolerance of fiber connector, the environmental variation such as vibration and changed temperature will lead to the radial, axial and angular offset. Especially for the optical link of short distance, the connection coupling loss is relatively large in the overall loss of the optical fiber link. And for fibers within in-vehicle networks, a high coupling offset tolerance is required due to the continuous vibration during the car driving and the frequent fiber dis-connect and re-connect during car maintenance. Therefore, it is particularly important to reduce the coupling loss of the fiber connector. It also needs to reduce the sensitivity of coupling loss to the coupling offset to let the link tolerable to some extent of coupling offset [4]-[6]. Consequently, a quantitative analysis of the relationship between coupling offset and coupling efficiency in fiber connection is necessary. Note that the coupling efficiency is same as the coupling loss here,

which is defined by the ratio of emitted optical power to received optical power for the connector under investigation.

Kangmei Li et al. proposed a new large-core MMF with 100μm core diameter recently. Such a new fiber has several advantages including low dispersion, small bending loss and large tolerable radial offset, compared with the traditional MMF with 50μm core diameter [7]. Koji Horiguchi et al. established a mathematical model of fiber coupling by detecting the mode field distribution of the transmitting fibers and receiving fibers. Their calculation results are similar to experiment results. However, the modeling process requires additional measured near field and far field patterns, which improves the complexity of modeling [8].

However, the coupling efficiency affected by axial and angular offsets during fiber connection have not been analyzed before. In this paper, we will use OpticStudio to model the coupling from a Gaussian light source to a length of fiber and the coupling between fibers. The coupling efficiency is analyzed by using the ray tracing method combined with a complete optical path during the coupling. Thus, the radial, axial and angular offsets tolerance with certain coupling efficiency will be compared for fibers with different core diameters.

II. PROPOSED METHOD OF FIBER COUPLING DESIGN

A. Fiber Coupling Model

We established a fiber coupling model with the OpticStudio non-sequential mode. One major advantage of OpticStudio is that it can use the ray tracing method to directly observe the light path, and then adjust the parameters. The near field model of fiber coupling is shown in Figure 1 (a), and the light source is set as a Gaussian beam with a wavelength of 850nm. It consists of a transmission fiber and a circular shading surface with a small hole in the center. Considering that the effect of the fiber refractive index distribution on the fiber coupling can be ignored. The transmission fiber is simulated by a hollow cylinder, and the hole in the center of the shielding surface is used to simulate the cross section of receiving fiber. The light through the hole is regarded as a successful coupling, and the light blocked by the shading surface is part of the coupling loss.

space for the tilt of the fiber, we set a certain axial offset in the simulated angular offset, and the subsequent conclusions show that this part of axial offset has little effect on the coupling efficiency.

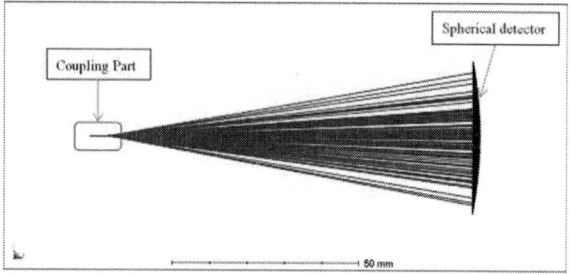

Figure 2.   Far field layout

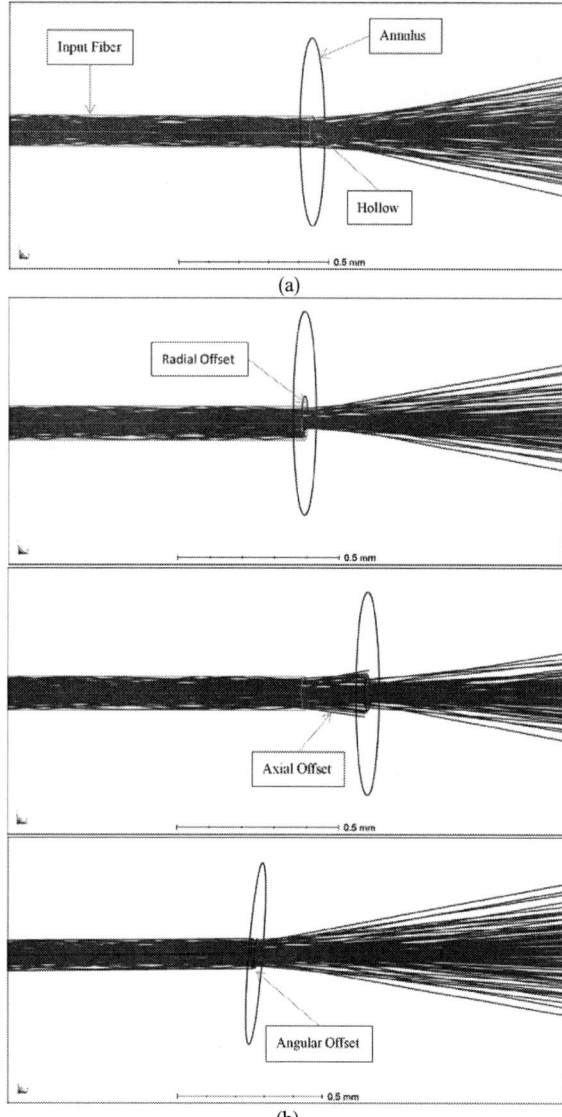

Figure 1.   1Fiber coupling model, (a) near field layout; (b) three kinds of fiber coupling misalignment

Three kinds of offsets during the optical fiber coupling are shown in Figure 1 (b). Moving the transmission fiber in the direction perpendicular to the optical axis can simulate the radial offset. Moving the transmission fiber in the direction parallel to the optical axis can simulate the axial offset. Changing the angle of the transmission fiber so that a certain angle between the two fibers can simulate the angular offset. It can be observed that a portion of the optical power in the transmitted fiber is absorbed by the circular shading surface when the fiber coupling appears radial and axial offset. Note that, in order to reserve some

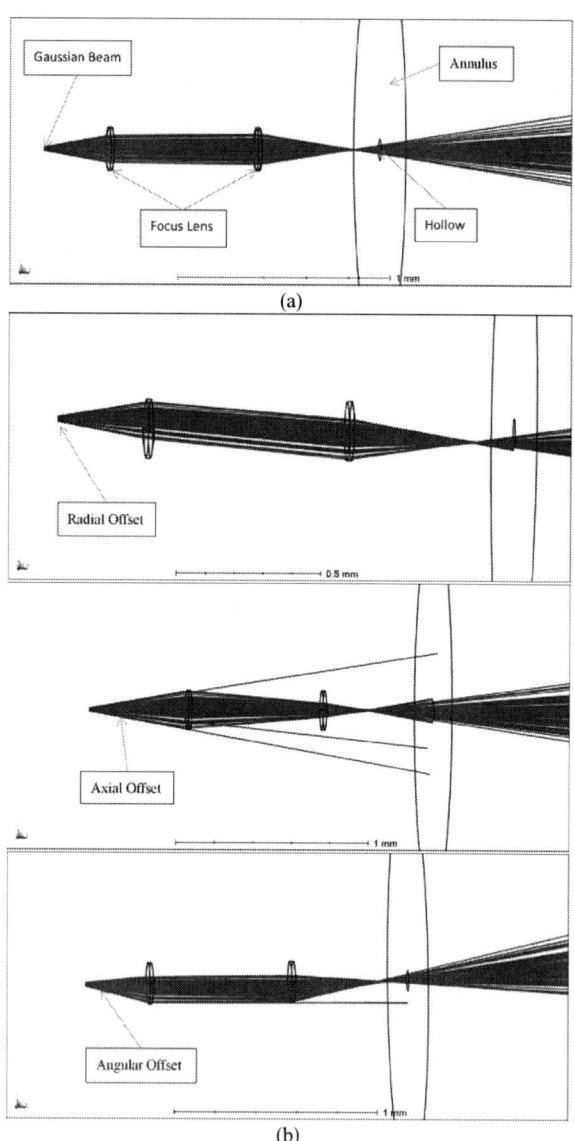

Figure 3.   Gaussian light source to fiber model, (a) near field layout; (b) three kinds of Gaussian light source to fiber coupling misalignment

978-1-6654-8699-6/22 $31.00 © 2022 IEEE

Figure 2 shows the far-field model of the optical fiber coupling. This structure consists of the coupling part shown in Figure 1 and a spherical detector. Because in the optical power received at the cross section of the receiving fiber, part of the optical power escapes from the cladding after entering the receiving fiber because of the large emitting angle. This is another part that constitutes the coupling loss. Therefore, we added a spherical detector far behind the torus, by calculating the numerical aperture (NA) of the optical fiber to adjust the position and rotation angle of the spherical detector, so as to filter the optical power not satisfying the receiving fiber's NA.

B. Gaussian light source to Fiber Coupling Model

The model of the Gaussian light source to fiber coupling is shown in Figure 3 (a). Gaussian light source passes through two lenses and focus into a small hole to simulate the reception of the fiber. Just as the fiber coupling, we add a spherical detector farther behind the shading surface to detect the receiving optical power.

As shown in Figure 3 (b), changing the emission position and emission direction of the Gaussian light source can also simulate three kinds of offsets of Gaussian light source to fiber coupling.

III. SIMULATION RESULTS AND DISCUSSION

In this paper, we compared the fibers with 50μm (Fiber A) and 100μm core diameter (Fiber B). Both Fiber A and Fiber B have the same NA of 0.2. For the fiber coupling model, the coupling efficiency is plotted against the axial and angular offsets in Figure 4 with and without radial offset.

As can be seen in Figure 4 (a), the decrease rate of the coupling efficiency increases for the coupling without the radial offset. The rate of Fiber A is higher than the rate of Fiber B. When the axial offset increases, without the radial offset, the coupling efficiency also decreases significantly. After introducing a radial offset of 10μm, there is no significant decrease in the coupling efficiency of Fiber A and Fiber B when the axial offset is small. With the axial offset increases, the decrease rate of the coupling efficiency will gradually increase to the situation when there is no radial offset. In Figure 4 (b), with the increasing angular offset, the coupling efficiency decreases significantly, and Fiber A and Fiber B perform almost the same. With a radial offset of 10μm, the coupling efficiency of Fiber A is slightly smaller than the one of Fiber B in the absence of angular offset, and the increase of angular offset does not affect the difference.

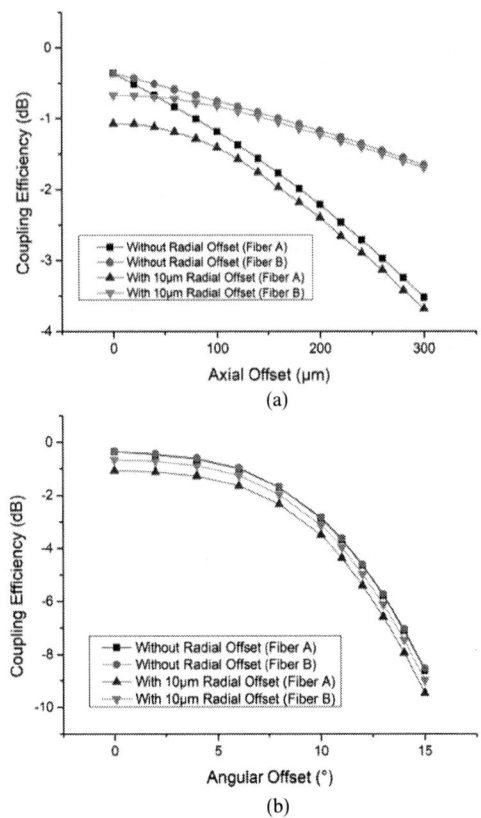

Figure 4. Three kinds of offsets in Fiber A and Fiber B coupling, (a) radial and axial offsets; (b) radial and angular offsets

The observed simulation results can be explained below. The Gaussian beam is centered emission. The mode field of fiber cross section is close to the Gaussian distribution. When a certain axial offset is set, the light power in outer layer of the fiber core will be leaked, leading to a decrease in the coupling efficiency. Adding a small amount of radial offset will further reduce the coupling efficiency, since the energy within the fiber core outer layer is relatively low compared with the one in the inner layer. As the axial offset increases, the loss from the axial offset gradually dominates, so for different radial offsets, the coupling efficiency eventually increases with the axial offset. In the model of this paper, the angular offset of the fiber coupling cause very few optical power leakage from the receiving fiber. Considering the light inputted into fiber is Gaussian distributed, the central optical power is larger. When the angular offset is small, the optical power of the edge is small. But with the increasing angular offset, the input angle of the light launched into the receiving fiber cross section also increases. The light with large input angle will be scattered when the input angle is larger than the numerical aperture of the receiving fiber. In the model, the output light gets out of the reception range of the spherical detector with an adjusted position. As the offset angle increases, the scattered optical power gradually approaches the central part of the Gaussian beam, that accelerates the decrease rate of coupling efficiency.

978-1-6654-8699-6/22 $31.00 © 2022 IEEE

The fiber coupling loss caused by offset is reflected in the amount of optical energy not emitted to the cross section of the receiving fiber. For the fiber with larger core diameter, the same radial and axial offset will bring less coupling loss, increasing the core diameter of the fiber will make the fiber less sensitive to radial and axial offset.

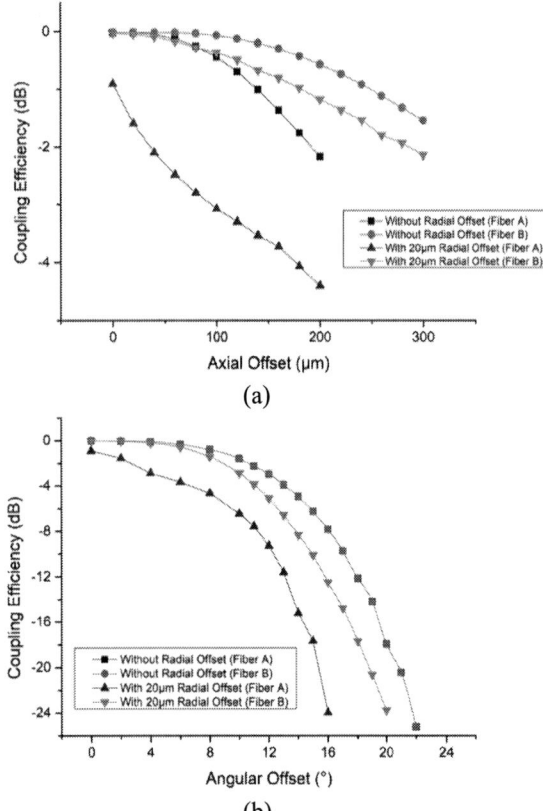

Figure 5. Three kinds of offsets in Gaussian light source to Fiber A and Fiber B coupling; (a) radial and axial offsets; (b) radial and angular offsets

Besides this, we simulate the coupling efficiency from Gaussian light source to Fiber A or Fiber B, with different axial and angular offsets, with and without radial offset. The results are shown in Figure 5. In Figure 5 (a), without radial offset, the coupling efficiency of Fiber A gradually decreases as the axial offset increases, with the pace faster than the one of Fiber B. Fiber B decreases with a slower speed, and the axial offset within 100 μ m does not cause significant loss. With a radial offset of 20μm, the coupling efficiency of the Gaussian light source to Fiber A decreased significantly, while Fiber B is almost unaffected. The axial offset is now increasing, Fiber A also showed a faster decrease rate in coupling efficiency than Fiber B. As can be seen in Figure 5 (b), in the absence of radial offset, the coupling efficiency of Gaussian light source to Fiber A and Fiber B accelerates with the increase of angular offset, with the silimar decrease law. After introducing a 20μm radial offset, the coupling efficiency of Fiber A decreases faster than Fiber B, and the difference in the coupling efficiency of two fibers grows when the angular offset increases.

The ability of Gaussian light source tolerating higher offset is mainly attributed to the lens coupling. Since the Gaussian light source does not cover the entire lens surface, only a small portion of light source is launched into the fiber cross section. Therefore, a small amount of radial, axial and angular offset has no significant effect on the coupling efficiency. And with the increase of offset, on the one hand, the Gaussian light source will get out of the receiving range of the lens. This is one part of optical power loss. On the other hand, with the movement or rotation changes of Gaussian light source, the focus on the cross section of receiving fiber will move. When the light spot moves beyond the range of the receiving fiber cross section, optical power will also lose. This is the other part of the loss. These two parts of loss cause that the coupling efficiency decrease with an increased speed when the radial and angular offset beyond a certain threshold.

When the Gaussian light source is coupled to the optical fiber, both the axial and angular offsets will make the light spot deviate from the original position. When the core diameter of the fiber is larger, the acceptable area for the moved light spot is also larger. Thus, larger axial and angular offsets can be tolerated.

IV. CONCLUSION

In order to simulate the coupling between two fibers and from a Gaussian light source to a fiber, the effects of axial and angular offsets on the coupling efficiency were investigated. We developed a new model by the OpticStudio. The radial offset simulation results are similar to previous results, prove the feasibility of this modeling method. Using this model, we found that in fiber coupling, fiber with larger core diameter has higher tolerance for axial offset and angular offset, thus reduce the requirements for mechanical fabrication accuracy and environmental stability. In Gaussian light source to fiber coupling, using fiber with larger core diameter can improve the tolerance of axial and angular offset, even the coupling efficiency is hardly affected by small offset. Choosing larger lenses and larger core diameter optical fibers can further widen the range of tolerable offset.

REFERENCES

[1] J. Lee et al., "Spectrum for 5G: Global Status, Challenges, and Enabling Technologies," in IEEE Communications Magazine, vol. 56, no. 3, pp. 12-18, (2018)

[2] Y. Koike and A. Inoue, "High-speed plastic optical fibers and their simple interconnects for 4K/8K video transmission," 2015 European Conference on Optical Communication (ECOC), pp.1-3, (2015)

[3] S. C. Mettler, "A general characterization of splice loss for multimode optical fibers," in The Bell System Technical Journal, vol. 58, no. 10, pp. 2163-2182, (1979)

[4] Itay Yahav, Nir Sheffi, Yacov Biofcic, and Dan Sadot, "Multi-Gigabit Spatial-Division Multiplexing Transmission Over Multicore Plastic Optical Fiber," J. Lightwave Technol.39, 2296-2304, (2021)

978-1-6654-8699-6/22 $31.00 © 2022 IEEE

[5] Ronald E. Freund, Christian-A. Bunge, Nikolay N. Ledentsov, D. Molin, and Ch. Caspar, "High-Speed Transmission in Multimode Fibers," J. Lightwave Technol.28, 569-586 (2010)

[6] Yasuhiro Koike and Azusa Inoue, "High-Speed Graded-Index Plastic Optical Fibers and Their Simple Interconnects for 4K/8K Video Transmission," J. Lightwave Technol.34, 1551-1555 (2016)

[7] K. Li, X. Chen, A. R. Zakharian, J. E. Hurley, J. S. Stone and M.-J. Li, "High Bandwidth Large Core Multimode Fibre with High Connector Tolerance for Short Distance Communications," 2020 European Conference on Optical Communications (ECOC), pp.1-4,(2020)

[8] K. Horiguchi, S. Kobayashi and O. Sugihara, "Calculation Model for Multimode Fiber Connection Using Measured Near- and Far-Field Patterns," IEEE Photonics Technology Letters, vol. 33, no. 6, pp. 285-288, 15 March15 (2021)

Proceedings of the 7th Optoelectronics Global Conference (OGC 2022)

Underwater Wireless Optical Communication Channel Characterization Using Machine Learning Techniques

Abdulaziz Al-Amodi
Department of Electrical Engineering
King Fahd University of Petroleum and Minerals
Dhahran 31261, Kingdom of Saudi Arabia
s201812280@kfupm.edu.sa

Mudassir Masood
Department of Electrical Engineering
Center for Communication Systems and Sensing
King Fahd University of Petroleum and Minerals
Dhahran 31261, Kingdom of Saudi Arabia
mudassir@kfupm.edu.sa

M. Z. M. Khan
Department of Electrical Engineering
Center for Communication Systems and Sensing
King Fahd University of Petroleum and Minerals
Dhahran 31261, Kingdom of Saudi Arabia
zahedmk@kfupm.edu.sa

Abstract—Recently Underwater Optical Wireless Communication (UOWC) has attracted major attention due to its high transmission rate, low link delay, high communication security, and low implementation cost. However, optical signals suffer from severe attenuation loss due to absorption and scattering effects, which impedes the establishment of an effective and reliable UWOC system. Hence, it is important to identify the characteristic of the underwater channel in order to overcome the mentioned challenges. In literature, the combination of the Exponential and the Generalized Gamma Distribution (EGG) has been shown to model the underwater channel environment with great accuracy. EGG is a comprehensive channel model incorporating the effect of temperature-induced turbulence in the presence of air bubbles, in both fresh and salty aqueous environments. In this work, we built a Machine Learning (ML) based system that utilizes Convolutional Neural Network (CNN) to estimate the parameters of the EGG channel model from the received signal. Furthermore, we take one more step and train a separate deep network to predict bubble level and temperature gradient in the UWOC channel using the estimated parameters. The two networks together form a pipeline enabling us to estimate the channel state from the received signal. The results confirm well with the experimental data from the literature.

Keywords-underwater optical wireless communication; channel characterization; machine learning; convolutional neural networks

I. INTRODUCTION

The transmission of wireless signals in the underwater environment is of great importance, as it can be utilized in exploring, monitoring, and researching offshore regions. Over the years, several approaches were implemented to establish a reliable underwater communication system (e.g., acoustic waves, optical waves, and radio frequency waves). To begin

with, acoustic waves are heavily implemented due to their long transmission range, and low attenuation rate. However, it suffers from severe limitations in bandwidth, along with high latency and significant transmission losses. As a result, RF waves were implemented to provide high bandwidth and hence, high data rate. In spite of that, as in the case of acoustic waves, RF waves have also been found to have major disadvantages, such as the short transmission range, and high attenuation rate. Furthermore, it requires large antenna size, since it operates at low frequency. These major drawbacks in both acoustic and RF waves boosted the research interest in optical waves. UWOC can provide very high bandwidth with low latency and power efficiency, but it also suffers from short transmission range, which stems from the absorption and scattering phenomena. Hence, limiting the performance of optical wireless signals and impeding the establishment of an effective and reliable UWOC system. The complexity of the underwater environment needs to be modeled accurately to characterize the absorption and scattering effects. Several models have been proposed in an attempt to accurately describe UWOC channel. A simple tuned Beer-Lambert's Law (BL) was proposed in [1,2] to model both the absorption and scattering effects. However, it is based on the unrealistic assumptions of having the transmitter and receiver in perfect line of sight (LOS). Furthermore, the model does not factor in the scattered light and it assumes that it is fully lost. Another attempt in this direction involves finding the approximated solution of the Radiative Transfer Equation (RTE) by utilizing the Monte Carlo simulation as in [3]-[5]. While this approach is considered as an accurate model, it requires thousands of iterations and hence, making it computationally expensive and impractical. The lognormal distribution was also proposed in [6] to model UWOC channel. However, while the lognormal distribution can be used to model optical signals in free space, it fails to accurately model the underwater channel as the refractive index differs greatly from free space due to

978-1-6654-8699-6/22 $31.00 © 2022 IEEE

temperature and air bubble levels [7]. Exponential-Lognormal distribution model was also proposed in [8]. While this model outperforms the Lognormal, it is not a comprehensive model as it does not factor in the temperature gradient and salinity effects. In [9], Weibull distribution was proposed to model salinity-induced turbulence, which showed an excellent fit with the experimentally measured data. As in the case of the Exponential-Lognormal, the Weibull distribution is limited as it cannot model the effect of both gradient temperature and bubble levels. The Generalized Gamma Distribution (GGD) was also proposed in [10] in order to model the underwater channel in the case of weak temperature-induced turbulence. It is worth noting that both Weibull and Exponential distributions are special cases of the GGD. The proposed distribution perfectly models the turbulence induced due to temperature gradient of the underwater channel, but it does not model the salinity-induced turbulence. A comprehensive model that considers the underwater channel under all different conditions in different water types was proposed in [11]. The proposed model employs the mixture of Exponential-Generalized Gamma Distributions (EGG).

In this paper, we extended the work of [11] that utilize the EGG model to characterize the underwater channel for variants scenarios of temperature gradient and bubble levels in fresh water. These channels are used to sensate received signals which are then used to train deep networks with the aim of estimating the unknown parameters. The parameters values that are used to simulate channels in the CNN training process are obtained experimentally from [11] by using the expectation maximization (EM) algorithm. In addition to that, we propose to add some more parameters settings as presented in Table I and described there. The proposed network is a deep CNN that exhibits an excellent fit with the experimentally collected data as well as the proposed scenarios.

The rest of the paper is organized as follows. Section II illustrates the signal and channel models along with the proposed network structure. Section III explains the process of the estimation of the unknown parameters, along with the channel intensity. Section IV discusses the simulation results of the CNN. In the end, some final remarks are shown in Section V.

II. PROBLEM FORMULATION

A. Signal Model

As mentioned earlier, we will use the EGG to model the turbulence in the UWOC channel. Based on this, the optical signal traveling through such a turbulent medium will be modeled at the receiver as follows [12]

$$y = \eta I x + n \quad (1)$$

Where η is the optical-to-electrical conversion coefficient, I is the normalized irradiance, and x is the transmitted information. Furthermore, n represent the additive noise. Note that I is the EGG distributed random variable that models the turbulent channel.

B. UWOC Channel Modeling with the EGG Distribution

Throughout this paper, the EGG distribution will be used to model gradient temperature and bubble levels in the underwater channel since it provides us with a comprehensive model for UWOC. The EGG is expressed in equation (2) as

$$f_I(I) = \omega f(I, \lambda) + (1 - \omega)g(I, a, b, c) \quad (2)$$

Here,

$$f(I, \lambda) = \frac{I}{\lambda} \exp\left(-\frac{I}{\lambda}\right) \quad (3)$$

and

$$g(I, a, b, c) = \frac{cI^{ac-1}}{b^{ac}} \frac{\exp\left(-\left(\frac{I}{b}\right)^c\right)}{\Gamma(a)} \quad (4)$$

Are respectively the exponential and generalized gamma distributions. ω is the parameter to form the mixture of the two distributions $(0 < \omega < 1)$. In this model, the exponential distribution models the bubbles-induced turbulence, while the GGD takes care of turbulence induced by temperature gradient. Hence, in the case of uniform temperature, the simple exponential distribution can perfectly model the underwater channel. This special case will aid us in the estimation of the parameters that are used to train the proposed deep network architecture. Furthermore, λ, a, b, and c are the distribution parameters that we are interested in estimating along with ω. The authors in [11] used different bubble levels and temperature variations to get the real data measurements. This data was then used with Expectation Maximization (EM) algorithm to obtain the corresponding Maximum Likelihood (ML) estimates of the parameters. We used these estimated parameters to generate different channel conditions in order to generate data that is used to train the deep learning network.

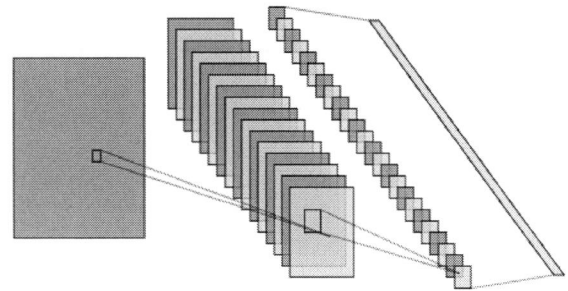

Figure 1. General structure of CNN Network.

C. Machine Learning Network

The network structure proposed in this paper is depicted in Fig.1 and is composed of four 1D Convolutional layers having the number of kernels equal to 16, 32, 64, and 128 respectively. This is followed by three fully connected layers of dimensions 120, 84, and 5 respectively. To illustrate, the input to the first convolutional layer is a vector of size 500, which is simply a collection of 500 samples of the received signal. The output of each convolutional layer is subject to ReLU activation function. Similar to the convolutional layers, ReLU activation is applied to the output of the fully connected layers except for the final layer whose output forms the estimated parameters values. Finally, mean square error (MSE) is used as the loss function.

III. CHANNEL ESTIMATION

A. Estimation of Channel Parameters

The parameters of the EGG distribution are estimated for different scenarios of bubble levels and temperatures gradient in fresh water environment. The estimation of these parameters allows us to monitor the intensity of bubbles and temperature. In other words, the turbulence induced by these factors can be measured by estimating these parameters. For training the Neural Network (NN), a number of scenarios listed in Table I are utilized to achieve a minimum error in the estimated value. The parameters values in the listed scenarios are based on the measurement done experimentally in [11]. Furthermore, some new scenarios have been added to the table. These are mainly the special scenarios where one of the controlling factors (i.e., bubbles and temperature variations) is missing. For example, $\omega = 0$ refers to the channel with temperature-induced turbulence only. For this case, λ does not matter. Hence, we need to estimate a, b, and c only. It was also observed that the pdf of GGD is highly concentrated around 1 when temperature gradient is zero. However, the pdf spreads out as the temperature gradient increases. Guided by these observations, additional scenarios were estimated to further enhance the training process since the variety of scenarios will ensure that the NN is trained properly for different water environments (e.g., harbor, coastal, and sea).

B. Estimation of Channel Intensity

Once the channel parameters are estimated, we can use these to accurately estimate the bubble level and temperature gradient. This step allows us to estimate the physical parameters of the channel using the statistical parameters (i.e., ω, λ, a, b, and c).

IV. SIMULATION RESULTS

The NN described earlier in Sec II-C is trained using the signal received after passing through channels corresponding to the various scenarios presented in Table I. We generate 50,000 samples of the received signal for each scenario to train the NN. Thus for 16 scenarios, we have 800,000 samples. The data used for testing is generated separately using 25,000 samples. In addition to that, to ascertain the efficiency of the network, we completely exclude some of the scenarios from the training phase. These are then used for testing the performance of the NN. We keep repeating this by excluding different scenarios every time, which allowed us to establish more confidence in the CNN performance. We used normalized mean-square-error to measure the accuracy of the predicted parameters. The simulation results of estimating the parameters are presented in Fig. 2 and summarized in Table II. Table III shows the NMSE in estimating the bubble level and temperature gradient. It is obvious from the results that all parameters as well as the channel state have been estimated with high accuracy.

Table I. EGG Parameters values for different underwater channel scenarios for NN training.

Scenarios		Exponential Generalized Gamma Distribution Parameters
Bubble Level (L/min)	Temperature Gradient (°C · cm^{-1})	$(\omega, \lambda, a, b, c)$
0	0	$(4.0628 \times 10^{-21}, 1.0225, 30.8432, 0.6993, 9.5461)$[11]
0	0.05	$(4.0628 \times 10^{-21}, 1.0225, 11.8268, 0.7709, 5.4146)$ This work
0	0.10	$(4.0628 \times 10^{-21}, 1.0225, 4.9787, 0.8347, 6.6622)$ This work
0	0.15	$(4.0628 \times 10^{-21}, 1.0225, 1.9303, 0.9264, 7.1128)$ This work
0	0.20	$(4.0628 \times 10^{-21}, 1.0225, 1.2894, 0.9926, 7.2019)$ This work
2.4	0	$(0.1953, 0.5273, 3.7291, 1.0721, 30.3214)$[11]
4.7	0	$(0.2109, 0.4603, 1.2526, 1.1501, 41.3258)$[11]
7.1	0	$(0.3489, 0.4771, 0.4319, 1.4531, 74.3650)$[11]
2.4	0.05	$(0.2130, 0.3291, 1.4299, 1.1817, 17.1984)$[11]
2.4	0.10	$(0.2108, 0.2694, 0.6020, 1.2795, 21.1611)$[11]
2.4	0.15	$(0.1807, 0.1641, 0.2334, 1.4201, 22.5924)$[11]
2.4	0.20	$(0.1665, 1.1207, 0.1559, 1.5216, 22.8754)$[11]
4.7	0.05	$(0.4589, 0.3449, 1.0421, 1.5768, 35.9424)$[11]
4.7	0.10	$(0.4539, 0.2744, 0.3008, 1.7053, 54.1422)$[11]
16.5	0.22	$(0.6238, 0.1094, 0.0111, 4.4750, 105.3550)$[11]
23.6	0.22	$(0.7210, 0.1479, 0.0121, 7.4189, 65.6983)$[11]

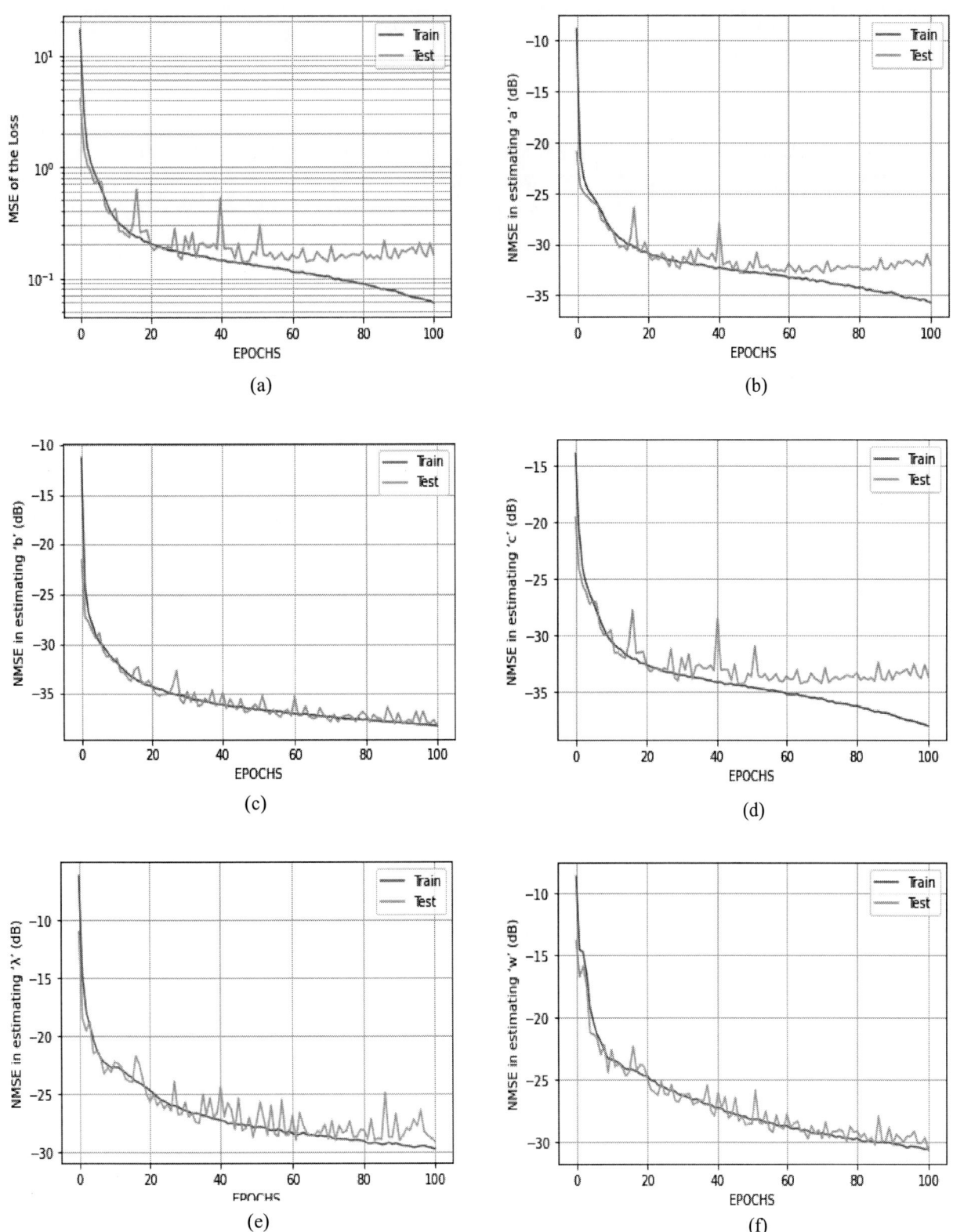

Fig. 2: The performance of the CNN in estimating the parameters of the EGG distribution. Training and validation error of (a) Loss, (b) a, (c) b, (d) c, (e) λ, and (f) ω. (a) is calculated in terms of MSE while the others are calculated in terms of NMSE.

Table II: Summary of the accuracy of the CNN in estimating parameters of EGG channel distribution.

CNN performance in estimating the parameters of the EGG

Training loss	NMSE 'a'	NMSE 'b'	NMSE 'c'	NMSE 'ω'	NMSE 'λ'
0.059079	0.000267	0.000150	0.000156	0.000876	0.001058
Validation loss	NMSE 'a'	NMSE 'b'	NMSE 'c'	NMSE 'ω'	NMSE 'λ'
0.160701	0.000627	0.000153	0.000430	0.000866	0.001224

Table III: Summary of the accuracy of the CNN in estimating the channel intensity.

CNN performance in estimating the channel intensity

Training loss	NMSE 'BL'	NMSE 'GT'
0.028476	0.001278	0.004994
Validation loss	NMSE 'BL'	NMSE 'GT'
0.024290	0.001077	0.004784

V. CONCLUSION

We utilized the EGG distribution, which was proposed in [11], to model the UWOC channel considering the temperature and bubble-induced turbulences. We exploited the experimentally measured data for different scenarios of gradient temperature and bubble levels in fresh water [11] to estimate four more scenarios. These are the special cases of the EGG distribution where we have uniform bubble levels. The CNN was trained first to estimate five different parameters of the EGG distribution. A small value of NMSE was observed that indicates an excellent fit with the experimental data. Furthermore, another CNN was also trained to estimate the channel intensity, which again showed an accurate estimation of both gradient temperature and bubble level. These obtained results with extended scenarios could potentially help in estimating the parameters of the EGG and the corresponding channel intensity, without the need of performing a series of complicated statistical analysis that involves the use of the expectation maximization algorithm. This can help accelerate the transition to UWOC by implementing a more effective and robust system, which will lead to significant improvement in bandwidth and transmission speed.

ACKNOWLEDGMENT

The authors would like to acknowledge the support provided by the Deanship of Research Oversight and Coordination (DROC) at King Fahd University of Petroleum and Minerals (KFUPM) for funding under the Research Center for Communication Systems and Sensing through project No. INCS2103.

REFERENCES

[1] J. H. Smart, "Underwater optical communications systems part 1: variability of water optical parameters," in MILCOM 2005 - 2005 IEEE Military Communications Conference, 2005, pp. 1140-1146 Vol. 2.

[2] J. W. Giles and I. N. Bankman, "Underwater optical communications systems. Part 2: basic design considerations," in MILCOM 2005 - 2005 IEEE Military Communications Conference, 2005, pp. 1700-1705 Vol.

[3] M.-A. Khalighi, C. Gabriel, T. Hamza, S. Bourennane, P. Leon, and V. Rigaud, "Underwater wireless optical communication; recent advances and remaining challenges," in Proceedings of IEEE 16th International Conference on Transparent Optical Networks (ICTON), 2014, pp. 1-4.

[4] C. Gabriel, M. A. Khalighi, S. Bourennane, P. Leon, and V. Rigaud, "Monte-Carlo-based channel characterization for underwater optical communication systems," IEEE J. Opt. Commun. Netw., vol. 5, no. 1, pp. 1-12, 2013.

[5] R. M. Lerner and J. D. Summers, "Monte carlo description of time and space-resolved multiple forward scatter in natural water," Appl. Opt., vol. 21, no. 5, pp. 861–869, Mar. 1982.

[6] L. C. Andrews, R. L. Phillips, and C. Y. Hopen, Laser Beam Scintillation with Applications. SPIE Press, 2001.

[7] E. Zedini, H. M. Oubei, A. Kammoun, M. Hamdi, B. S. Ooi, and M. S. Alouini, "A New Simple Model for Underwater Wireless Optical Channels in the Presence of Air Bubbles," in GLOBECOM 2017 - 2017 IEEE Global Communications Conference, 2017, pp. 1-6.

[8] M. V. Jamali, P. Khorramshahi, A. Tashakori, A. Chizari, S. Shahsavari, S. AbdollahRamezani, M. Fazelian, S. Bahrani, and J. A. Salehi, "Statistical distribution of intensity fluctuations for underwater wireless optical channels in the presence of air bubbles," in 2016 Iran Workshop on Communication and Information Theory (IWCIT'16), Tehran, Iran, May 2016, pp. 1–6.

[9] H. M. Oubei, E. Zedini, R. T. ElAfandy, A. Kammoun, T. K. Ng, M.-S. Alouini, and B. S. Ooi, "Efficient weibull channel model for salinity induced turbulent underwater wireless optical communications," in 22nd OptoElectronics and Communications Conference, ser. Oral 2-3K-2, Singapore, 2017.

[10] Oubei, Hassan M. "Underwater Wireless Optical Communications Systems: from System-Level Demonstrations to Channel Modeling." PhD diss., 2018.

[11] Zedini, E., Oubei, H. M., Kammoun, A., Hamdi, M., Ooi, B. S., & Alouini, M.-S. (2019). Unified Statistical Channel Model for Turbulence-Induced Fading in Underwater Wireless Optical Communication Systems. IEEE Transactions on Communications, 67(4), 2893–2907. doi:10.1109/tcomm.2019.28.

[12] W. Liu, Z. Xu, and L. Yang, "SIMO detection schemes for underwater optical wireless communication under turbulence," Photon. Res., vol. 3, no. 3, pp. 48–53, Jun. 2015.

Proceedings of the 7th Optoelectronics Global Conference (OGC 2022)

Joint Optimization of Multidimensional Resources Allocation in Cloud Networking

Jialong Li[1,2,3], Kangqi Zhu[1,2], Nan Hua[1,2], Chen Zhao[1,2], Yanhe Li[4], Xiaoping Zheng[1,2], and Bingkun Zhou[1,2]

[1]Beijing National Research Center for Information Science and Technology (BNRist), Beijing 100084, China
[2]Department of Electronic Engineering, Tsinghua University, Beijing 100084, China
[3]Max Planck Institute for Informatics, Germany
[4]Tsinghua Unigroup, Tsinghua University, Beijing 100084, China
huan@mail.tsinghua.edu.cn*, xpzheng@mail.tsinghua.edu.cn

Abstract—Cloud networking enables flexible service deployments and agile function managements. The convergence of cloud and networking requires the integration of transmission, computation, and storage resources. In this paper, a unified mathematical model on joint optimization of transmission, computation, and storage resources allocation is established, aiming at reducing the job completion time and improving the resource utilization efficiency. To so do, computation resources are first virtualized and then discretized in the time domain. Then we realize the coordination of routing process and transmission, computation resources in cloud networking. Simulation results demonstrate that the propose method could reduce the job completion time by 33%.

Index Terms—Cloud networking, multidimensional resources, joint optimization

I. Introduction

Networking and cloud computing are responsible for communications and data process/storage respectively, working separately and lack collaborations. The convergence of networking and cloud computing, namely cloud networking [1], enables faster service deployment and higher resource utilization efficiency. Unlike the legacy network, cloud networking requires the allocation of multidimensional resources (transmission, computation, and storage) simultaneously [2], which plays a vital role in the convergence of cloud and networking. Furthermore, the service modelling is different from the traditional network architecture, where data are processed in a centralized way and the networks act as pipes connecting end users and data centers. Benefit from the deployment of computation and storage resources at the network edge, task executions among geographically distributed nodes are possible, which is essential for some applications, such as federated learning (FL) [3], [4] and intelligent video surveillance (IVS) [5]. It is crucial to take into account the availability of the multidimensional resources as well as the distributed tasks' interdependence when assigning distributed tasks. If the task assignment and multidimensional resource allocation could be jointly conducted, the entire job completion time could be reduced greatly.

In terms of the collaborative allocation of multidimensional resources, Ref. [6] proposed a strategy by jointly adjusting the size of allocated computation and transmission resources. In this way, multiple computation results generated by distributed

tasks could be transmitted to the next node simultaneously, avoiding the buffer time among these results. However, the buffer time reduction is limited, and its performances are highly dependent on the composition characteristic of the job. Ref. [7] introduced the resource utilization balance factor to allocate multidimensional resources evenly. Unexpected long buffer time caused by the shortage of resources could be avoided. Ref. [8] designed a multi-state adjustment algorithm for multidimensional resource allocation. The task mapping procedures are adjusted according to the availability of the transmission resources as well as the location of the raw data. However, this strategy is based on greedy methods and could not obtain the theoretical optimal value. Furthermore, this work does not take into account the availability of computation resources.

The traditional method based on single-dimensional resource optimization can not realize the collaboration of multidimensional resources, which will cause higher completion time and lower resource utilization efficiency. A more efficient and practical strategy is required in distributed task assignments and allocations of multidimensional resources. In our previous work, we proposed a mathematical model aiming at minimizing the job completion time [9]. However, this optimization model is too complex and requires plenty of time to obtain optimal results. In this work, we focus on minimizing the completion time of jobs with sequential task interdependence. To this end, we design a strategy based on the joint optimization of NEtwork and COmputation resources (NECO). Simulation results validate the effectiveness of NECO compared with other benchmark algorithms.

The rest of the paper is organized as follows. Multidimensional resources and service modelling are shown in Sec. II. In Sec. III, NECO algorithm is presented in detail. We demonstrate the simulation performances in Sec. IV. We conclude this work in Sec. V.

II. Multidimensional Resources and Service Modelling

Fig. 1 is a schematic diagram of multidimensional resources in cloud networking. It can be seen from the figure that distributed task executions involve the participation of three types of resources, namely transmission, computation, and

978-1-6654-8699-6/22 $31.00 © 2022 IEEE

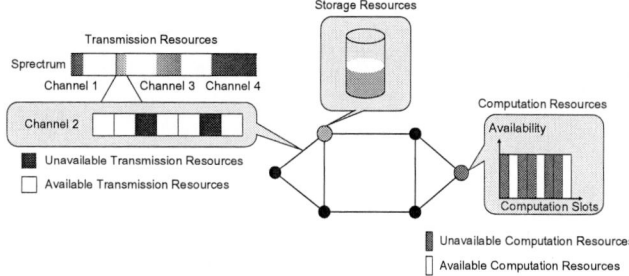

Fig. 1. Multidimensional resources in the network.

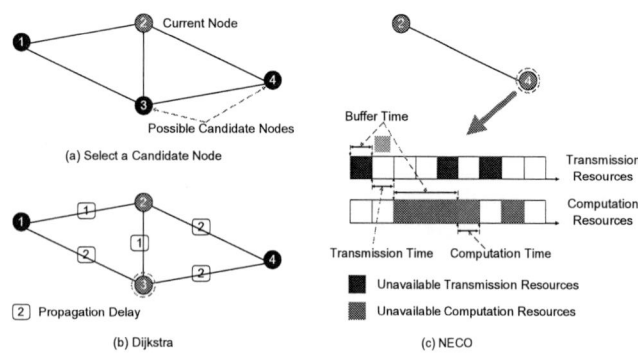

Fig. 2. Illustration of Dijkstra and NECO strategy.

storage resources. Transmission resources are used for transmitting raw data and computation results generated by tasks. Computation and storage resources are responsible for task executions and data storage. In this work, we mainly focus on transmission and computation resources. A unified abstraction for multidimensional resources is critical in cloud networking. In this work, transmission and computation resources are discretized in the time domain. For transmission resources, the optical networks are based on optical time slices switching (OTSS) [10], where optical channels are divided into repetitive frames and each frame is further slicing to multiple slots. Under such circumstances, data transmission could occupy several continuous slots in a selected wavelength. As for computation resources, we assume that a task will occupy all computation resources in the node for execution, while other tasks will be queued until resources are released. In this way, the availability of transmission and computation resources could be measured by time slots, and we could design the NECO strategy based on this assumption.

In terms of the service model, a job is decomposed into several tasks under interdependence constraints. The interdependence among computation tasks could be modelled by a directed acyclic graph (DAG). Three typical interdependence models are mentioned in Ref. [11], namely sequential, parallel, and general interdependence. In this paper, we mainly consider the sequential interdependence, i.e., the DAG is a chain.

III. NECO STRATEGY

A. Discussion

Fig. 2 presents an example of Dijkstra and NECO algorithm. In Fig. 2(a), Node 2 is the current node, and the next step is to select a candidate node. Node 3 and 4 are both possible candidate nodes. Fig. 2(b) shows how Dijkstra algorithm determines the candidate node. Dijkstra method tries to minize the propagation delay and does not take computation and transmission resources into consideration. It adopts propagation delay as the cost, and Node 3 is chosen as the candidate node since the path $2 \rightarrow 3$ is the shortest.

NECO deals with this problem in another way. It tries to find the optimal completion time by considering the buffer time, transmission/computation time, and the propagation delay. From Fig. 2(c), we could learn that the time required by the task execution and the result transmission is both 1 time slice. The buffer time caused by the shortage of transmission and computation resources are 1 and 3 time slices, respectively. It is a one-step strategy and jointly takes computation and transmission resources into consideration, which will select the candidate node with the minimum completion time.

Algorithm 1 describes the NECO algorithm in detail. NECO consists of three stages, namely Initialization, Update, and Relaxation. In the Initialization stage, the distances from the source node s to other nodes are infinity and all nodes are unvisited. Then the first task is executed in source node s. In the second stage, the distances from sources node s to its neighbour nodes are updated. In the Relaxation stage, if a visiting node could realize a shorter distance than the current distance, update the distance and set the visiting node as a predecessor. We could find the shortest paths and minimum distances from the source node s to other nodes by the three stages.

Then we analyze the time complexity of NECO. The time complexity of Dijkstra algorithm is $O(|N|^2)$, where N is the number of nodes. In NECO algorithm, latency is used to replace the physical length of the edge. The time complexity of calculating the latency for an edge is $O(|W||S|^2)$, where W and S denote the number of wavelengths and the number of minimum time slices in an OTSS frame, respectively. The number of nodes in the DAG is L, so there are $L - 1$ edges that requires latency calculation. The overall time complexity of NECO is $O(|W||L||N|^2|S|^2)$.

B. Benchmark Algorithms

Besides Dijkstra algorithm, we further design another two benchmark algorithms, namely Optimal and depth-first search (DFS). In Optimal strategy, all possible mapping cases will be checked and the case with minimum completion time will be selected. In other words, optimal results could be obtained by Optimal strategy.

As for DFS scheme, it could be regarded as a random assignment policy. First, we select a physical node randomly as a root node and adopt DFS to search the following nodes as far as possible. Then tasks are mapped into these nodes

978-1-6654-8699-6/22 $31.00 © 2022 IEEE

Algorithm 1 : NECO Algorithm

Require:
$\mathcal{N} \leftarrow$ set of nodes;
$dis \leftarrow$ distance vector; **dis[v]** $\leftarrow$ distance from source **s** to node **v**;
$pre \leftarrow$ indicate a node is already visited or not;
$ns(v) \leftarrow$ start time of the task execution in node **v**;
$es(s, d) \leftarrow$ start time of the computation result transmission between edge **(s, d)**;
$pl(s, d) \leftarrow$ propagation latency for edge **(s, d)**;
$ct(v) \leftarrow$ task execution time required in node **v**;
$tt(s, d) \leftarrow$ computation result transmission time required between edge **(s, d)**;

1: // Initialization
2: **for** each node **v** in $\mathcal{N}$ **do**
3:    **dis[v]** = Infinity
4:    **pre[v]** = Unvisited
5: **end for**
6: calculate **ns(s)**;
7: **dis[s]** = **ns(s)** + **ct(s)**
8: **pre[s]** = Visited
9: // Update the distance vector
10: **for** each neighbor **v** of **s do**
11:    calculate **es(s, v)**;
12:    calculate **ns(v)**; **ns(v)** should satisfy: **ns(v)** $\geq$ **es(s, v)** + **tt(s, v)** + **pl(s, v)**;
13:    **dis[v]** = **ns(v)** + **ct(v)**
14: **end for**
15: // Relaxation operation
16: **while** not all elements in pre are Visited **do**
17:    **u** $\leftarrow$ node with **min dis[u]** and **pre[u]** == Unvisited
18:    **for** each neighbor **v** of **u do**
19:      **if pre[v]** is Visited **then**
20:        **continue**
21:      **end if**
22:      calculate **es(u, v)**;
23:      calculate **ns(v)**; **ns(v)** should satisfy: **ns(v)** $\geq$ **es(u, v)** + **tt(u, v)** + **pl(u, v)**;
24:      **if dis[v]** > **ns(v)** + **ct(v) then**
25:        **dis[v]** = **ns(v)**+ **ct(v)**
26:      **end if**
27:    **end for**
28:    **pre[u]** = Visited
29: **end while**

in order. In DFS scheme, the task assignment and resource allocations are separated.

IV. SIMULATION EVALUATION

A. Simulation Setup

A 6-node mesh topology is used to evaluate the performances in this paper. The distance between two adjacent nodes is 20 km. One wavelength is deployed in each link. The length of an OTSS frame is 100 ms. Each frame is further divided into 1000 minimum slices, which means the length of a minimum slice is 100 μs. As for the parameters of DAGs, the minimum and maximum length of the DAG are 2 and 4, respectively. The required time for task execution in the physical node ranges from 100 to 200 μs. And the transmission time of the computation results varies from 100 to 300 μs.

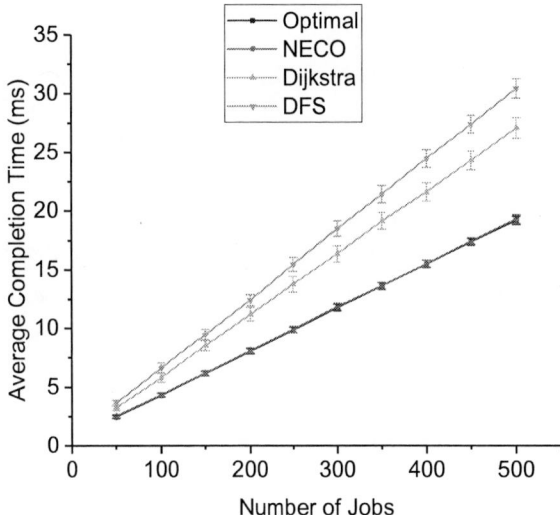

Fig. 3. Average completion time comparison between Optimal, NECO, Dijkstra, and DFS under different number of jobs.

B. Average Completion Time Performance

Fig. 3 shows the average completion time comparison between Optimal, NECO, Dijkstra, and DFS under different number of jobs. From this figure, we could learn that average completion time increases with the number of jobs. Among these four algorithms, Optimal algorithm and NECO could realize lower average completion time than Dijkstra and DFS methods. When the number of jobs is 500, the average completion time for DFS and Dijkstra are around 31 ms and 27 ms, respectively. Optimal algorithm and NECO achieve 18 ms average completion time under the same condition, which is 13 ms and 9 ms lower than DFS and Dijkstra, respectively. Furthermore, Optimal and NECO performances are almost identical, and it is hard to tell the differences.

Fig. 4(a) demonstrates average completion time performance under the number of wavelengths of 1 and 2, respectively. W1 and W2 denote the number of wavelengths deployed in each link is 1 and 2, respectively. We could notice that NECO-W2 realizes lower average completion time than NECO-W1. It is the same case with the other three algorithms. The wavelength that provides the lowest buffer time before the transmission will be selected in priority when multiple wavelengths are deployed, which is why W2 could achieve better performances than W1.

Then we study the results under different average lengths of the DAG in Fig. 4(b). SL (short length) denotes the length of the DAG ranges from 2 to 3, while the length ranges from 2 to 4 for LL (long length). In our simulation, the length of the DAG is generated uniformly from the minimum to the maximum value. The average lengths of SL and LL are 2.5 and 3, respectively. The average completion time of LL is higher than that of SL. DAGs with longer average lengths require more physical nodes to execute tasks and traverse more links, leading to higher average completion time.

978-1-6654-8699-6/22 $31.00 © 2022 IEEE

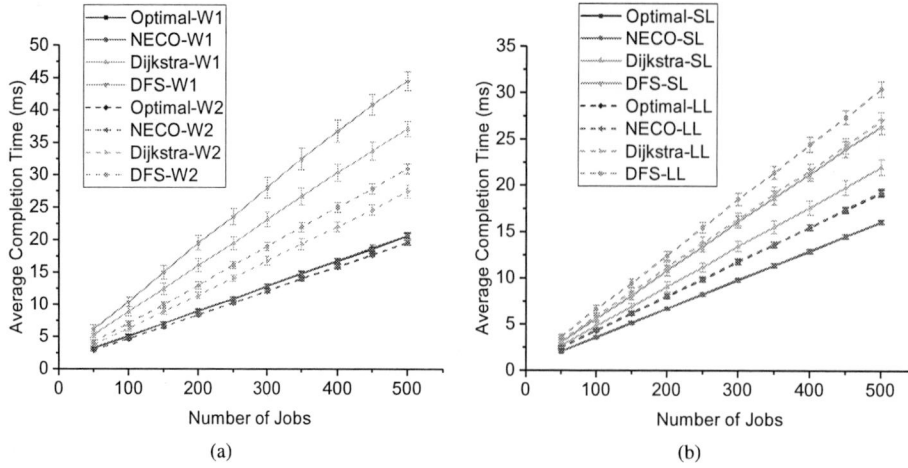

Fig. 4. (a) Average completion time performance under the number of wavelengths of 1 and 2, respectively. (b) Average completion time performance under different average lengths of the DAG, respectively.

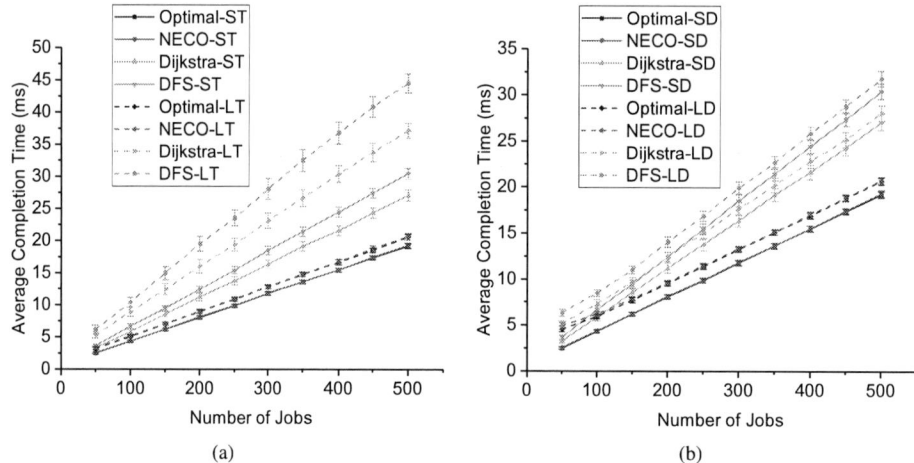

Fig. 5. (a) Average completion time performance under the maximum computation result transmission time of 300 and 1000 μs, respectively. (b) Average completion time performance under different propagation delays.

Fig. 5(a) shows the average completion time performances under different computation result transmission time. ST (short transmission time) and LT (long transmission time) represent the computation result transmission time are 300 and 1000 μs, respectively. From the figure, we could learn that the average completion time of LT is higher than that of ST. The longer the transmission time is, the more transmission resources are demanded. The buffer time before the result transmission becomes larger since it is harder to find more available transmission resources, resulting in higher job completion time.

Finally, the performances under different propagation delays are discussed here. SD denotes short distances and LD represents long distances. SD and LD denote the distances between two adjacent nodes are 20 and 200 km, respectively. Optical signals propagates at a speed of 200 km/ms. The propagation delay in SD for each link is 100 μs, while the value is 1000 μs for LD. Fig. 5(b) indicates that the average

completion time of LD is larger than that of SD. The reason is that the propagation delay is also considered in the completion time. It is worth noting that the range of the metro network is generally less than 100 km, and the distance between two adjacent nodes is less than this value.

V. CONCLUSION

In this work, we focus on the integration of transmission and computation resources in cloud networking. To this end, transmission and computation resources are first discretized in the time domain. Based on this, we design a NECO strategy aiming at minimizing the job completion time. Simulation results are demonstrated and validate the effectiveness of our proposed policy.

ACKNOWLEDGMENT

This work was supported in part by the National Key R&D Program (No.2020YFB1805602), and the National Natural Science Foundation of China (No.61871448).

REFERENCES

[1] Q. Duan, S. Wang, and N. Ansari, "Convergence of networking and cloud/edge computing: Status, challenges, and opportunities," *IEEE Network*, vol. 34, no. 6, pp. 148–155, 2020.

[2] M. Ruffini, "Multidimensional convergence in future 5G networks," *Journal of Lightwave Technology*, vol. 35, no. 3, pp. 535–549, 2016.

[3] T. Li, A. K. Sahu, A. Talwalkar, and V. Smith, "Federated learning: Challenges, methods, and future directions," *IEEE Signal Processing Magazine*, vol. 37, no. 3, pp. 50–60, 2020.

[4] B. Shariati, P. Safari, A. Mitrovska, N. Hashemi, J. Fischer, and R. Freund, "Demonstration of federated learning over edge-computing enabled metro optical networks," in *2020 European Conference on Optical Communications (ECOC)*. IEEE, 2020, pp. 1–4.

[5] W. Shi, J. Cao, Q. Zhang, Y. Li, and L. Xu, "Edge computing: Vision and challenges," *IEEE Internet of Things Journal*, vol. 3, no. 5, pp. 637–646, 2016.

[6] J. Zhang, L. Cui, Z. Liu, and Y. Ji, "Demonstration of geo-distributed data processing and aggregation in mec-empowered metro optical networks," in *2020 Optical Fiber Communications Conference and Exhibition (OFC)*. IEEE, 2020, pp. 1–3.

[7] H. Yang, J. Zhang, Y. Zhao, J. Han, Y. Lin, and Y. Lee, "Cross stratum optimization for software defined multi-domain and multi-layer optical transport networks deploying with data centers," *Optical Switching and Networking*, vol. 26, pp. 14–24, 2017.

[8] Y. Sahni, J. Cao, and L. Yang, "Data-aware task allocation for achieving low latency in collaborative edge computing," *IEEE Internet of Things Journal*, vol. 6, no. 2, pp. 3512–3524, 2018.

[9] J. Li, N. Hua, C. Zhao, Y. Li, X. Zheng, and B. Zhou, "Towards low-latency distributed tasks collaboration by joint optimization of transmission, computation and storage resources allocation in edge computing," in *Asia Communications and Photonics Conference*. Optical Society of America, 2020, pp. M4A–210.

[10] N. Hua and X. Zheng, "Optical time slice switching (OTSS): An all-optical sub-wavelength solution based on time synchronization," in *Asia Communications and Photonics Conference*. Optical Society of America, 2013, pp. AW3H–3.

[11] Y. Mao, C. You, J. Zhang, K. Huang, and K. B. Letaief, "A survey on mobile edge computing: The communication perspective," *IEEE Communications Surveys & Tutorials*, vol. 19, no. 4, pp. 2322–2358, 2017.

978-1-6654-8699-6/22 $31.00 © 2022 IEEE

Proceedings of the 7th Optoelectronics Global Conference (OGC 2022)

Optical Labels Enabled Optical Network Performance Monitoring

Tao Yang
State Key Laboratory of Information Photonics and Optical Communications
Beijing University of Posts and Telecommunications, Beijing, 100876, China
e-mail: yangtao@bupt.edu.cn

Abstract—**Optical performance monitoring (OPM) technology, especially the optical power and OSNR of each WDM channel, are of great importance and significance that need to be performed to ensure stable and efficient operation and maintenance of network itself. In this paper, we present our recent research activities and progresses on optical-label enabled OPM. Simulation results of WDM system show that the maximum error of channel power monitoring and OSNR estimation are less than 1dB after 20-span transmission. It confirms that the optical-label enabled OPM has lower cost and higher efficiency for mass deployment in practical WDM networks.**

Keywords-WDM networks; optical label; performance monitoring; channel power; OSNR

I. INTRODUCTION

With the rapid progress of various high bandwidth services and emerging internet applications, the capacity demand has been increasing dramatically [1]. To comply with the demand, large-capacity and long-haul wavelength division multiplexing (WDM) systems have been widely deployed in metro and backbone optical networks nowadays. Accordingly, with the wide deployment of reconfigurable optical add-and-drop multiplexers (ROADMs) based on wavelength selective switches (WSSs), the architecture of WDM optical networks is becoming more complex and the wavelength connection is more dynamic, flexible and reconfigurable. Therefore, to ensure its working stability and operation efficiency, low cost and highly reliable multi-channel optical performance monitoring (OPM) technology is indispensable and attract a lot of attention [2]. Especially the optical power and optical signal noise ratio (OSNR) of each WDM channel, as key channel performance indicators, are the most important and meaningful parameters that need to be accurately monitored/estimated [3]. Besides, if the OPM can monitor the node location (hereinafter called "node ID") where the service signal is initially added and obtain specific surveillance information (such as symbol rate, modulation format, route configuration, etc., hereinafter called "wavelength ID"), it will be beneficial to the efficient management and optimization of WDM optical networks. However, the majority of critical monitoring module of existing solutions as shown in Fig. 1, such as the conventional method using optical spectral analysis by spectrometer or optical filter, generally are too expensive, operationally inconvenient and functionally limited to be applied in performance monitoring of all channels over a WDM system with numerous nodes [4].

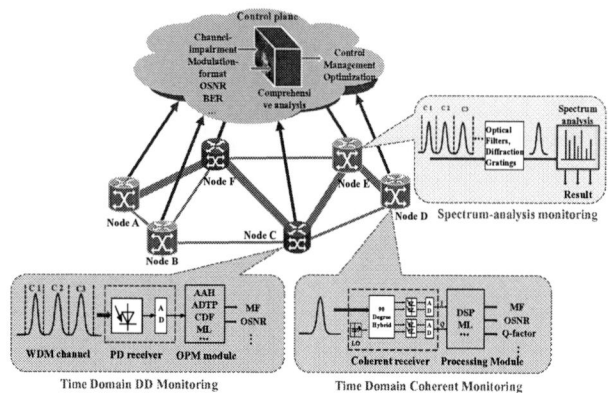

Fig.1. Existing mainstream OPM solutions.

In this paper, we present our recent research activities and progresses on optical-label enabled OPM. A low cost and high efficiency optical power monitoring and OSNR estimation scheme based on DPSK digital Optical Labels is proposed and demonstrated. The scheme adopts pilot tones with frequency of tens of MHz to be modulated by a KHz low-speed (~K Baud) digital surveillance information corresponding to the target channel, then the pilot-tone carried signal is loaded on the service signal as a label [5]. At monitoring nodes, a low bandwidth PD receives the label signal through direct detection to realize the performance monitoring of all wavelength channels simultaneously. Because that the optical labels here are tied up to their corresponding optical channels anywhere in the network, it can not only monitor the wavelength channel performance, but also can reliably deliver any interested surveillance information to further sense the working state and optimize resource utilization of WDM networks. In our proposed scheme, the channel optical power is monitored using the method of spectral integration, and the OSNR is estimated by calculating the ASE noise accumulation in all amplifiers. To verify the proposed scheme, we have performed simulations on WDM systems within 20 spans fiber transmission, and the results show that the maximum error of power monitoring is less than 1dB while the error of OSNR estimation is also less than 1dB. Therefore, the implementation of the proposed scheme is of lower cost and higher efficiency, which has great potential for large-scale deployment in practical WDM networks.

978-1-6654-8699-6/22 $31.00 © 2022 IEEE

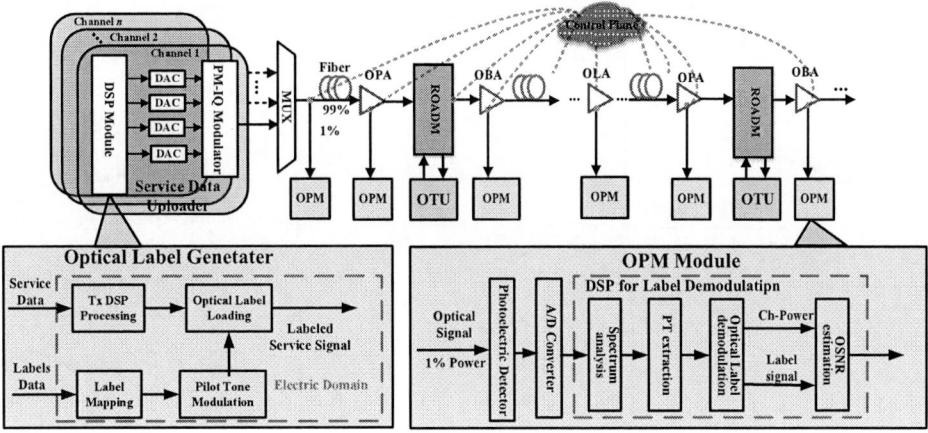

Fig. 2. Schematic diagram of the optical-label enabled OPM scheme.

II. OPERATION PRINCIPLES

Optical label is known as top modulation or a small amounts amplitude modulation, also called as low-frequency perturbation or pilot tones in some lectures. The schematic diagram of the scheme is shown in Fig. 2. After the digital surveillance information of node ID and wavelength ID of each wavelength channel is mapped by DPSK coding, it is modulated (~K Baud rate) to the pilot tone with a specific frequency to be the final optical label. It is noteworthy that the pilot tone frequency is larger than 40MHz in order to avoid the impact of stimulated Raman scattering (SRS) crosstalk. Thus the power monitoring error caused by the SRS, resulting in the non-correspondence between the label power and the channel power, can be reasonably ignored.

Then the optical label is loaded into the high-speed service signal, and it is modulated by the optical I/Q modulator and transmitted to the destination node. when arriving at optical pre-amplifier (OPA) or optical boost amplifier (OBA) in a service node, or optical line amplifier (OLA) in relay node, the monitoring signal is separated out from the monitoring port of the EDFA, and enters the OPM module to monitor the optical power, estimate the OSNR and recovery the delivered node/wavelength ID of each channel. In the OPM module, the low-bandwidth PD and low-sampling-rate analog to digital converter (ADC) are used to detect the labeled optical signals, so as to reduce the cost and bulk. After photoelectric conversion and digital sampling by PD and ADC, a DSP unit is used to calculate the power and OSNR. Firstly, FFT-based spectrum analysis is used to obtain the receiving spectrum of optical labels. Then, spectrum integration of each optical label spectrum from the beginning PT frequency to the ending PT frequency is carried out respectively. For example, the channel-1 power is equal to the spectrum integration value from [39.9~40.1] MHz range. Next, a proportional coefficient of photoelectric conversion of PD is used to calculate the final channel optical power of the service signal. Finally, the monitoring results of all network monitoring nodes are transmitted to the control plane through optical supervising channel (OSC), and the control plane analyzes the performance of the whole optical network and optimizes resource utilization comprehensively.

In a multi-span WDM transmission system, OSNR will gradually deteriorate due to the accumulation of ASE noise generated by the EDFA. Therefore, the OSNR could be estimated by calculating the ASE noise accumulation. In practical optical transmission system, the actual gain of the amplifier may be different from the configured gain, so we can calculate the actual gain of each optical amplifier with the help of the optical power monitored by the optical labels. At the same time, because the noise figure of optical amplifier will change with different wavelengths, here we can obtain the value of the NF by measuring it in advance. Finally, according to the routing of service signals, control plane obtains the amplification case and the corresponding NF parameters, and based on the EDFA amplification and ASE noise accumulation of each wavelength channel, the OSNR can be estimated by the 58-Formula.

III. SIMULATION AND DISCUSSION

We have constructed a simulation platform of WDM transmission with digital optical labels to investigate the performance of channel power monitoring and OSNR estimation. Optical labels with 4 MHz grid is arranged to 16G Baud polarization multiplexing (PM) QPSK and 16QAM transmission systems. Here we simulate WDM transmission with 8/64 channels and 20 spans. The modulation of optical label is set to 10%. A 300MHz PD and 600 MHz 10-bit ADC are used to detect the optical labels of 8/64 channels.

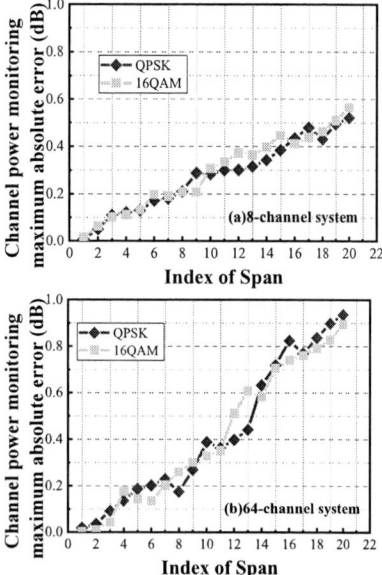

Figure 3. The maximum monitoring error under 10% optical label modulation depth of (a) 8-channel PM-QPSK/16QAM, (b) 64-channel PM-QPSK/16QAM systems.

A. Optical Power

The figure 3(a) and (b) show that the QPSK system and the 16QAM system have similar maximum power monitoring errors. The maximum monitoring error is similar at different label modulation depths, so the monitoring accuracy of the two systems can be considered to be roughly the same. The maximum channel power monitoring error does not exceed 0.6/1 dB after 20-span transmission in 8/64-channel system.

B. OSNR

In order to validate practically the performance of OSNR estimation in PM-QPSK/16QAM systems, the noise figure (NF) value of EDFA for each optical channel is 5dB. Meanwhile, the length of each span is set to 100 km. The input-power and output-power are monitored based on optical labels, then the OSNR in each channel is calculated by using the 58-Formula. The result of OSNR estimation and the estimated error after 20-span transmission is shown in Figure 4.

It shows that the OSNR monitoring error is less than 1dB after 20-span transmission in 64-channel PM-QPSK/16QAM systems respectively. The main monitoring error occurs because of the error that occurred during the previous power monitoring. It is also found that when the transmission distance increases, the power monitoring and OSNR estimation error and its fluctuations caused by noise, dispersion and other factors during the fiber transmission process are also relatively enhanced.

IV. SUMMARY

In this paper, we present our recent research activities and progresses on optical-label enabled OPM. Simulation

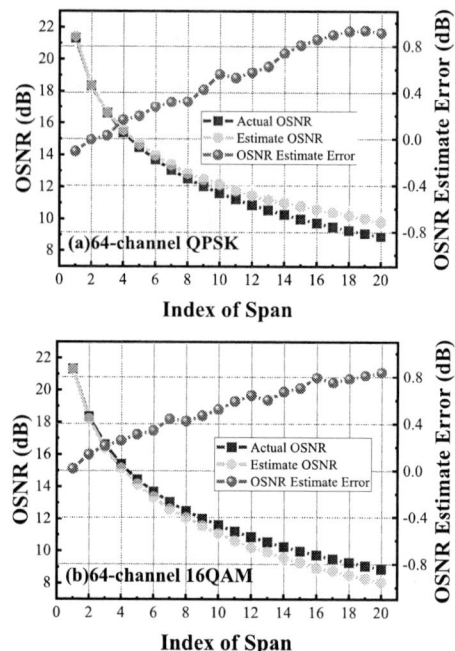

Fig. 4. Performance of OSNR estimation in 64-channel (a) PM-QPSK and (b) PM-16QAM systems.

results of WDM system show that the maximum error of channel power monitoring and OSNR estimation are both less than 1dB after 20-span transmission. It confirms that the optical-label enabled OPM has adequate channel power monitoring performance, and the optical-label enabled OPM has lower cost and higher efficiency for mass deployment in practical WDM networks.

ACKNOWLEDGMENT

This work was supported by the National Natural Science Foundation of China (No. 62001045), Fund of State Key Laboratory of IPOC (BUPT) (No. IPOC2021ZT17), Fundamental Research Funds for the Central Universities (No. 2022RC09), and Beijing Municipal Natural Science Foundation (No. 4214059).

REFERENCES

[1] C. C. K. Chan, Optical performance monitoring: Advanced techniques for next-generation photonic networks. Burlington, MA: Academic Press/Elsevier, 2010.

[2] Z. Jiang, T. Liu; H. Yang, D. Jin, M. Si, "Method and apparatus for providing a pilot tone," United States Patent Application 20170373748. United States Patent Application 20170373748

[3] J. Yuan, L. Xi, D. Zhao, H. Xu, X. Tang, W. Zhang, J. Li, X. Zhang, "OSNR monitoring in presence of fiber nonlinearities for coherent Nyquist-WDM system," Opt. Commun. 380, 10-14 (2016).

[4] Z. Jiang and X. Tang, "Low-Cost Signal Spectrum Monitoring Enabled by Multiband Pilot Tone Techniques," in 2018 European Conference on Optical Communication (ECOC) (2018), pp. 1-3.

[5] S. B. Jun, H. Kim, P. K. J. Park, J. H. Lee and Y. C. Chung, "Pilot-tone-based WDM monitoring technique for DPSK systems," IEEE Photonics Technol. Lett. 18(20), 2171-2173(2006)

Proceedings of the 7th Optoelectronics Global Conference (OGC 2022)

Equalization for Optical PAM Data Center Interconnects

Gordon Ning Liu, Lin Sun, Caoyang Liu, Jiawang Xiao, Yi Cai, Gangxiang Shen

Suzhou Key Laboratory of Advanced Optical Communication Network Technology, and School of Electronic and
Information Engineering, Soochow University, Suzhou, Jiangsu 215006, China
e-mail: gordonnliu@suda.edu.cn

Abstract—With the rapid development of optical data center interconnect, the requirement for data rate of client-side transceivers increases drastically. To meet the transmission rate demand, the recent pulse-amplitude modulation optical interconnects are discussed. To lower down the cost of the optical interconnects, low cost opto-electronic devices with strong bandwidth limitation and nonlinear distortion are employed. Moreover, the interplay between the fiber dispersion and square law photodetection leads to additional nonlinear inter-symbol interference. To handle the problem, various digital equalizers are introduced, which are reviewed in this paper.

Keywords- pulse-amplitude modulation; equalization; neural network

I. INTRODUCTION

To support data-intensive and latency-sensitive demands of emerging services such as 5G, self-driving collaborative vehicles and the industrial Internet, the cloud data center and edge data center are growing rapidly, accompanied by a high demand for traffic within cloud/edge data centers. Such a huge data traffic is normally achieved through the optical data center interconnect (DCI) between many client side transceivers. However, due to the low spectrum efficiency (SE), the traditional on–off keying (OOK) format is difficult to achieve the data rate of 100 Gbps and above. While line side coherent transceivers with high SE have been deployed in long-haul systems, they are expensive, large size, and high power consumption for DCIs, especially for intra-data center interconnects.

Recently, with the development of CMOS process and high-speed analog-to-digital/digital-to-analog converters, several intensity modulation and direct detection (IM/DD) techniques with powerful digital signal processing (DSP) capability are introduced to increase the line rate of optical interconnects with the limited bandwidth of opto-electronic devices. They are four-level pulse-amplitude modulation (PAM-4) [1], discrete multitone (DMT) [2], and carrierless amplitude phase modulation (CAP) [3] etc. Among them, PAM4 is generally superior to the other two due to its simplicity. Currently, PAM4 has been adopted in 100 Gbps per wavelength DCI by 400 Gbps standard or multi-source agreements (MSA) groups for 400GE-DR4 and 400G-FR4. Within the current discussion for future 800 GbE or 1.6 TbE standards or MSAs [4], PAM4 or PAM-N are also overwhelming for beyond 200 Gbps per wavelength DCI

links. However, to lower down the transceivers cost, current opto-electronic devices fail to accommodate the required high baud rates. The narrow bandwidth introduces strong inter symbol interference (ISI). In addition, the nonlinear responses of opto-electronic devices such as modulators, drivers and amplifiers also limit the achievable line rate. Moreover, the chromatic dispersion or modal dispersion of fiber combined with the square law photodetection in IM/DD systems can severely generate nonlinear distortions and limit the transmission distance [5]. Therefore, some digital equalization techniques were proposed to address these issues, which will be reviewed in this paper.

II. PAM AND TRADITIONAL DIGITAL EQUALIZERS

Different from the traditional OOK format with two levels, the PAM signal uses multiple amplitude modulated levels to represent the data information. The data rate R of PAM-N can be expressed as $R = B \cdot \log_2(N)$, where N represents the number of signal levels and B is the signal Baud rate. With the introduction of PAM-N, the designated data rate of client side transceiver can be achieved by low cost opto-electronic devices with narrow bandwidth since the bandwidth requirement is related to the signal symbol rate. The required devices bandwidth decreases with the modulation levels. However, the receiver sensitivity decreases significantly with the modulation levels. When the modulation level N rises to 16 or above, the eye opening becomes very small, and the signal to noise ratio (SNR) requirement becomes very high. Meanwhile, the nonlinear impairments introduced by opto-electronic devices and fiber is intolerable since the signal becomes more similar to analog signals. Therefore, because of the trade-off between the system performance and device, the selection of modulation level is a very important issue for a certain application scenario. Only up to PAM6 and PAM8 are commonly investigated in the optical data center interconnects [6,7]. Moreover, to reduce the bandwidth and then the cost of opto-electronic devices, the duobinary (or polybinary) and faster-than-Nyquist (FTN) signaling are employed to further increase the spectrum efficiency [8,9]. The filtering in the duobinary / polybinary or FTN introduces strong ISI. To mitigate the impairments, various digital equalization algorithms have been studied.

A common and simple equalization algorithm is feed-forward equalizer (FFE), which is a symbol-by-symbol linear equalizer by inversing the channel transfer function to mitigate the linear impairment effectively [10]. The inversing

978-1-6654-8699-6/22 $31.00 © 2022 IEEE

is implemented by a number of taps with finite impulse response (FIR) filters and the tap weights are determined by several algorithms such as zero-forcing (ZF), least mean squares (LMS), or recursive least squares (RLS) etc. It improves the high frequency components of signal which is filtered due to the bandwidth limitation. However, FFE is only optimized to suppress the precursor ISI. When the bandwidth is severely limited, FFE increases the noise along with the signal at high frequencies. To reduce the noise at high frequencies, FFE can be followed by a nonlinear equalizer, decision feedback equalizer (DFE) [11,12]. It can be utilized to cancel the postcursor ISI. But it suffers from the error propagation and decision feedback delay because of its structure. To avoid the error propagation of DFE, Tomlinson-Harashima precoding (THP) is applied at the transmitter side working as a transmitter side DFE [13-15]. It consists of a feedback filter and a modulo operator, which was proposed by M. Tomlinson, H. Harashima and H. Miyakawa [16,17]. It can greatly increase the transmission distance and data rate of the signal with duobinary or FTN filtering.

FFE, DFE, and THP can mitigate most of the bandwidth limitation induced inter symbol interference. However, the nonlinear interferences induced by the low linearity of opto-electronic devices and the interplay between the fiber dispersion and the square law photodetection cannot be completely equalized. Volterra nonlinear equalizer (VNLE) is one of the most popular nonlinear equalizers to mitigate nonlinear effects in both IM/DD and coherent systems [18]. However, the high computational complexity of VNLE limits its commercial deployment. To reduce its implementation complexity, some simplification techniques are investigated by replacing the square terms with absolute operation [19,20], or truncating the terms with low impact to performance from a full VNLE. There are many different truncating methods, such as the diagonally pruned/grouped VNLE [21,22], Schmidt orthogonalization searching based sparse VNLE [23] and threshold-based pruned retraining VNLE [24] etc.

III. Machine Learning Based Equalization Schemes

In recent years, the machine learning (ML) has been introduced to simplify the VNLE complexity. The weighted principal component analysis (WPCA) was proposed to decrease the number of VNLE taps by employing unsupervised learning to lower down the complexity and exploring supervised learning to maintain the equalization performance [25].

Meanwhile, machine learning algorithms such as artificial neural network (ANN) can be used separately as nonlinear equalizers [26]. Fully connected neural network (FCNN) based equalizer was demonstrated to support stronger equalization performance, especially for equalizing the nonlinear impairments [27]. Many other neural networks were explored to work as nonlinear equalizers, such as deep neural network (DNN) [28], convolutional neural network (CNN) [29], Radial Basis function neural network (RBFNN) [30], and Recurrent neural network (RNN) [31] etc. Because

RNN has an efficient capability in processing sequential data, its variant long short-term memory (LSTM) shows an excellent nonlinear equalization performance [32] and the cascaded RNN enhances the 100 Gbps PAM4 transmission performance [33].

However, the neural network based equalizer (NNE) suffers from the high complexity so that some investigations are conducted to compare NNEs with their traditional counterparts [34,35]. Some other studies are carried out to lower down the algorithm complexity. A pruning technique with sparse connection was applied to reduce the complexity of ANN [36]. We have also investigated the relationship between the complexity and the hidden layer numbers of an RNN equalizer [37].

Moreover, another disadvantage of NNE is that its neural network is acquired by a training of long time with large data volume. Once the transmission channel condition changes, the training has to be restarted from the beginning, which slows down the convergence and induces implementation difficulty in rapidly changing transmission environment. To handle the issue, the adaptive neural network (AdaNN) based equalizer was proposed [38]. Meanwhile, the transfer learning is introduced to train the neural network from the factors of another similar scenario to expedite the training. The scheme was demonstrated in the nonlinear distortion variation [39] and optical switching [40] scenarios. By combining both pruning and transfer learning techniques, we can get a fast-convergence and low-complexity RNNE for PAM4 data center optical interconnects [41].

IV. Conclusions

The paper has reviewed one popular IM/DD modulation format, PAM-N, for the high-speed optical data center interconnects. With the bandwidth limitation caused by low cost opto-electronic devices, duobinary/polybinary and FTN, and the nonlinear impairment induced by low linearity of opto-electronic devices and the interplay between fiber dispersion and square law photodetection, the inter symbol interference strongly affects the transmission performance of PAM-N signals. Consequently, the digital equalizers are important to mitigate the interference and improve the transmission performance. FFE is widely used since its simple and efficient structure, but it may increase the noise at high frequencies. DFE will not increase the high frequency noise, but it suffers from the error propagation issue. THP solves the error propagation problem by placing at the transmitter side. VNLE is widely used for equalizing the nonlinear distortions and many efforts were conducted to lower down its complexity without too much performance compromise recently. Various machine learning based digital equalization schemes are demonstrated in recent years for its strong performance. Among them, the neural network based equalizer attracts huge attention. To make those neural network based equalizers practically and commercially available, the investigations on reducing their complexity and training time are conducted, by introducing the pruning and transfer learning techniques, and choosing the proper hidden layer numbers.

978-1-6654-8699-6/22 $31.00 © 2022 IEEE

REFERENCES

[1] C. Yang et al, "IM/DD-Based 112-Gb/s/lambda PAM-4 Transmission Using 18-Gbps DML," IEEE Photonics Journal, vol. 8, no. 3, 2016.

[2] G. N. Liu et al., "IM/DD Transmission Techniques for Emerging 5G Fronthaul, DCI and Metro Applications," J. Lightw. Technol., vol. 36, no. 2, pp. 560-567, Jan. 15, 2018.

[3] M. I. Olmedo et al., "Multiband Carrierless Amplitude Phase Modulation for High Capacity Optical Data Links," J. Lightw. Technol., vol. 32, no. 4, pp. 798-804, Feb. 15, 2014.

[4] Tom Palkert et al, "400G-FR4 Technical Specification," 100G Lambda MSA Group, Jan, 2018.

[5] K. Zhong et al., "Digital Signal Processing for Short-Reach Optical Communications: A Review of Current Technologies and Future Trends," J. Lightw. Technol., vol. 36, no. 2, pp. 377-400, Jan. 15, 2018.

[6] Y. Xu et al., "260-Gb/s PAM-6 Transmission Using Joint Optical Pre-equalization and a Low-complexity Volterra Equalizer for Short-Reach Optical Interconnects," 2018 Asia Communications and Photonics Conference (ACP), 2018.

[7] D. Zou et al., "100G PAM-6 and PAM-8 Signal Transmission Enabled by Pre-Chirping for 10-km Intra-DCI Utilizing MZM in C-band," J. Lightw. Technol., vol. 38, no. 13, pp. 3445-3453, July 2020.

[8] T. Zuo et al., "Single Lane 112-Gbps Analog Small Form-factor Pluggable Module with Only 4-GHz End-to-end 3-dB Bandwidth Employing Duobinary 4-PAM," in Proc. Optical Fiber Communication Conference (OFC) , paper Th1G.5, 2016.

[9] K. Zhong et al., "Experimental demonstration of 500Gbit/s short reach transmission employing PAM4 signal and direct detection with 25Gbps device," in Proc. Optical Fiber Communication Conference (OFC), paper Th3a.3, 2015.

[10] X. Pang et al., "200 Gbps/Lane IM/DD Technologies for Short Reach Optical Interconnects," J. Lightw. Technol., vol. 38, no. 2, pp. 492-503, Jan. 15, 2020.

[11] M. E. Austin, "Decision-feedback equalization for digital communication over dispersive channels," MIT Res. Lab. Electron., Cambridge, MA, USA, Tech. Rep. 461, Aug. 1967.

[12] G. Glentis et al., "Performance evaluation of decision feedback equalizers in fiber communication links," in Proc. Int. Symp. Commun., Control Sig. Proc. (ISCCSP), 2014.

[13] Q. Hu et al., "High Data Rate Tomlinson-Harashima-Precoding-Based PAM Transmission," in Proc. European Conference and Exhibition on Optical Communication (ECOC), Tu3D.3, 2019.

[14] Q. Hu et al., "IM/DD beyond bandwidth limitation for data center optical interconnects," J. Lightw. Technol., vol. 37, no. 19, pp. 4940–4946, Oct. 1, 2019.

[15] T. Wettlin et al., "Comparison of PAM Formats for 200 Gb/s Short Reach Transmission Systems," in Proc. Optical Fiber Communication Conference (OFC) , paper Th2A.37, 2020.

[16] M. Tomlinson, "New automatic equaliser employing modulo arithmetic," Electron. Lett., vol. 7, no. 5–6, pp. 138–139, 1971.

[17] H. Harashima et al., "Matched-transmission technique for channels with intersymbol interference," IEEE Trans. Commun., vol. 20, no. 4, pp. 774–780, Aug. 1972.

[18] N. Stojanovic et al., "Volterra and wiener equalizers for short-reach 100G PAM-4 applications," J. Lightw. Technol., vol. 35, no. 21, pp. 4583–4594, Nov. 2017.

[19] Y. Yu et al., "Nonlinear Equalizer Based on Absolute Operation for IM/DD System Using DML," IEEE Photon. Technol. Lett., vol. 32, no. 7, pp. 426-429, April 1, 2020.

[20] T. Zuo et al., "850-nm VCSEL-Based Single-Lane 200-Gbps PAM-4 Transmission for Datacenter Intra-Connections," IEEE Photon. Technol. Lett., vol. 33, no. 18, pp. 1042-1045, 15 Sept.15, 2021.

[21] E. Batista et al., "On the performance of adaptive pruned Volterra filters," Signal Process., vol. 93, no. 7, pp. 1909–1920, 2013.

[22] Y. Yu et al., "Low-Complexity Second-Order Volterra Equalizer for DML-Based IM/DD Transmission System," J. Lightw. Technol., vol. 38, no. 7, pp. 1735-1746, April 1, 2020.

[23] H. Ying et al., "Sparse volterra model based on single side-band optical NPAM-4 direct-detection system," in Proc. Conference on Lasers and Electro-Optics Pacific Rim (CLEO-PR), 2017.

[24] L. Ge et al., "Threshold-Based Pruned Retraining Volterra Equalization for 100 Gbps/Lane and 100-m Optical Interconnects Based on VCSEL and MMF," J. Lightw. Technol., vol. 37, no. 13, pp. 3222-3228, July 1, 2019.

[25] Z. Yang et al., "Hardware-efficient Nonlinear Equalizer based on Joint Unsupervised Learning and Supervised Weights," in Proc. Optical Fiber Communication Conference (OFC), paper M5F.5, 2021.

[26] S. Gaiarin et al., "High Speed PAM-8 Optical Interconnects with Digital Equalization Based on Neural Network," in Proc. Asia Communications and Photonics Conference (ACP), 2016.

[27] L. Yi et al., "Machine Learning for 100 Gb/s/λ Passive Optical Network," J. Lightw. Technol., vol. 37, no. 6, pp. 1621-1630, March 15, 2019.

[28] C.Y. Chuang, et al., "Employing deep neural network for high speed 4-PAM optical interconnect," in Proc. European Conference and Exhibition on Optical Communication (ECOC), 2017.

[29] J. Zhang et al., "Convolutional Neural Network Equalizer for Short-reach Optical Communication Systems," in Proc. Asia Communications and Photonics Conference (ACP), 2020.

[30] Z. Yang et al., "Radial basis function neural network enabled C-band 4×50 Gb/s PAM-4 transmission over 80 km SSMF," Opt. Lett., vol. 43, no. 15, pp. 3542-3545, 2018.

[31] C. Ye et al. "Recurrent Neural Network (RNN) Based End-to-End Nonlinear Management for Symmetrical 50Gbps NRZ PON with 29dB+ Loss Budget," in Proc. European Conference and Exhibition on Optical Communication (ECOC), 2018.

[32] X. Dai et al., "LSTM networks enabled nonlinear equalization in 50-Gb/s PAM-4 transmission links," Applied optics, vol. 58, no.22, pp. 6079-6084, 2019.

[33] Z. Xu et al., "Cascade Recurrent Neural Network Enabled 100-Gb/s PAM4 Short-Reach Optical Link Based on DML," in Proc. Optical Fiber Communications Conference and Exhibition (OFC), 2020.

[34] Z. Xu et al., "Computational complexity comparison of feedforward / radial basis function/recurrent neural network-based equalizer for a 50-Gb/s PAM4 direct-detection optical link," Opt. Express, vol.27, no. 25, pp. 36953-36964, 2019.

[35] L. Huang et al., "Performance and complexity analysis of conventional and deep learning equalizers for the high-speed IMDD PON," J. Lightw. Technol., vol. 40, no. 14, pp. 4528-4538, July 2022.

[36] Z. Wan et al., "Nonlinear equalization based on pruned artificial neural networks for 112-Gb/s SSB-PAM4 transmission over 80-km SSMF," Opt. Express, vol. 26, no. 8, pp. 10631-10642, 2018.

[37] C. Liu et al., "Low-complexity RNN equalizer with multiple hidden layers for optical PAM-4 signals," in Proc. 27th OptoElectronics and Communications Conference (OECC), 2022.

[38] Q. Zhou et al., "AdaNN: Adaptive Neural Network-Based Equalizer via Online Semi-Supervised Learning," J. Lightw. Technol., vol. 38, no. 16, pp. 4315-4324, Aug.15, 2020.

[39] J. Zhang et al., "Fast remodeling for nonlinear distortion mitigation based on transfer learning," Opt. Lett., vol. 44, no. 17, pp. 4243-4246, 2019.

[40] Z. Xu et al., "Transfer learning aided neural networks for nonlinear equalization in short-reach direct detection systems," in Proc. Optical Fiber Communications Conference and Exhibition (OFC), 2020.

[41] J. Xiao et al., "Fastly converged transfer learning using neuron-pruning nonlinear equalizer for intra data center networking," in Proc. 27th OptoElectronics and Communications Conference (OECC), 2022.

Research on 5.5 μm Infrared Filter Applied to Infrared Thermometer

Suotao Dong
School of OptoElectronic Engineering
Changchun University of Science and Technology
Changchun, China
e-mail: dongsuotao@163.com

Xiuhua Fu
School of OptoElectronic Engineering
Changchun University of Science and Technology
Changchun, China
e-mail: fuxiuhua@cust.edu.cn

Abstract—**With the expansion of novel coronavirus pneumonia's influence on the world, people's dependence on infrared thermometer guns is increasing. In order to improve the measurement accuracy of the infrared temperature measuring gun and meet the requirements of rapid and accurate measurement of human body temperature, the core components for the infrared temperature measuring gun are developed and prepared in this paper. The film fogging phenomenon caused by the anisotropy of metal germanium and semiconductor properties is analysed and solved by measuring the atomic force microscope image and infrared spectrum of the film, the 5.5-micron infrared filter with high transmittance and good film quality was prepared by electron beam evaporation, resistance evaporation and ion source assisted deposition.**

Keywords-COVID-19; infrared thermometer; anisotropic; infrared filter

I. INTRODUCTION

Because of the rapid spread of COVID-19 in the world, medical testing equipment is also an indispensable anti epidemic substance besides essential medical materials. Among them, the infrared temperature measuring gun with wide application and high demand has become a necessary material for public places and families. In nature, any object above absolute zero (- 273 ℃) has irregular movement of molecules and atoms, and objects are constantly releasing infrared radiation energy. The size and wavelength of radiation energy are closely related to the temperature of the object surface. For example, the normal temperature of the human body is between 36 ~ 37 ℃, and the infrared wavelength of radiation is 9 ~ 13 μm。 In the infrared thermometer, the core component is the infrared temperature sensor. At present, infrared temperature sensors can be divided into the following five types according to the conversion technology used after receiving infrared energy: pyroelectric type: triglyceride sulfate, lithium tantalate, etc; Thermopile type: n-type and p-type polysilicon; Diode type: single crystal or polycrystalline PN junction; Thermoelectric capacity type: double material film; Thermistor type: vanadium oxide, amorphous silicon, etc. [1]

The infrared temperature sensor is the core component of the infrared thermometer, while the infrared filter is the core of the temperature sensor. Therefore, the preparation of high-precision infrared filter is the key to improve the accuracy of infrared thermometer. The infrared band includes a wide range of wavelengths, ranging from 780nm to 14um,

including near infrared, mid infrared and far infrared. Near infrared (code IR-A, wavelength 780-1500nm, NIR), medium infrared (IR-B, 1500-6000nm, MIR) and far infrared (IR-C, 6000-14000nm, FIR). In terms of materials, there are optical glass coating, colored glass and plastic infrared filter. This refers to the near-infrared filter. If the medium and far infrared is involved, the materials include Si, ZnSe, Sapphire, CaF2, quartz glass, etc. In terms of optical characteristics, there are long wave pass IR filter and bandpass IR bandpass filter. Long wave pass type, made of colored glass, usually black, such as IPG-800. It absorbs visible light and allows transmission of infrared light. If you look at the sun through this infrared filter, you can still see a red sun. Suitable for infrared imaging. (2) An infrared filter, such as IPGC-720, is made on optical white glass by vacuum coating. This type of filter is also called an optical cold mirror, which reflects visible light and allows infrared light to pass through. The appearance looks silver, like a mirror. If it is a medium and far infrared filter, it should be coated on Si, Sapphire and quartz glass. The infrared plastic filter made of special plastic has a black appearance. If you look at the sun through this infrared filter, you can also see a red sun. The material of this plastic can be either PC or PMMA. Besides, band-pass filters are made of vacuum coating. For near-infrared bandpass, it is coated on white glass. If it is medium and far infrared, it will be coated on Sapphire. The corresponding light source of the near-infrared band-pass filter is mainly infrared IR LED and infrared laser, so the main wavelengths are 808nm, 850nm, 905nm, 940nm, 1064nm, and 780nm, which are not commonly used. Therefore, these are the main wavelengths of the commonly used near-infrared filters. The medium and far infrared filters mainly include 1550nm, 2300nm, 3800nm, 5500nm, 7800nm, etc. This paper mainly studies 5500 nm infrared filter. [2]

Most of the sensors used in common frontal temperature gun and ear temperature gun belong to thermopile type, that is, the corresponding temperature feedback is provided by means of bimaterial film. The traditional infrared temperature measuring gun cannot accurately measure the actual temperature of patients because of the low transmittance of the infrared filter in the core component. A high-precision infrared filter film is prepared in this paper, which can increase the accuracy of infrared temperature measuring gun and achieve more remarkable effect of epidemic prevention. [3], [4].

978-1-6654-8699-6/22 $31.00 © 2022 IEEE

II. Experiment and Analysis

A. Selection of Materials

Common infrared band film materials include Ge, Si, ZnS, YbF3, ZnSe, etc. According to the transmittance of the required band and the influence of film adhesion, Ge is finally selected as the high refractive index material and ZnS as the ground refractive index material. Ge and ZnS are usually used in the preparation of infrared films. [5, 6]

Ge is a rare metal, non-toxic and non-radioactive. It is mainly used in semiconductor industry, plastic industry, infrared optical devices, aerospace industry, optical fibre communication, etc. The light transmission range is 2000Nm --- 14000 nm, n = 4 or more. It melts at 937 (°C) and forms a liquid in the electron gun, and then evaporates easily at 1400 (°C). When evaporated by electron gun, its density is lower than the overall bulk density, while it can be close to the loose density by ion assisted plating or laser evaporation. Tens of layers of 8000-12000nm band-pass filters are prepared on germanium substrate with thf4. If the chamber temperature is too high, the absorption will change significantly. In the range of 240-280 (°C), GE has a critical point in the process of transformation from amorphous to crystal. Germanium doesn't work with carbon, so it melts in a graphite crucible and won't be polluted by carbon. Due to the anisotropy of metal Ge, irregular protrusions appear on the surface of the film after melting. [7]

ZnS has a refractive index of 2.35 and a light transmission range of 400-13000 nm, which has good stress and good environmental durability. ZnS is easy to sublimate when steamed at high temperature. In this way, it forms a non-adsorptive film on the substrate before the required film is attached. Therefore, it is necessary to thoroughly clean the furnace and dry it at the highest temperature. It takes several hours to eliminate the adverse effect of zinc. Therefore, the part baked in advance is added in the subsequent film preparation to ensure the smooth evaporation of ZnS. [8][9].

B. Membrane Stack Design

The basic theory is that multilayer film is equivalent to monolayer film, and the equivalent layer is used to match the admittance with the substrate

$$\eta = \frac{H_t^+}{(S_0 \times E_t^+)} = -\frac{H_t^-}{(S_0 \times E_t^-)}, Y = \frac{C}{B}$$

According to boundary conditions, the combined characteristic matrix of substrate and m-layer film is

$$E_0 \begin{bmatrix} 1 \\ Y \end{bmatrix} = \begin{bmatrix} B \\ C \end{bmatrix} = \left\{ \prod_{j=1}^{M} \begin{bmatrix} \cos\delta_j & i/\eta_j \sin\delta_j \\ i\eta_j \sin\delta_j & \cos\delta_j \end{bmatrix} \right\} \begin{bmatrix} 1 \\ \eta_{M+1} \end{bmatrix}$$

So, the overall transmittance is

$$T = \frac{4\eta_0\eta_{M+1}}{(\eta_0 B + C)(\eta_0 B + C)*}$$

where η_i is the effective admittance of layer i, H_t is the tangential component of magnetic field strength, E_t is the tangential component of electric field strength, S_0 is the

energy flow density; Value of P component $\eta_i = n_i/cos\theta_i$. Value of S component $\eta_i = n_i cos\theta_i$; Y is the admittance of the combination of film layer and substrate, and M is the number of film layers; δ_i is the phase thickness of layer i.

The filter studied in this paper belongs to the long wave pass band-pass model. According to the above formula, sub| (0.5hl0.5h) s| air is selected as the basic membrane unit. Where sub represents the base material, H represents the high refractive index material, Ge, l represents the low refractive index material, ZnS, s represents the number of cycles, and air represents air; Then the essential MacLeod is used to optimize the film. Finally, the main film structure and film thickness are as follows: Sub|4.43H, 3.24L, 9.43H, 4.40L, 8.11H, 7.62L, 5.42L, 10.36L, 5.61H, 7.78L, 8.53L, 6.07L, 7.53H, 9.71L, 4.15H,12.05L, 5.71H, 5.08L, 16.34L, 2.10L, 7.71H, 0.25L, 0.26H, 0.58L, 0.06H, 16.17L | Air. The design spectrum was shown in the figure 1.

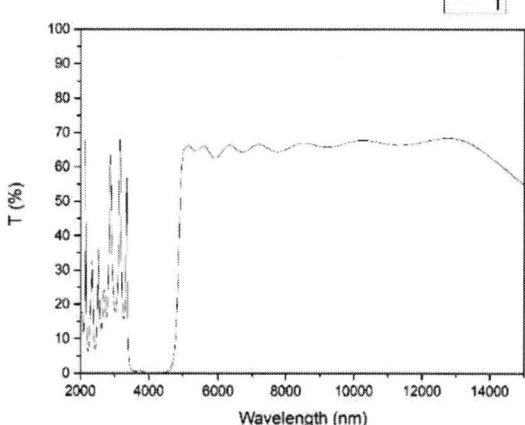

Figure 1. Design spectrum of main film system of 5.5 μm infrared filter film

The secondary film is infrared AR film, and the design spectrum is shown in Figure 2.

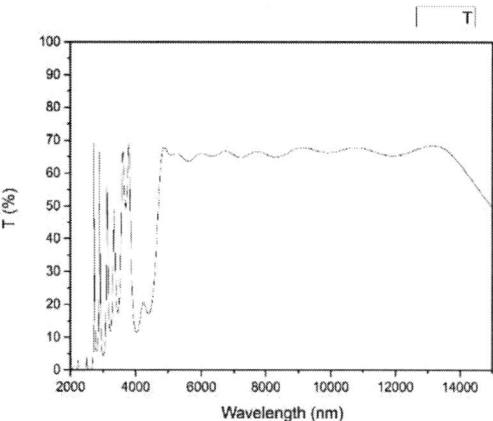

Figure 2. design spectrum of secondary film system of 5.5 μm infrared filter film

C. Preparation of Infrared Filter Film

In this experiment, MF1550 vacuum coating machine is selected, and the equipment is equipped with 4 Shimadzu molecular pumps and cryogenic. Evaporation source 1 is an independently developed resistance evaporation system, evaporation source 2 is an E-type electron gun, and is equipped with Hall ion source for auxiliary deposition. The low refractive material ZnS is prepared by resistance evaporation, and the metal Ge is evaporated by electron gun.

In order to improve the control accuracy of film thickness, the optical film thickness monitoring system developed independently was used in this experiment. High precision film thickness control system is needed to control the optical thickness of the film. Based on the actual production requirements, a set of high-precision and wide spectrum optical film thickness monitoring system has been developed, which can cover the visible light to the mid infrared part, and the control accuracy is up to 0.05%. The system is based on the combination of Finier formula and dielectric film substrate optical admittance, and the reflectivity formula applicable to the optical film thickness monitoring system is deduced using topology analysis and other algorithms. By combining the Savitzky Golay fitting method with the least square method, the fitting and filtering processing of the actual acquisition curve is realized. According to the required wave band and responsiveness requirements of the system, we independently developed and manufactured a high stability light source luminous device combining visible light and near-infrared and mid infrared light, which achieved the index requirements of high precision and wide spectrum. In order to meet the collimation characteristics of the optical path system, the combination of fiber optical waveguide and free-form curved lens group is used to achieve the optical path imaging mode that meets the requirements of the device. In the detector part, the traditional photomultiplier tube scheme is abandoned, and the spectral width is delayed by combining the silicon photodetector with the indium arsenic tellurium amplification photodetector. Finally, the system is built based on mechanical principle and optical design principle. An electron beam evaporator equipped with the system is used for film preparation. Specific process parameters are shown in Table 1. [10]

TABLE I. PROCESS PARAMETERS OF 5.5 MM INFRARED FILTER FILM

	ZnS	Ge
Rate (Å/s)	20	6
Pressure (Pa)	9.0E-4	9.0E-4
IBS	N	N

In the evaporation process, the ion source is used to clean the substrate for 10 minutes. The first and second layers are deposited with the aid of ion source, and other layers are not selected. In order to increase the firmness of the initial layer and prevent the film from peeling off.

D. Fogging on the Surface of Infrared Filter Film

After the preparation of the main film system, it was found that there was obvious fogging on the film surface and the hand felt dull. The surface is tested by onlookers, and the sample AFM test results are shown in the Figure 3.

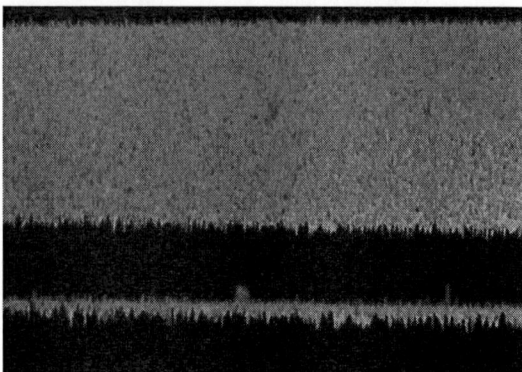

Figure 3. AFM test spectrum of main film of 5.5 μm infrared filter film

As shown in Fig. 3, AFM test map of main film of infrared filter film, the surface roughness of the film layer is high, and there is an obvious gap between particles. Conduct experiments in two directions to judge which of the two materials causes the fogging phenomenon. Firstly, ZnS was tested, and a regular film system with a large number of ZnS layers was selected to prepare and observe the surface phenomenon of the substrate. It was found that there was no obvious fog on the surface.

Then test the material Ge, steam 10 layers of test film system, and the thickness of each layer of Ge is 1 μm. The thickness of ZnS is 50 nm. There is obvious fog on the surface of the experimental piece. Based on the analysis of the material properties of Ge, it is suspected that it is related to the crystal properties of Ge. Ge is in the range of 240-280 (℃), and GE has a critical point in the process of transformation from amorphous to crystal. According to the anisotropic property of the nonspecific protrusion produced by Ge in the melting process, it is judged that the mixed evaporation of crystalline germanium and amorphous germanium may lead to the uneven surface of the film, resulting in fogging. It is analyzed that the co evaporation of crystal and amorphous may be caused by the uneven heating of crucible. The original graphite crucible with poor thermal conductivity is replaced by a molybdenum crucible with good thermal conductivity. Due to the uneven heating of the graphite crucible, there is sufficient water cooling at the bottom, the evaporation rate is fast, there is no water cooling at the edge, and the evaporation rate is slow, resulting in amorphous Ge evaporated at the edge of the crucible. After replacing the crucible with molybdenum crucible, according to the shape of the remaining material in the crucible, only the crystal germanium in the middle part evaporated to the substrate. Observe that there is no fogging phenomenon on the evaporated test piece [11], [12].

III. TEST AND ANALYSIS

As shown in the figure, the final test curve of the main film system has an average transmittance of more than 80% from 5.5 microns to 13 microns, and there is no fogging on the substrate surface, which meets other environmental test indexes of the film.

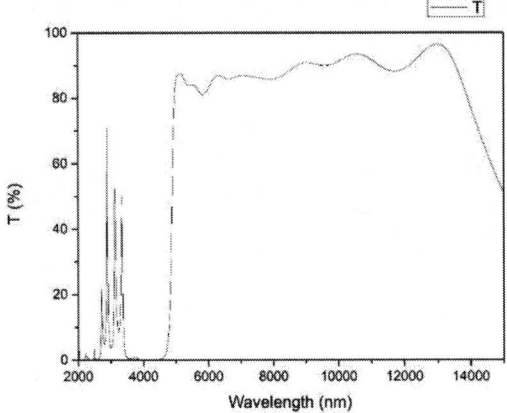

Figure 4. test spectrum of 5.5 μm infrared filter film

IV. SUMMARY

According to the principle of photoelectric extreme value method, a set of infrared optical film thickness monitoring system is designed and manufactured by using Macleod programming software. The preparation accuracy of infrared narrow-band filter film can be improved. The infrared narrow-band filter film is prepared by using the system. The test results are good and meet the requirements of spectrum and other optical parameters. The research and development of this system is of great significance to the promotion of military industry and aerospace industry. In the future research, it will further refine the research and improve the spectral monitoring range and accuracy.

REFERENCES

[1] Li Xiang, Zhang Min, Li Bing, et al. Study on a new infrared temperature measurement method [J] Journal of Tianjin University of technology, 2010, 26 (1): 58-61.

[2] Tokunaga A T , Simons D A . The Mauna Kea Observatories Near-Infrared Filter Set for 1—5 Microns[C]// Instrument Design and Performance for Optical/Infrared Ground-based Telescopes Pt.1. Institute for Astronomy, Univ. of Hawaii, Honolulu, Hawaii, USA, 2003.

[3] SU. R, ABHI. V, SPINELLO. D, NECSULESCU. D. Infrared fever body identification using shape and temperature filters[J]. Instrumentation and Measurement Technology Conference. IEEE, 2012, 1556-1560.

[4] Zhu Zhenyi Optical temperature measurement technology and physical principal analysis [J] Mathematics learning and research, 2019 (16): 127-127

[5] LAHIRI. B, BAAAVATHIAPPAN. S, Jayakumar T, et al. medical applications of infrared thermography: A review [J]. Infrared Physics and Technology, 2012, 55(4): 221-235.

[6] Liu Henghui, Yin Yong, Li Yu Research and implementation of high precision infrared temperature measurement system based on FPGA [J] Electronic devices, 2009, 32 (2): 452-454

[7] L. Persichetti, A. Sgarlata, M. Fanfoni, A. Balzarotti. Heteroepitaxy of Ge on singular and vicinal Si surfaces: elastic field symmetry and nanostructure growth[J]. Journal of Physics: Condensed Matter. 2015 (25)

[8] Yinhua Zhang,Shengming Xiong,Wei Huang,Kepeng Zhang,Xiaoxi Tian. Six-spectral antireflection coating on zinc sulfide simultaneously effective for the visible, near-IR, and long wavelength IR regions[J]. Surface & Coatings Technology. 2019.

[9] M R Kozlowski, IM Thomas. High power op tical coatings for a mega joule class ICF laser [A]. Proc. SP IE, 1992, 1782: 105 - 119.

[10] Dong, S.; Fu, X.; Li, C. Noble Infrared Optical Thickness Monitoring System Based on the Algorithm of Phase-Locked Output Current–Reflectivity Coefficient. Coatings 2022, 12, 782.

[11] J. A. Dobrowolski, F. Ho. High-performance step-down AR coatings for high refractive index IR materials [J]. Appl. Opt. 1982, 21 (2): 288-292 [10] H. A. Macleod, E. Pelletier. Error compensation mechanisms in some thin film monitoring systems [J]. Opt. Acta, 1977, 24 (9): 907-930.

[12] K. GangaReddy, P. Nagaraju, G.L.N. Reddy, Partha Ghosal, M.V. Ramana Reddy, Growth and characterization of electron beam evaporated NiO thin films for room temperature formaldehyde sensing, Sensors and Actuators A: Physical, Volume 346, 2022, 113876, 0924-4247.

Dark Current Analysis in Type-II InAs/GaSb Superlattice LWIR Detector with M-structure Barrier

Dongpei Shen[1,2], Tong Sun[1], Pengfei Zhu[1], Xiaoning Guan[1], Baonan Jia[1], Haizhi Song[3,4], Pengfei Lu[1,2]*

[1]State Key Laboratory of Information Photonics and Optical Communications, Beijing University of Posts and Telecommunications, Beijing 100876, China

[2]School of Electronic Engineering, Beijing University of Posts and Telecommunications, Beijing 100876, China

[3]Southwest Institute of Technical Physics, Chengdu 610041, China

[4]Institute of Fundamental and Frontier Science, University of Electronic Science and Technology of China, Chengdu 610054, China

e-mail: photon.bupt@gmail.com

Abstract—We designed a long-wave infrared detector using InAs/GaSb and InAs/GaSb/AlSb/GaSb superlattices and further studied the effect of some sensitive parameters on dark current characteristics. We utilize the numerical model to analyze the dark current characteristics of the contact layer and the absorption layer at different doping levels, and also calculate the dark current characteristics of the absorption layer and barrier layer at different thicknesses. By designing different absorption layer and barrier layer, we found that the detector has a hole barrier in the valence band, which effectively reducing the dark current level. Under the optimal detector structure, the dark current at low temperature is maintained at a relatively ideal level about 2.25×10^{-5} A/cm^2 and the quantum efficiency is close to 42%.

Keywords-LWIR; T2SL; M structure; dark current

I. INTRODUCTION

III-V binary composites have made great progress in the growth of their own materials and the manufacture of devices through the development of nearly half a century [1]. Traditional HgCdTe (MCT) materials of group II-VI compounds have been developed and researched for nearly 60 years, the disadvantages of high cost of epitaxial substrates have gradually emerged. InAs/GaSb type II superlattices (T2SL), as a new material that can replace HgCdTe and develop infrared detectors, have attracted great attention [2,3]. In 1970, IBM laboratories Esaki and Tsu proposed the concept of superlattices and prepared them on GaAs substrates [4]. They theoretically analyzed the material characteristics of InAs/GaSb T2SL and calculated its energy band structure [5]. In 1987, Smith and Mailhiot published the theoretical performance prediction of InAs/GaSb type II superlattices as infrared detectors. The results showed that the InAs/GaSb type II superlattices had better optical properties than MCT materials and the high lattice matching of the two compounds also reduced the overall defects and dislocation density of materials. This has greatly promoted the research of InAs/GaSb type II superlattices as infrared detector materials [6,7]. On the one hand, due to the low

auger recombination rate and high effective mass of T2SL, the G-R current component in the dark current is greatly reduced and the tunneling effect can be well suppressed. On the other hand, the InAs/GaSb heterojunction presents the second-class staggered arrangement in energy band arrangement, which makes the energy band structure of the superlattice flexible and adjustable. The theory predicts that T2SL have better performance than HgCdTe, but so far, the superior performance has not been fully realized. Because the number of SRH recombination centers is not well controlled and the minority lifetime is small, the dark current remains at a very high level [8,9,10].

Therefore, various novel barrier structures have been proposed to reduce the dark current density, such as pπMn [11,12,13], CBIRD [14,15], nBn [16] and pBp structures. In these structures, the band offset of the heterojunction material mainly falls in the conduction band or the valence band through reasonable structural design, so that the barrier structure can prevent the transmission of the majority carrier in the contact layer, but cannot affect the photogenerated carrier in the absorption layer. The photocurrent is formed by applying a certain reverse voltage bias to help the photogenic minority carriers of the absorber layer to cross the smaller barrier. For barrier detectors, the depletion region can be transferred from the narrow-band absorption region to the wide-band barrier region as much as possible by controlling doping, while there is almost no SRH current in the wide-band gap barrier layer, so the G-R dark current can be significantly reduced [17].

This paper is arranged as follows. In section 2, the material properties of each layer and the growth structure of device are introduced, meanwhile the existence mechanism and the model analysis method of dark current are simply described. In section 3, the Apsys software developed by Crosslight company is used to simulate and calculate the dark current. The influence of a series of sensitive parameters on the dark current is analyzed. In section 4, we summarize the overall performance of the device.

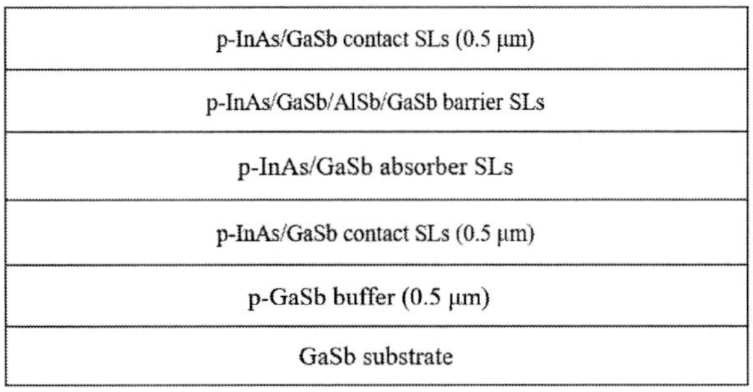

| p-InAs/GaSb contact SLs (0.5 μm) |
| p-InAs/GaSb/AlSb/GaSb barrier SLs |
| p-InAs/GaSb absorber SLs |
| p-InAs/GaSb contact SLs (0.5 μm) |
| p-GaSb buffer (0.5 μm) |
| GaSb substrate |

Figure 1. The structure of pπMp LWIR detector.

II. DETECTOR STRUCTURES AND THEORETICAL

A. Detector Structures

Fig. 1 shows the device structure of the long-wave infrared (LWIR) detector. The buffer layer is p-type GaSb layers with doping concentration of 1×10^{18} cm^{-3}, and the buffer layer thickness is 0.5 μm. The bottom contact layer composed of InAs/GaSb superlattice grown on the buffer layer. Above the bottom contact layer is the absorber layer and the top contact layer of the same structure. Between the absorber layer and the top contact layer is the InAs/GaSb/AlSb/GaSb superlattice barrier layer. The barrier layer structure is 6 ML (Monolayer) InAs/3 ML GaSb/5 ML AlSb/3 ML GaSb. The bottom and top contact layer and absorption layer structure is 13 ML InAs/7 ML GaSb. Both the bottom and top contact layer thickness is 0.5 μm. The barrier layer, the contact layer and the absorption layer are all p-type doping. In this paper, InSb-like interface layers are added between InAs and GaSb layers to ensure good lattice matching. This phenomenon can also be found between InAs and AlSb layers that there are similar InSb-like interface layers.

B. Dark Current

In narrow band gap semiconductor devices, dark current mechanisms for electrons and holes passing through the depletion region can be divided into two categories. One is intrinsic mechanism, which only depends on material parameters and device design. The other is the defect dependent mechanism, which requires the defect as an intermediate state. The mechanisms that determine the dark current of LWIR detector mainly include: diffusion current, generation-recombination current (G-R current), direct tunneling current and trap-assisted tunneling current. These four current mechanisms are often used to describe and predict the performance limits of LWIR detectors. In addition, there is surface leakage current which is not the dark current caused by the device structure itself, but related to the surface passivation effect of the device, it can be suppressed by optimizing passivation technology. It can be said that dark current is a combination of the above four currents.

At present, the dark current research of infrared detectors also has many challenges, and the dark current suppression

method is still not very thorough under different conditions. In terms of experiments, the fabrication process of infrared detectors is very complex, with long cycle and high price, which makes device simulation technology an important tool for studying dark current composition and mechanism [19,20]. The device simulation of infrared detector includes numerical model and analytical model. Although the numerical model takes a long time and the results are not easy to converge, it is more suitable for the design and optimization of device structure because drift diffusion is the most common physical model of infrared detectors. In the following chapters, we will use a series of software based on numerical analysis to simulate the dark current properties of devices. By modifying a series of sensitive parameters, the dark current components in different device structures and operating environments are analyzed to achieve the suppression of dark current, improve device performance and provide theoretical support and reference for the structural optimization of barrier infrared detectors.

III. RESULTS AND DISCUSSIONS

The absorption layer, barrier layer and contact layer affect the dark current characteristics of LWIR detector. In the pπMp LWIR detector structure, we focus on the diffusion current, G-R current, direct tunneling current and trap assisted tunneling current. Among them, radiation recombination, Auger recombination and trap-assisted recombination are the important components that affect the recombination process, while the minority carrier lifetime affects the diffusion, trap recombination and defect assisted tunneling process. The key parameters affecting dark current are summarized in Table I.

TABLE I. MATERIAL AND DEVICE PARAMETERS USED IN THE SIMULATION

Parameters	pπMp LWIR detector	
	13/7 MLs InAs/GaSb	6/3/5/3 MLs InAs/GaSb/AlSb/GaSb
Energy band gap at T=77 K [eV]	0.121	0.395
Electron effective Mass [/m$_0$]	0.022	0.067
Hole effective mass [/m$_0$]	0.30	0.11
Radiative recombination rate [m^3/s]	1×10^{-15}	1×10^{-13}
Auger rate [m^6/s]	1×10^{-36}	1×10^{-36}
SRH lifetime [ns]	7	7

978-1-6654-8699-6/22 $31.00 © 2022 IEEE

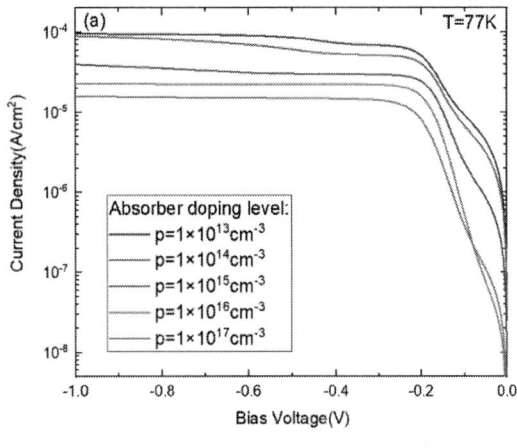

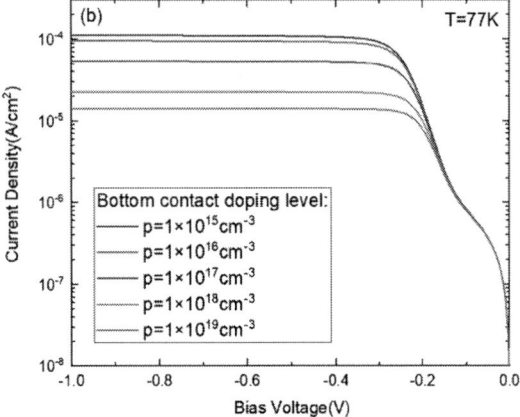

Figure 2. Dark current characteristics of pπMp LWIR detector at different doping concentrations of 77 K: (a) absorb layer doping concentration, (b) contact layer doping concentration.

A. Doping Concentration

The dark current characteristic curves of the bottom contact layer and the absorption layer under different doping concentrations at 77 K are showed in Fig. 2 respectively. Fig. 2 (a) shows that the dark current level decreases when the absorber layer doping concentration increases. This is because when pπMp detector operates under working bias, the dark current is mainly dominated by diffusion current. The key factor affecting the diffusion current value is the absorber layer doping concentration and the minority carrier lifetime, that is, the diffusion current value is inversely proportional to the product of absorber layer doping concentration and minority carrier lifetime. When fixing the minority carrier lifetime, increasing the absorber layer doping concentration will reduce the dark current level. However, it is worth noting that increasing the absorption layer doping concentration will also increase depletion region width and the G-R current will increase accordingly. Therefore, the absorber layer doping concentration cannot be continuously increased.

Fig. 2 (b) shows saturated dark current decreases when the bottom contact layer doping concentration increases. This is because diffusion current becomes the main component at a high bias voltage and the material selected for the bottom contact layer is the same as the absorption layer. When exposed to a certain wavelength of light, the bottom contact layer will produce the same carrier generation process as the absorber layer. It can be seen from the above relationship between the absorber layer doping concentration and dark current that increasing doping concentration will cause the decrease of diffusion dark current. At different contact layer doping concentrations, the G-R current is almost the same at a low voltage bias.

B. Layer Thickness

The dark current characteristic curves corresponding to different barrier layer and absorption layer thickness at 77 K are plotted in Fig. 3 respectively. Fig. 3 (a) shows that saturation dark current is almost the same, but the saturation voltage increases when the barrier thickness increases. This is due to the fact that both diffusion current and tunneling current are dominant at lower temperatures. At a low voltage bias, tunneling current becomes the main component. As the barrier layer thickness increases, electrons and holes are difficult to tunnel through the thick barrier layer, which avoids the diffusion of exhausted electric field from the barrier layer to the absorption layer under reverse bias, so the tunneling current is effectively suppressed. However, the quantum efficiency of the detector decreases when the barrier thickness increases. Therefore, the influence of dark current level and quantum efficiency should be considered comprehensively in the selection of the barrier thickness. At high bias voltage, diffusion current becomes the main mechanism of dark current and barrier thickness has little influence on diffusion current, so saturated dark current is almost the same under different barrier thickness.

Fig. 3 (b) shows that the device's dark current increases when the thickness of the absorption layer increases at low bias voltage. This is because the dark current will be affected by the diffusion mechanism at low working voltage bias. With the increase of the absorber layer thickness, the width of depletion region will also increase and the dark current level will increase accordingly. However, the quantum efficiency will be improved when the absorber layer thickness increases. Therefore, the selection of the absorber layer thickness needs to balance the two change mechanisms of the absorption layer photogenerated carrier concentration and the built-in electric field.

C. Structure Optimization

Through the above researches on the changes of dark current curves under different sensitive parameters and the influence of the integrated device quantum efficiency, we selected the best detector combination structure at 77 K temperature. Due to the above research process adopts control variable method, so that herein only need to select contact layer, absorbing layer and barrier layer with the optimal doping and thickness level. The optimized detector structure is summarized below.

978-1-6654-8699-6/22 $31.00 © 2022 IEEE

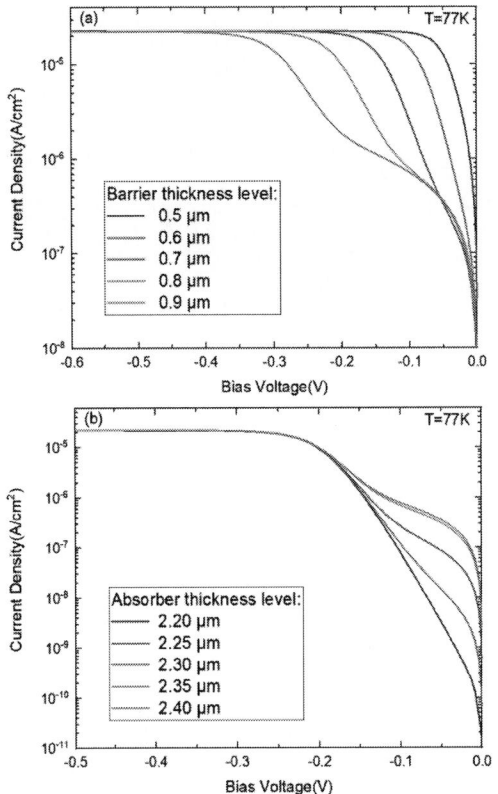

Figure 3. Dark current characteristics of pπMp LWIR detector at different thickness level of 77 K: (a) barrier thickness level, (b) absorber thickness level.

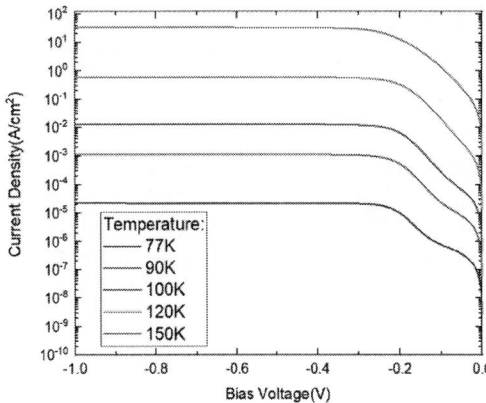

Figure 4. Dark current characteristics of pπMp LWIR detector at different temperature. The barrier thickness is 0.8 μm. The absorber is p-type doping and the concentration is $1×10^{16}$ cm^{-3}, the thickness of absorber is 2.4 μm.

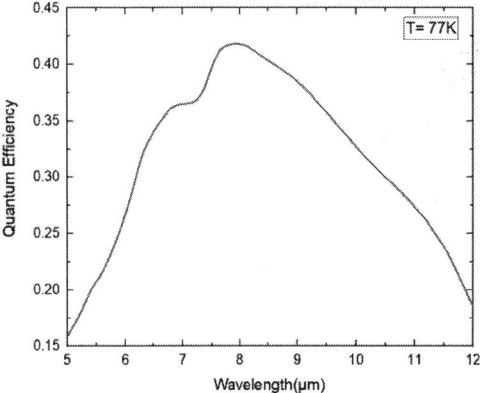

Figure 5. The quantum efficiency of pπMp LWIR detector at 77 K. The peak value of QE can reach nearly 42% at 8 μm.

In order to achieve better optical detection performance, the doping concentration of the two contact layers is $1×10^{18}$ cm^{-3} and the thickness is 0.5 μm. For the absorption layer of the device, p-type slight doping is selected and the doping concentration is $1×10^{16}$ cm^{-3}, so as not to increase the G-R current while reducing the diffusion current. Considering the change of dark current with the absorber layer thickness and the light absorption effect, the absorber layer thickness was selected as 2.4 μm. Under the premise of restraining the direct tunneling current and improving the quantum efficiency, the barrier layer thickness is set to 0.8 μm and the doping concentration is set to $1×10^{15}$ cm^{-3}.

The dark current curves of the detector at different temperatures are showed in Fig. 4. When temperature rises, all the dark current components will increase, especially the diffusion dark current. Therefore, diffusion dark current becomes the main mechanism of the dark current of the dominant device at high temperature. At low temperature, the dark current of the detector can be maintained at a good level, about $2.25×10^{-5}$ A/cm^2 at 77 K, but at high temperature, the dark current suppression needs to be further improved and perfected. Fig. 5 shows the quantum efficiency change curve with wavelength. It can be seen that there is an obvious efficiency peak at 8 μm, and the quantum efficiency peak can reach nearly 42%.

IV. CONCLUSIONS

In summary, we propose and design a pπMp LWIR detector with M barrier structure. The energy band structures of InAs/GaSb and InAs/GaSb/AlSb/GaSb superlattices are calculated. The result shows that the band gap of absorption layer superlattice is 0.121eV, which is suitable for long wave absorption. And the barrier layer band gap reaches 0.395eV, which meets the requirement of wide band gap. We analyze the mechanism of dark current and calculate the influence of different sensitive parameters on the dark current in LWIR detector.

(1) At 77 K, the dark current is the coexisting result of diffusion current, G-R current and tunneling current. At lower voltage bias, tunneling current and G-R current are the main dark current mechanisms, while at higher voltage bias, diffusion current is the main component.

(2) The dark current decreases with the absorption layer doping concentration increases, the same rule also appears in the relationship between the saturated dark current and the contact layer doping concentration.

(3) Increasing the absorption layer thickness will affect the carrier diffusion mechanism, leading to the increase of the dark current level at lower voltage bias, but the increase

of the barrier layer will significantly reduce the tunneling current of the detector. However, above two parameters have no significant effect on saturated dark current at higher voltage bias.

We select the best combination of detector structural parameters and calculate the change of dark current level at different temperatures. At a 77 K lower temperature, the dark current level of the detector can be maintained at 2.25×10^{-5} A/cm^2 while the quantum efficiency can reach nearly 42%. But the dark current suppression work at a higher temperature needs further improvement and perfection.

ACKNOWLEDGMENT

This work was supported by the Open-Foundation of Key Laboratory of Laser Device Technology, China North Industries Group Corporation Limited (No. KLLDT202103).

REFERENCES

[1] Plis. E. A, "InAs/GaSb Type-II Superlattice Detectors," Advances in Electronics, vol. 2014, pp. 1-12, 2014.

[2] Mailhiot. C, Smith. D. L, "Long-wavelength infrared detectors based on strained InAs-Ga$_{1-x}$In$_x$Sb type-II superlattices," J. V ac. Sci. Technol, vol. A7(2), pp. 445-449, 1989.

[3] Rogalski. A, "New material systems for third generation infrared detectors," Opto-Electron. Rev, vol.16(4), pp. 458-482, 2008.

[4] Esaki. L, Tsu. R, "Superlattice and negative differential conductivity in semiconductors," IBM J. Res. & Dev, vol. 14(1), pp. 61-65, 1970.

[5] Sai-halasz. G. A, Esaki. L, Harrison. W. A, "InAs/GaSb superlattice energy structure and its semiconductor-semimetal transition," Phys. Rev. B, vol. 18, pp. 2812, 1978.

[6] Smith. D. L, Mailhiot. C, "Proposal for strained type II superlattice infrared detectors, " J. Appl. Phys, vol. 62(6), pp. 2545-2548, 1987.

[7] Razeghi. M, Nguyen. B. M, "Band gap tunability of Type II Antimonide-based superlattices," Phys. Procedia, vol. 3(2), pp. 1207-1212, 2010.

[8] Connelly. B. C, Steenbergen. E. H, Metcalfe. G. D, Shen. H. Wraback. M, "Significantly improved minority carrier lifetime observed in a long-wavelength infrared III-V type-II superlattice comprised of InAs/InAsSb," Appl. Phys. Lett, vol. 99(25), pp. 1815-500, 2011.

[9] Pellegrino. J, DeWames. R, "Minority carrier lifetime characteristics in type II InAs/GaSb LWIR superlattice n$^+\pi$p$^+$ photodiodes," Proc. SPIE 7298, 72981U-72981U-10, 2009.

[10] Razeghi. M, "Focal plane arrays in type II-superlattices," US 6864552 B2, 2004.

[11] B.−M. Nguyen, M. Razeghi, V. Nathan, G. J. Brown, "Type−II 'M' structure photodiodes: an alternative material design for midwave to long wavelength infrared regimes," Proc. SPIE 6479, 2007.

[12] B.−M. Nguyen, D. Hoffman, P.−Y. Delaunay, M. Razeghi, "Dark current suppression in type II InAs/GaSb superlattice long wavelength infrared photodiodes with M−structure," Appl. Phys. Lett, 2007.

[13] B.−M. Nguyen, D. Hoffman, P.−Y. Delaunay, E. K. Huang, M. Razeghi, J. Pellegrino, "Band edge tunability of M−structure for heterojunction design in Sb based type II superlattice photodiodes," Appl. Phys. Lett. 93, 2008.

[14] D. Z. Ting, C. J. Hill, A. Soibel, J. Nguyen, S. A. Keo, M. C. Lee, et al, "Anti− monide−based barrier infrared detectors," Proc. SPIE 7660, 76601R−1−12, 2010.

[15] D. Z. Ting, C. J. Hill, A. Soibel, S. A. Keo, J. M. Mumolo, J. Nguyen, S. D. Gunapala, "A high−performance long wavelength superlattice complementary barrier infrared de− tector," Appl. Phys. Lett, vol. 95, 2009.

[16] Rodriguez. J. B, Plis. E, Bishop. G, Sharma. Y. D, Kim. H, Dawson. L. R, et al, "NBn structure based on InAs/GaSb type-II strained layer superlattices," Appl. Phys. Lett, vol. 91(4), 2007.

[17] Hu Weida, Li Qing, Chen Xiaoshuang, Lu Wei, "Recent progress on advanced infrared photodetectors," Acta Physica Sinica, vol. 68, 2019.

[18] Klipstein. P. C, Avnon. E, Benny. Y, Fraenkel. R, Glozman. A, Grossman. S, Weiss. E, "InAs/GaSb Type II superlattice barrier devices with a low dark current and a high-quantum efficiency," Infrared Technology and Applications XL, 2014.

[19] G. M. Williams, R. E. DeWames, "Numerical simulation of HgCdTe detector characteristics," Journal of Electronic Material, vol. 24, pp. 1239-1248, 1995.

[20] K.Kosai, "Status and application of HgCdTe device modeling," J. Electronic Mat, vol. 24, pp. 635-640, 1995.

Proceedings of the 7th Optoelectronics Global Conference (OGC 2022)

A Single-laser System for Mobile Cold Atom Gravimeter

[1]Pei Dongliang, [1, 2]Kong Delong, [1]Wang Jieying, [1]Chen Weiting, [1, 2]Lu Xiangxiang, [1]Wei Junxin, [1, 2]Liu Weiren

[1]Tianjin Navigation Instrument Research Institute；
[2]Maritime Support and Technology Laboratory of China State Shipbuilding Corporation
Tianjin, China
Email: 1710877856@qq.com

Abstract: Improving the environmental adaptability of laser system is the key to realize field measurement for cold atom interference gravimeter. In this paper, a laser system based on a 1560nm fiber laser is described in detail. The output power of trapping laser and Raman laser are 1.1W and 1.9W at 780.2nm, with near-ideal Gaussian mode of TEM$_{00}$, respectively. Using this compact optical scheme, we achieved the atom gravimeter interference with 2T=140ms, corresponding to an acceleration sensitivity of 610μGal/√Hz, which is limited by vibration conditions. The results of optical frequency locking and power stabilization for laser system meet requirements of high precision mobile gravimeter.

Keywords-atom gravimeter; laser system; frequency-doubling; 780nm laser

I. INTRODUCTION

In recent years, atom interferometer has been gradually used in the measurement of gravity [1-4], gravity gradient [5,6], rotation[7,8] and basic physical constants[9,10]. In order to adapt to the outfield environment, most sensors must have the characteristics of high reliability and convenient to transport performance, therefore, several teams such as Muquans of France, Wuhan Institute of Physics and Mathematics, Zhejiang University of Technology and Huazhong University of Science and Technology have carried out related research works in the past few years. The laser system of commercial gravimeter developed by Muquans use full fiber structures, and all function lasers are provided by four laser sources. The phase between lasers is controlled by beating frequency locking. The full fiber devices constitute the optical system, hence the system stability is excellent and suitable for field environment measurement. However, using of multiple lasers is inconducive to the miniaturization of the system[11]. The latest gravimeter laser system developed by Wuhan Institute of Physics and Mathematics is produced by a 780nm fiber laser modulated by an electro optic modulator (EOM). The system is compact and miniature, but the maturity of device corresponding to the 780nm wavelength is relative poor compared with the communication wavelength[12]. The newly reported laser system of Zhejiang University of Technology is based on two 1560nm distributed-feedback Bragg (DFB) lasers, which is frequency doubled to 780nm, and the two lasers achieved phase locked by optical phase-locking loop [13]. In the early stage, the research team

of Huazhong University of Science and Technology used three external cavity diode lasers (ECDL) to build the miniature system[14]. In 2019, this group reported a compact gravimeter laser system based on a single 1560 nm DFB seed laser[15]. The German research team reported a single-laser doubling frequency system for the [87]Rb atomic interferometer in 2015, which achieved a high-stability marine interferometer laser system using the rapid and large-scale frequency-tuning technology[16].

At present, most atom interference gravimeters operate with Rb atoms having a wavelength of 780nm. The laser source of 780nm ECDL do not need additional components, while the stability of ECDL deteriorate in the field. Also, the measurement in field needs to use additional temperature control system and anti-vibration system. Compared to 780nm laser, the 1560nm laser is in the conventional band of optical fiber communication, which is in the low loss window of optical fiber transmission, hence its performance is more stable. Moreover, Raman light is generated by phase modulation technology, which avoids the use of optical phase-locking loop and reduces the number of optical components, so the optical fiber system is more suitable for field measurement.

With a 1560.4nm laser (fiber external cavity laser, FECL) as the laser source, we achieve a compact and robust laser system for the [87]Rb atom interference gravimeter based on the laser frequency doubling technology[17]. The frequency jumping process from cooling laser to Raman laser in the above single laser system scheme requires complex frequency locking and controlling loop. Through the design of frequency shifting based on an acousto-optic modulator (AOM), we eliminate the unstable factors caused by the frequency jumping at GHz and simplify requirements of the frequency control system. Consequently, a compact single-laser system is achieved and implement for mobile gravity measurement.

II. THE METHOD OF FREQUENCY DOUBLING

The generation of 1560nm laser frequency-doubling to 780nm laser needs to meet the specific phase matching conditions, and the highly efficient conversion of the frequency doubling needs to design the special mode matching between the laser beam and the doubling crystal. The nonlinear doubling crystal for the frequency conversion selected in the experiment is the periodically poled lithium niobate (PPLN) waveguide crystal. By the theoretic model of the polarization interaction between electric field component and the medium when laser wave propagation in crystal, dividing a certain length of waveguide into N regions length l_i respectively, each region refractive indexes is n_i, single-pass nonlinear conversion efficiency is η_i, quasi-phase matching(QPM) temperature is T_P, the power of frequency doubling laser conversion can be expressed as[18]:

$$P_{2\omega} = \sum_{i=1}^{N} \eta_i P_\omega^2 \frac{\sin(\Delta k_i l_i/2)}{(\Delta k_i l_i/2)^2} + \\ 2\sum_{i=1}^{N-1}\sum_{j=i+1}^{N} \sqrt{\eta_i \eta_j} P_\omega^2 \frac{\sin(\Delta k_i l_i/2)}{(\Delta k_i l_i/2)^2} \frac{\sin(\Delta k_j l_j/2)}{(\Delta k_j l_j/2)^2} \cos\left[\frac{\Delta k_i l_i/2}{2} - \frac{\Delta k_j l_j/2}{2}\right] \tag{1}$$

$P_{2\omega}$ and P_ω are harmonic and fundamental power respectively; $\Delta k_{i,j}$ is phase mismatched. Δk is a function of the phase matching temperature T_p, expanded in a Taylor series with the first-order:

$$\Delta k = \frac{4\pi}{\lambda} \frac{\partial \Delta n}{\partial T}(T - T_P) \tag{2}$$

For a low-loss waveguide, with ideal quasi-phase matching conditions of $\Delta k = 0$, the power of frequency doubled laser power P_{SHG} and fundamental laser power $P_{coupled}$ can be theoretically related as:

$$P_{SHG} = P_{coupled} \tanh^2\left(\sqrt{\eta P_{coupled}}\right) \tag{3}$$

The normalized conversion efficiency is:

$$\eta = \left. P_{SHG} \middle/ (P_{coupled})^2 \right. \tag{4}$$

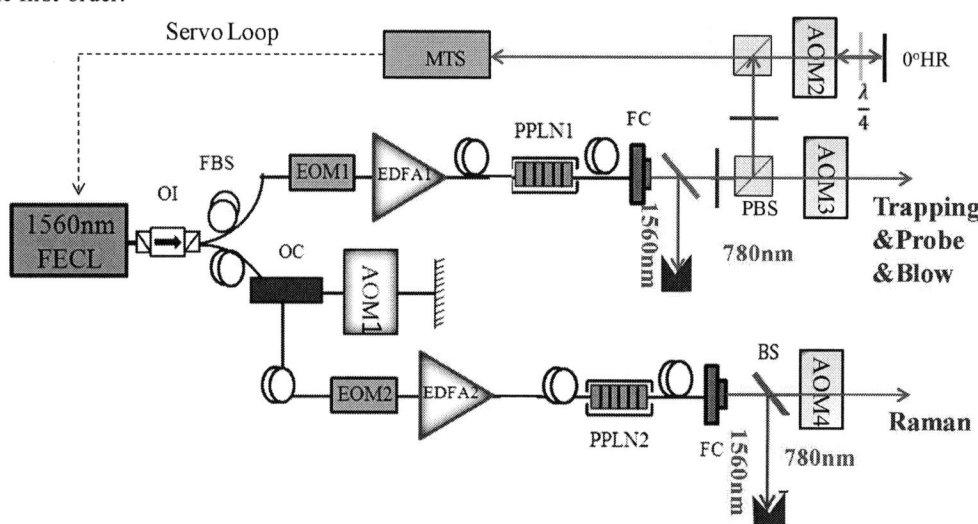

Figure 1. Frequency doubling schematic diagram of cold atom interference gravimeter.

III. SINGLE-LASER SYSTEM OF COLD ATOM GRAVIMETER

A. Design of Laser System

The general measurement process of a cold atom interference gravimeter is as follows: firstly, cooling and trapping ^{87}Rb atoms in the 3D-MOT. The frequency of cooling laser is red detuning 12 MHz to ^{87}Rb atomic $5S_{1/2}$, F=2→$5P_{3/2}$, F'=3 transition, the repumping laser is resonant with $5S_{1/2}$, F=1→$5P_{3/2}$, F'=2 transition. After trapping the atoms, it is necessary to further reduce the atom temperature by the polarization gradient cooling process, the frequency of cooling laser needs to jump to red detuning of 100MHz in this time. Secondly, the atoms fall freely, and are prepared to a magnetic-insensitive state $|5S_{1/2}, F=1, m_F=0\rangle$ by acting two microwave pulses and one Raman pulse. After entering the interference zone, the interference process is realized by a series of π/2-π-π/2 Raman pulses, and the Raman laser frequency are red detuning ~576.5MHz relatively to transition of $5S_{1/2}$, F=2→$5P_{3/2}$, F'=1 and $5S_{1/2}$, F=1→$5P_{3/2}$, F'=1. Finally, the atomic information is extracted by acting the probe laser of frequency resonate to $5S_{1/2}$, F=2→$5P_{3/2}$, F'=3 transition. The interference fringe is obtained by scanning the chirp rate of Raman laser.

The schematic diagram of the optical path used for the cold atomic gravimeter is shown in Figure 1, in which one of the key technology is the frequency-doubling of 1560.4nm laser to 780.2nm laser. A narrow linewidth FECL with a center wavelength of 1560.4nm first pass through a fiber optical isolator (FOI) to prevent the optical feedback damage from subsequent optical components. Then the light is divided into two beams by a fiber beam splitter (FBS), one of the beams is modulated by EOM1, and used for seed laser of erbium-doped fiber amplifier (EDFA1). Then the 1560.4nm laser pass through the PPLN1 waveguide and dichroic mirror (DM) to generated the pure 780.2nm laser. Finally, we obtained cooling laser, probe laser and blow laser by frequency shifting with AOM3. The maximum output power of EDFA1 is 3W. A few part of the 780nm laser is used to build a modulation transfer spectroscopy (MTS) after frequency-doubling of PPLN1. Another seed laser firstly pass through an optical circulator (OC) and AOM1 to realize a frequency shift of 400MHz. Then the laser pass through the EOM2 to modulate the Raman laser frequency component and used as the seed laser of EDFA2. The maximum output power of the EDFA2 is 5W. The 780.2nm harmonic laser can be used as the Raman laser in laser system after passing through AOM4. The input and output fibers of amplifiers are both single-mode polarization-maintaining fiber. In order to match the AOM aperture and the fiber coupling efficiency in the subsequent optical path, the Gaussian diameter of the EDFA output beams are 0.95mm and 0.93mm.

The frequency-doubling PPLN waveguide used in our system is a waveguide structure made on the matrix of the PPLN crystal, so that the laser transverse mode entered into the crystal is limited to a very narrow waveguide. Compared to transmitting in the PPLN crystal, the laser mode matches better in the PPLN waveguide, so it has higher conversion efficiency.

B. Optimization of Frequency Doubling Parameters

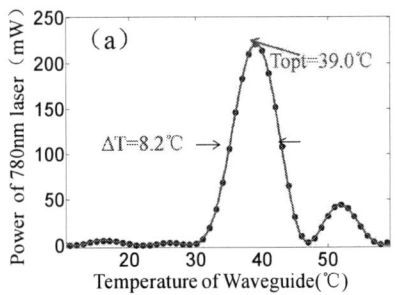

(a) Waveguide of trapping laser

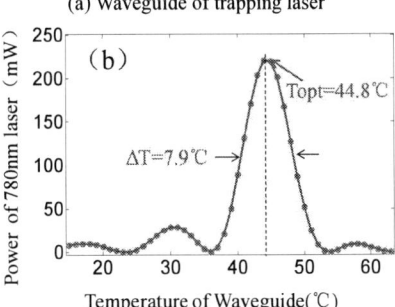

(b) Waveguide of Raman laser

Figure 2. The temperature tuning curves of the PPLN waveguide. The dots in the figure are the experimental data, and the red line are the theoretical fitting curves.

The results shown in Fig. 2 are the experimentally measured temperature tuning curves of both QPM waveguides. Fig. 2(a) corresponds to the trapping laser path waveguide, the 1560.4nm fundamental injection power is fixed at 1W, the highest 780.2nm output power is obtained when the temperature of PPLN1 is tuning to 39.0°C; the temperature tuning curves of PPLN2 is shown in Fig. 2(b), the fundamental power is fixed at 1.5W, and the highest

780.2nm laser output is obtained when the temperature of PPLN2 is tuning to 44.8°C. The dots in the figure are the experimental data, and the red line are the theoretical fitting curves(N=4) obtained from Equation (1). The asymmetry of the temperature curve coincides with the theoretical fitting curve, which is mainly derived from the optical inhomogeneity inside crystal. According to the theoretical curves fitting of the waveguide, the phase-matching acceptable bandwidths of temperature are 8.2°C and 7.9°C, respectively. The phase-matching temperature bandwidth of the crystal can be represented as:

$$\Delta T \cdot L = \frac{2\lambda_1}{2.25} \left| 2\frac{\partial n_1}{\partial T} - \frac{\lambda_1}{\lambda_2}\frac{\partial n_2}{\partial T} \right| \quad (5)$$

Here, ΔT indicates the temperature bandwidth, L is the crystal length, $\lambda_i(i=1, 2)$ represents wavelength of fundamental and harmonic wave, respectively, $n_i(i=1, 2)$ is the refractive indexes of crystal, and $\partial n_i / \partial T$ is the temperature coefficient of fundamental and harmonic wave component in the crystal. For a certain nonlinear interaction process, when the phase-matching condition is satisfied, Equation (5) shows that the temperature bandwidth is inversely proportional to the crystal length, coincided with Fig. 2.

Since the EDFA is directly fused with the fiber of waveguides, the coupling and transmission loss of the fundamental and the frequency doubling laser in the waveguides can be ignored. The frequency doubling conversion efficiency can be defined as the ratio of the frequency doubling laser power separated by DM to the power of input fundamental before the waveguide. Fig. 3 shows the change of 780.2nm frequency doubling laser power with the power of 1560.4nm fundamental. The dots in Fig. 3 are experimental results, and the solid lines represent the theoretical curves fitting. In order to maximize the conversion efficiency of frequency doubling, the phase matching temperature should be optimized at each set of fundamental. For trapping laser path (Fig. 3(a)), the maximum power of frequency doubled laser at 780.2nm is 1.1W when the fundamental laser power is 3W, indicating the single-pass optical conversion efficiency is 36.7%. For Raman laser path (Fig. 3(b)), when the fundamental laser power is 5W, the maximum power at 780.2nm is 1.9W, indicating the optical conversion efficiency is 38.0%.

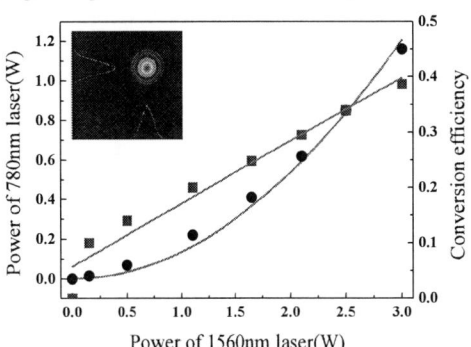

Power of 1560nm laser(W)

(a) Trapping laser

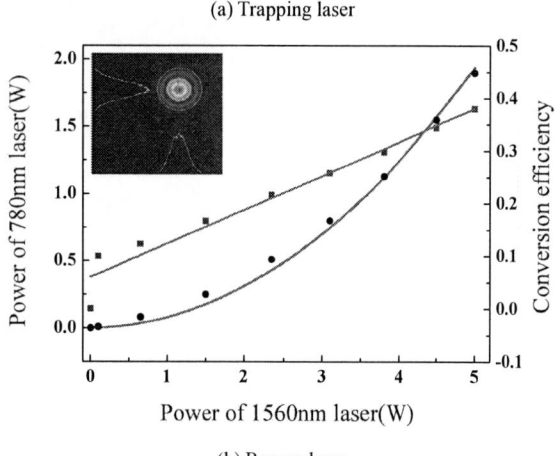

(b) Raman laser

Figure 3. Doubling output power verse power of fundamental laser. The black dots are the power of 780nm laser, and the blue dots are the conversion efficiency.

Fig. 3 shows the output laser spot of the frequency doubled laser at 780.2nm and the distribution curve of the laser intensity in two vertical directions of X and Y. Figure 3(a) shows the measurement result of the trapping path laser spot. When the output power is 1.1W, the laser spot diameter is 0.95mm. Fig. 3(b) shows the measurement result of the laser spot diameter in Raman path, it is 0.93mm with the laser output power of 1.9W. The M^2 factors of two laser beams were measured simultaneously, 1.03 and 1.05, respectively. Obviously, the optical intensity distribution in both X and Y directions are near-ideal Gaussian distribution, proving that the 780.2nm laser generated by frequency doubling is near-ideal Gaussian mode of TEM_{00}.

C. Frequency and Power Stabilization

The seed laser is frequency locked by MTS, optical paths schematic diagram is shown in Fig. 4. By double-passing through AOM2, the frequency shift of laser is 200MHz, then the MTS is set up based on 780.2nm laser up to 2mW. The pumping laser is modulated by radio frequency (RF) signal at 12.5MHz. The laser frequency is locked to transition $5S_{1/2}$, $F=2 \rightarrow 5P_{3/2}$, $F'=3$, as shown in Fig. 5.

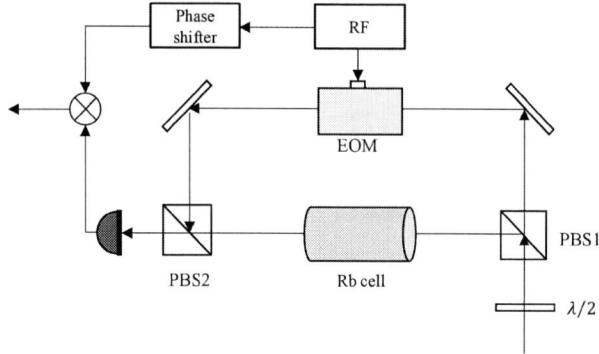

Figure 4. The schematic diagram of MTS

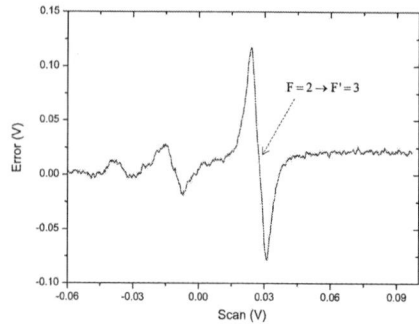

Figure 5. MTS for ^{87}Rb $5S_{1/2}$, $F=2 \rightarrow 5P_{3/2}$, $F'=1,2,3$ transition

Fig. 6 shows the change of error signal before and after frequency locking. The fluctuation of error signal is 50mV in free-running and 1mV after frequency-locking, the laser frequency stability improve from 15MHz to 0.3MHz correspondingly. The contribution to the measuring error of gravimeter is less than 1μGal.

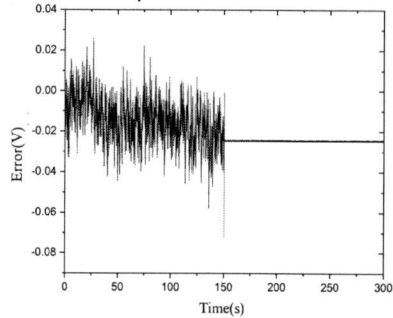

Figure 6. The error of frequency locking

As the polarization fluctuation of output laser at 780.2nm is measured to be higher than 10%, optical power stabilization is necessary to laser system. As shown in Fig. 7, a power stabilization loop is set up based on AOM. The diffractive laser of AOM is collected by a photodetector(PD₃) after beam splitter, then the signal fed back to the variable-gain amplifier (VGA), which is used to tune the driving power of RF signal to AOM.

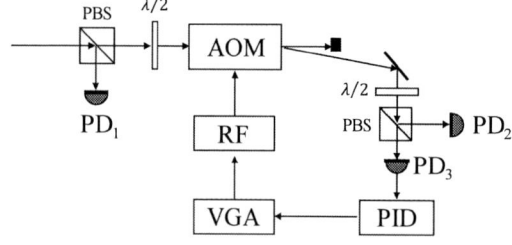

Figure 7. The schematic diagram of optical power stabilization loop

By monitoring the output of PD₂, the stabilization of diffractive laser is measured. As shown in Fig. 8, the optical power fluctuation is 0.34%@20 hours after optical power stabilization, far better than 13.2%@20 hours before power stabilization.

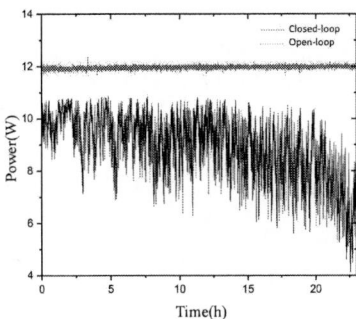

Figure 8 The power stability of closed-loop

IV. EXPERIMENTAL VERIFICATION

In the interference process, the cooling laser and repumping laser component are generated by frequency modulating on EOM1 with 6.583 GHz. By jumping the frequency modulated on AOM2 after cooling atoms ~490ms, we increase the frequency detuning of cooling laser to 100 MHz, reduce the atomic temperature to 7μK. By acting two microwave π-pulses and one Raman π-pulse, the atoms are populated to state $5S_{1/2}$, F=1, m_F=0. The probe and blow laser are realized by jumping the frequency of modulated on AOM3. The Raman laser is red detuning 576.5MHz to the relative state $5P_{3/2}$, F'=1, and it produces the other Raman laser component by frequency modulating on EOM2 with 6.834GHz. A π/2-π-π/2 pulses series acts to falling freely atoms. Finally, it's time to probe the state of atoms. The entire measurement period is 800ms. Fig. 9 shows the gravimeter interference fringes derived from the scanning of frequency chirp rate to Raman laser. The interference pulses interval T is 70ms. The dots are the mean of measurements and the solid lines are the theoretical fitted curve. Generally, the gravimeter phase $\Phi = k_{eff} \cdot g \cdot T^2$, where the k_{eff} presents the effective wave vector to Raman laser. The measured 120-point fringe phase-uncertainty $\Delta\Phi$ is 22mrad, indicating to a noise measurement resolution of $610\mu Gal/\sqrt{Hz}$.

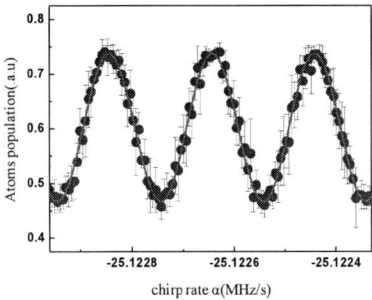

Figure 9. Interferometry fringe at T=70ms

Compared to the results of other research groups, there is still a certain gap, which is limited by poor vibration conditions, where is next to street and the subway goes frequently.

V. CONCLUSION

An all-fiber single laser system for cold atom interferometer is built, and achieved a continuous laser output with excellent performance. When the 1560.4nm EDFA output power is 3 W and 5 W, respectively, up to 1.1W and 1.9W of 780.2nm laser output can be obtained. In addition, the optical frequency and power stabilization loop are studied with good results. The frequency-uncertainty error is less than 1μGal. The noise resolution is $610\mu Gal/\sqrt{Hz}$ based on the laser system.

Moreover, there are also some aspects of this laser system need to be optimized and improved. The accurate temperature control of whole laser system can further improve the anti-environmental disturbance; and the precise measurement and evaluation of phase shift introduced by modulated sideband are required.

REFERENCES

[1] Hauth M, Freier C, Schkolnik V, et al. First gravity measurements using the mobile atom interferometer GAIN. Applied Physics B, 2013, 113(1): 49-55.

[2] Freier C, Hauth M, Schkolnik V, et al. Mobile quantum gravity sensor with unprecedented stability. Journal of physics: cinference series, 2016, 723(1): 012050.

[3] Wu X, Pagel Z, Malek B S, et al. Gravity surveys using a mobile atom interferometer. Science Advances, 2019, 5(9): 1-9.

[4] Y. Bidel, N. Zahzam, C. Blanchard, et al. Absolute marine gravimetry with matter-wave interferometry. Nat. Commun, 2018 9(1): 627.

[5] B.Stray, A.Lamb, A. Kaushik, et al. Quantun sensing for gravity cartograph. Nature, 2022 602(7898):590-594.

[6] Wang Y, Zhong J, Song H, et al. Location dependent Raman transition in gravity-gradient measurements using dual atom interferometers. Physics Review A, 2017, 95(5): 053612.

[7] Dutta I, Savoie D, Fang B, et al. Continuous cold-atom inertial sensor with 1 nrad/sec rotation stability. Physics Review Letters, 2016, 116(18): 183003.

[8] D. Savoie, M. Altorio, B. Fang, et al. Interleaved atom interferometry for high-sensitivity inertial measurement. Science Advance, 2018 4(12).

[9] Rosi G, Amico G D, Cacciapuoti L, et al. Quantum test of the equivalence principle for atoms in coherent superposition of internal energy states. Nature Communication, 2017, 8(1): 1-6.

[10] Parker R H, Yu C, Zhong W, et al. Measurement of the fine-structure constant as a test of the standard model. Science, 2018, 360(6385): 191-195.

[11] Sabulsky D O, Junca J, Lefevre G, et al. A fibered laser system for the MIGA large scale atom interferometer. Scientific Reports, 2020, 10(1): 1-14.

[12] Fang J, Hu J, Chen X, et al. Realization of a compact one-seed laser system for atom interferometer-based gravimeters. Optics Express, 2018, 26(2): 1586-1596.

[13] Wang Q, Wang Z, Fu Z, et al. A compact laser system for the cold atom gravimeter. Optics Communication, 2016, 358(1): 82-87.

[14] Song H. Precise measurement of gravity gradient based on the cold atom interferometer. Wuhan: Huazhong University of science & technology, 2017.

[15] Luo Q, Zhang H, Zhang K, et al. A compact laser system for a portable atom interferometry gravimeter. Review of Scientific Instruments, 2019, 90(4): 043104.

[16] Theron F, Carraz O, Renon, et al. Narrow linewidth single laser source system for onboard atom interferometry. Applied Physics B, 2015, 118(1): 1-5.

[17] Wang Jieying, Kong Delong, et al. Method of single-laser system based on frequency doubling for cold atom gravimeter. Journal of Chinese Inertial Technology, 2021, 29, 6.

[18] Jiang H, Li G, Xu X. Highly efficient single-pass second harmonic generation in a periodically poled $MgO:LiNbO_3$ waveguide pumped by a fiber laser at 1111.6nm. Optics Express, 2009, 17(18): 16073-16080.

Proceedings of the 7th Optoelectronics Global Conference (OGC 2022)

Intensity Compensation of Echo Pulses for Fiber Interferometers Based on uwFBG Reflectors

Yandong Pang, Junbin Huang, Hongcan Gu, Su Wu, Zhiqiang Zhang*

Department of Weaponry Engineering, Naval Engineering University, Wuhan 430032, China
*Corresponding author e-mail: 732508739@qq.com

Abstract—**We report on current theoretical and experimental results of echo pulses based on ultra-weak Fiber Bragg Gratings (uwFBGs), using intensity compensation to solve the signal distortion caused by light pulse delay. Cubic Spline Interpolation is applied, leading to better robustness for the interference intensity. Then the linear compensation value is obtained by comparing the peak intensity in order to correct attenuation in three-way optical path. We report improvements over the conventional demodulation algorithm based on 3×3 coupler with circulator. The experimental results have shown that, when the intensity reduction is existed due to pulse delay, the hydroacoustic field could be reconstructed with high fidelity by using the proposed method. Leading to much lower noise density level for the interference pulse intensity, more than 7 dB when the reference signal is 1rad/√Hz in 5~50 Hz, and the SNR is improved approximately 10 dB@10 Hz. The proposed method exhibits a particularly applicability in passive optical path, which is of great importance for increasing the practicality of the system in complicated environments.**

Keywords-Fiber optic Hydrophone; uwFBG; Fiber Interferometers

I. INTRODUCTION

Fiber-optic sensors have been widely studied in recent years [1], for advantages of high sensitivity, anti-electromagnetic interference, and large-scale reuse. Among various types of fiber-optic sensors, FBG-based sensor networks can be configured differently according to practical scenarios and multiplexing of FBGs has been demonstrated [2], such as vibration sensing based on simultaneous vector demodulation [3], hot spot detection system [4], OTDR (Optical Time Domain Reflectometry) sensing network [5], and DAS (Distributed Acoustic Sensing) system with long sensing distance [6]. For underwater environment applications, large-scale reuse and excellent resolution are the keys of fiber-optic hydrophones, especially in surface towed array sonar system [7], because of increasing the sensor number of arrays can improve the detection range, and good spatial resolution can further enhance the detection accuracy.

In order to achieve high spatial resolution of quasi-distributed time division multiplexing (TDM) array, we generally reduce the grating interval and improve the system sampling rate. At this point, the pulse width should be reduced as much as possible to ensure that the adjacent light pulses do not overlap. Besides that, the reflectivity of uwFBG is between -40 dB and -50 dB [8], so detecting the weak light pulse with high precision and selecting a stable demodulation algorithm of interference signal have an important influence on the system performance. Previous study has shown that [9],

algorithms based on a 3×3 coupler are proved to be able to demodulate ultra-weak Fiber Bragg Gratings (uwFBGs) signals due to simple architecture and immunity for signal fading. But when we use a 3×3 coupler to obtain the interference signal, the insertion loss of the circulator will reduce the signal intensity, on the other side, the arrival time for the three-way pulse to the detector will not be consistent, owing to the optical path extension. Especially when the detection accuracy of fiber-optic delay is not enough or has difficulty to measure, so fluctuations in pulse amplitude will be unpredictable, the combined effect for the two aspects will further increase the asymmetric results of the three-way signal. Experiments on TDM interference system based on uwFBG have been proved that [3], when the uwFBG spacing is about 5 m, the pulse width should be kept under 20 ns to ensure that the adjacent pulses do not overlap. For the same reason, high sampling rate is also required, though it is up to 250 MSa/s, only about 5 sampling points form a single pulse. Therefore, under the circumstances for the optical pulse drift, it is very likely to result in the peak position offset, further, the intensity of interference signal is greatly reduced. Based on this question, Ding [10] attempted to demodulate only through Hilbert transform to reduce the influence of the circulator, however, the experimental results show a low signal noise ratio (SNR). In order to eliminate the intensity noise, Zhang [11] reported an improved phase generated carrier (PGC) demodulation algorithm to eliminate the influence of light intensity disturbance (LID), then they further proposed an asymmetric division demodulation algorithm based on fundamental frequency mixing [12], the SNR and distortion ratio of the sensor achieves a gain of 7.77 dB over the traditional algorithm. However, the above methods only apply to PGC and its improved algorithms. For FBG interference scheme with 3×3 coupler, the non-interferential pulse is used as reference signal to suppress signal distortion caused by optical intensity [13], the total harmonic distortion can be decreased by 30 dB, but it cannot suppress the change of interference pulse intensity caused by optical pulse drift. It is worth noting that light intensity based on the 3×3 coupler can form elliptic curve [14], inspired by it, Qu [15] proposed an ellipse fitting algorithm (EFA) to compensate the nonlinear distortion, however, when the phase change less then π due to the mildly intrusive, the interference signal cannot form a complete ellipse, which leads to the error of parameter estimation. To solve this problem, Shi [16] proposed a phase-shifted demodulation technique with a 3×3 coupler and EFA for the interrogation of interferometric sensors, and experimental result shows a high accuracy and stability for measuring small phase signals. Although this method can

978-1-6654-8699-6/22 $31.00 © 2022 IEEE

solve many problems resulting from parameter inconsistency perfectly, it has high complexity. Focus on TDM systems, when the sensor spacing magnitude is meter, the sampling rate must reach hundreds of megahertz to get more accurate interference signals [17], but for high-speed acquisition system, EFA method undoubtedly increases the burden of data processing.

In this paper, we propose an intensity compensation method of fiber interferometers array based on uwFBG reflectors. At first, in order to locate the trajectory for each pulse peak more accurately, cubic spline interpolation is used to resampling the distorted pulse, then based on the peak coordinates of the normal two-way pulses, the intensity compensation of the signal offset by the third pulse is made. This scheme can solve the amplitude attenuation of interference signal caused by introducing the circulator fundamentally. So that, the demodulation stability of interference signal is improved efficiently.

Having these ideas in mind, in the following sections, firstly, we discuss the source of interference light intensity distortion caused by circulator and present the mathematical background of the demodulation technique used in our scheme in section 2. Section 3 is dedicated to the experimental setup used to demonstrate the technique. In section 4, results of time-domain and frequency-domain obtained by the proposed method, the SNR enhancement results at different frequencies are calculated.

II. PRINCIPLE

A. Interference Scheme

Fig. 1 shows the structure of uwFBG-based interferometric fiber-optic system. The continuous light emitted from a narrow linewidth laser is modulated into pulse light by an acousto-optic modulator (AOM), whereafter, it is injected into circulator-1 (CIR-1) after amplification by erbium-doped fiber amplifier (EDFA), the pulse enters the uwFBGs array, and then the pulse train reflected by the uwFBGs are transmitted to CIR-2. The arm length difference of the nonequilibrium interferometer constructed by Faraday rotating mirror (FRM) is the same with the spacing of uwFBGs. Therefore, the optical path compensation can be realized in the interferometer by the successive reflected pulses, thus complete pulse interference. Finally, the interference pulse is converted by photoelectric detectors (PDs) into FPGA module for data acquisition and processing.

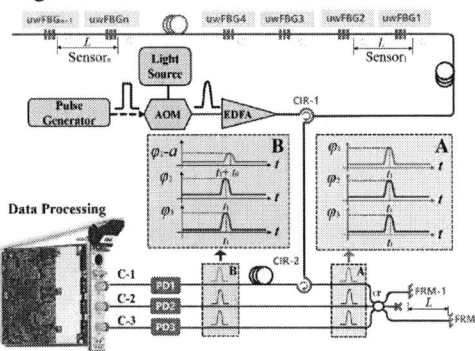

Figure 1. Structure of uwFBG-based interferometric fiber-optic system.

In practice, even if the driving signal is a square pulse, the light pulse still has nonlinear rising edge and falling edge after modulation by AOM. By this time, the theoretical pulse will become the Gaussian signal analogously [16], so the trajectory of pulse peak for interference signal reflects the change of interference light intensity. In order to satisfy the condition of 3×3 demodulation algorithm, we usually introduce CIR-2 to transmit optical signal of C-1.

Concrete analysis, the changes of light intensity are shown in Part A and Part B in Fig. 1. The pulse train from 3×3 coupler is equiamplitude and synchronization during the same time, but after passing through CIR-2, the pulse of C-1 has decrease for a in intensity, and delay of t_0. At the same time, lower sampling rates will lead to peak fluctuation, so that the interference intensity changes, and finally results in the difference of interference signal for output channels. The three-way interference intensity can be represented as

$$\begin{cases} I_{C-1} = b\left[\left(E_m^2 + E_k^2 - a + 2E_mE_k \cos\left(\varphi_s(t) + \varphi_0\right)\right)\right] \\ I_{C-2} = c\left[E_m^2 + E_k^2 + 2E_mE_k \cos\left(\varphi_s(t) + \varphi_0 + 2\pi/3\right)\right] \\ I_{C-3} = d\left[E_m^2 + E_k^2 + 2E_mE_k \cos\left(\varphi_s(t) + \varphi_0 + 4\pi/3\right)\right] \end{cases}, \quad (1)$$

where E_m and E_k respectively express the amplitudes of the two fields, and $\varphi_s(t)$ is the measured phase change due to the external impact, φ_0 being constants denoting initial phases, a represents the DC offset caused by light attenuation of CIR-2. b, c, and d respectively express the variation coefficient of due to pulse delay for C-1, C-2, and C-3. On the other hand, when the sampling rate of TDM system is insufficient, the trajectory of pulse peak fluctuates greatly, which leads to the loud interference intensity noise.

B. Revised Scheme

The signal compensation scheme is shown in Fig. 2. First, we adjust the DC bias signal, second, the intensity of I_{C-1}, I_{C-2}, and I_{C-3} is also correlated with the trajectory fluctuation of the interference pulse, the low sampling density is the key leading to the fluctuation obviously. Considering that the pulse distribution does not satisfy the specific mathematical law, we refer to the appropriate methods for fitting nonlinear curves [18], the pulses are resampled based on cubic spline interpolation algorithm to obtain the stable interference intensity. For the C-1 interference pulse composed of discrete signals, the cubic spline $y_i(\delta)$ is obtained by the form

$$y_i(\delta) = \sum_0^i N_{i,k}(\delta)y_k + \sum_0^i M_{i,k}(\delta)y'_{k+1}, \quad (2)$$

where δ is an auxiliary variable, $k=0,1,2\cdots i-1$, and in each discrete time interval $[t_k, t_{k+1}]$, for $1 \leq k \leq i-1$, the shape functions $N_{i,k}(\delta)$ and $M_{i,k}(\delta)$ will be given by

$$N_{i,k}(\delta) = \begin{cases} \left(1 - 2\dfrac{\delta - t_k}{t_k - t_{k+1}}\right)\left(\dfrac{\delta - t_{k+1}}{t_k - t_{k+1}}\right)^2, t_{k-1} \leq \delta \leq t_k \\ \left(1 - 2\dfrac{\delta - t_{k+1}}{t_{k+1} - t_k}\right)\left(\dfrac{\delta - t_k}{t_{k+1} - t_k}\right)^2, t_k \leq \delta \leq t_{k+1} \end{cases}, \quad (3)$$

$$M_{i,k}(\delta) = \begin{cases} (\delta - t_k)\left(\dfrac{\delta - t_{k+1}}{t_k - t_{k+1}}\right)^2, & t_{k-1} \le \delta \le t_k \\[2ex] (\delta - t_{k+1})\left(\dfrac{\delta - t_k}{t_{k+1} - t_k}\right)^2, & t_k \le \delta \le t_{k+1} \end{cases}, \quad (4)$$

in case of $\delta < k\text{-}1$ or $\delta > k\text{+}1$, $N_{i,k}(\delta) = 0$, $M_{i,k}(\delta) = 0$. The calculation results of C-2 for $y_j(\delta)$ and C-3 for $y_h(\delta)$ can be also obtained by using this method. Since the transformation of pulse waveform is caused by interference changes, $I_{C\text{-}z} = 0$ ($z = 1$, 2, 3) is likely to occur when AC signal reaches the lowest point. But we know by the formula, $I_{C\text{-}2} + I_{C\text{-}3} > 0$, therefore, the above relationship can accurately find the corresponding moment δ_0 of C-2, C-3, so the C-1 signal could be obtained at time δ_0. $y_j(\delta)$, $y_h(\delta)$, $y_i(\delta)$ for C-2, C-3 and C-1 can be expressed as

$$\begin{cases} \max\left(y_j\right) = y_j(\delta)\big|_{\delta = \delta_0} \\[1ex] \max\left(y_h\right) = y_h(\delta)\big|_{\delta = \delta_0} \\[1ex] y_i(\delta)\big|_{\delta = \delta_0} = \mathrm{fix}(y_i) \end{cases}. \quad (5)$$

We obtain the maximum value of the interpolation curve by using the peak-seeking algorithm, where $\max(y_j)$ and $\max(y_h)$ is the maximum of $y_j(\delta)$, $y_h(\delta)$ in the moment of $\delta = \delta_0$, and so that we can obtain $y_i(\delta) = \mathrm{fix}(y_i)$ at this point. When the amplitude of phase change caused by external physical quantity exceeds $\pi/2$, the numerical relation is used to compensate the intensity. The three-way corrected signal can be represented as

$$\begin{cases} I'_{C\text{-}1} = \dfrac{\max(y_j)}{\mathrm{fix}(y_i)} y_i(\delta)\bigg|_{\delta = \delta_0 + nT} \\[2ex] I'_{C\text{-}2} = y_j(\delta)\big|_{\delta = \delta_0 + nT} \\[1ex] I'_{C\text{-}3} = y_h(\delta)\big|_{\delta = \delta_0 + nT} \end{cases}, \quad (6)$$

where T is the period of a modulated pulse signal for AOM. After the above pretreatment, the problem of interference intensity variation caused by CIR-2 can be solved. Finally, we use the arctangent demodulation algorithm [9], the correct original signal will be obtained.

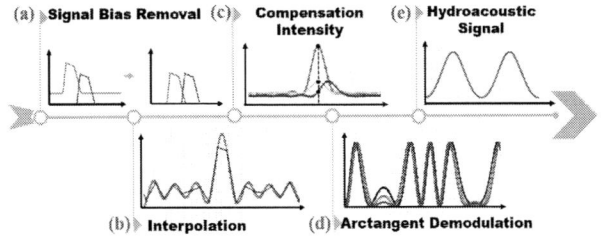

Figure 2. Signal compensation scheme.

III. EXPERIMENT SETUP AND RESULTS

The experimental principle of matched interferometric fiber-optic hydrophone system is shown in Fig. 3. To satisfy the interference conditions in light source system, narrow linewidth laser (DFB-M-1550-150-F-10-09MPF-FC/APC)

has a bandwidth of 3 kHz, and a central wavelength of 1550 nm. AOM (T-M200-0.1C2J-3-F2S) generates a pulse period of 2 kHz, pulse width for 20 ns. The interference output of the 3×3 coupler is composed of three-way parts, C-2 and C-3 routes directly enter PD, whereas route C-1 reaches PD through port 3 of CIR-2. The detection bandwidth of PDs (KG-200M-APR) is 200 MHz. In order to distinguish each narrow-pulse of sensor, the sampling rate of the data acquisition card (NI PXIe5170R) is 250 MSa/s. Finally, cubic spline interpolation and compensation are used to complete the pretreatment to achieve stable demodulation.

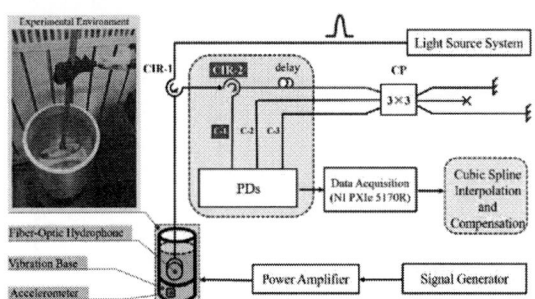

Figure 3. Experimental principle of matched interferometric fiber-optic hydrophone system.

We use vibratory fluid column method to simulate the propagation rules of hydroacoustic wave. The sound pressure field in the liquid column is generated by an excitation table with a linear power amplifier (VT500), and the voltage sensitivity of accelerometer is 5.76 pC/m·s⁻². Limited by output power, the exciter can produce large amplitude only at low frequency, so in order to satisfy the requirement of generating phase variation that greater than π amplitude, we select hydroacoustic signals of low-frequency band 5-50 Hz for experiment. At the same time, for obtaining the correct signal of standing wave, the inner diameter of the round tube is set as 7.50 cm, and liquid column height H is 18.75 cm. The optic-fiber hydrophone in the experimental system uses uwFBGs array with a central wavelength of 1550 nm and a reflectivity between -40 and -50 dB. The spacing between adjacent gratings is 5 m.

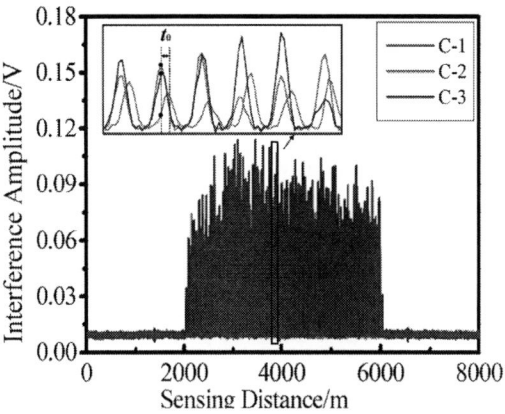

Figure 4. Three-way signal of uwFBG array with 800 sensors.

978-1-6654-8699-6/22 $31.00 © 2022 IEEE

Fig. 4 shows the three-way pulse signal of uwFBG TDM array with 800 sensors, and its sensing distance is 4 km. By amplifying the partial pulse signal, comparing with C-2 and C-3, the C-1 pulse train has overall offset of t_0, so the amplitude of C-1 interference signal indexed at the same position will be greatly reduced, which verifies the theoretical analysis mentioned above.

Previous studies have shown that we can use EFA to compensate the nonlinear distortion caused by LID [19], in the case of light intensity fluctuates greatly, the edge of the ellipse is more discrete. For verifying the effectiveness of cubic spline interpolation, we choose the same light intensity, so that select interference signals of C-2 and C-3 to obtain the Lissajous figure.

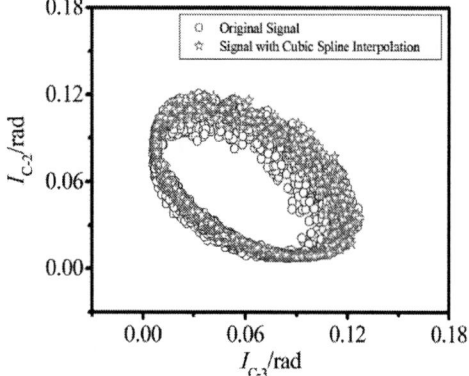

Figure 5. Lissajous figure of two interference signals for I_{C-2} and I_{C-3}.

In Fig. 5, The blue symbol represents the formation of original signal, and the red symbol represents the result of cubic spline interpolation. The comparison results show that, the cubic spline interpolation lattice is more concentrated than the untreated lattice, so the intensity noise is lower, indicating that when the interference pulse sampling rate is low, the cubic spline interpolation algorithm can effectively increase the stability for trajectory of interference signal. And it can also explain that when the delay distance caused by the CIR-2 is smaller than the sampling resolution of the acquisition card, we cannot accurately find the position of the peak by estimating the delay, which is an important reason of intensity noise.

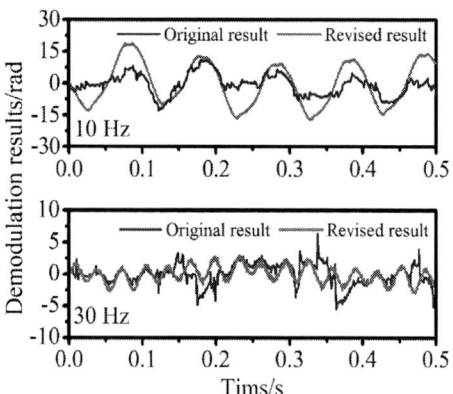

Figure 6. Comparison of time-domain demodulation results in 10 Hz and 30Hz.

With cubic spline interpolation and linear compensation, the demodulation results are analyzed in experiment. Fig.6 shows the comparison of time-domain demodulation results for 10 Hz and 30 Hz. Due to the influence of intensity noise, the time domain signals produce obvious distortion and the demodulation amplitudes are also reduced. The results of algorithm application are improved effectively by using the proposed method.

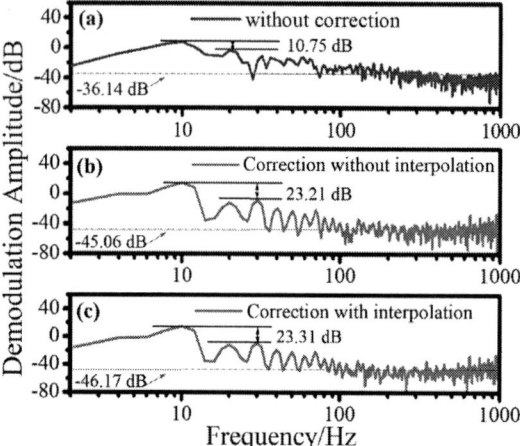

Figure 7. Comparison of PSD with different methods.

In order to analyze the results of the proposed algorithm qualitatively, we use three methods to obtain the 10 Hz demodulation signal and calculate the power spectral density (PSD). In Fig. 7, (a) represents the results obtained without correction, (b) represents the results obtained by direct compensation without interpolation, (c) represents the results obtained by linear compensation after interpolation, the SNR is 10.75 dB, 23.21 dB, 23.31 dB, respectively. The signal without any processing has a low SNR, and attenuation of interference signal caused by the pulse delay is the main factor for signal distortion. After linear compensation, the result is significantly improved. On the other hand, we analyze the average value for PSD of noise when the reference signal is 1 rad/√Hz, which are severally -36.14 dB, -45.06 dB and -46.17 dB. It can be reduced by 1.11 dB after cubic spline interpolation, The linear compensation can further increase the noise level by about 8.92 dB. The results are also consistent with the results of the Lissajous figure shown in Fig. 4, which prove that the proposed algorithm for dealing with the pulse delay problem is very effective.

IV. CONCLUSION

In this paper, for the intensity noise caused by the light delay introduced of the circulator in 3×3 optical path, a linear compensation method based on cubic spline interpolation fitting is proposed for echo pulses. It provides a significant over direct demodulation method. The proposed method can reduce the background noise of low-frequency band by more than 7 dB and improve the SNR by up to 10 dB. Compared with other compensation algorithms, the proposed algorithm is more efficient and has less computation. However, the disadvantage is that the intensity compensation method is only effective in the monopulse period, so when sensor spacing of 5 m with 20 ns optical pulse, it can be immune to the optical path delay within about 2 m caused by the circulator. As a result, the proposed pretreatment method is greatly suitable for practical applications in uwFBGs array based on dense TDM systems.

ACKNOWLEDGEMENT

This work was supported by National Natural Science Foundation of China (11774432; 51509252; 41874091; 41774021). We also thank for my girlfriend Xiaoshu Quan with school of architecture and urban planning in Guizhou university for her suggestions on the color matching of the pictures in this paper.

REFERENCES

[1] S. W. Tyler, S. A. Burak, J. P. Mcnamara, A. Lamontagne, J. S. Selker, and J. Dozier, "Spatially distributed temperatures at the base of two mountain snowpacks measured with fiber-optic sensors,". J. Glaciol. 54(187), 2008, pp. 673.

[2] H. Y. Guo, J. G. Tang, X. F. Li, Y. Zheng, H. Yu, and H. H. Yu, "On-line writing identical and weak fiber Bragg grating arrays," Chin. Opt. Lett. 11(3), 2013, pp. 030602.

[3] Y. Li, L. Qian, C. M. Zhou, D. Fan, Q. N. Xu, Y. D. Pang, X. Chen, and J. G. Tang, "Multiple-Octave-Spanning vibration sensing based on simultaneous vector demodulation of 499 Fizeau interference signals from identical ultra-weak fiber Bragg gratings over 2.5 km," Sensors 18(1), 2018, pp. 210

[4] J. Q. Wang, Z. Y. Li, X. L. Fu, J. Zhan, H. H. Wang, and D. S. Jiang, "High-sensing-resolution distributed hot spot detection system implemented by a relaxed pulsewidth," Opt. Express 28, 2020, pp. 16045.

[5] C. Wang, Y. Shang, X. H. Liu, C. Wang, H. H. Yu, D. S. Jiang, and G. D. Peng, "Distributed OTDR-interferometric sensing network with identical ultra-weak fiber Bragg gratings," Opt. express 23(22), 2015, pp. 29038.

[6] Y. Y. Shan, W. B. Ji, X. Y. Dong, M. Zabihi, Q. Wang, Y. X. Zhang, and X. P. Zhang, "An Enhanced Distributed Acoustic Sensor Based on UWFBG and Self-Heterodyne Detection," J. Lightw. Technol. 37(11), 2019, pp. 2700.

[7] H. Qayyum, and M. Ashraf, "Performance Comparison of Direction-of-Arrival Estimation Algorithms for Towed Array Sonar System,"

[8] C. M. Zhou, Y. D. Pang, L. Qian, X. Chen, Q. N. Xu, C. G. Zhao, H. R. Zhang, Z. W. Tu, J. B. Huang, H. C. Gu, and D. Fan, "Demodulation of a Hydroacoustic Sensor Array of Fiber Interferometers Based on Ultra-Weak Fiber Bragg Grating Reflectors Using a Self-Referencing Signal," J. Lightw. Technol. 37(11), 2018, pp. 2568.

[9] Y. D. Pang, C. M. Zhou, D. Fan, J. B. Huang, H. C. Gu, Q. N. Xu, D. S. Jiang, and L. X. Wang, "Research of very low-frequency (VLF) hydroacoustic detection based on Fizeau interferometry," Optical Fiber Sensors, 2018, pp. TuE37.

[10] P. Ding, W. Liu, H. C. Gu, Y. Y. Wang, J. Wu, J. B. Huang, and J. S. Tang, "Demodulation of a weak fiber Bragg grating array using a fiber delay line," Appl. Opt. 59(8), 2020, pp. 2325.

[11] A. L. Zhang, and S. Zhang, "High stability fiber-optics sensors with an improved PGC demodulation algorithm," IEEE Sens. J. 16(21), 2016, pp. 7681.

[12] A. Zhang, and D. Li, "Interferometric sensor with a PGC-AD-DSM demodulation algorithm insensitive to phase modulation depth and light intensity disturbance," Appl. Opt. 57(27), 2018, pp. 7950.

[13] H. Y. Zhang, D. N. Wang, Q. P. Shi, C. D. Tian, L. W. Wang, M. Zhang, and Y. B. Liao, "Optical intensity compensating method for time division multiplexing of fiber-optic hydrophone using 3×3 coupler," Chin. J. Lasers 38(11), 2011, pp. 1105006.

[14] N. Chang, M. Zhang, Y. Zhu, C. X. Hu, S. Q. Ding, and Z. Jia, "Sinusoidal phase-modulating interferometer with ellipse fitting and a correction method," Appl. Opt. 56(13), 2017, pp. 3895.

[15] Z. Y. Qu, S. Guo, C. B. Hou, J. Yang, and L. B. Yuan, "Real-time self-calibration PGC-Arctan demodulation algorithm in fiber-optic interferometric sensors," Opt. Express 27(16), 2019, pp. 23593.

[16] J. H. Shi, D. Guang, S. L. Li, X. Q. Wu, G. S. Zhang, C. Zuo, G. Zhang, R. Wang, Q. Ge, and B. L. Yu, "Phase-shifted demodulation technique with additional modulation based on a 3× 3 coupler and EFA for the interrogation of fiber-optic interferometric sensors," Opt. Lett. 46(12), 2021, pp. 2900.

[17] Y. Muanenda, S. Faralli, C. J. Oton, C. Cheng, M. H. Yang, and F. D. Pasquale, "Dynamic phase extraction in high-SNR DAS based on UWFBGs without phase unwrapping using scalable homodyne demodulation in direct detection," Opt. express 27(8), 2019, pp. 10644.

[18] S. Yaghoobi, B. P. Moghaddam, and K. Ivaz, "An efficient cubic spline approximation for variable-order fractional differential equations with time delay," Nonlinear Dyn. 87(2), 2017, pp. 815.

[19] C. Ni, M. Zhang, Y. Zhu, C. X. Hu, S. Q. Ding, and Z. Jia, "Sinusoidal phase-modulating interferometer with ellipse fitting and a correction method," Appl. Opt. 56(13), 2017, pp. 3895.

Proceedings of the 7th Optoelectronics Global Conference (OGC 2022)

Strain and Temperature Discrimination by Fourier Analyzing Transmission Spectrum of an In-fiber Mach-Zehnder Interferometer

Shiying Xiao, Beilei Wu*, Zixiao Wang, Youchao Jiang

Institute of Lightwave Technology
Beijing Jiaotong University
Beijing, China
e-mail: syxiao@bjtu.edu.cn, wubeilei@bjtu.edu.cn,
zixiaow@bjtu.edu.cn, jiangyc@bjtu.edu.cn

Chunran Sun

System Integration Department
China North Industries Corp.
Beijing, China
e-mail: sunchunran92@163.com

Abstract—An approach for strain and temperature discrimination is proposed and demonstrated by Fourier analyzing the transmission spectrum of a Mach-Zehnder interferometer (MZI) which consists of two peanut tapers and a polarization maintaining fiber (PMF). The high resolution together with the ease-of-fabrication and robustness of the sensor allow the MZI sensor combining with Fourier analysis a competitive approach for simultaneous strain and temperature sensing applications.

Keywords-fourier analysis; Mach-Zehnder interferometer; Peanut Taper; Simultaneous Measurement

I. INTRODUCTION

In-fiber Mach-Zehnder interferometers have been studied in recent years for various sensing applications with their advantages of flexibility, compact size and immunity to electromagnetic interference. A typical in-fiber MZI sensor incorporates a mode splitter, a functional fiber and a mode combiner. The mode splitter/combiner and the functional fiber can be very diverse. Both common fibers such as single mode fibers [1, 2] and specialty fibers such as photonic crystal fibers [3] can be utilized as the functional fiber. As to the mode splitter and the mode combiner, structures of core-offset splicing [4], core mismatch splicing [5], down-tapers [6], up-tapers [7] and long period gratings [8] are used. Among these structures, up-tapers have the advantages of robustness, ease of fabrication and compatibility with fibers that have either matched or mis-matched cores. In-fiber MZI sensors based on up-tapers are employed for the sensing of multiple physical quantities, such as strain [9], temperature [10], curvature [3], humidity [5], refractive index [7] and so on.

As with other optical fiber based sensors, in-fiber MZI sensors suffer the susceptibility to two or more physical quantities and therefore simultaneous sensing of dual or more quantities attracts much attention [11-18]. For instance, many MZI sensors are proposed for simultaneous measurement of strain and temperature [1, 11-15]. Most of these sensors utilize traditional wavelength tracking method to track the wavelength shift of two or more dips in the transmission spectrum of the sensor to monitor the strain and temperature variations. However, the transmission spectrum

of an in-fiber MZI sensor is the superstition of interference between core mode and multiple cladding modes. Though most of the cladding modes are weak modes, they can induce different wavelength shifts to the overall spectrum. Therefore, the wavelength shift of a single spectrum dip is affected by all cladding modes, which influences the measurement accuracy of the sensor [19]. Fourier analysis is a method which can solve this problem. It can help extract the cosine spectrum components contributed by different cladding modes from the overall transmission spectrum. By monitoring the extracted spectrum components, the sensing information carried by each cladding mode can be obtained and used for dual even multiple parameter sensing [20].

In this paper, an in-fiber MZI sensor combined with a Fourier analysis method is proposed for the discriminative sensing of temperature and strain. The MZI sensor incorporates two peanut-shaped up-tapers and a section of panda-type PMF. The two peanut tapers respectively serve as the mode splitter and the mode combiner whereas the PMF in-between serves as the functional fiber. The peanut tapers help to excite multiple cladding modes whereas the functional PMF allows temperature and strain to exert different influences on different cladding modes. Further, the Fourier analysis method helps to extract different spectrum components contributed by different cladding modes and therefore allows the possibility to monitor their strain and temperature responses to achieve strain and temperature discrimination.

II. SENSOR STRUCTURE AND PRINCIPLE

The proposed MZI sensor structure is shown in Fig. 1(a). It is shown that the sensor is composed of two peanut tapers and a section of 10 cm PMF. The PMF is a commercial panda-type polarization maintaining fiber (PM1017-C from Yangtze Optical Fibre and Cable Co., Ltd.). As shown in the figure, the light from the core of the lead-in single mode fiber (SMF) is distributed into the core and cladding of the PMF at the first peanut taper. At the second taper, the light from the core and cladding of the PMF are coupled to the core of the lead-out SMF. The peanut tapers are easily manufactured by a fusion splicer (Erisson FSU 975) with two arc-discharges. Firstly, the ends of the SMF and PMF are arc discharged into spherical tapers, as shown in Fig. 1(b).

978-1-6654-8699-6/22 $31.00 © 2022 IEEE

Secondly, the SMF and the PMF with spherical taper ends are arc discharged for a second time into the peanut taper, as presented in Fig. 1(c). Both of the two arc-discharges are carried out with the SMF-to-SMF splicing mode set in the splicer. The second fuse time and second fuse current defined in the fusion mode are altered into 2s and 25mA for the first discharge whereas 5s and 15mA for the second discharge. As shown in Fig. 1(c), the length, belly and waist of the peanut taper are respectively 625 μm, 243 μm and 152 μm.

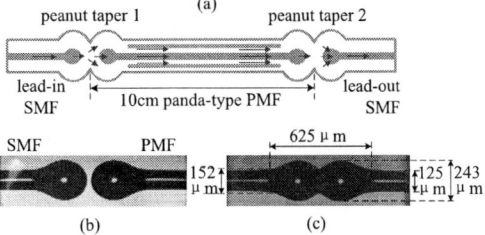

Figure 1. (a) The proposed MZI sensor structure; microscopic images of (b) SMF and PMF with spherical end, (c) peanut taper.

The proposed sensor is a PMF-based MZI. The output intensity of the MZI is governed by [21]

$$I = I_{01} + \sum_{mn} I_{mn} + 2\sum_{mn} \sqrt{I_{01}I_{mn}} \cos\left(\frac{2\pi\Delta n_{eff}^{mn}L}{\lambda}\right) \quad (1)$$

where I represents the intensity of the interference signal, I_{01} and I_{mn} respectively represent the intensity of the core mode LP_{01} and the cladding mode LP_{mn}, Δn_{eff}^{mn} is the effective index difference between LP_{01} and LP_{mn}, L is the length of the functional fiber and λ is the optical wavelength. The interference dips in the transmission spectrum appear at

$$\lambda_j^{mn} = \frac{2\Delta n_{eff}^{mn}L}{2j+1} \quad (2)$$

different, λ_j^{mn} of different cladding modes differ from each other. By taking fast Fourier transform (FFT) to the transmission spectrum, a spatial spectrum is obtained with the spatial frequencies of the cladding modes appear at [22]

$$\xi_{mn} \approx \frac{\Delta n_{eff}^{mn}L}{\lambda^2} \quad (3)$$

When strain and temperature changes in the environment, the optical path difference $\Delta n_{eff}^{mn}L$ changes and therefore the wavelength λ_j^{mn} changes. By differentiating (2), the relationship between the wavelength shift λ_j^{mn} and the strain

variation ($\Delta\varepsilon$) and the temperature variation (ΔT) can be expressed by [2]

$$\frac{\Delta\lambda_j^{mn}}{\lambda_j^{mn}} = \left[1 + \frac{\partial(\Delta n_{eff}^{mn})/\partial\varepsilon}{\Delta n_{eff}^{mn}}\right]\Delta\varepsilon + \left[\frac{\partial(\Delta n_{eff}^{mn})/\partial T}{\Delta n_{eff}^{mn}} + \frac{\partial L/\partial T}{L}\right]\Delta T \quad (4)$$

From the equation, λ_j^{mn} of different cladding modes differ from each other if their Δn_{eff}^{mn} response to strain and temperature differently.

From (1), it is known that the transmission spectrum of the proposed MZI sensor is a superstition of multiple cosine curves each of which is contributed by the interference between a certain cladding mode and the core mode. On the other hand, by using Fourier analysis, a certain transmission spectrum can be decomposed into different cosine spectrum components contributed by different cladding modes [19]. Based on this concept, it is possible to use Fourier analysis to obtain two spectrum components contributed by two different cladding modes to monitor their dips for strain and temperature sensing.

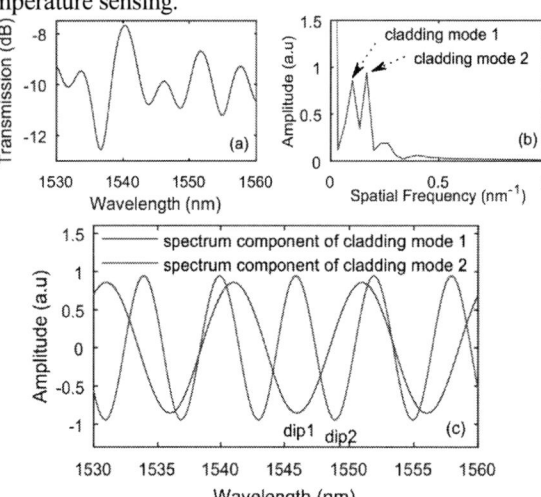

Figure 2. (a) optical transmission spectrum and (b) spatial spectrum of the proposed sensor; (c) optical spectrum components of two distinct cladding modes

The transmission spectrum and the spatial spectrum of the proposed MZI sensor are respectively shown in Fig. 2(a) and Fig. 2(b). The spatial spectrum is obtained by taking FFT to the transmission spectrum. As can be seen in Fig. 2(b), two cladding modes with spatial frequencies of 0.0999 nm⁻¹ and 0.1666 nm⁻¹ stand out in the spatial spectrum. By inversely fast Fourier transforming the two spatial frequency components, the optical spectrum components contributed by the two cladding modes are obtained, as displayed in Fig. 2(c). For the proposed MZI sensor, the PMF utilized is a panda-type PMF whose cladding has two stress areas. Since higher cladding modes have larger effective mode field areas, they are more vulnerable to the two stress areas. When temperature or strain changes, the stress areas exert different influences on the effective index of different cladding modes. Therefore, the spectrum components contributed by the two distinctive cladding modes mentioned ahead can response to strain and temperature differently. It is possible to monitor the wavelength shift of the dips in their spectrum components to discriminate temperature from strain. As shown in Fig. 2(c), dip1 in the spectrum component of cladding mode 1 and dip2 in that of cladding mode 2 are chosen in the following strain and temperature sensing experiments.

III. EXPERIMENT RESULTS AND DISCUSSION

The experiment setups for the measurement of strain and temperature are shown in Fig. 3(a) and Fig. 3(b) respectively. An optical broadband source (BBS) is utilized as the light source. The resolution of the optical spectrum analyzer (OSA) is 20 pm. In the strain sensing experiment, the proposed sensor is placed between a fixed stage and a translation stage to simulate different strain environment. In the temperature sensing experiment, the proposed sensor is placed in a temperature chamber to simulate different temperature environment.

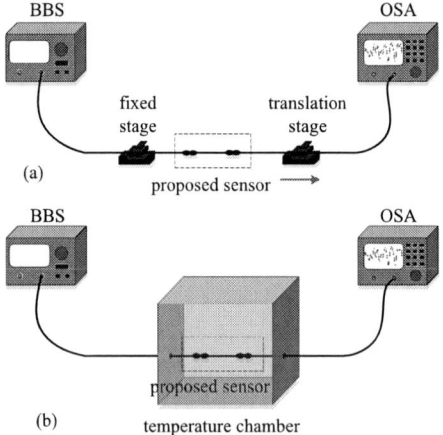

Figure 3. Experiment setups for the measurement of (a) strain and (b) temperature.

In the strain measurement, the experiment setup is kept at room temperature of 19 °C. The initial distance between the fixed stage and the translation stage (L) is set to be 20 cm. Different strain environment is implemented by modulating the translation stage away from the fixed stage. When an increment of the distance (ΔL) is applied, the strain exerted on the proposed sensor is calculated as $\Delta L/L$. ΔL is increased at a step of 0.02 mm which corresponds to 100 $\mu\varepsilon$ added to the proposed sensor. The strain ranges from 0 $\mu\varepsilon$ to 800 $\mu\varepsilon$ in the experiment. The strain response of the overall transmission spectrum of the proposed sensor is shown in Fig. 4(a). Fig. 4(b) and Fig. 4(c) respectively show the responses of the spectrum component contributed by cladding mode 1 and cladding mode 2. As can be seen, dip1 of cladding mode 1 shows a red shift whereas dip2 of cladding mode 2 shows a blue shift with the increasing strain. Fig. 5 displays the wavelength shifts of the two dips versus

the strain applied to the sensor. The fitted linear curves show that the strain sensitivity of dip1 is 1.103 pm/$\mu\varepsilon$ with R-square of 0.9809 whereas that of dip2 is -3.343 pm/$\mu\varepsilon$ with R-square of 0.9586.

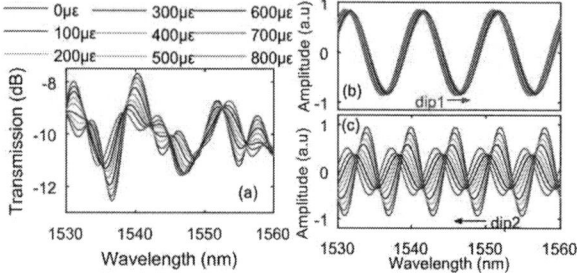

Figure 4. (a) transmission spectrum response of the proposed MZI sensor and the spectrum component responses of (b) cladding mode 1 and (c) cladding mode 2 to strain variations.

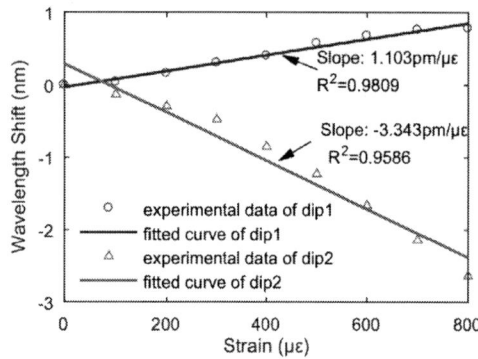

Figure 5. The wavelength shift of dip1 of cladding mode 1 and dip2 of cladding mode 2 under different strain.

In the temperature measurement, the proposed sensor is placed into the temperature chamber with a resolution of 0.1 °C. The temperature changes from 30.0 °C to 64.3 °C at a step of about 5 °C. Fig. 6(a) exhibits the response of the overall transmission spectrum of the sensor as temperature increases. Fig 6(b) and Fig 6(c) respectively shows the responses of the spectrum components contributed by the two cladding modes. As shown in the figures, both dip1 and dip2 present a red shift with the increasing temperature. Fig. 7 displays the wavelength shift of the two dips versus different temperature, where linear relationships are seen. The two fitted linear curves show that the temperature sensitivity of dip 1 is 127.8 pm/°C with R-square of 0.9643 and that of dip2 is 121.3 pm/ °Cwith R-square of 0.9617.

The discrimination of strain and temperature can be obtained by monitoring the wavelength shift of the two dips using the following matrix [1]

$$\begin{bmatrix} \Delta\varepsilon \\ \Delta T \end{bmatrix} = \frac{1}{C_{\varepsilon 1}C_{t2} - C_{\varepsilon 2}C_{t1}} \begin{bmatrix} C_{t2} & -C_{t1} \\ -C_{\varepsilon 2} & C_{\varepsilon 1} \end{bmatrix} \begin{bmatrix} \Delta\lambda_1 \\ \Delta\lambda_2 \end{bmatrix} \tag{5}$$

978-1-6654-8699-6/22 $31.00 © 2022 IEEE

where $\Delta\lambda_1$ and $\Delta\lambda_2$ are respectively the wavelength shifts of dip1 and dip2, $C_{\varepsilon 1}$ and $C_{\varepsilon 1}$ are the strain sensitivities whereas C_{t1} and C_{t2} are the temperature sensitivities. Based on the experiment results, (5) can be rewritten for the proposed sensor as follows

$$\begin{bmatrix} \Delta\varepsilon \\ \Delta T \end{bmatrix} = \frac{1}{561.029} \begin{bmatrix} 121.3 & -127.8 \\ 3.343 & 1.103 \end{bmatrix} \begin{bmatrix} \Delta\lambda_1 \\ \Delta\lambda_2 \end{bmatrix} \quad (6)$$

Since the wavelength resolution of OSA used in the experiment is 20 pm, the strain resolution and temperature resolution are calculated to be ± 0.23 $\mu\varepsilon$ and $\pm 0.16\ ^\circ$C based on (6).

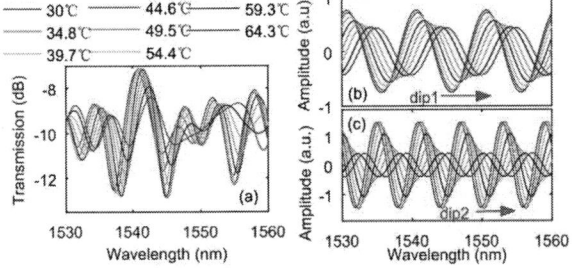

Figure 6. (a) transmission spectrum response of the proposed MZI sensor and the spectrum component responses of (b) cladding mode 1 and (c) cladding mode 2 to temperature variations.

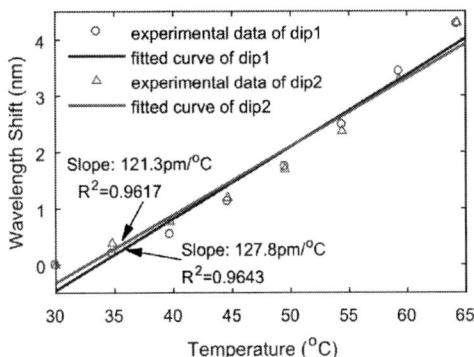

Figure 7. The wavelength shift of dip1 of cladding mode 1 and dip2 of cladding mode 2 under different temperature.

Table I exhibits the sensing properties of the proposed MZI sensor with Fourier analysis and some other MZI sensors with traditional wavelength tracking method (TWTM). As shown in the table, the proposed method possesses both high sensitivity and high resolution. It is worth noticing that, the strain resolution of the proposed method is at least one order higher than that of other methods. The high sensitivity and high resolution permit the approach combining PMF-based MZI with Fourier analysis a competitive candidate for strain and temperature discrimination.

TABLE I. SENSING PROPERTIES OF THE PROPOSED APPROACH AND SOME OTHER MZI SENSORS WITH TWTM

Sensing method	Sensor properties			
	SS^a (pm/$\mu\varepsilon$)	TS^a (pm/$^\circ$C)	SR^a ($\mu\varepsilon$)	TR^a ($\mu\varepsilon$)
SMF-based MZI and FBG^a [1]	1.07; 0.891	55.35; 10.85	± 3.104	± 0.149
SCF^a-based MZI and FBG[11]	0; 0.627	93.11; 11.46	± 27.95	± 0.21
Micro-cavity based MZI and FPI^a [12]	-0.98; 4.24	72.5; 0.26	± 4.73	± 0.34
FMF^a-based MZI [13]	-13; 9	-212; 262	± 4.7	± 0.2
MZI and FBG written in parallel [14]	0.70; 0.55	10.2; 20.4	not given	not given
the proposed approach	1.103; -3.343	127.8; 121.3	± 0.23	± 0.16

a. SS: strain sensitivity; TS: temperature sensitivity; SR: strain resolution; TR: temperature resolution; FBG: fiber Bragg grating; SCF: seven core fiber; FPI: Fabry-Perot interferometer; FMF: few mode fiber.

IV. CONCLUSION

A Fourier analysis method is utilized to analyze the transmission spectrum of an in-fiber MZI sensor. The MZI sensor is based on a PMF and two peanut tapers. The peanut tapers help to excite multiple cladding modes and the PMF helps to exert different influences on different cladding modes when strain and temperature changes. By taking FFT and inverse FFT, individual spectrum components contributed by two different cladding modes are obtained and monitored for the discriminative sensing of strain and temperature. The wavelength shifts of dip1 in the spectrum component of cladding mode 1 and dip2 in that of cladding mode 2 are monitored to achieve the dual parameter sensing. It is shown in the experiment that the strain and temperature sensitivities are 1.103 pm/$\mu\varepsilon$ and 127.8 pm/$^\circ$C for dip1 whereas -3.343 pm/$\mu\varepsilon$ and 121.3 pm/$^\circ$C for dip2. In the meanwhile, the temperature and strain discrimination resolutions are ± 0.23 $\mu\varepsilon$ and $\pm 0.16\ ^\circ$C, respectively. The high resolution along with the ease of fabrication and robustness of the sensor allow the proposed approach a competitive alternative for the simultaneous measurement of strain and temperature.

ACKNOWLEDGMENT

This work is supported by the National Natural Science Foundation of China (No. 62005011, No. 61801017 and No. 62005013) and the Beijing Municipal Natural Science Foundation (No. 4212009).

REFERENCES

[1] L. Lv, S. Wang, L. Jiang, F. Zhang, Z. Cao, P. Wang, Y. Jiang, and Y. Lu, "Simultaneous measurement of strain and temperature by two peanut tapers with embedded fiber Bragg grating," Applied Optics, vol. 54, no. 36, pp. 10678-83, Dec 20, 2015.

[2] D. Wu, T. Zhu, K. S. Chiang, and M. Deng, "All Single-Mode Fiber Mach–Zehnder Interferometer Based on Two Peanut-Shape Structures," Journal of Lightwave Technology, vol. 30, no. 5, pp. 805-810, 2012.

[3] X. Zhou, S. Li, X. Li, X. Yan, X. Zhang, and T. Cheng, "A Vectorial Analysis of the Curvature Sensor Based on a Dual-Core Photonic Crystal Fiber," IEEE Transactions on Instrumentation and Measurement, vol. 69, no. 9, pp. 6564-6570, 2020.

[4] Y. Wang, Y. Zhou, Z. Liu, D. Chen, C. Lu, and H. Y. Tam, "Sensitive Mach-Zehnder interferometric sensor based on a grapefruit microstructured fiber by lateral offset splicing," Optics Express, vol. 28, no. 18, pp. 26564-26571, Aug 31, 2020.

[5] C. Bian, Y. Cheng, W. Zhu, R. Tong, M. Hu, and T. Gang, "A Novel Optical Fiber Mach–Zehnder Interferometer Based on the Calcium Alginate Hydrogel Film for Humidity Sensing," IEEE Sensors Journal, vol. 20, no. 11, pp. 5759-5765, 2020.

[6] S. Gu, D. Feng, T. Zhang, S. Deng, M. Li, Y. Hu, W. Sun, and M. Deng, "Highly Sensitive Magnetic Field Measurement With Taper-Based In-Line Mach-Zehnder Interferometer and Vernier Effect," Journal of Lightwave Technology, vol. 40, no. 3, pp. 909-917, 2022.

[7] Z. Zhou, Z. Gong, B. Dong, S. Ruan, and C. C. Chan, "Enhanced Sensitivity Refractometer Based on Spherical Mach–Zehnder Interferometer With Side-Polished Structure," IEEE Sensors Journal, vol. 21, no. 2, pp. 1548-1553, 2021.

[8] M. Singh, S. K. Raghuwanshi, and O. Prakash, "Ultra-Sensitive Fiber Optic Gas Sensor Using Graphene Oxide Coated Long Period Gratings," IEEE Photonics Technology Letters, vol. 31, no. 17, pp. 1473-1476, 2019.

[9] S. Yuan, Z. Tong, J. Zhao, W. Zhang, and Y. Cao, "High temperature fiber sensor based on spherical-shape structures with high sensitivity," Optics Communications, vol. 332, pp. 154-157, 2014.

[10] J. Wang, A. Wang, X. Chen, S. Liu, X. Xu, C. Sun, Y. Yan, Q. Yan, S. Wang, T. Geng, and W. Sun, "An All Fiber Mach-Zehnder Interferometer Based on Tapering Core-Offset Joint for Strain Sensing," IEEE Photonics Technology Letters, vol. 34, no. 1, pp. 11-14, 2022.

[11] Y. Liu, X. Song, B. Li, J. Dong, L. Huang, D. Yu, and D. Feng, "Simultaneous measurement of temperature and strain based on SCF-based MZI cascaded with FBG," Applied Optics, vol. 59, no. 30, pp. 9476-9481, Oct 20, 2020.

[12] B. Huang, S. Xiong, Z. Chen, S. Zhu, H. Zhang, X. Huang, Y. Feng, S. Gao, S. e. Chen, W. Liu, and Z. Li, "In-Fiber Mach-Zehnder Interferometer Exploiting a Micro-Cavity For Strain and Temperature Simultaneous Measurement," IEEE Sensors Journal, vol. 19, no. 14, pp. 5632-5638, 2019.

[13] C. Lu, J. Su, X. Dong, T. Sun, and K. T. V. Grattan, "Simultaneous Measurement of Strain and Temperature With a Few-Mode Fiber-Based Sensor," Journal of Lightwave Technology, vol. 36, no. 13, pp. 2796-2802, 2018.

[14] K. Tian, M. Zhang, Z. Zhao, R. Wang, D. Liu, X. Wang, E. Lewis, G. Farrell, and P. Wang, "Ultra-compact in-core-parallel-written FBG and Mach-Zehnder interferometer for simultaneous measurement of strain and temperature," Optics Letters, vol. 46, no. 22, pp. 5595-5598, Nov 15, 2021.

[15] J. Zhou, C. Liao, Y. Wang, G. Yin, X. Zhong, K. Yang, B. Sun, G. Wang, and Z. Li, "Simultaneous measurement of strain and temperature by employing fiber Mach-Zehnder interferometer," Optics Express, vol. 22, no. 2, pp. 1680-6, Jan 27, 2014.

[16] M. Zhang, L. Hou, H. Zhang, X. Zhang, and J. Yang, "Composite Modal Interferometer for Simultaneous Triple-Parameter Measurement Based on Cascaded No-Cladding Fiber and Thin-Core Fiber Structure," IEEE Access, vol. 8, pp. 180133-180139, 2020.

[17] L. Hou, J. Yang, X. Zhang, J. Kang, and L. Ran, "Bias-Taper-Based Hybrid Modal Interferometer for Simultaneous Triple-Parameter Measurement With Joint Wavelength and Intensity Demodulation," IEEE Sensors Journal, vol. 19, no. 21, pp. 9775-9781, 2019.

[18] N. Zhang, W. Xu, S. You, C. Yu, C. Yu, B. Dong, and K. Li, "Simultaneous measurement of refractive index, strain and temperature using a tapered structure based on SMF," Optics Communications, vol. 410, pp. 70-74, 2018.

[19] H.-W. Fu, X. Yan, M. Shao, H. Li, H. Gao, Z. Jia, and X. Qiao, "Fourier Analysis Applied on MZI Transmission Spectrum for Refractive Index Measurement," IEEE Photonics Technology Letters, vol. 27, no. 6, pp. 658-660, 2015.

[20] X. Y. Xu Yan, H. F. Haiwei Fu, H. L. Huidong Li, and a. X. Q. and Xueguang Qiao, "Simultaneous refractive index and temperature measurements by using dual interference in an all-fiber Mach–Zehnder interferometer," Chinese Optics Letters, vol. 14, no. 3, pp. 030603-30607, 2016.

[21] F. Yu, P. Xue, X. Zhao, and J. Zheng, "Investigation of an in-line fiber Mach–Zehnder interferometer based on peanut-shape structure for refractive index sensing," Optics Communications, vol. 435, pp. 173-177, 2019.

[22] J. Harris, P. Lu, H. Larocque, Y. Xu，L. Chen，X. Bao "Highly sensitive in-fiber interferometric refractometer with temperature and axial strain compensation," Optics Express, vol.21, no.8, pp.9996-10009, 2013.

978-1-6654-8699-6/22 $31.00 © 2022 IEEE

High Brightness Ultra Wideband Fiber Source

Tongle Yuan, Yu Cheng, Ming Chen, Libo Yuan
Guangxi Key Laboratory of Optoelectronic Information Processing;
School of Optoelectronic Engineering, Guilin University of Electronic Technology, 541004, China
e-mail: yuantl2020@163.com, chengyu@guet.edu.cn, mchenqq2011@163.com, lbyuan@vip.sina.com

Sumei Huang, Jing Li
Han's Laser Technology Industry Group Co., Ltd.
Shenzhen, China
e-mail: hsm@hanslaser.com, lij111946@hanslaser.com

Abstract—Novel bismuth-erbium co-doped fibers (BEDFs) are prepared by the modified chemical vapor deposition process, and its fluorescence spectra covers 1380~1700nm. The structure, fiber length, pump power and Erbium ion doping concentration are discussed in this paper to explore the properties of this ultra-wideband fiber light source based on BEDF. The experiments show that the optimization of the above parameters is beneficial in improving the fluorescence intensity of this fiber light source from 1450 nm to 1700 nm. The concentration of erbium ions affects the luminescence efficiency of the fiber. It is demonstrated that the decrease of the fluorescence luminescence efficiency is due to the presence of a strong upconversion luminescence via experiments described in this paper. Finally, a high-brightness ultra-broadband light source with a full spectral width of 320 nm is obtained through a multi-parameter equalization method, and its 5db bandwidth near 1550 nm can reach 51 nm, and the output power of the light source can reach 1.95 mW.

Keywords-bismuth-erbium co-doped fiber; upconversion luminescence; ion doping concentration; ultra-wideband light source

I. INTRODUCTION

The technological progress of light sources, as key instruments for scientific research and social applications, is of great importance. From laser diodes (LD), light-emitting diodes (LED), superradiant light-emitting diodes (SLED), to Erbium-doped superfluorescent fiber optic light sources (SFS) have been developed rapidly within a few decades [1-3]. However, the current fiber optic light sources generally suffer from the defects of difficult coordination of luminous intensity, bandwidth and flatness to meet the demand of practical applications. Therefore, BEDF light sources that can theoretically achieve coverage from 1100 to 1700 nm have good prospects for development [4, 5].

For this reason, many studies have been conducted at home and abroad. For example, Dianov systematically studied the fluorescence characteristics of Bismuth-doped optical fibers with four basic matrix materials, including Bismuth-doped phosphate fiber (PSBi), Bismuth-doped aluminosilicate fiber (ASBi), Bismuth-doped germaninosilicate fiber (GSBi) and Bismuth-doped silicate fiber (SiBi) [6]. Wu obtained a fluorescence spectrum with a central wavelength of 1536 nm and a 3 dB bandwidth of 29

nm by pumping BEDF with 980 nm laser source [7]. Yuan achieved a 105 nm broadband spectrum with 5 dB bandwidth near 1550 nm using Bismuth/Erbium co-doped fiber[8].

To achieve a high bright, broadband fiber source, new BEDFs are fabricated using the liquid phase doping method via the modified chemical vapor deposition (MCVD) process. A broadband fluorescence spectrum from 1300 to 1700 nm is measured with the fiber. The difference of the BEDF fluorescence spectrum is linked to erbium concentration in different fibers. The overall brightness of the broadband light source can be enhanced by reducing the Er ion doping concentration and increasing the Bi ion doping concentration at the same time [9]. Finally, A 5dB bandwidth of 51nm near 1550nm and the overall output power of 1.95mW can be achieved in the fiber source that made in GUET.

II. EXPERIMENT

A. BEDF Preparation

Three BEDFs with bismuth and erbium co-doping recipes were fabricated by MCVD process, named 1#, 2# and 3# BEDFs, respectively. In these BEDFs, the doping concentration of Er^{3+} decreased sequentially, where the Er^{3+} concentration were 6800 ppm in 1# BEDF, 1816 ppm in 2# BEDF and 1535 ppm in 3# BEDF via INCAX-act measurement. The doping concentration of Bi ions increased sequentially in the fabrication recipes. However, due to the high temperature (>2100°C) during the collapse process of the preform, there was volatilization of bismuth that make the concentration unknown to all. The bismuth doping concentration was too lower to be measured, so only the Er ion concentration was available in the actual test. At 980nm, 1#, 2# and 3# BEDF core absorption was 91dB/m, 24.3dB/m and 20.55dB/m respectively.

B. Absorption Spectrum of BEDF

For BEDF performance test, these instruments were employed: an optical spectrum analyzer (OSA) (from Agilent, model 86142B), a miniature spectrometer (from ideaoptics, model FX4000), a 974 nm semiconductor laser (from MC Fiber Optics, model MCSPL-974-520-0-1-1-FA-T1), a halogen light source (from ideaoptics, model HL2000-12), a supercontinuum light source (from YSL Photonics, model SC-5).

In this paper, the absorption spectrum of 3# BEDF was tested by the cut back method, as shown in Fig. 1.

It could be seen that the absorption spectrum of BEDF had typical absorption peaks of Er^{3+} ions, whose main peak wavelengths were at (482 nm and 498 nm), 514 nm, 548 nm, (645 nm and 667 nm), 800 nm, 980 nm, (1530 nrm and 1550 nm), named a-j in Fig. 1, respectively. They could be known from the energy level diagram of Er^{3+} as $^4I_{15/2} \rightarrow ^4F_{7/2}$, $^4I_{15/2} \rightarrow ^2H_{11/2}$, $^4I_{15/2} \rightarrow ^4S_{3/2}$, $^4I_{15/2} \rightarrow ^4F_{9/2}$, $^4I_{15/2} \rightarrow ^4I_{9/2}$, $^4I_{15/2} \rightarrow ^4I_{9/2}$, $^4I_{15/2} \rightarrow ^4I_{11/2}$, $^4I_{15/2} \rightarrow ^4I_{13/2}$ of electron leap. And the small absorption peaks with wavelengths located near 425 nm, 1147 nm, and 1629 nm are the absorption peaks produced by Bismuth [10].

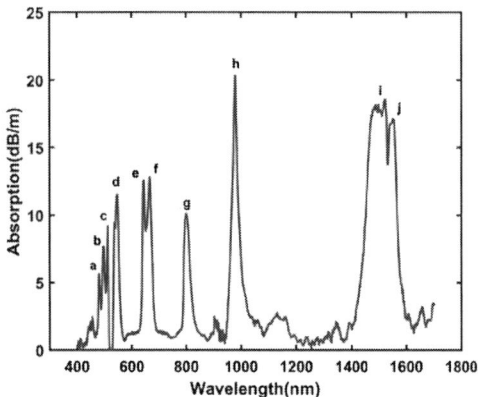

Figure 1. Absorption spectrum of BEDF

C. Emission Spectrum of BEDF

In this paper, the emission spectrum of BEDF was obtained by using the forward pumping structure. The BEDF length was 30 cm. The 974 nm semiconductor laser output power was 100 mW. The experiment setup was shown in Fig. 2(a).

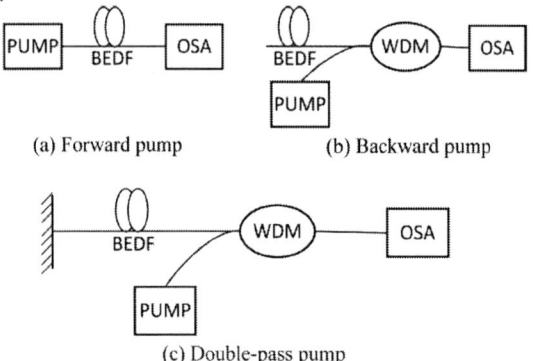

Figure 2. BEDF sources with different structures

The emission spectrum of BEDF from 237 to 600 nm was acquired from FX4000 Spectrometer, and the emission spectrum of BEDF from 600 to 1700 nm was acquired from the OSA. The total emission spectrum of BEDF from 200 to 1700 nm was shown in Fig. 3.

The fluorescence peak near 974 nm indicated that there was a large residual of pump power and the BEDF had been

adequately pumped. BEDF was found to have two sets of upconversion luminescence peaks at 438.5 nm and 536.1 nm, which were $^4F_{3/2} \rightarrow ^4I_{15/2}$ and $^4S_{3/2} \rightarrow ^4I_{15/2}$ upconversion luminescence. Meanwhile, BEDF also had a fluorescence peak with high fluorescence intensity and good flatness near 1530 nm, which was the Er^{3+} electron leap of $^4I_{13/2} \rightarrow ^4I_{15/2}$.

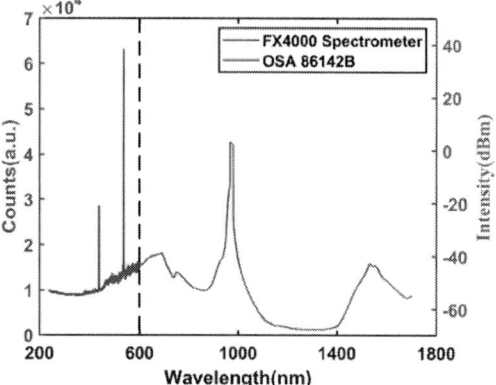

Figure 3. Emission spectrum of BEDF

D. Different BEDF Source Structures

To obtain the promising fluorescence intensity, bandwidth and flatness of the BEDF source, three different structures were employed, namely forward pumping structure, backward pumping structure and dual-range pumping structure, the output spectrum was acquired from the OSA, as shown in Fig. 2.

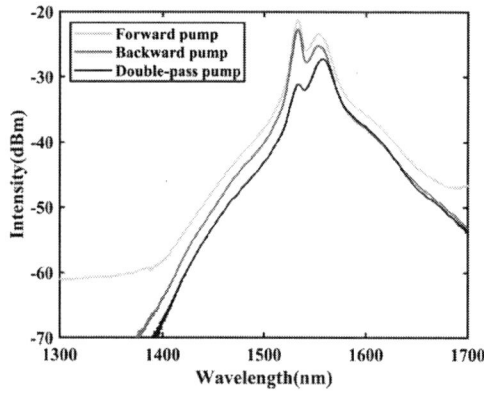

Figure 4. Influence of different BEDF source structure on output spectrum

The same length of BEDF and the same pump power were fixed at these experiments in order to compare different output spectrum with different structures. The effects of different light source structures on the output spectrum were shown in Fig. 4. It could be clearly observed from Fig. 4 that the overall brightness and bandwidth of the fluorescence spectrum obtained by forward pumping was the best at the same BEDF length and pumping power, and its flatness was slightly worse than that of the two-way pumping structure, which was because the WDM and fiber mirror in the backward pumping and two-way pumping structures had certain insertion loss, which made the overall intensity of the fluorescence spectrum decreased, and part of the fluorescence of the two-way pumping needed to pass through the BEDF twice. Due to the absorption of BEDF at 1530 nm, the overall fluorescence spectrum was red-shifted. It was also observed that the fluorescence intensity of the reverse pumping and two-way pumping structures in the range of 1300 ~ 1450 nm and 1670 ~ 1700 nm was abruptly reduced compared with that of the forward pumping structure, which was due to the bandwidth limitation of the WDM used in the experiment, and all the wavelength beyond the bandwidth limitation was filtered out by the WDM. On balance, we chose the forward-pumped structure for the subsequent experiments.

E. Different BEDF Length

To explore the relation between the length of BEDF and the fluorescence intensity, the BEDF was spliced to pump laser source and a fiber pigtail. The fiber pigtail was connected to the OSA. The pump power was fixed at 100mW. The fiber length was cut-back from 88cm to 7.5cm, and the fluorescence spectrum of BEDF at different length was acquired from the OSA.

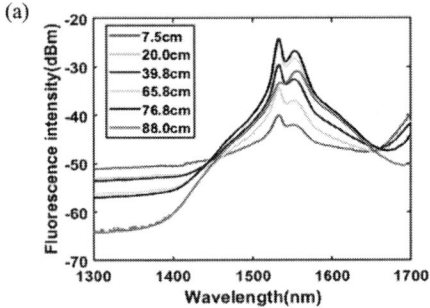

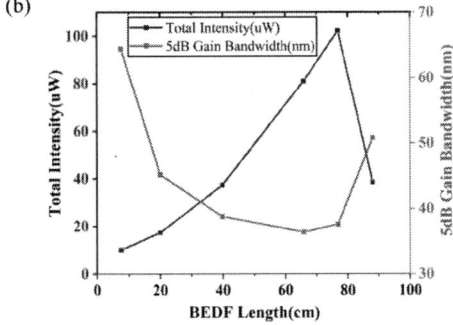

Figure 5.   (a) Fluorescence spectra of BEDF at different lengths; (b) Total fluorescence power and 5dB bandwidth of BEDF Influence of different light source structure on the output spectrum

As shown in Fig. 5(a), if the length of BEDF was less than 76.8 cm, the intensity of the fluorescence peak at 1530 nm and 1550 nm increased gradually with the increase of fiber length. If the BEDF length was larger than 76.8 cm, the intensity of the two fluorescence peaks at 1530 nm and 1550 nm decreased with the increase of fiber length. And the decay rate of the fluorescence intensity at 1530 nm was faster than that at 1550 nm. The reason was the absorption peaks at 1530 nm in the redundancy BEDF. So there must be

an optimized fiber length in BEDF that made the spontaneous radiation and absorption to be balanced. As shown in Fig. 5(a), if the length of BEDF was less than 76.8 cm, the spontaneous radiation of Er3+ ions in the fiber dominated, and the intensity of the fluorescence peak gradually increased with the increasing fiber length. If the length of BEDF was greater than 76.8 cm, the absorption of BEDF near 1530 nm surpassed the spontaneous radiation. So the intensity of the 1530 nm fluorescence peak decreased with increasing fiber length, while the 1550 nm fluorescence peak showed a red shift. The reason of the red shift was the redundancy BEDF absorbed the C band photon and emitted at longer wavelength. Meanwhile, the intensity of the fluorescence intensity near 1450 nm could be observed unchanged because this emission was attribute to bismuth silicate [11].

To compare the effect of different lengths of BEDF on the fluorescence spectra intuitively, the 5 dB bandwidth of the fluorescence spectrum and the total power of the fluorescence were shown in Fig. 5(b). If the length of BEDF was less than 76.8 cm, the 5 dB bandwidth of fluorescence spectrum gradually decreased while the total power gradually increased. If the length of BEDF was less than 76.8 cm, the 5 dB bandwidth of fluorescence spectrum gradually increased while the total power rapidly decreased as the fiber length increased. The intensity of the fluorescence peaks at 1530 nm and 1550 nm increased while the waveform steepened, increasing in the overall intensity of the fluorescence spectrum. If the length of BEDF was larger than 76.8 cm, the total fluorescence intensity decreased rapidly although the 5 dB bandwidth increased slightly. This indicated that there was an optimal fiber length to balance the 5 dB bandwidth and the overall fluorescence intensity of the BEDF fluorescence spectrum.

F. Upconversion Luminescence of BEDF

BEDF with highest doping concentration was employed to produce the light source. It was observed that the BEDF emitted pronounced green light which was much bright than the laboratory natural light, as shown in Fig. 6. The reason was the upconversion in the BEDF.

Figure 6.   Green upconversion from BEDF

To know more about the pump power to the upconversion luminescence and fluorescence characteristics, BEDF with a fixed length of 88 cm was employed to build the light source with the forward pumping structure. By varying the output power of the 974 nm semiconductor laser, the relation between pump power and the upconversion luminescence intensity was shown in Fig. 7(a), and the relation of the fluorescence intensity vs. the pump power was shown in Fig. 7(b).

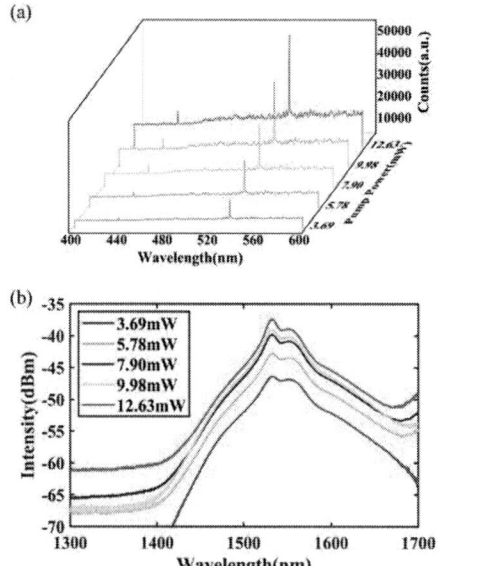

Figure 7. (a) Effect of pump power on BEDF up conversion; (b) Effect of pump power on BEDF fluorescence spectrum

The pump power increased from 3.69 mW to 12.63 mW, the luminescence intensity of the 438.5 nm and 536.1 nm increased obviously. More Er3+ ions were excited in BEDF that increased the probability of two-photon absorption jumps as the pump power increased, which finally enhanced the upconversion luminescence. The overall fluorescence intensity of BEDF at 1400 ~ 1700 nm also increased with the increase of the pump power as could be seen in Fig. 7(b).

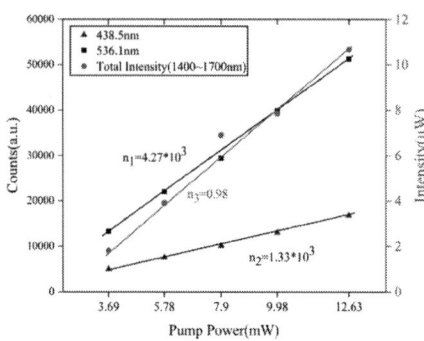

Figure 8. Relation of the pump power with upconversion intensity in the miniature spectrometer and total fluorescence intensity in the OSA

Nearly linear relationship of the upconversion intensity and fluorescence intensity vs. the pump power could be seen in Fig. 8. It could be found that the intensity of the upconversion at 438.5 nm and 536.1 nm produced by BEDF was proportional to the pumping power. The intensity and slope efficiency of the upconversion at the 536.1 nm were

larger than that at the 438.5 nm, which was consistent with the observation in Fig.6 that the BEDF showed green color as it was pumped at 974nm. At lower pumping power, the total power of the fluorescence spectrum from 1400 nm to 1700 nm was approximately proportional to the pumping power.

G. Research on Upconversion Luminescence and Fluorescence Properties of Erbium Bismuth ion Doping Concentration

Different BEDFs with different doping concentration were fabricated for our experiments. The obvious upconversion luminescence in BEDF could be reduced by changing the doping concentration of erbium bismuth ions in BEDF. The overall fluorescence intensity of BEDF also improved as the doping concentration decreased.

Three BEDFs with different doping concentration were forword pumping at 974nm, while the fiber length was fixed at 88cm and the pump power was fixed at 12.63mW. The fluorescence spectra of different BEDF was shown in Fig. 9(a). And the fluorescence spectrum of BEDF enhanced in 1400 ~ 1700 nm significantly as the decrease of Er^{3+} ion concentration.

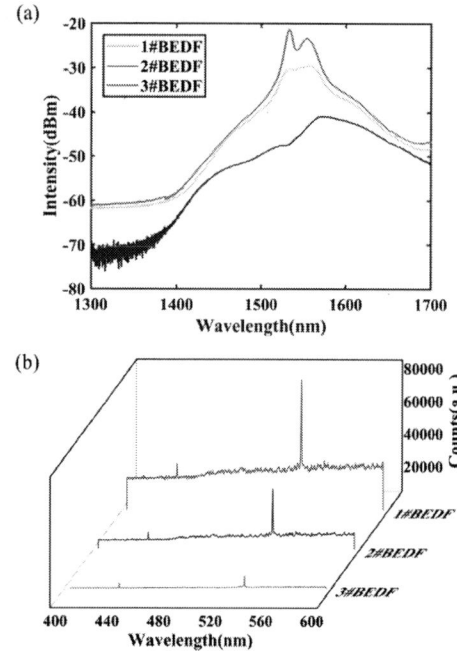

Figure 9. (a) The fluorescence spectra of different doping concentration in BEDF; (b) The upconversion of different doping concentration in BEDF

It could be seen that the decrease of Er^{3+} ion doping concentration made the intensity of the upconversion of BEDF decreased in Fig. 9 (b). This is because the reduced concentration of Er^{3+} ions in BEDF reduces the number of excited particles, and most of the excited Er^{3+} ions directly returned to the ground state and emit fluorescence at 1532 nm after the radiation-free leap to the sub-stable state, making it difficult to generate upconversion luminescence by the two-photon approach.

III. RESULTS & DISCUSSION

It could be seen that a high brightness fiber light source based on BEDF was achieved with appropriate bismuth-erbium doping concentration. The forward pumping structure seems better than other structure as the results showed in our experiments. The BEDF upconversion luminescence decreases while the fluorescence spectrum intensity increases if the Er3+ ions concentration appropriately reduces in our BEDF sample. The main reason is that higher Er^{3+} ions doping concentration made it easier to form two-photon jumps under higher pumping power, while the emission of Er^{3+} ions near 1530 nm ($^4I_{13/2} \rightarrow {}^4I_{15/2}$) is suppressed. If the concentration of Er^{3+} ions decreases, most upper energy level Er^{3+} ions relaxes to the sub-stable energy level to suppress the upconversion luminescence and enhances the overall fluorescence spectrum intensity of BEDF near 1530 nm. Finally, through the multi-parameter equalization method, the ultra-broadband light source are achieved whose bandwidth is over 320nm. And its 5dB bandwidth near 1550nm can reach 51nm, and the total output power of fluorescence reaches 1.95mW. However, the flatness of the fluorescence spectrum still needs to be optimized, and the pump power is only up to 300 mW due to the limitation of the existing 974 nm laser source in our laboratory. The spectral flatness can be improved if the pump power could be further increased. An ultra-broadband light source with greater brightness can be obtained in our further research.

IV. CONCLUSION

In this paper, the relationship between upconversion luminescence and fluorescence spectra in bismuth-erbium co-doped fiber is explored with the novel bismuth-erbium co-doped fiber made for this paper. Different factors, such as fiber length, pumping power and doping concentration in BEDF, are explored in detail and report in this paper to extend more knowledge about the novel BEDFs. It is found that the fluorescence spectrum of bismuth-erbium co-doped fiber is generally better than that of erbium-doped fiber. Meanwhile, appropriately reducing the Er^{3+} ions doping concentration and increasing the bismuth ions doping concentration in the fiber can decrease the upconversion luminescence and improve the total fluorescence intensity. A high brightness ultra-wideband fiber light source based on BEDF with a 5 dB bandwidth up to 51 nm and a total output power up to 1.95 mW is achieved. At the same time, this fiber light source has the advantages of low cost and low manufacturing difficulty. The fiber light source is all fiberized, which has very promising value due to its robust and miniature size.

ACKNOWLEDGMENT

The authors thank the National Key Research and Development Program of China (2019YFB2203901) and the National Natural Science Foundations of China (61975038, 62065006, and 62075047) for their support.

REFERENCES

[1] R Sumathy, A Muthukumar, "Wavelength division multiplexing transmission using multimode Erbium doped fiber amplifier with elevated refractive index profile," *Optical and Quantum Electronics*, vol. 53, no. 2, pp. 1-11, 2021.

[2] J Zhang, Z M Sathi, Y Luo, et al., "Toward an ultra-broadband emission source based on the Bismuth and Erbium co-doped optical fiber and a single 830nm laser diode pump," *Optics express*, vol 21, no. 6, pp. 7786-7792, 2013.

[3] A Nassiri, A Boulezhar, H Idrissi_Saba, "Enhanced data transmission Erbium doped fiber amplifier," *Materials Today: Proceedings*, vol. 30,pp. 1043-1045, 2020.

[4] B. B. Yan, Y. H. Luo, A Zareanborji, et al., "Performance comparison of Bismuth/Erbium co-doped optical fibre by 830 nm and 980 nm pumping," *Journal of Optics*, vol. 18, no. 10, pp. 105705, 2016.

[5] E M Dianov, S V Firstov, M A Melkumov, "Bismuth-doped fiber lasers covering the spectral region 1150–1775 nm," in *Laser Science*, 2015.

[6] I A Bufetov, M A Melkumov, S V Firstov, et al., "Bi-doped optical fibers and fiber lasers," *IEEE Journal of Selected Topics in Quantum Electronics*, vol.20, no. 5, pp. 111-125, 2014.

[7] X. J. WU, B. B. YAN, C. S. LI, et al., "Experimental Investigation on the Luminescence Properties of Bismuth/Erbium Co-doped Optical Fiber," *Acta Photonica Sinica*, vol. 45, no. 3, pp. 0306001, 2016.

[8] T. L. Yuan, M Xiong, Y Cheng, et al., "Study of ultra-wideband light source with Erbium-Bismuth co-doped fiber," in *Eighth Symposium on Novel Photoelectronic Detection Technology and Applications*, 2022,vol. 12169,pp. 1541-1547.

[9] S V Firstov, E G Firstova, A V Kharakhordin, et al., "Anti-Stokes luminescence in Bismuth-doped high-germania core fibres," *Quantum Electronics*, vol. 49, no. 3, pp. 237, 2019.

[10] B. N. Jia, P. F. Lu, Z. X. Peng, et al., "Near-IR luminescence characteristics of monovalent Bismuth in Bi-doped pure silica optical fiber: First-principle study," *Journal of Luminescence*, vol. 198, pp. 384-388, 2018.

[11] A N Romanov, E V Haula, D P Shashkin, et al., "On the origin of near-IR luminescence in SiO2 glass with Bismuth as the single dopant. Formation of the photoluminescent univalent Bismuth silanolate by SiO2 surface modification," *Journal of Luminescence*, vol. 183, pp. 233-237, 2017.

Proceedings of the 7th Optoelectronics Global Conference (OGC 2022)

BOTDR Denoising by Sparse Representation Algorithm with Preformed Dictionary

Yuting Liu[2], Zhijie Sun[1], Ning Cui[2]

[1.] Shanxi Transportation Technology Research & Development Co., Ltd
Taiyuan, China
e-mail: liuyuting1232@link.tyut.edu.cn,
46395114@qq.com, cuining_tyut@163.com

Qing Bai[1,2*], Yu Wang[1,2], Baoquan Jin[2]

[2.] Key Laboratory of Advanced Transducers and Intelligent Control Systems, Ministry of Education and Shanxi Province
Taiyuan University of Technology
Taiyuan, China
e-mail: baiqing@tyut.edu.cn, wangyu@tyut.edu.cn,
jinbaoquan@tyut.edu.cn

Abstract—**In Brillouin optical time domain reflectometers, the signal-to-noise ratio is a key factor restricting the sensor performance. Using redundancy and correlation of 3D-Brillouin gain spectrum in multi-dimensional domain, sparse representation algorithm can be used to improve signal-to-noise ratio. According to basic principle of sparse representation, a dictionary can be designed to reconstruct valid signals. During reconstruction, random noise will be discarded as residuals. In this paper, discrete cosine transform algorithm is used to design the dictionary, orthogonal matching pursuit algorithm is used to extract the coefficient matrix, and the signal is finally reconstructed to achieve the purpose of noise reduction. The simulation results show that when 5dBm random noise is added, signal-to-noise ratio in the non-temperature-change region is increased by 24.3dB, which provides a new idea for improving signal-to-noise ratio of BOTDR sensor.**

Keywords-distributed optical fiber sensing; BOTDR; signal-to-noise ratio; sparse representation; discrete cosine transform

I. INTRODUCTION

Brillouin Optical Time Domain Reflectometer (BOTDR) technology is a distributed optical fiber sensing technology that uses the relationship between Brillouin gain signals (BGS) (such as power, frequency) and properties of the fiber itself to detect temperature or strain changes at each location along the fiber [1]. It has advantages of single-ended pumping, anti-electromagnetic interference, and so on. It has been widely used in many fields of engineering practice.

However, due to threshold limit and attenuation of the optical pulse, spontaneous Brillouin scattering signal in the fiber is relatively weak. In addition, BOTDR sensor will introduce random noises during measurement process. And harsh environments in practical applications will also deteriorate signal-to-noise ratio (SNR). Therefore, it is important for BOTDR sensor to improve SNR.

In the field of SNR improvement of BOTDR sensor, existing methods, such as multi-wavelength detection [2], pulse coding [3], etc., can effectively improve SNR. Although these methods, whether changing optical path structure or changing the detection pulse, have achieved an improvement in SNR. It also leads to a more complex

hardware structure and increases overhead. Or it is necessary to introduce an additional demodulation algorithm to obtain Brillouin frequency shift (BFS), resulting in longer signal processing time and reduced real-time performance. Furthermore, a study find that these methods do not fully utilize redundancy and correlation of signals in multi-dimensional domain, which limits ability to improve SNR [4]. In recent years, many experts have applied methods of digital image processing to the field of distributed optical fiber sensing, and achieved rich results. Some of these methods treat the acquired BFS signal as an image to be processed, utilize an image filter [5, 6], and filter it to reduce noise. And experiments show that they can enhance SNR in different degrees of, which prove a feasibility of digital image processing methods.

In this paper, 3D-Brillouin gain spectrum (3D-BGS) is regarded as an image to be processed. Pixelating the top view of 3D-BGS, and a sparse representation denoising algorithm for BOTDR sensor is proposed. Based on the basic principle of this algorithm, by constructing a dictionary matrix and extracting a coefficient matrix, a reconstructed signal can be obtained. The residual generated by the algorithm is discarded as noise to achieve the purpose of noise reduction. After simulation and analysis of the algorithm, in the case of adding 5dBm noise, SNR in the non-temperature-change region is improved by 24.3dB.

II. ALGORITHM PRINCIPLE

A. The Principle of Sparse Representation Algorithm

According to the basic principle of sparse representation [7-9], images without noise has its certain inherent regularity. After being projected onto a dictionary, they can be sparsely represented by a linear combination of dictionary atoms, while noise is random and cannot be sparsely represented by dictionary atoms.

The ideal 3D-BGS measured by BOTDR sensor has certain rules, and Brillouin center frequency at different positions is roughly the same. Only when properties of the fiber itself (such as temperature, strain, etc.) change, center frequency will change. Therefore, the 3D-BGS matrix composed of BGS measured along optical fiber has

978-1-6654-8699-6/22 $31.00 © 2022 IEEE

similarity and redundancy, which consistent with the requirements of sparse representation principle for effective signals, and can be sparsely represented by linear combination of the dictionary atoms. The noise (such as thermal noise, shot noise, etc.), which introduced by the sensor during measurement process, is random and cannot be sparsely represented by dictionary atoms. The actual measured 3D-BGS can be regarded as a superposition of the effective signal and the noise signal. Therefore, based on the difference between effective signal and random noise, the separation of effective signal and random noise can be accomplished by using the algorithm. The noise will be discarded as residual between original 3D-BGS signal and reconstructed signal, so as to achieve the purpose of separating noise and improving SNR of BOTDR sensor.

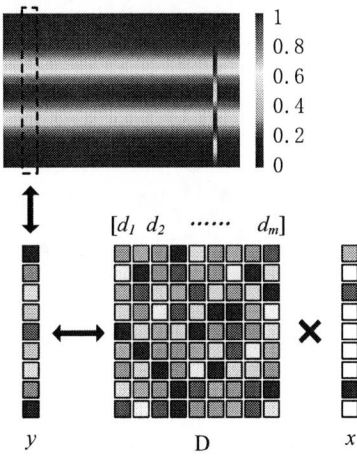

Figure 1. Fundamentals of sparse representation algorithm

The basic principle of sparse representation algorithm is shown in Fig. 1. Take original BGS data at a certain position of the fiber, and map power data at different frequencies to each pixel of color patch matrix one-to-one. The color of color block matrix represents size of the value, and the lighter the color, the smaller the value. The column vector y represents color block column vector corresponding to BGS data at a certain position along the fiber. The matrix D represents designed dictionary matrix, and d_i ($i=1, 2, ..., m$) is column vector of the dictionary D. A sparse representation of y can be obtained by linearly combining the column vectors of D. Its coefficients are coefficient vector x. The reconstruction of y is done by multiplying x by D. By sequentially selecting BGS data at each position of optical fiber and performing the above operations, the above operations can be extended to entire 3D-BGS image, and reconstruction of entire 3D-BGS image can be completed.

According to the above, the design goal of sparse representation algorithm is to complete noise reduction work on premise that sparseness of the algorithm and effective information of the original data are not affected.

Mathematically, the above goal is:

$$\begin{cases} \left\| Y - DX \right\|_2^2 \leq \varepsilon \\ min \left\| x_i \right\|_0 \ (i = 1, 2, ..., m) \end{cases} \tag{1}$$

Eq. (1) consists of two parts. In the above equation, Y refers to the original signal matrix, D refers to the dictionary matrix, X refers to the coefficient matrix, and ε refers to the target error. DX represents the reconstruction matrix obtained from D and X. By setting $\left\| Y - DX \right\|_2^2 \leq \varepsilon$, the reconstructed matrix maintains a certain accuracy compared with the original data. In the following equation, the l_0 norm $\left\| \cdot \right\|_0$ is used to represent the number of non-zero elements in the vector. Making $\left\| x_i \right\|_0$ the smallest can ensure that there are few non-zero elements in the column vector of the coefficient matrix, thus ensuring sparsity of the algorithm. Therefore, solving Eq. (1) can achieve the purpose of noise reduction on premise of satisfying sparseness and accuracy of the algorithm. There are two variables to be solved in Eq. (1), the dictionary matrix D and the coefficient matrix X. In this paper, discrete cosine transform (DCT) algorithm is used to construct the dictionary matrix D, and orthogonal matching pursuit (OMP) algorithm is used to solve the coefficient matrix X.

B. Algorithm Flow

The sparse representation algorithm flow with preformed DCT dictionary is as follows:

1) Construct a preformed dictionary D_{DCT} using discrete cosine transform. The DCT is a special form of the discrete Fourier transform. It can transform signal from time domain to frequency domain. By filtering out noise which is considered to be in high-frequency part and retaining effective information in low-frequency part, it can achieve the purpose of noise reduction. In this paper, the DCT is extended to two-dimensions, as shown in Eq. (2):

$$F(u,v) = g(u)g(v)\sqrt{\frac{2}{MN}} \sum_{m=0}^{M-1} \sum_{n=0}^{N-1} x(m,n) cos(\frac{(n+\frac{1}{2})\pi u}{N})$$
$$cos(\frac{(n+\frac{1}{2})\pi v}{M}) \tag{2}$$

$$(u = 0,1,2,...,M-1)(v = 0,1,2,...,N-1)$$

Among them, $x(m,n)$ represents the data in time domain before transformation, and $F(u,v)$ represents the data in frequency domain after transformation. $g(k) = \begin{cases} \frac{1}{\sqrt{2}}, k = 0 \\ 1 \ , k \neq 0 \end{cases}$.

According to Eq. (2), the discrete cosine transform dictionary D_{DCT} can be calculated.

2) Project original data Y into space formed by the dictionary D_{DCT}, and use OMP algorithm [10] to extract the coefficient matrix X_{DCT}. First, a column vector of the dictionary which can acquire a largest inner product with original data is calculated to represent original data to obtain a closest expression to the original data Y, that is, Eq. (3):

$$Y = x_1 d_{q_1} + Y_1 \tag{3}$$

Among them, d_{q_1} represents the result of seeking approximate expression of Y in the dictionary space for the first time, x_1 is the corresponding coefficient, and Y_1 is residual after the first approximate expression. Perform Schmidt orthogonalization to all the column vectors currently solving D. Then use the dictionary space to seek approximate expression of the residual Y_1 again, and generate a new residual Y_2, repeating continuously. Finally, the approximate expression of the original data Y can be completed under condition that the residual is less than the threshold, as shown in Eq. (4):

$$\begin{aligned} Y &= x_1 d_{q_1} + x_2 d_{q_2} + \cdots + x_k d_{q_k} + \varepsilon \\ &= D_{DCT} X_{DCT} + \varepsilon \end{aligned} \tag{4}$$

In this process, the orthogonalization process prevents the same dictionary column vector from being selected repeatedly. Therefore, the whole solution process can be quickly converged, and operation speed and complexity of the algorithm can be reduced.

3) Reconstruct target matrix Y' by using the dictionary D_{DCT} and the coefficient matrix X_{DCT} to complete noise reduction.

III. SIMULATION EXPERIMENT AND RESULT ANALYSIS

A. Simulation Data Construction

In order to verify effectiveness of the algorithm in this paper, a simulation experiment is designed as follows.

First, BOTDR simulation data based on the heterodyne frequency sweep detection scheme is constructed. In range of 10.600GHz~10.840GHz, frequency sweep is simulated with 5MHz steps. Set the simulated sensing fiber range to 10.15km, Brillouin center frequency to 10.721GHz. Set the simulated temperature-change region at 9.156km~9.295km, and move center frequency to 10.752GHz to simulate local BFS change when the fiber temperature changes. The simulated ideal data without noise of BOTDR system is shown in Fig. 2(a) and (b).

Fig. 2(a) is an ideal 3D-BGS image without noise. Fig. 2(b) is the top view corresponding to ideal 3D-BGS image. It can be seen from Fig. 2(a) and (b) that BFS of the simulated temperature-change region is significantly shifted from that of the non-temperature-change region. BFS of the simulated non-temperature-change region is 10.721GHz, and BFS of the simulated temperature-change region is moved to 10.752GHz.

In actual measurement process of BOTDR system, random noise will be introduced in propagation process of the signal in optical domain link and electrical domain link. These random noises are well-distributed in power and frequency, much like white Gaussian noise. Therefore, this paper adopts a method of adding 5dBm white Gaussian noise to the ideal data to simulate BOTDR original data. 3D-BGS image after adding noise is shown in Fig. 2(c) and (d).

Fig. 2(c) is 3D-BGS image after adding 5dBm white Gaussian noise. Fig. 2(d) is the corresponding top view of

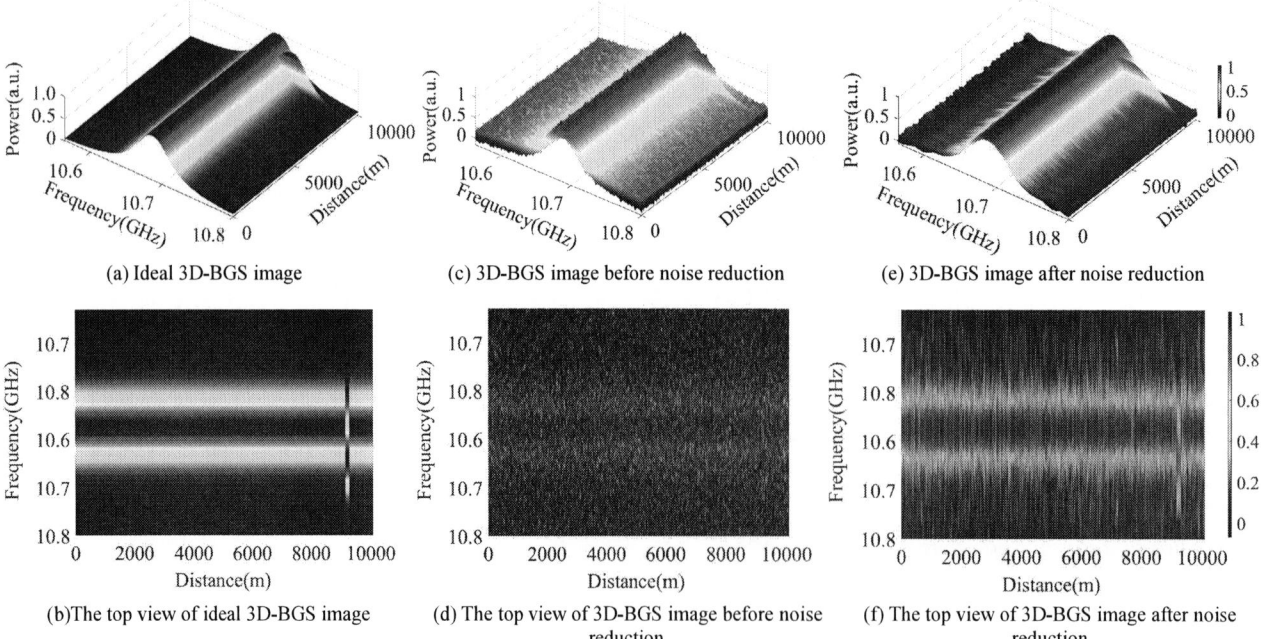

Figure 2. 3D-BGS image and its top view before and after noise reduction

978-1-6654-8699-6/22 $31.00 © 2022 IEEE

3D-BGS image. It can be seen from Fig. 2(c) and (d) that 3D-BGS data after adding noise becomes obviously messy, which is very unfavorable for subsequent demodulation of temperature and strain information.

B. Simulation Result Analysis

The simulated original data is denoised using sparse representation algorithm with preformed DCT dictionary. 3D-BGS images after noise reduction are shown in Fig. 2(e) and (f).

It can be seen that after the algorithm processing, random noise in Fig. 2(e) is significantly reduced as a whole, and the data smoothness is significantly improved. In Fig. 2(f), the overall data becomes clear, and location of the temperature-change region in the simulation data can be clearly distinguished. The effectiveness of sparse representation algorithm with preformed dictionary is proved intuitively.

Arbitrarily extract a certain point along the fiber, and draw BGS curve before and after noise reduction at this point, as shown in Fig. 3.

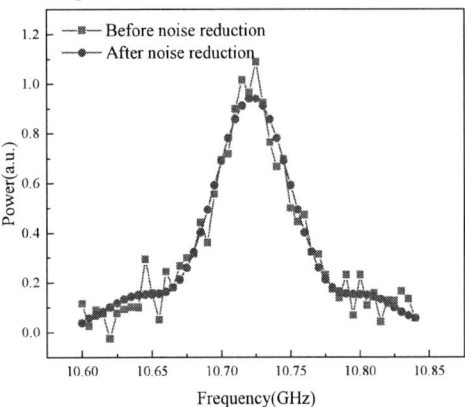

Figure 3. BGS curve before and after noise reduction at a certain position in the non-temperature-change region

As shown in Fig. 3, red curve in figure represents simulated original BGS curve of a certain point in the non-temperature-change region before noise reduction, and blue curve represents BGS curve after noise reduction by the algorithm. It can be seen from Fig. 3 that, after processing of sparse representation algorithm with preformed dictionary, fluctuation of BGS curve is obviously suppressed, and smoothness is greatly improved. The center frequency point of BGS curve after noise reduction can be found more accurately, which is convenient for demodulation of subsequent frequency shift data of BOTDR system.

Quantitatively analyze noise reduction performance of the algorithm, and draw BFS curve before and after noise reduction, as shown in Fig. 4.

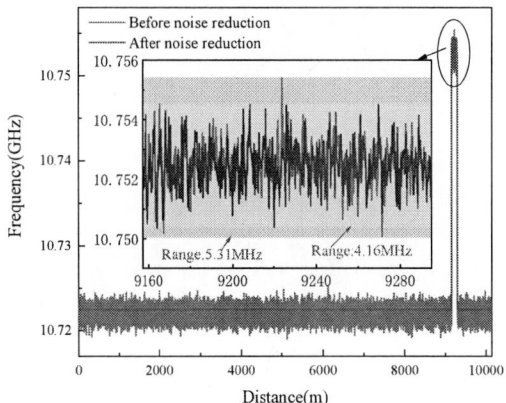

Figure 4. BFS curve before and after noise reduction

As shown in Fig. 4, red curve is BFS curve before noise reduction of the simulated original data, and blue curve is BFS curve after noise reduction. It can be seen from Fig. 4 that fluctuation range of BFS curve after noise reduction in the non-temperature-change region is significantly narrower than that before noise reduction, and data accuracy is significantly improved. Fluctuation range of BFS curve after noise reduction in the temperature-change region is also narrowed, from 5.31MHz to 4.16MHz, reduced by 1.15MHz.

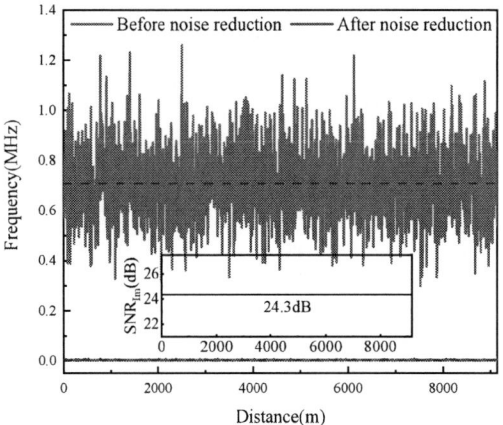

Figure 5. RMSE curve before and after noise reduction

SNR is defined as ratio of signal power to noise power, as shown in Eq. (5):

$$SNR = 10\lg\frac{P_{\text{S}}}{P_{\text{N}}} \quad (5)$$

Among them, P_{S} refers to the signal power, and P_{N} refers to the noise power. Since noise reduction processing does not affect P_{S} of the system, SNR improvement in the system is defined as Eq. (6):

$$SNR_{Im} = SNR_{Af} - SNR_{Be}$$

$$= 10\lg \frac{P_s}{P_{N-Af}} - 10\lg \frac{P_s}{P_{N-Be}} \qquad (6)$$

$$= 10\lg \frac{P_{N-Be}}{P_{N-Af}}$$

Among them, SNR_{Af} and SNR_{Be} refer to SNR of the system before and after noise reduction, and P_{N-Af} and P_{N-Be} refer to noise power before and after noise reduction. According to the data in Fig. 5 and Eq. (6), RMSE before and after noise reduction can be calculated. Using the sparse representation algorithm with preformed DCT dictionary, SNR of BOTDR system is improved by 24.3dB, which proves the effectiveness of the algorithm proposed in this paper.

IV. CONCLUSION

This paper proposes a sparse representation algorithm with preformed DCT dictionary, which uses the dictionary matrix and the coefficient matrix to reconstruct signal, discards residual, and achieves the purpose of noise reduction. In order to verify effectiveness of the algorithm, this paper regards measurement data of BOTDR system as an image to be processed, and studies noise reduction performance of the algorithm by designing a simulation experiment. The experimental results show that the algorithm has a significant noise reduction effect on the simulation data, and SNR in the non-temperature-change region is improved by 24.3dB, which proves effectiveness of the algorithm. This also provides a new idea and theoretical basis for application of image processing methods in field of SNR improvement of BOTDR.

It should be noted that, the noise reduction method in this paper is only suitable for random noise in signal of BOTDR system. In actual application of BOTDR system, a variety of non-random noises will also be introduced, and these noises have certain rules. Therefore, more in-depth research is still needed on how to filter out this part of noise. In addition, after noise reduction by the algorithm, BFS in the temperature-change region is lower than ideal signal, and improvement of SNR in the temperature-change region is not significant. These problems also need to be resolved.

ACKNOWLEDGMENT

This work was supported in part by the National Natural Science Foundation of China under Grants 61975142 and 62005190, in part by the Natural Science Foundation of Shanxi Province, China, under Grant 201901D211072, and in part by the Postdoctoral Science Foundation of China, under Grant 2021M691989.

REFERENCES

[1] Q. Bai, Q. L. Wang, et al, "Recent Advances in Brillouin Optical Time Domain Reflectometry,"Sensors (Basel), vol. 19, pp. 18-62, April 2019.

[2] N. Lalam, W. P. Ng, X. W. Dai, Q. Wu, and Y. Q. Fu, "Performance analysis of Brillouin optical time domain reflectometry (BOTDR) employing wavelength diversity and passive depolarizer techniques", Measurement Science & Technology, vol. 29, December 2018.

[3] Q. L. Wang, Q. Bai, C. S. Liang, Y. Wang, Y. T. Liu, and B. Q. Jin, "Random coding method for SNR enhancement of BOTDR", Optics Express, vol. 30, pp. 11604-11618, March 2022.

[4] H. J. He, L. Y. Shao, et al, "SNR Enhancement in Phase-Sensitive OTDR with Adaptive 2-D Bilateral Filtering Algorithm", IEEE Photonics Journal, vol. 9, June 2017.

[5] S. Le Floch, and F. Sanser, "New improvements for Brillouin optical time-domain reflectometry", presented at 25th International Conference on Optical Fibre Sensors, South Korea, 2017.

[6] Y. Y. Zhang, Y. G. Lu, Z. L. Zhang, J. M. Wang, C. J. He, and T. Wu, "Noise reduction by Brillouin spectrum reassembly in Brillouin optical time domain sensors", Optics and Lasers in Engineering, vol. 125, February 2020.

[7] M. Elad, and M. Aharon, "Image denoising via sparse and redundant representations over learned dictionaries", IEEE Transactions on Image Processing, vol. 15, pp. 3736-3745, December 2006.

[8] L. Q. Xu, J. L. Chen, and X. Y. Liu, "Sparse representation DOA estimation based on spatial smoothing", Electronic Measurement Technology, vol. 42, pp. 7-11, August 2019.

[9] S. S. Zhang, and J. H. Zhao, "Simulation analysis of reconstruction algorithm based on compressed sensing", Foreign Electronic Measurement Technology, vol. 38, pp. 44-48, October 2019.

[10] T. T. Cai, and L. Wang, "Orthogonal Matching Pursuit for Sparse Signal Recovery With Noise", vol. 57, pp. 4680-4688, July 2011.

Proceedings of the 7th Optoelectronics Global Conference (OGC 2022)

Characterization of Various Bound State Solitons Using Linear Optical Sampling Technique

Jingwen Li
School of Mechanical and Electronic Information
China University of Geosciences
Wuhan, China
llljw@cug.edu.cn

Zhichao Wu
School of Mechanical and Electronic Information
China University of Geosciences
Wuhan, China
wuzhichao@cug.edu.cn

Zhe Yu
School of Optics and Electronic Information
Huazhong University of Science and Technology
Wuhan, China
yuzhe@hust.edu.cn

Chaoyu Xu
School of Mechanical and Electronic Information
China University of Geosciences
Wuhan, China
cug-xiaoxu-2009@cug.edu.cn

Tianye Huang
School of Mechanical and Electronic Information
China University of Geosciences
Wuhan, China
tianye_huang@163.com

Songnian Fu
School of Information Engineering
Guangdong University of Technology
Guangzhou, China
songnian@gdut.edu.cn

Abstract—**The linear optical sampling (LOS) technique enables the full-field information acquisition of ultra-high speed optical signals with low bandwidth optoelectronic devices. Compared to conventional coherent detection method, LOS is a more powerful tool to monitor ultra-high speed optical signals, since it gets rid of the bandwidth limitation of electronic bottleneck. In this paper, the LOS is proposed to implement the characterization of bound state solitons generated from a passively mode-locked fiber laser. According to our experimental results, it is confirmed that the LOS enables more accurate measurements with much higher resolution in both time and spectral domains than conventional measurement devices.**

Keywords-linear optical sampling; soliton; bound state soliton

I. INTRODUCTION

Over the past decades, linear optical sampling (LOS) has shown its capability to characterize advanced modulation format signals with increasing demands for fiber optical communication[1-5]. Generally, it is challenging for conventional coherent detection, due to the bandwidth constraint of commercially available optoelectronic devices. With the help of LOS, the ultra-high speed optical signals can be accurately characterized with the recovered eye diagram and constellation, while avoiding the stringent requirement for ultra-wide bandwidth electronic devices[6-11].

Owing to the unique advantages for distortionless transmission, the optical soliton is regarded as a perfect candidate for high-speed fiber optical transmission. For decades, one of the ideas in fiber laser research is that soliton.

Moreover, soliton characterization has always been a research hotspot due to its significance for both physical mechanisms and practical uses[12, 13]. However, the acquisition of its full-field information is the prerequisite of soliton characterization. Simultaneous implementation of dispersive Fourier transform (DFT) and time-lens have been used to comprehensively characterize the spectral and temporal evolutions[14]. Nevertheless, the DFT and time-lens will greatly complicate the experimental setup. On the other hand, coherent detection is a straightforward method for obtaining the full-field information of ultrafast optical pulses[15]. However, the operation bandwidth of the photodetector (PD) and analog-to-digital converter (ADC) must meet the Nyquist sampling theorem in relation to the ultrafast optical pulse to ensure pulse characterization without distortion during coherent detection. The narrower the optical pulse width, the greater the bandwidth of the required optoelectronics. Therefore, the requirements of the balanced photodetector (BPD) with ultra-wide bandwidth and ADC with ultra-high speed and precision tremendously restrict the characterization of picosecond or sub-picosecond optical pulses[16, 17]. In 2019, Turitsyn et al carried out pulse dynamics analysis in mode-locked fiber lasers by acquiring the full-field information of ultrafast pulses through a full-bandwidth coherent receiver, and then calculated their eigenvalue distributions by nonlinear Frontier transform (NFT) to reveal the soliton dynamics using the evolution of eigenvalues in the nonlinear frequency domain[18]. In the same year, they also demonstrated the NFT-based characterization of dissipative solitons in the positive

978-1-6654-8699-6/22 $31.00 © 2022 IEEE

dispersion region[19]. In these experiments, they investigated the use of NFT and coherent reception to characterize 200-ps pulses with a BPD bandwidth of 50 GHz. However, the system has already reached its detection limit and it would be unable to cope with narrower pulses. Therefore, this inspires us to introduce the LOS technique into soliton characterization.

In this paper, we have reported, for the first time, on the experimental demonstration of the LOS enabled characterization for various bound state solitons. By finely adjusting the repetition rate difference between the solitons under test and the pulsed laser source as the local oscillator (LO), 11.62-TSa/s equivalent sampling rate arising in the LOS can be secured, while only 400-MHz BPDs and 5-GSa/s ADC are used. Finally, the time and spectral domain features of both loosely and tightly bound state solitons are accurately characterized with excellent resolutions.

II. PRINCIPLE AND EXPERIMENTAL SETUP

The mechanism of LOS is a combination of all-optical sampling and coherent detection, as shown in Fig 1. A 90-degree hybrid implements the mixing between soliton under test (SUT) and LO. The pulse-train which has a slight repetition rate difference compared to the SUT is served as the LO to guarantee the high equivalent sampling rate.

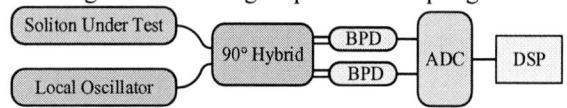

Figure 1. Linear optical sampling system. SUT: soliton under test; LO: local oscillator; BPD: balanced photodetector; ADC: analog-to-digital converter; DSP: digital signal processing.

$$\Delta T = \frac{1}{f_{LO}} - \frac{1}{f_P} \tag{1}$$

$$f_{sample} = \frac{1}{\Delta T} = \frac{f_{LO}f_P}{f_P - f_{LO}} = \frac{f_{LO}f_P}{\Delta f} \tag{2}$$

The f_{LO} and f_P are the repetition rates of the LO and the SUT, respectively. ΔT is the sampling interval, and the sampling rate f_{sample} of LOS can be cauculated from (2). Δf is the difference between f_{LO} and f_P. From (2), it is obvious that the equivalent sampling rate of LOS is inversely proportional to the repetition rate difference.

The structure of the fiber laser to generate SUT is shown in Fig 2. A commercial 976-nm laser diode (LD) with a maximum power of 400 mW introduces the pump light into the laser cavity via a 980/1550 nm wavelength division multiplexer (WDM). A 1-m EDF (Nufern, SM-ESF-7/125) is used as the gain medium. A polarization insensitive isolator (ISO) is used to ensure the unidirectional operation and suppress the detrimental reflections. The self-started mode locking can be achieved by a homemade carbon nanotube (CNT) saturable absorber, which is attached between the end surface of two standard fiber connectors. A polarization controller (PC) is used to adjust the cavity birefringence. Particularly, an optical delay line is used to precisely manipulate the repetition rate of soliton pulse-train, which can be tuned from 19.88 to 20.12 MHz with 19-Hz tuning resolution. The 3-dB spectral width of SUT can be compressed to less than 1 nm with the help of a filter. The commercial tunable pulse laser (Calmar Laser, FPL-M3CFT) is served as LO with a 99.98-MHz pulse repetition rate, and a 3-dB spectral bandwidth of 1.76 nm, which can completely cover the spectrum of the SUT, in order that all frequency information can be entirely extracted and characterized.

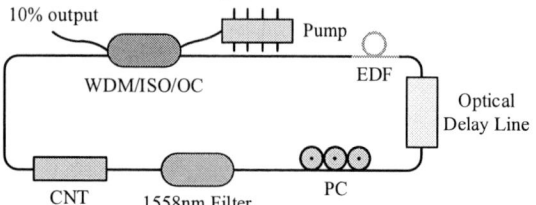

Figure 2. Passively mode-locked fiber laser to generate SUT. EDF: erbium-doped fiber; PC: polarization controller; CNT: carbon nanotube; WDM: wavelength division multiplexer; ISO: isolator; OC: optical coupler.

After coherent mixing in the 90-degree hybrid, the optical signals enter the BPDs which transfer the optical signals into analog electrical signals, which then are converted to digital signals via ADC. Finally, the effective information related to the SUT can be obtained by digital signal processing (DSP).

III. EXPERIMENTAL RESULTS

A. Weakly bound state from soliton laser

When the pump power is increased to 144 mA, by finely adjusting the PC, we observe the loosely bound state solitons from the fiber laser, as shown in Fig 3. The pulse-train, as shown in Fig 3(a), presents dual-pulse features with a repetition rate of 19.99 MHz, which corresponds to the 10-m cavity length. Normally a bound state soliton cannot be directly observed from an oscilloscope due to the narrow separation of two solitons. However, in this condition, the soliton separation is relatively large so that a 10-GHz PD could help distinguish two solitons. Therefore, it is regarded as a weakly bound state soliton. The corresponding spectrum with a central wavelength of 1558.49 nm shown in Fig 3(b) shows a typical modulation fringe which is formed by the interference between the two solitons. Actually, the entirely destructive interference should be observed on the spectrum of the two solitons with the same pulse intensity. Due to the limitation of the 0.02-nm resolution of optical signal analyzer (OSA), the fringe dips are not as deep as expected, which fully exposes the shortcoming of conventional measurement. The spectral spacing and pulse separation satisfy the basic relationship (3), where Δt is the pulse interval in the time-domain, λ is the central wavelength, $\Delta\lambda$ is the spectral modulation period. According to the inset of Fig 3(b), the spectral modulation period is 0.033 nm, corresponding to

$$\frac{1}{\Delta t} = \Delta f = \frac{c}{\lambda^2}\Delta\lambda \tag{3}$$

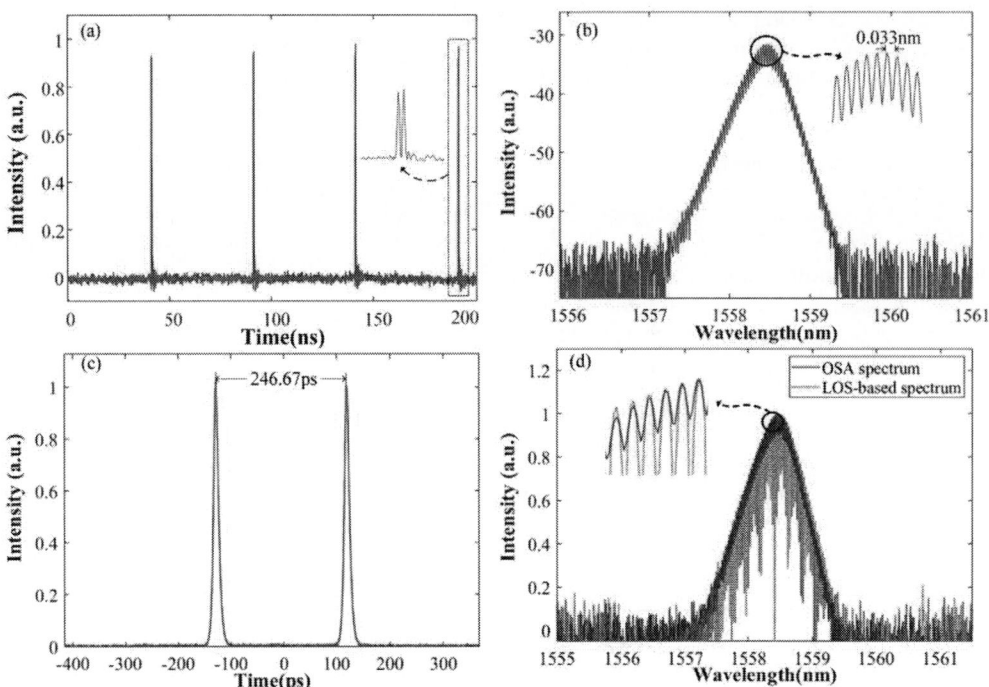

Figure 3. Weakly bound state solitons: (a) pulse-train on oscilloscope, (b) optical spectrum on OSA;

(c)(d) temporal profile and spectrum characterized by LOS.

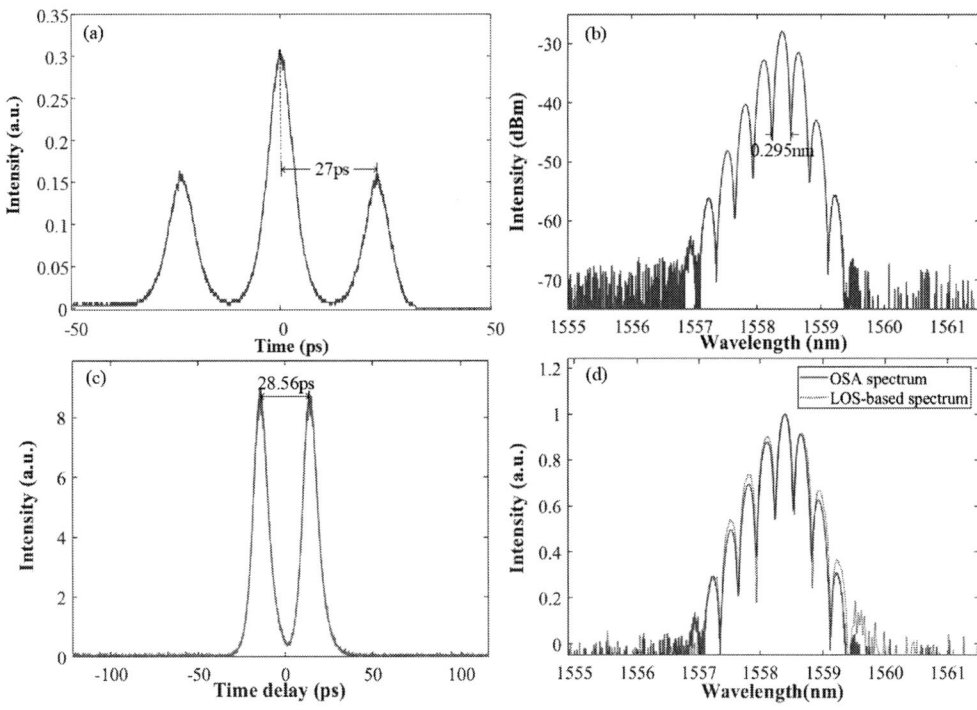

Figure 4. Strongly bound state solitons: (a) pulse profile on autocorrelator, (b) optical spectrum on OSA;

(c)(d) temporal profile and spectrum characterized by LOS

the pulse time-domain interval of 245.15 ps. Then, we implement LOS to acquire the full-field information of SUT. The obtained temporal profile and spectrum by LOS are shown in Fig 3(c) and (d). Fig 3(c) shows that the LOS with 2.73-TSa/s equivalent sampling rate can accurately characterize the time-domain features. By comparing Fig 3(d) and Fig 3(b), the spectra recovered by using the LOS technique have more obvious interference fringes, which indicates the LOS has excellent spectral resolution than OSA. The spectral resolution of LOS is 0.005 nm which is much higher than the 0.02-nm resolution of OSA. Therefore, it is verified that the LOS enables more precise spectral measurements than OSA with good accuracy.

B. Strongly Bound State from Soliton Laser

Fixing the pump power, another bound state from the fiber laser can be achieved by slightly adjusting the PC. In this condition, the two bounded solitons are close to each other with quite a small separation, resulting that the bound state cannot be directly observed on the oscilloscope. With the help of a commercial autocorrelator, a typical double-humped feature is obtained, which confirms the generation of strongly bound state solitons, as shown in Fig 4(a). The pulse time-domain interval is 27.44 ps, corresponding to the 0.295-nm spectral modulation period in Fig 4(b). Fig 4(c) and (d) show the characterization results obtained by LOS, which effectively reproduce the time and spectral domain results. Here, LOS fully reveals the unique advantage to measure ultra-fast signals, while avoiding the dependence on high-speed devices. Moreover, the equivalent sampling rate is up to 11.62-TSa/s. The time-domain resolution is as high as 0.086 ps, equivalent to 5.81-THz BPD which is far beyond commercially available optoelectronic devices.

IV. CONCLUSION

The LOS enables full-time information acquisition with low-speed and low-bandwidth devices, which effectively avoids the inevitable electronic bottlenecks in coherent detection. Here, we successfully utilize LOS to characterize weakly and strongly bound state solitons. By finely adjusting repetition rate difference between SUT and LO, up to 11.62-TSa/s equivalent sampling rate can be experimentally secured, which gives rise to 0.086-ps and 0.005-nm resolution in time and spectral domain, respectively. We believe these results provide an effective methodology for the investigation of more complex soliton dynamics.

REFERENCES

[1] C. Haffner, W. Heni, J. Fedoryshyn, J. Niegemann, A. Melikyan, D. L. Elder et al., "All-plasmonic Mach–Zehnder modulator enabling optical high-speed communication at the microscale," Nature Photonics, vol. 9, July. 2015, pp. 525-528, 2015, doi: 10.1038/nphoton.2015.127.

[2] F. E. Becerra, J. Fan and A. Migdall, "Photon number resolution enables quantum receiver for realistic coherent optical communications," Nature Photonics, vol. 9, November. 2014, pp. 48-53, doi: 10.1038/nphoton.2014.280.

[3] J. Cho, X. Chen, G. Raybon, D. Che, E. Burrows, S. Olsson et al., "Shaping lightwaves in time and frequency for optical fiber communication," Nature communications, vol. 13, February. 2022, pp. 1-11, doi: 10.1038/s41467-022-28349-x.

[4] V. Giovannetti, R. Garcia-Patron, N. J. Cerf and A. S. Holevo, "Ultimate classical communication rates of quantum optical channels," Nature Photonics, vol. 8, September. 2014, pp. 796-800, doi: 10.1038/nphoton.2014.216.

[5] S. Koenig, D. Lopez-Diaz, J. Antes, F. Boes, R. Henneberger, A. Leuther et al., "Wireless sub-THz communication system with high data rate," Nature photonics, vol. 7, October. 2013, pp. 977-981, doi: 10.1038/nphoton.2013.275.

[6] I. Coddington, W. C. Swann and N. R. Newbury, "Coherent linear optical sampling at 15 bits of resolution," Optics letters, vol. 34, July. 2009, pp. 2153-2155, doi: 10.1364/OL.34.002153.

[7] J. Song, S. Fu, B. Liu, M. Tang, P. Shum, and D. Liu, "Impact of sampling source repetition frequency in linear optical sampling," IEEE Photonics Technology Letters, vol. 28, September. 2015, pp. 15-18, doi: 10.1109/LPT.2015.2478877.

[8] C. Dorrer, D. C. Kilper, H. R. Stuart, G. Raybon, and M. G. Raymer, "Linear optical sampling," IEEE Photonics Technology Letters, vol. 15, December. 2003, pp. 1746-1748, doi: 10.1109/LPT.2003.819729.

[9] S. Wang, B. Xu, X. Fan, and Z. He, "Linear optical sampling technique for simultaneously characterizing WDM signals with a single receiving channel," Optics Express, vol. 26, January. 2018, pp. 2089-2098, doi: 10.1364/OE.26.002089.

[10] B. Liu, Z. Wu, S. Fu, Y. Feng, and D. Liu, "On-field measurement trial of 4× 128 Gbps PDM-QPSK signals by linear optical sampling," Optics Communications, vol. 384, February. 2017, pp. 36-40, doi: 10.1016/j.optcom.2016.10.018.

[11] M. Westlund, H. Sunnerud, M. Karlsson, and P. A. Andrekson, "Software-synchronized all-optical sampling for fiber communication systems," Journal of lightwave technology, vol. 23, March. 2005, p. 1088, doi:10.1109/JLT.2004.838875.

[12] Y. Wang, S. Fu, J. Kong, A. Komarov, M. Klimczak, R. Buczyński et al., "Nonlinear Fourier transform enabled eigenvalue spectrum investigation for fiber laser radiation," Photonics Research, vol. 9, June. 2021, pp. 1531-1539, doi: 10.1364/PRJ.427842.

[13] Y. Wang, S. Fu, C. Zhang, X. Tang, J. Kong, J. H. Lee et al., "Soliton distillation of pulses from a fiber laser," Journal of Lightwave Technology, vol. 39, April. 2021, pp. 2542-2546, doi: 10.1109/JLT.2021.3051036.

[14] P. Ryczkowski, M. Närhi, C. Billet, J. M. Merolla, G. Genty, and J. M. Dudley, "Real-time full-field characterization of transient dissipative soliton dynamics in a mode-locked laser," Nature Photonics, vol. 12, March. 2018, pp. 221-227, doi: 10.1038/s41566-018-0106-7.

[15] Z. Yu, S. Fu, H. He, Z. Wu, T. Huang, M. Tang et al., "Biased Balance Detection for Fiber Optical Frequency Comb Based Linear Optical Sampling," Journal of Lightwave Technology, vol. 39, June1. 2021, pp. 3458-3465, doi: 10.1109/JLT.2021.3069197.

[16] R. Liao, Z. Wu, S. Fu, S. Zhu, Z. Yu, M. Tang et al., "Fiber optics frequency comb enabled linear optical sampling with operation wavelength range extension," Optics letters, vol. 43, February. 2018, pp. 439-442, doi:10.1364/OL.43.000439.

[17] Z. Yu, Y. Wang, H. He, Z. Wu, D. Liu, L. Zhao et al., "Linear Optical Sampling Enabled Eigenvalue Analysis of Fiber Laser Radiation," 2021 Asia Communications and Photonics Conference (ACP), 2021, pp. 1-3, doi:10.1364/ACPC.2021.T4A.239.

[18] S. Sugavanam, M. K. Kopae, J. Peng, J. E. Prilepsky, and S. K. Turitsyn, "Analysis of laser radiation using the Nonlinear Fourier transform," Nature communications, vol. 10, December. 2019, pp. 1-10, doi: 10.1038/s41467-019-13265-4.

[19] I. S. Chekhovskoy, O. V. Shtyrina, M. P. Fedoruk, S. B. Medvedev, and S. K. Turitsyn, "Nonlinear Fourier transform for analysis of coherent structures in dissipative systems," Physical review letters, vol. 122, April. 2019, p. 153901, doi: 10.1109/CLEOE-EQEC.2019.8872485.

Proceedings of the 7th Optoelectronics Global Conference (OGC 2022)

Highly Sensitive Multi-coating Photonic Crystal Fiber Biosensor at Near-Infrared Waveband

Duanming Li

National key Laboratory of Science and Technology on Underwater Acoustic Antagonizing
Shanghai Marine Electronic Equipment Research Institute
Shanghai, China
duangming.li@hotmail.com

Wei Zhang

Shenzhen Institute of Information Technology
Shenzhen, China
zhwv@foxmail.com

Jiangfei Hu, Minxue Gu

Shanghai Marine Electronic Equipment Research Institute
Shanghai, China
jf330178465@163.com, gmxmx726@163.com

Abstract—A multi-coating photonic crystal fiber with a trapezoid-shaped slot (TS-PCF) for highly refractive index (RI) sensing is proposed. TiO_2 and Indium tin oxide (ITO) are coated on the bottom of the polished-area, ITO is used as the plasmonic material. Through the full-vector finite element method (FV-FEM), the wavelength sensitivity of 5000~17000 nm/RIU in the analyte RI range of 1.31 to 1.36 is obtained. Moreover, the maximum amplitude sensitivity of 358.94 RIU^{-1} is also achieved with the relevant resolution of 2.79E-5 RIU. The proposed fiber sensor can be the suitable candidate for real time detecting in medical diagnostics and biomolecules applications.

Keywords- Photonic crystal fiber, surface plasmon resonance (SPR), optical fiber sensors

I. INTRODUCTION

Surface plasmon resonance (SPR) fiber sensor for analyte detecting becomes a commonly and effectively method in recent years, comparing with the traditional prism ones [1-3]. J. Zhao et al reported a silver-coated side-polished single mode fiber (SMF) with the sensitivity of 4365 nm/RIU in the liquid RI range of 1.32 to 1.4, and the relevant resonance wavelength changed from 500 to 670 nm [4]. Photonic crystal fiber (PCF), firstly fabricated by J. C. Knight and P. J. Russell at 1996 [5], had been wildly used as an ideal platform for gas or liquid analyzing, benefitting from its flexible cladding structure [6-7]. Z. K. Fan et al demonstrated a RI sensor based on analyte-filled PCF with the average sensitivity of 7040 nm/RIU in the range of 1.4 to 1.42 [8]. Luan et al obtained the sensitivity of 2900 nm/RIU through a D-shaped micro-structured optical fiber with a hollow-core [9].

According to the coating position of active plasmonic material, the numerous SPR-PCF sensors can be divided into the nanowire or layer inside air-hole types [7], [10], [11], and

coating film on the outer silica cladding like the traditional fiber or D-shaped ones [6, 9, 12]. For the former sensors, the stronger SPR intensity can be obtained by the optimization evanescent field between fiber core and plasmonic material layer. On the other hand, the analyte can only be detected through the micro-fluid way, which results in a slow and complex processing. Considering the real time detecting applications, the outer cladding coated sensors can offer a more effective solution with its larger reaction area and the more convenient coating fabrication. However, the weak polarization maintaining ability and SPR intensity etc. are also exist.

In this paper, a multi-coating photonic crystal fiber with a trapezoid-shaped slot (TS-PCF) is designed to improve the problems mentioned above with its birefringence cladding structure and the dual coating layers of TiO_2 and ITO. The materials are deposited on the slot bottom at the vertical direction, approaching the pure silica core. Simulation work had been accomplished by the FEM with anisotropic Perfectly Matched Layers (PML), it shows that the maximum sensing resolution of 5.88E-6 RIU is obtained with the relevant sensitivity of 17000 nm/RIU, and the average spectral sensitivity is 9400 nm/RIU with the analyzing RI range of 1.31 to 1.36.

II. STRUCTURAL DESIGN AND NUMERICAL MODELING

Fig. 1(a) and (b) display the cross-section of the proposed fiber sensor, which consists of the trapezoid-shaped birefringent cladding structure. Unlike the traditional detecting ways mentioned above, the TS reaction area can efficiently provide a higher flow rate detection project, without the slow micro-hole injecting process. Meanwhile, the cladding structure has less confinement at the vertical axis, resulting in the increase of the energy leaking from the sensor core to the reaction area. Then, the surface plasmon

978-1-6654-8699-6/22 $31.00 © 2022 IEEE

resonance between the fiber core-guided mode and the SPPs mode is enhanced.

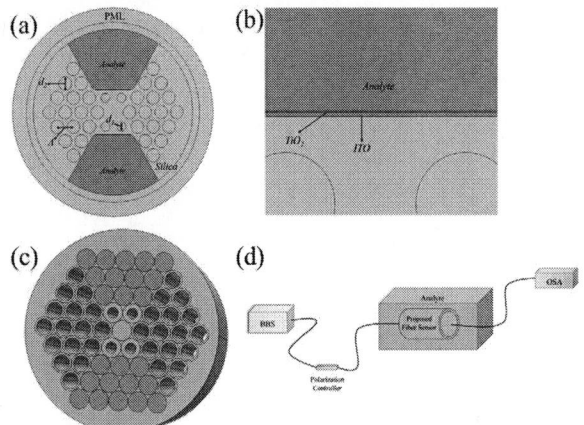

Figure 1. (a)-(c) Cross-section of the proposed fiber. (d) Schematic of the experimental set-up.

The lattice constant of Λ is 3.5 μm. The birefringence cladding air-holes are d_1 and d_2 with the sizes of 0.55Λ and 0.8Λ, respectively. The background material is pure silica. The active plasmonic material of ITO is applied, which the tunable plasmonic resonance wavelength is benefited from its broad transparency range [13, 14]. TiO$_2$ will be coated on the surface layer to improve the sensing region, which was also known in Refs. 15 and 16. TiO$_2$ and ITO coating layers are 40 and 70 nm, respectively, which are displayed as the red and green layers in Fig. 1(b).

The relevant preparation can be as: firstly, the birefringent PCF's fabrication is based on the stack and drawing procedure, while the two parts of trapezoid-shaped area will be replaced with the silica micro-stick at the vertical axis, as shown in Fig. 1(c); then, micromachining and polishing treatment are applied in these solid areas; finally, the ITO and TiO$_2$ can be coated at the bottom of the polished slot.

The schematic of experimental set-up is shown in Fig. 1(d), depending on the trapezoid-shaped slot, the real-time detecting process can be implemented without pumping analyte through the air-holes. The input light comes from a mid-IR supercontinuum light source, and the polarization state is maintained by a polarization controller. The transmission spectrum containing the SPR signals is received by the mid-IR optical spectrum analyzer (OSA). Then, the wavelength shifts resulting from the changing of analyte RI can be observed.

The material dispersion of pure silica is calculated by the Sellmeier formula [17]. Furthermore, the permittivity of the ITO coating layer is determined by Drude model,

$$\varepsilon_m = \varepsilon_{mr} + i\varepsilon_{mi} = \varepsilon_\infty - \frac{\lambda^2 \lambda_c}{\lambda_P^2(\lambda_c + i\lambda)} \qquad (1)$$

Where ε_∞ is the infrared dielectric constant [18], approach to 3.80, λ_c is 11.21076E-6 m, and λ_p is 5.6497E-7 m. Additionally, the refractive index of TiO$_2$ is calculated by [19],

$$n_{TiO_2} = \sqrt{5.913 + \frac{0.2441}{\lambda^2 - 0.0843}} \qquad (2)$$

Where the λ is the input wavelength (unit: nm). Besides, the confinement loss (CL), which depends on the imaginary part of the effective RI (n_{eff}) is calculated as,

$$CL = 8.686 \times \frac{2\pi}{\lambda} \mathrm{Im}(n_{eff}) \times 10^4 \ (dB/cm) \qquad (3)$$

III. RESULTS AND DISCUSSION

A. Coupling Characteristics

Based on the designed birefringent structure, the stronger coupling strength and the larger CL gap are obtained, comparing with the n_{eff} along to x-axis, as shown in Fig. 2(a), with the fixed parameters of Λ=3.5 μm, d_1/Λ=0.55, d_2/Λ=0.8, t_{ITO}=70 nm, t_{TiO2}=40 nm, n_a=1.34. The extinction ratio is introduced, which is defined as:

$$ER = 10 * \log\left(\frac{Loss\_y}{Loss\_x}\right) \qquad (4)$$

(a)

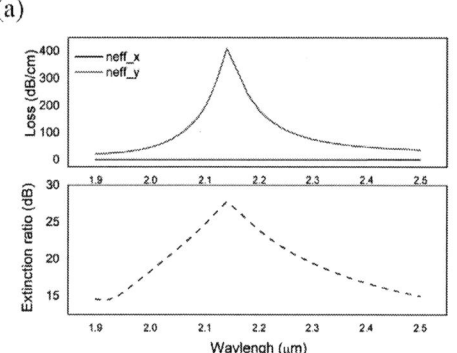

(b)

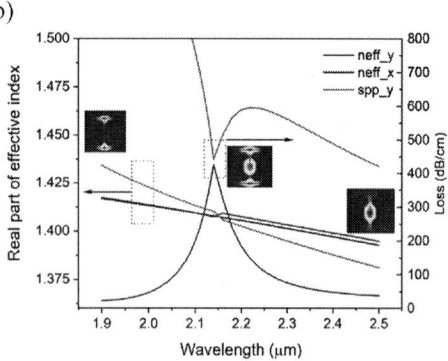

Figure 2. (a) Wavelength dependence on confinement loss (up) at two directions and the restrain ratio (down). (b) Wavelength dependence on the dispersion relations of the fundamental mode and SPP mode with the relevant confinement loss of the fundamental mode.

The maximum value of 27.7 dB is achieved at the SPR wavelength, and the bandwidth of 240 nm can be maintained during the restrain ratio of over 20 dB. So, the paper works will concentrate on the analysis at *y*-axis in the later chapters. The mode coupling evolution between the fundamental mode $n_{eff\,y}$ and the surface plasmon polariton (SPP) modes is displayed in Fig. 2(b). As the refractive index difference of the fundamental mode and the SPP mode becomes narrowing, the SPR effect gradually increases, leading to a significantly rising of loss curve till the peak (the phase matching point).

B. Sensing Response for Analyte Types

The SPR peak wavelengths with the analyte RI range of 1.31 to 1.36 are demonstrated in Fig. 3(a), with the parameters as: Λ=3.5 μm, d_1/Λ=0.55, d_2/Λ=0.8, t_{ITO}=70 nm, t_{TiO2}=40 nm. The simulation data curves show that, the small variation of analyte RI can lead to the obviously changing of corresponding SPR peak wavelength. It presents a red shift with the increasing of RI, as 1.95, 2.0, 2.06, 2.14, 2.25, 2.42 μm, and the corresponding wavelength shifts are 50, 60, 80, 110, 170 nm, respectively. Through these two data, it seems that the maximum value of wavelength shift depends on the higher RI region. However, the relevant SPR strength becomes weak and flat simultaneously, along with the increasing of analyte RI. However, gradually broadening loss curve and attenuating SPR peak would reduce the signal noise ratio, so the upper limit of the analyte RI detecting is set to 1.36 in this paper.

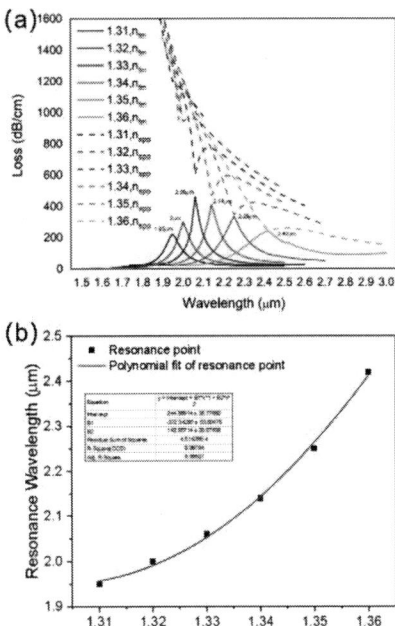

Figure 3. Fundamental and SPP loss spectrums with the variation of analyte RI from 1.31 to 1.36. (b) Polynomial Fit of the SPR wavelength on the analyte of RI variation.

The Fig. 3(b) shows a polynomial fit for the analyte RI with the relevant SPR wavelength, the smooth changing

curve is obtained, which can be a forecast method in working range. The sensitivity is from 5000 to 17000 nm/RIU, with the relevant analyte RI range of 1.31 to 1.36, indicating that the proposed sensor is more sensitive than the ones in Refs. 6, 7, 8, 16, 20. And the corresponding resolution can be as high as 5.88E-6 RIU by assuming $\Delta\lambda_{min}$ = 0.1 nm.

Meanwhile, amplitude interrogation is also applied to determine the sensor sensitivity, as known as another economy and convenient detecting method. The power is measured at the specific wavelength, the relevant formula can be expressed as:

$$S_\lambda(\lambda) = \frac{1}{\alpha(\lambda, n_a)} \frac{\partial \alpha(\lambda, n_a)}{\partial n_a} [\text{RIU}^{-1}]$$

(5)

where the $\alpha(\lambda, n_a)$ is the overall confinement loss, the RI is n_a, and $\delta\alpha(\lambda, n_a)$ is the loss change between the adjacent analyte RI. From the Fig. 4, the amplitude sensitivity depending on the RI and the thickness of ITO are illustrated. The proposed sensor reaches 273.4 RIU^{-1} with the analyte RI of 1.32. And it achieves 358.94 RIU^{-1} with the ITO thickness of 60 nm, which can be equivalent to the resolution of 2.79E-5 RIU by assuming the change in transmitted intensity of 1%.

(a)

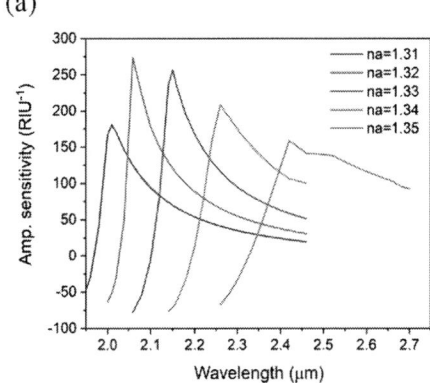

(b)

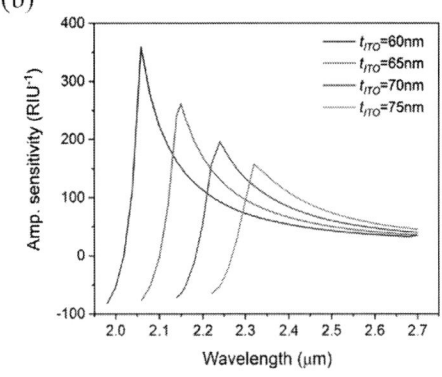

Figure 4. Variations of analyte RI (a) and ITO thickness (b on the sensor sensitivity.

C. Coating Layers Influence

As the active plasmonic material, the influence depending on the thickness of ITO is also considered in the work. Fig. 5 displays that, the SPR wavelength has the red shift with the thickening ITO layer, and the other parameters of fiber are as: Λ=3.5 μm, d_1/Λ=0.55, d_2/Λ=0.8, t_{TiO2}=40 nm, na=1.34. Based on the CL curves' comparison, the ITO thickness significantly affects the coupling wavelength and SPR strength. From the Fig. 5, it seems that the thinner of ITO, the higher of relevant peak is obtained, and the blue shift of SPR wavelength can be observed at the range of 2.32 μm to 1.98 μm. However, considering the complexity and accuracy of coating process, the thickness of ITO is set to 70 nm, which the corresponding value (411dB) is 85.5% of the maximum loss at 60nm, and 109.42% of the minimum loss at 75 nm. The coating-tolerance of ±10 nm can be obtained by the proposed sensor.

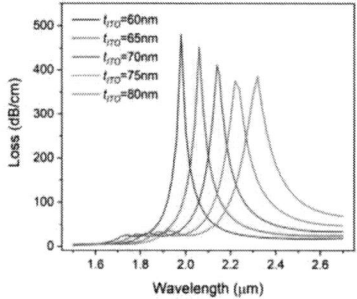

Figure 5. Variations of ITO thickness on the fundamental mode confinement loss.

Furthermore, the influence of TiO$_2$ layer is also investigated, due to its contribution in the mode coupling. As shown in Fig. 6, fixing other parameters of fiber as: Λ=3.5 μm, d_1/Λ=0.55, d_2/Λ=0.8, t_{ITO}=70 nm, na=1.34, it illustrates that the peak wavelength has the red shift, and the coupling strength reaches the peak, then decreasing gradually along with the increasing of TiO$_2$. The relevant maximum and minimum values are 486.74 dB and 331.7 dB, respectively. Even though the higher loss can be achieved at the thickness of 35 nm, the thickness of 40 nm (corresponding value: 410 dB) is chose as the final parameter, considering the fabrication accuracy.

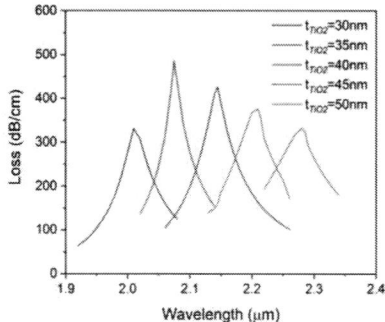

Figure 6. Variations of TiO$_2$ thickness on the fundamental mode confinement loss.

IV. CONCLUSIONS

In summary, with the dual-coating layers of ITO and TiO$_2$, the TS-PCF for RI sensing is designed and analyzed.

The simulation results show that, the proposed sensor has the wavelength sensitivity of 5000~17000 nm/RIU at the analyte RI range of 1.31 to 1.36. The maximum value is obtained with the relevant resolution of 5.88E-6 RIU, as well as the amplitude sensitivity of 358.94 RIU^{-1} with the resolution of 2.79E-5 RIU. Based on the trapezoid-shaped slot and dual-coating layers, the TS-PCF sensor can be utilized for the convenient real-time detecting application at the near-infrared band.

ACKNOWLEDGMENT

The authors wish to thank the anonymous reviewers for their valuable suggestions.

REFERENCES

[1] A. Tubb, F. Payne, R. Millington, C. Lowe, "Single-mode optical fibre surface plasma wave chemical sensor," *ens. Actuators B*, vol.41, 1997, pp. 71–79.

[2] B. Gupta and R. Verma, "Surface plasmon resonance-based fiber optic sensors: principle, probe designs, and some applications," *J. Sens.*, vol. 2009, 2009, pp. 979761.

[3] J. Homola, S. S. Yee, and G. Gauglitz, "Surface plasmon resonance sensors: review," *Sens. Actuat. B Chem.*, vol. 54, 1999, pp. 3-15.

[4] J. Zhao, S. Q. Cao, C. R. Liao, Y. Wang, G. J. W, X. Z. Xu, C. L. Fu, G. W. Xu, J. R. Lian and Y. P. Wang, "Surface plasmon resonance refractive sensor based on silver-coated side-polished fiber," *Sensors and Actuators B: Chemical*, vol. 230, 2016, pp. 206-211.

[5] J. C. Knight, T. A. Birks, and P. St. J. Russell, "All-silica single-mode optical fiber with photonic crystal cladding." *Opt. Lett.*, vol. 21, 1996, pp. 1547-1549.

[6] G. W. An, S. G. Li, H. Y. Wang and X. N. Zhang, "Metal oxide-graphene-based quasi-D-shaped optical fiber plasmonic biosensor," *IEEE Photonics Journal*, vol. 9, 2017, 6803309.

[7] D. M. Li, W. Zhang, H. Liu, J. F. Hu and G. Y. Zhou. "High sensitivity refractive index sensor based on multi-coating photonics crystal fiber with surface plasmon resonance at near-infrared wavelength." *IEEE Photonics Journal*, vol. 9, 2017, pp. 6801608.

[8] Z. Fan, S. G. Li, Q. Liu, G. An, H. Chen, J. Li, D. Chao, H. Li, J. Zi, and W. Tian, "High sensitivity of refractive index sensor based on analyte-filed photonic crystal fiber with surface plasmon resonance," *IEEE Photonics Journal*, vol. 7, 2015, pp. 4800809.

[9] N. Luan, R. Wang, W. Lv, and J. Yao, "Surface plasmon resonance sensor based on D-shaped microstructured optical fiber with hollow core," *Opt. Express*, vol. 23, 2015, pp. 8576–8582.

[10] B. Shuai, L. Xia, and D. Liu, "Coexistence of positive and negative refractive index sensitivity in the liquid-core photonic crystal fiber based plasmonic sensor," *Opt. Express*, vol. 20, 2012, pp. 25858–25866.

[11] A. Rifat, G. Mahdiraji, D. Chow, Y. Shee, R. Ahmed, and F. Adikan, "Photonic crystal fiber-based surface plasmon resonance sensor with selective analyte channels and graphene-silver deposited core," *Sensors*, vol. 15, 2015, pp. 11499–11510.

[12] E. K. Akowuah, T. Gorman, H. Ademgil, S. Haxha, G. K. Robinson, and J. V. Oliver, "Numerical analysis of a photonic crystal fiber for biosensing applications," *Quantum Electron. IEEE J.*, vol. 48, 2012, pp. 1403–1410.

[13] J. N. Dash and R. Jha, "SPR biosensor based on polymer PCF coated with conducting metal oxide," *IEEE Photonics Technol. Lett.*, vol. 26, 2014, pp. 595–598.

[14] S. Singh and B. D. Gupta, "Simulation of a surface plasmon resonance-based fiber-optic sensor for gas sensing in visible range using films of nanocomposites." *Meas. Sci. Technol.*, vol. 21, 2010, pp. 115202.

[15] D. Gao, C. Guan, Y. Wen, X. Zhong, and L. Yuan, "Multi-hole fiber based surface plasmon resonance sensor operated at near-infraed wavelengths," *Opt. Commun.*, vol. 313, 2014, pp. 94-98

[16] A. A. Rifat, G. A. Mahdiraji, Y. M. Sua, R. Ahmed, Y. G. Shee, and F. R. M. Mahamd Adikan, "Highly sensitive multi-core flat fiber surface plasmon resonance refractive index sensor," *Opt. Express*, vol. 24, 2016, pp. 2485-2495.

[17] G. P. Agrawal, *Nonlinear Fiber Optics*, San Diego, CA, USA: Academic (1989).

[18] C. Rhodes, M. Cerruti, A. Efremenko, M. Losego, D. E. Aspens, J. P. Maria and S. Franzen, "Dependence of plasmon polaritons on the thickness of ITO thin films," *J. Appl, Phys.*, vol. 103, 2008, pp. 1-6,.

[19] J. Dash and R. Jha, "On the performance of graphene-based D-shaped photonic crystal fibre biosensor using surface plasmon resonance," *Plasmonics*, vol. 10, 2015, pp. 1123-1131.

[20] Z. Tan, X. Li, Y. Chen and P. Fan, "Improving the sensitivity of fiber surface plasmon resonance sensor by filling liquid in a hollow core photonic crystal fiber," *Plasmonics*, vol. 9, 2014, pp. 167–173.

Proceedings of the 7th Optoelectronics Global Conference (OGC 2022)

A Sensitive Material for Optical Fiber Sensor

——$Dy_8Fe_{16-x}Co_x$ (x=0,2,3): First-principles calculations

Yue Yuan [a,c], Tao Shen* [a,b,c], Chi Liu [a,b], Tianyu Yang [a,c], Ai Na Gong [a,c]

[a]Heilongjiang Province Key Laboratory of Laser Spectroscopy Technology and Application, Harbin University of Science and Technology, Harbin 150080, China.
[b]Key Laboratory of Engineering Dielectrics and Its Application, Ministry of Education, Harbin University of Science and Technology, Harbin, 150080, China.
[c]Heilongjiang Provincial Key Laboratory of Quantum Control, Harbin University of Science and Technology, Harbin 150080, China.
*Corresponding author
e-mail: taoshenchina@163.com

Abstract—In this paper, the structural, electrical, magnetic and optical properties of $Dy_8Fe_{16-x}Co_x$ (x=0,2,3) compounds have been studied by first principles calculation. Among them, the structural properties show that the atoms of $DyFe_2$, $Dy_8Fe_{14}Co_2$ and $Dy_8Fe_{13}Co_3$ are symmetrically located, and each layer forms a triangle. In addition, the calculation of electrical properties shows that the three compounds have obvious metal properties and ferromagnetism. At the same time, an interesting phenomenon was found in the magnetic calculation, The magnetic moment of the same electron orbit of the same element is the same on the premise of ensuring the triangle symmetry of each layer. Finally, in order to study the optical properties of $Dy_8Fe_{16-x}Co_x$ (x=0,2,3), We calculated the optical absorption spectra of the three compounds with photon energy. It is found that the absorption spectrum is red shifted in the ultraviolet range. The above basic physical property calculations could provide theoretical guidance for the optical elements of $Dy_8Fe_{16-x}Co_x$ (x=0,2,3). It also lays a foundation for improving the sensitivity of $Dy_8Fe_{16-x}Co_x$ (x=0,2,3) compound optical fiber sensing.

Keywords-electrical properties; magnetic; optical properties; optical fiber sensing

I. INTRODUCTION

The magnetic field is everywhere. The earth itself is a huge magnet, and we live in the magnetic field every day. The method of magnetic field detection can be applied to many fields. For example, in medical [1], power detection [2], aerospace [3] and other fields. Optical fiber sensor has been applied in the detection of magnetic field environment because of its small size, strong anti-interference ability, digitization and high sensitivity. However, in order to improve the performance of the sensor, the magnetostrictive effect of $DyFe_2$ has been concerned by many researchers. As a photosensitive material, $DyFe_2$ can be combined with optical fiber to form an optical fiber sensor, which can be used to measure the magnetic field and other parameters.

As early as 1978, A. Clark et al. Found that the saturation magnetization of the compounds doped with Tb in $DyFe_2$ is much stronger than before in the temperature range of 0~300k. The concept of anisotropic compensation was

introduced. It can reduce the anisotropy of the compound while maintaining good magnetostrictive properties [4]. Verhoeven J D et al. Discussed the values of x and y of $Tb_xDy_{1-x}Fe_y$ metal compounds at 6.9 MPa in 1988. It was found that the magnetostrictive strain A of the alloy will change obviously when the applied magnetic field increases to a suitable value. Since then, researchers have started to doping different concentrations of Tb in $DyFe2$ to improve the magnetostrictive properties of this metal compound [5]. In the same year, Clark A E team observed the magnetostrictive "jump" in the research. In practical application, if only one trigger magnetic field centered on the optimal bias field is applied, a lot of work can be completed [6]. In the past decades, researchers have devoted themselves to studying the influence of different ratios of Tb and Dy on magnetostriction [7,8]. Although the magnetostriction of this metal compound is very practical，but Tb and Fe are rare earth elements with high cost, which are difficult to be widely used in magnetic field detection. Therefore, we should take some measures to reduce the content of rare earth elements and have excellent magnetostrictive properties.

In recent years, in order to understand this compound clearly, many researchers began to study its basic physical properties. In 2016, A. Bentouaf et al. analyzed the physical properties of binary Laves phase $TbCo_2$ and $TbFe_2$ metal compounds. The state types associated with each electron orbit are calculated by the first principles. It emphasizes the role of relevant electronic processing in accurately describing binary Laves phase compounds [9]. In 2018, Li f et al. Turned their attention to non-rare earth elements, the team used X-ray and XRD to study the structure, magnetic and magnetostrictive properties of $Tb_{0.2}Dy_{0.8-x}Pr_x$ $(Fe_{0.8}Co_{0.2})_{1.93}$ compound. The results show that, $Tb_{0.2}Dy_{0.8-x}Pr_x(Fe_{0.8}Co_{0.2})_{1.93}$ compound has excellent magnetostrictive properties at room temperature. In the compound, Co replaces Fe to move the anisotropy compensation point to the Pr side [10,11]. Although there are many studies on metal compounds, it is still to be studied to find a compound with simple chemical formula and high stability that non-rare

978-1-6654-8699-6/22 $31.00 © 2022 IEEE

earth elements replace rare earth elements. What's more，there is a lack of basic research on this compound, especially the optical properties. Therefore, in this article, we proposed a single over Metal Co doped dyfe2 intermetallic compound, and comprehensively studied the physical properties of two concentration Co doped $DyFe_2$ compounds and intrinsic $DyFe_2$ compounds by using the first principle.

II. COMPUTATIONAL DATAILEDT

In this article, we used Vienna Ab-initio Simulation Package (VASP) software to calculate the structure, electrical, magnetic and optical properties of $Dy_8Fe_{16-x}Co_x$ (x=0,2,3) by using density functional theory. In the calculation process, we have used the Generalized Gradient Approximation (GGA) combined with the U-Hubbard Hamiltonian (GGA+U). As everyone knows, the calculation of high electron orbitals of rare earth elements by GGA method ignores the Coulomb repulsion between electrons in the open shell. The method of GGA+U can achieve a correction effect [12]. At the same time, the calculated value is closer to the experimental value. In this study, we determined a set of U values based on previous studies [9,12], U=5.8 and J=0.7 for Dy, U=6.8 and J=6.8 for Fe, U=7.8 and J=8.92 for Co. In order to obtain high precision convergence, we set the EDIFF value to 1E-6. At the same time, the cut-off energy parameter (ENCUT) was tested regarding the model of $Dy_8Fe_{16-x}Co_x$ (x=0,2,3). Finally, we have selected ENCUT=300.

III. RESULTS AND DISCUSSION

A. Structural Properties

First, we doped Co into $DyFe_2$ compound to obtain $Dy_8Fe_{16-x}Co_x$ (x=0,2,3) compound. Fig. 1 (a) shows the intrinsic $DyFe_2$ compound when x=0. There are 8 Dy atoms and 16 Fe atoms in the cell. The intrinsic $DyFe_2$ compound can be regarded as a layered structure, and each layer forms a triangle. Fig. 1 (b) and (c) show $Dy_8Fe_{16-x}Co_x$ (x=0,2,3) compounds when Co was doped in the same layer. We optimized the structures of the three models and obtained the steady-state structure with the lowest energy.

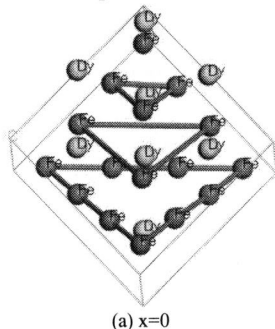 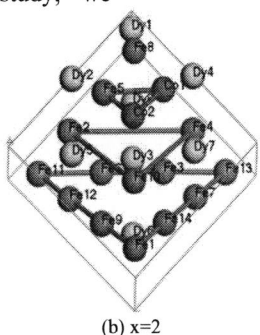 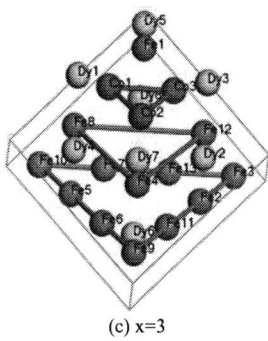

(a) x=0   (b) x=2   (c) x=3

Figure 1. The crystal structure of $Dy_8Fe_{16-x}Co_x$ (x=0,2,3) (a) x =0: DyFe2, (b) x =2: $Dy_8Fe_{14}Co_2$, (c) x =3: $Dy_8Fe_{13}Co_3$.

B. Electronic Properties

Electronic properties are the key to analyze the basic physical properties of materials. In this paper, the energy bands and density of states of $Dy_8Fe_{16-x}Co_x$ (x=0,2,3) are calculated. Firstly, according to the calculation results of energy band properties, it can be found that the conduction and valence bands overlap at the Fermi level, as shown in Fig. 2 (a~f). This is an important manifestation of metallic properties. Secondly, comparing the energy band of $Dy_8Fe_{16-x}Co_x$ (x=0,2,3) in Fig. 2 (b) and (c) compound with the energy band $DyFe_2$ in Fig. 2 (a), it can be seen that the energy band at -25ev~-20ev in Fig. 2 (b) and (c) is significantly widened. At the same time, the energy band becomes dense at -10ev~-5ev, which indicates that a new impurity band was generated due to the addition of Co. Similarly, Fig. 2 (d~f) has the same phenomenon, After Co was doped, the electronic activity of some orbitals will be enhanced, and the electronic activity of some orbitals will also be suppressed. The interaction between new electrons will also lead to the change of magnetism [13].

In order to clarify how the contribution of each electronic orbital to the band structure is represented, we calculated the total density of states (DOS) and partial density of states (TDOS) of the three compounds, as shown in Fig. 3 (a~c). Obviously, the total density of states of the three compounds as a whole is antisymmetric, indicating that the compounds are ferromagnetic compounds. Obviously, we can see from Fig. 3 (a) that the intrinsic $DyFe_2$ has three obvious strong peaks at -45ev, -20ev and 0ev. The energy contribution of the total density of states at -45ev and -20ev mainly comes from the s and p orbitals of Dy. In Fig. 3 (b) and (c), the strong peaks of s, p and f orbitals of Dy disappeared obviously. In addition, in Figure 3 (a), the s orbit of Dy only contributes at -45ev. The energy contribution is relatively small, but the s orbital energy contribution of Dy increases obviously after the addition of Co was doped, and other orbital energy contributions of Dy are also enhanced. This is because the addition of Co promotes the activity of the three orbital electrons of Dy elements s, p and f. As a result, the electron energy contribution of each orbital doped with Co element is stronger than that of intrinsic $DyFe_2$. The DOS value at Fermi level is not zero, which indicates that this compound has metallic properties, which is consistent with the conclusion obtained from electronic properties [14].

978-1-6654-8699-6/22 $31.00 © 2022 IEEE

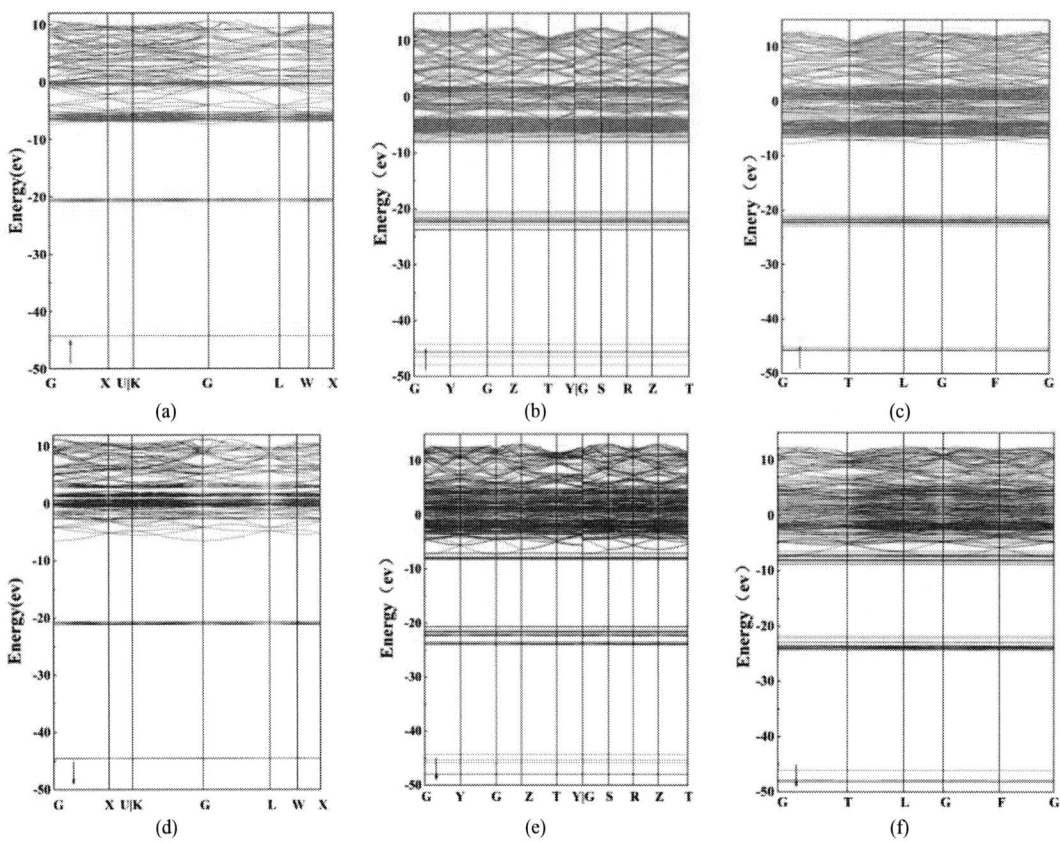

Figure 2. The calculated band structure of the $Dy_8Fe_{16-x}Co_x$ (x=0,2,3) as obtained within the (a-f) GGA+U for spin-up and spin-down.

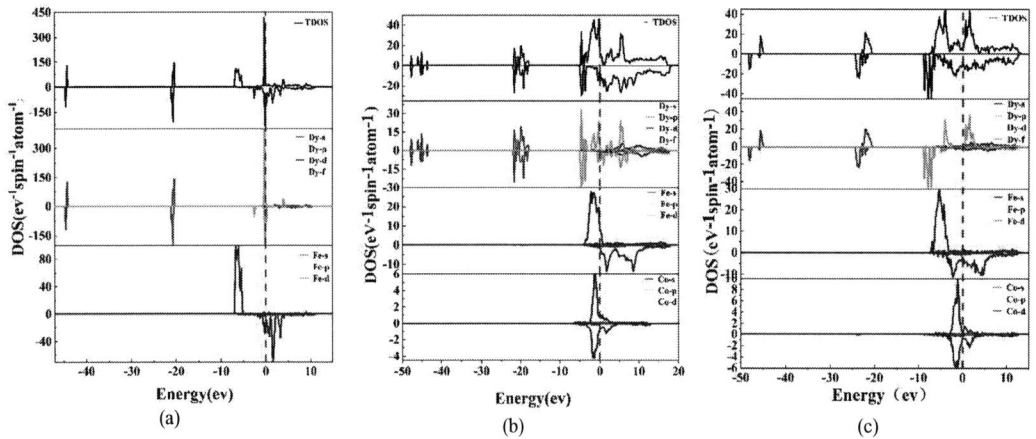

Figure 3. The TDOS and PDOS about $Dy_8Fe_{16-x}Co_x$ (x=0,2,3) (a) x =0: DyFe2, (b) x =2: $Dy_8Fe_{14}Co_2$, (c) x =3: $Dy_8Fe_{13}Co_3$.

C. Magnetic Properties

In order to understand how the atomic magnetic moments at different positions contribute, we have carried out magnetic calculations. Table I shows the magnetic moment values of intrinsic $DyFe_2$, it can be seen that the magnetic moment of each atom is very regular. The magnetic moment of the s orbital of all Dy elements is -0.034, P orbital of all Dy elements is 0, d orbital of all Dy elements is -0.036, f orbital of all Dy elements is 2.104. In addition, the magnetic

moment of the s orbital of all Fe elements is 0.009, p orbital of all Fe elements is -0.024, dorbital of all Fe elements is 3.100. All orbital magnetic moments of each atom are equal, which can be clearly seen from Fig. 1 (a). All atoms in the intrinsic $DyFe_2$ compound present a very symmetrical state, so the interaction between electrons is equal, so the magnetic moment of the same electron orbit of the same element is also equal.

$Dy_8Fe_{16-x}Co_x$ (x=2,3) compound can be found in Table II, and the law of magnetic moment value is obviously broken.

This phenomenon can also be found from the structural analysis. As shown in Fig. 1 (b), the positions of the two doped atoms are on the same layer, which obviously breaks the principle of triangular symmetry of each layer of Fe. However, in Table III, an interesting phenomenon is found from the magnetic moment values of $Dy_8Fe_{16-x}Co_x$ (x=,3) compounds. There are three atoms in the same layer, such as $Dy_{1/3/5}$, So the magnetic moment of their same orbit is the same. Similarly, the three atoms $Dy_{2/4/8}$ are in the same position. However, Dy_6 and Dy_7 are in two separate layers, so their magnetic moments are different. For Fe and Co atoms, the atoms at the three vertices of the triangle on each layer are symmetrical, and the magnetic moment values of the same orbital of the atoms at these three positions are the same. The atoms at $Fe_{3/9/10}$ and $Fe_{4/12}$ are the same. However, the total magnetic moment of doping in the case of triangular symmetry is smaller than that in the case of breaking symmetry. This is an interesting discovery that we can further study in the future.

TABLE I.   THE MAGNETIC MOMENT OF TOTAL AND PARTIAL ABOUT $DyFe_2$

	s	p	d	f	total
Dy_{1-8}	-0.034	-	-0.036	2.104	2.034
Fe_{1-16}	0.009	-0.024	3.100	-	3.086
tot	-0.124	-0.381	49.315	16.831	65.641

TABLE II.   THE MAGNETIC MOMENT OF TOTAL AND PARTIAL ABOUT $Dy_8Fe_{14}Co_2$

	s	p	d	f	total
$Dy_{1/2}$	-0.020	-0.007	-0.118	2.640	2.495
$Dy_{3/4}$	-0.049	-0.017	-0.371	-4.978	-5.415
$Dy_{5/8}$	-0.009	-0.016	-0.087	4.936	4.824
Dy_6	-0.029	-0.003	-0.147	0.055	-0.125
Dy_7	-0.049	-0.021	-0.424	-4.979	-5.473
$Fe_{1/7/13/14}$	0.011	-0.031	3.116	-	3.095
$Fe_{2/11}$	0.005	-0.007	2.983	-	2.982
$Fe_{3/4/9/10}$	0.012	-0.012	3.020	-	3.012
$Fe_{5/8}$	0.013	-0.007	3.029	-	3.035
$Fe_{6/12}$	0.013	-0.007	3.035	-	3.042
$Co_{1/2}$	-0.020	-0.079	0.382	-	0.283
tot	-0.119	-0.511	41.680	0.272	41.321

TABLE III.   THE MAGNETIC MOMENT OF TOTAL AND PARTIAL ABOUT $Dy_8Fe_{13}Co_3$

	s	p	d	f	total
$Dy_{1/3/5}$	-0.051	-0.024	-0.402	-4.983	-5.459
$Dy_{2/4/8}$	-0.048	-0.030	-0.384	-4.980	-5.442
Dy_6	-0.035	-0.020	-0.276	-1.451	-1.783
Dy_7	-0.071	-0.023	-0.447	-4.988	-5.539
Fe_1	0.013	-0.032	3.113	-	3.093
$Fe_{2/5/6/7/11/13}$	0.013	-0.052	3.052	-	3.013
$Fe_{3/9/10}$	0.004	-0.049	3.061	-	3.016
$Fe_{4/8/12}$	0.009	-0.027	3.032	-	3.014
$Co_{1/2/3}$	-0.019	-0.096	0.455	-	0.340
tot	-0.331	-1.063	37.984	-36.338	0.252

D. Optical Properties

In this study, we also calculated the relationship between the absorption spectrum of the intermetallic compound of $Dy_8Fe_{16-x}Co_x$ (x=0,2,3) and the photon energy, as shown in Fig. 4. Intrinsic $DyFe_2$ has three very obvious absorption peaks, and there is a strongest absorption peak at 4.4ev, and there are also two weak absorption peaks at 0.5ev and 2.5ev. Obviously, the three absorption peaks are suppressed after Co was doped. After doping two co atoms, there is only the strongest absorption peak at 4ev, while after doping three Co atoms, the strongest absorption peak is at 2.3ev, and another less obvious absorption peak appears at 6ev. More importantly, the absorption spectrum of the doped system will produce red shift in the ultraviolet range, and the red shift of the absorption spectrum of the doped two Co is stronger. The above phenomena can provide theoretical guidance for the optical fiber sensor of $Dy_8Fe_{16-x}Co_x$ (x=0,2,3 compound. The optical element of this compound can produce red shift in the ultraviolet region [15,16].

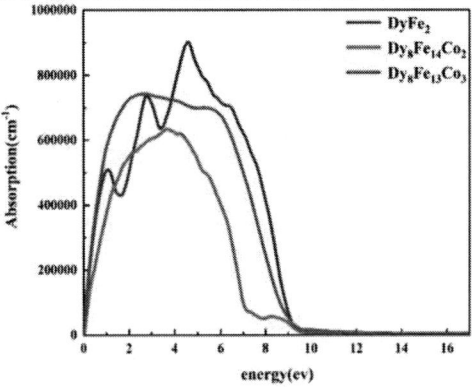

Figure 4. Absorption spectra of Dy8Fe16-xCox (x=0,2,3).

IV. CONCLUSION

The structural, electrical and magnetic properties of $Dy_8Fe_{16-x}Co_x$ (x=0,2,3) compounds were studied by GGA+U method. Firstly, through the structural models of the three compounds, we can intuitively see the symmetric relationship between atoms. Secondly, in the calculation of electrical properties, the energy band and density of states show the same metal characteristics, and it is found that Co-doping will improve the activity of electrons, resulting in new impurity bands. More importantly, in the study of magnetism, we found an interesting phenomenon. Doping Co on the basis of breaking the triangle symmetry will produce different magnetic moments for the same electron orbit of the same element. Finally, in the study of optical properties, we calculated the optical absorption spectrum. From the change between the absorption intensity and energy, it can be found that the compound doped with Co will inhibit the two absorption peaks of the intrinsic $DyFe_2$. And the absorption spectrum in the ultraviolet region is red shifted. The above research has laid a foundation for understanding the basic physical properties of $Dy_8Fe_{16-x}Co_x$ (x=0,2,3). Researchers can take corresponding measures to modify the optical fiber sensor of $Dy_8Fe_{16-x}Co_x$(x=0,2,3) compound, so as to improve the performance of the sensor and broaden the application field of the sensor.

ACKNOWLEDGMENT

The research was supported by the National Natural Science Foundation of China (52102164).

REFERENCES

[1] Fernando T, Gammulle H, Denman S, et al. Deep learning for medical anomaly detection–a survey[J]. ACM Computing Surveys (CSUR), 2021, 54(7): 1-37.

[2] Wang S, Tang J, Wang W, et al. Electrical detection of magnetic skyrmions[J]. Journal of Low Temperature Physics, 2019, 197(3): 321-336.

[3] Ismagilov F R, Papini L, Vavilov V E, et al. Design and performance of a high-speed permanent magnet generator with amorphous alloy magnetic core for aerospace applications[J]. IEEE Transactions on Industrial Electronics, 2019, 67(3): 1750-1758.

[4] Clark A, Abbundi R, Gillmor W. Magnetization and magnetic anisotropy of $TbFe_2$, $DyFe_2$, $Tb_{0.27}Dy_{0.73}Fe_2$ and $TmFe_2$[J]. IEEE Transactions on Magnetics, 1978, 14(5): 542-544.

[5] Verhoeven J D, Ostenson J E, Gibson E D, et al. The effect of composition and magnetic heat treatment on the magnetostriction of $Tb_x Dy_{1-x} Fe_y$ twinned single crystals[J]. Journal of applied physics, 1989, 66(2): 772-779.

[6] Clark A E, Teter J P, McMasters O D. Magnetostriction ''jumps'' in twinned $Tb_{0.3}Dy_{0.7}Fe_{1.9}$[J]. Journal of applied physics, 1988, 63(8): 3910-3912.

[7] Yu H, Xiao J, Schultheiss H. Magnetic texture based magnonics[J]. Physics Reports, 2021, 905: 1-59.

[8] Wang N, Liu Y, Zhang H, et al. Fabrication, magnetostriction properties and applications of Tb-Dy-Fe alloys: a review[J]. China Foundry, 2016, 13(2): 75-84.

[9] Bentouaf A, Mebsout R, Rached H, et al. Theoretical investigation of the structural, electronic, magnetic and elastic properties of binary cubic C15-Laves phases TbX_2 (X= Co and Fe)[J]. Journal of Alloys and Compounds, 2016.

[10] Li F, Liu J J, Zhu X Y, et al. Composition anisotropy compensation and magnetostriction of Co-doped Laves compounds $Tb_{0.2}Dy_{0.8-x}PrxFe_{1.93}$ ($0 \leq x \leq 0.40$)[J]. Solid State Communications, 2018, 275: 63-67.

[11] Miyake T, Harashima Y, Fukazawa T, et al. Understanding and optimization of hard magnetic compounds from first principles[J]. Science and Technology of Advanced Materials, 2021, 22(1): 543-556.

[12] Liu C, Shen T, Feng Y, et al. First-principles calculations to investigate electronic, magnetism, elastic properties of $Tb_xDy_{1-x}Fe_2$ (x= 0, 0.25, 0.5, 1)[J]. Journal of Magnetism and Magnetic Materials, 2022, 547: 168953.

[13] Jia X F, Hou Q Y, Xu Z C, et al. Effect of Ce doping on the magnetic and optical properties of ZnO by the first principle[J]. Journal of Magnetism and Magnetic Materials, 2018, 465: 128-135.

[14] Bentouaf A, Benmedjahed T, Mebsout R, et al. Electronic structure of $REFe_2$ (RE= Dy, Ho and Er) intermetallic compounds: Ab initio spin-density functional theory[J]. Solid State Communications, 2019, 296: 42-48.

[15] Wang X, Huang S, Deng L, et al. Enhanced optical absorption of Fe-, Co-and Ni-decorated Ti_3C_2 MXene: A first-principles investigation[J]. Physica E: Low-dimensional Systems and Nanostructures, 2021, 127: 114565.

[16] Berdiyorov G R. Optical properties of functionalized $Ti_3C_2T_2$ (T= F, O, OH) MXene: First-principles calculations[J]. Aip Advances, 2016, 6(5): 055105.

Proceedings of the 7th Optoelectronics Global Conference (OGC 2022)

Isopropanol-sealed Cascaded-Peanut Taper fiber Structure for Temperature Sensing Incorporated Fiber Laser

Weihao Lin
Department of Electrical and Electronic Engineering
Southern University of science and technology
Shenzhen, China
e-mail: 11510630@mail.sustech.edu.cn

Jie Hu
Department of Electrical and Electronic Engineering
Southern University of science and technology
Shenzhen, China
e-mail: 12031313@mail.sustech.edu.cn

Siming Sun
Department of Electrical and Electronic Engineering
Southern University of science and technology
Shenzhen, China
e-mail: 12132015@mail.sustech.edu.cn

Perry Ping Shum
Department of Electrical and Electronic Engineering
Southern University of science and technology
Shenzhen, China
e-mail: shenp@sustech.edu.cn;

Fang Zhao
Department of Electrical and Electronic Engineering
Southern University of science and technology
Shenzhen, China
e-mail: 12031197@mail.sustech.edu.cn

Changyuan Yu
Department of Electronic and Information Engineering
The Hong Kong Polytechnic University
Hong Kong, China
e-mail: changyuan.yu@polyu.edu.hk

Siming Sun
Department of Electrical and Electronic Engineering
Southern University of science and technology
Shenzhen, China
e-mail: 12132015@mail.sustech.edu.cn

Liyang Shao
Department of Electrical and Electronic Engineering
Southern University of science and technology
Shenzhen, China
e-mail: shaoly@sustech.edu.cn

Abstract—**In this paper, a new method for ultrasensitive temperature measurement is proposed and demonstrated. The sensor is based on two cascaded peanut structures to form a Mach Zehnder interferometer (MZI) in single mode fiber, and a cone is formed by repeat discharge in the interference region to enhance the evanescent field on its surface. At the same time, isopropanol with high thermal optical coefficient is filled for effective sensitization. The MZI was placed in a fiber ring cavity for filtering and temperature sensing. The results show that the fiber ring laser (FRL) has a narrower 3 dB band width (~0.15nm) and a high signal-to-noise ratio (SNR) (~50dB). In the temperature range of 20°C to 50°C, the sensitivity reached -285 pm/°C. The sensor has the advantages of high sensitivity, simple structure and low cost. It is expected to be widely used in ocean temperature measurement.**

Keywords-temperature measurement; cascaded peanut structure; isopropanol; fiber ring laser

I. INTRODUCTION

Fiber optic sensors [1-3] have been extensively studied in the measurement of various physical parameters, including temperature, refractive index, strain, stress, etc., due to their advantages of small size, easy fabrication, immunity to electromagnetic interference and low cost. Temperature is an important parameter to characterize the normal operation of the system, which means that the measurement and monitoring of temperature plays a crucial role in many application fields such as battery charge and discharge analysis, biological activity detection, aerospace equipment operation and maintenance. In recent years, optical fiber temperature sensors such as fiber grinding and polishing technology [4], fiber tapering technology [5], fiber Bragg grating (FBG) [6] and liquid-filled optical fiber MZI [7] have been widely studied. In the structural design of various optical fiber temperature sensors, the core mode is used as one of the arms in the structure of MZI, interference with the cladding modes, eliminating the interference of the huge structure system. Therefore, a fiber-optic MZI-based temperature sensor can be easily realized by a commercial welding machine. It has small size, simple production and other unique advantages.

In 2021, Lin et al. proposed a fiber temperature sensor based on erbium-doped fiber peanut structure, and obtained a temperature sensitivity of 0.158nm/°C [8]. However, the

978-1-6654-8699-6/22 $31.00 © 2022 IEEE

temperature sensitivity could not be further improved due to the material characteristics. In the same year, Du et al. proposed a tapered two-mode fiber MZI temperature sensor based on isopropanol filling. Due to the high thermo-optical coefficient of the material, the temperature sensitivity is 140.5nm/°C [9]. However, tapered fiber is very sensitive to the changes of the surrounding environment, resulting in the phenomenon of cross sensitivity.

In this paper, we propose and demonstrate a temperature sensor based on a cascade peanut structure. A cone region is formed by repeated discharge between the two structures to enhance the contact between the evanescent field and the outside world. Isopropanol and the designed MZI are encapsulated together through capillary tubes. Due to the high thermo-optical coefficient of the material, the sensing system is sensitive to temperature, reaching a sensitivity of −285 pm /°C over a temperature range of 20-50 °C.

II. EXPERIMENT SETUP AND WORKING PRINCIPLE

Fig. 1 shows the schematic diagram of the sensor structure. The two fiber peanut structures function as a fiber beam splitter and a synthesizer respectively. After being modulated by the first interference structure, part of the optical light excitation transmitted in the core forms a cladding mode,and the rest of the light continues forward in the core as a fundamental mode. Since the refractive index difference between different modes, the optical path of light transmission between fiber core and cladding is different, giving rise to phase difference. When the light is transmitted to the cone region, part of the light field is transmitted along the isopropyl alcohol with high thermal-optical coefficient, resulting in a larger phase difference with the core mode and improving the sensitivity. Then, different modes of light pass through the sensor unit to reach the designed beam combination, interference will be formed due to the coupling of cladding mode and the core mode.

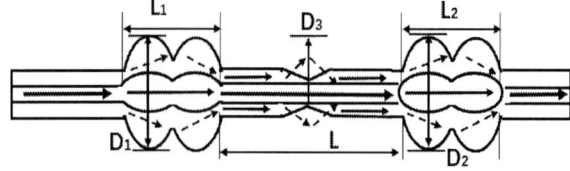

Figure 1. The schematic of designed fiber sensor based on tapered fiber with cascade peanut structure

The cut single-mode fiber is discharged repeatedly for 10 times through the welding machine (parameter: 1000 ms; 100 bit) to form a spherical structure, and then weld the two ends to form two peanut structures. D_1, D_2=190 μm. L_1, L_2= 450 μm. The cone region was discharged by 20 repeated discharges (1200ms; 110bit), D_3=50μm, L=15cm. After the interference between the core mode and the cladding mode, the expression of the output light intensity at the beam combiner is [10]:

$$I = I_1 + I_2 + 2\sqrt{I_1 I_2} COS\big[2\pi L\big(n_{cl}^{eff} - n_{cm,n}^{eff}\big)/\lambda\big] \quad (1)$$

where I_1 is the light field intensity of the core mode, I_2 is the light intensity of cladding mode. L is the spacing between two peanut structures; The initial wavelength of the incident light is λ. n_{cl}^{eff} and $n_{cm,n}^{eff}$ represents the effective refractive index of core mode and cladding mode respectively. The phase difference between the structure ca be expressed as [8]:

$$\Delta\varphi = 2\pi L\big(n_{cl}^{eff} - n_{cm,n}^{eff}\big)/\lambda \quad (2)$$

The free spectral range (FSR) can be simplified as:

$$\lambda_{dip} = 2L\big(n_{cl}^{eff} - n_{cm,n}^{eff}\big)/(2m + 1) \quad (3)$$

When m is rounded, the minimum light intensity of the corresponding output spectrum is the trough of the interference peak. The influence expression between transmission spectrum and temperature of interference structure filled with isopropyl alcohol is as follows [8]:

$$\frac{\Delta\lambda_{dip}}{\Delta T} \approx \lambda(\alpha + \zeta) \quad (4)$$

Among, α is the thermal expansion coefficient of the sensing system, and ζ is the thermal optical coefficient of the system.

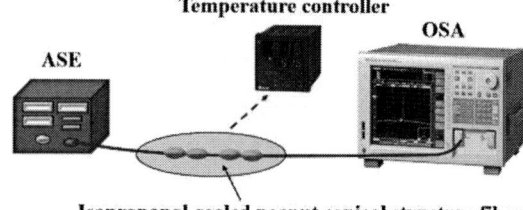

Figure 2. Experimental setup of the proposed temperature sensor.

Fig. 2 shows the experimental setup of the temperature sensor based on cascade peanut shaped structure filled with isopropanol. It consists of a broadband source (BBS), a 1.5cm long interferometer sandwiched between two SMFS and an optical spectrum analyzer (OSA Yokogawa AQ6370D). The refractive index changes with the filling temperature of the capillary to improve the sensitivity.

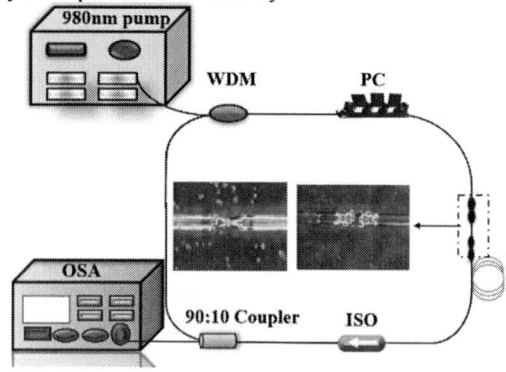

Figure 3. The schematic diagram of temperature sensor in FRL.

978-1-6654-8699-6/22 $31.00 © 2022 IEEE

After verifying the temperature sensing characteristics of the interferometer, a ring laser is used instead of an amplified spontaneous emission for monitoring. This is based on the fact that the laser has narrower linewidth, high output power and less burrs. As shown in Fig. 3, the 980nm laser enters the laser cavity through WDM as the pump light source. The polarization state in the cavity is controlled by an inserted polarization controller leading the output laser with the highest intensity. 1.5m erbium-doped fiber is used as gain medium. The isolator prevents backscattered light. Finally, 10% of the light is input into OSA through coupler for signal.

III. EXPERIMENTAL RESULTS

The temperature sensing spectrum based on amplified spontaneous emission source is shown in Fig. 4. Taking 5 °C as the step, with the increase of temperature, the refractive index of isopropanol changes due to the high thermal optical coefficient of its own material, the interference wavelength decreases with the increase of ambient temperature, and the spectrum has a blue shift in the range of 20°C-50 °C. Fig. 5 shows the change of the interference spectrum of the temperature sensor near 1500nm-1520nm. It can be clearly found that the interference peak shows good linearity with the change of temperature.

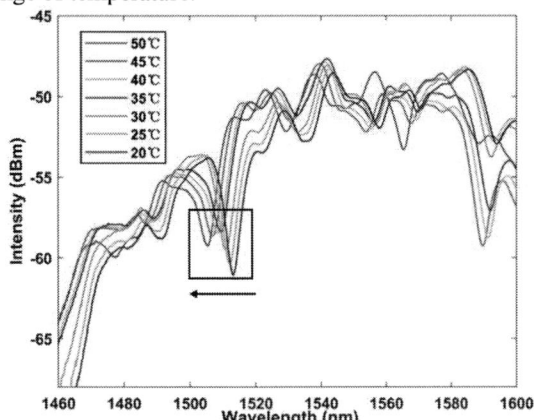

Figure 4. Transmission spectrum of sensor under temperature change of 20 °C - 50 °C

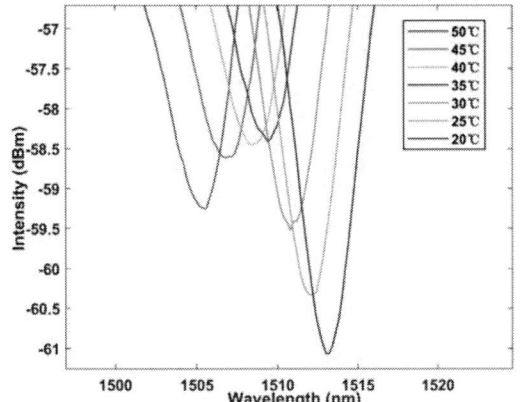

Figure 5. Transmission spectrum of the sensor under the temperature change of 20 °C ~ 50 °C (around 1500nm-1520nm)

The linear fitting curve of temperature sensing characteristics of the designed interferometer is shown in Fig. 6. The figure shows that the temperature sensitivity is -247pm/°C and the linear fitting coefficient is 0.996, indicating the good linearity of the sensor.

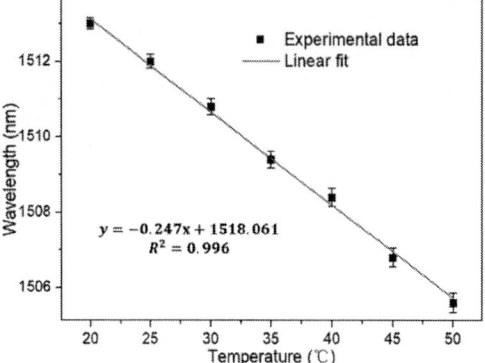

Figure 6. Linear fitting of wavelength shift versus temperature for the broadband light source

Fig. 7 indicates the spectral accordance between the broadband light source and the ring laser in 50 °C. As shown in the figure, the wavelength of the laser corresponds to the peak of the interference, which shows the good working effect of the interferometer as a filter in the laser cavity. The 3-db bandwidth of laser output is 0.15nm and the SNR ratio is up to 50dB, which is much better than that of broadband light source.

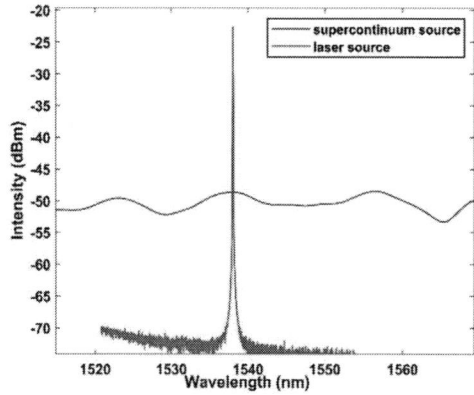

Figure 7. Transmission spectrum (blue line) of peanut structure interferometer under amplified spontaneous emission source and fiber ring laser system (Orange Line).

The wavelength shift of the output laser spectrum caused by the filtering action of the interferometer is shown in Fig. 8. The wavelength moves toward the shorter wavelength as the temperature rises, producing a blue shift. A tunable laser shift of about 10nm was generated.

The linear fitting curve of laser output with temperature is shown in Fig. 9. Within the limits of 20 °C -50 °C, the wavelength shifts blue with the increase of temperature. The temperature sensitivity is up to -285pm/ °C, and the linear fitting coefficient is 0.997, which shows the good

978-1-6654-8699-6/22 $31.00 © 2022 IEEE   117

linearity of the sensor. Compared with the convention fiber temperature sensor, the designed peanut structure has the advantages of narrow line width, low insertion loss, simple structure and low cost. It is conducive to the application of sensors in practical application scenarios.

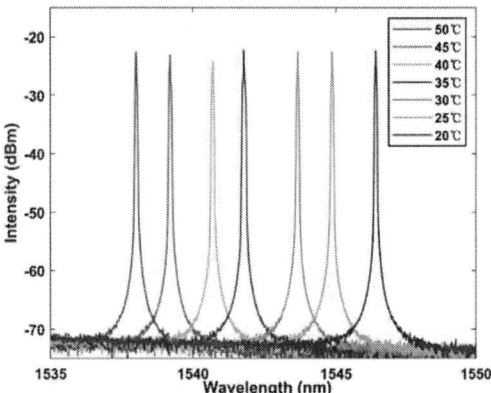

Figure 8. The output spectrum of the laser temperature sensor system increases with the temperature from 20℃ to 50℃.

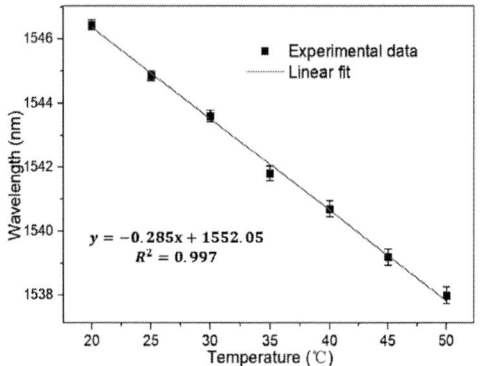

Figure 9. Linear fitting of wavelength shift versus temperature for the fiber ring laser sensor

IV. CONCLUSION

An interferometric high sensitivity temperature sensor based on cascade peanut SMF micro-structure fiber sealed with isopropanol is introduced. In this paper, the peanut structure acts as a fiber splitter and combinator in the sensor; the cone region is used to enhance the interaction between the evanescent field and the outside environment. Using the ultra-high thermal optical coefficient of isopropanol. Real time on-line high sensitivity temperature detection is realized. In the experiment, it is found that the sensor shows good linear fit in the temperature range of 20 ℃ ~50 ℃, and its sensitivity is -285 pm/ ℃. At the same time, the system also has a narrow linewidth (~0.15nm), high signal-to-noise ratio (~50dB) combined with the benefits of easy to manufacture, compact size and low cost. The proposed sensor has a potential application prospect in perimeter security, battery in-situ monitoring and other fields.

REFERENCES

[1] W. Lin, Y. Liu, L. Shao, and M. I. Vai, "A Fiber Ring Laser Sensor with a Side Polished Evanescent Enhanced Fiber for Highly Sensitive Temperature Measurement," *Micromachines (Basel),* vol. 12, no. 5, May 20 2021.

[2] A. Masoudi, J. A. Pilgrim, T. P. Newson, and G. Brambilla, "Subsea cable condition monitoring with distributed optical fiber vibration sensor," *Journal of Lightwave Technology*, vol. 37, no. 4, pp. 1352–1358, 2019.

[3] Z. Liu et al., "PDMS-assisted microfiber M-Z interferometer with a knot resonator for temperature sensing," *IEEE Photonics Technology Letters*, vol. 31, no. 5, pp. 337–340 , 2019.

[4] J. Woong-Gyu, K. Sang-Woo, K. Kwang-Taek, K. Eung-Soo, and K. Shin-Won, "High-sensitivity temperature sensor using a side-polished single-mode fiber covered with the polymer planar waveguide," *IEEE Photonics Technology Letters,* vol. 13, no. 11, pp. 1209-1211, 2001.

[5] H. Deng *et al.*, "A Temperature Sensor Based on Composite Optical Waveguide," *Journal of Lightwave Technology,* vol. 40, no. 8, pp. 2663-2669, 2022.

[6] B. Zhang and M. Kahrizi, "High-Temperature Resistance Fiber Bragg Grating Temperature Sensor Fabrication," *IEEE Sensors Journal,* vol. 7, no. 4, pp. 586-591, 2007.

[7] Y. Liu *et al.*, "Fiber Optic Electric Field Intensity Sensor Based on Liquid Crystal-Filled Photonic Crystal Fiber Incorporated Ring Laser," *IEEE Photonics Journal,* vol. 14, no. 1, pp. 1-5, 2022.

[8] W. Lin *et al.*, "In-Fiber Mach–Zehnder Interferometer Sensor Based on Er Doped Fiber Peanut Structure in Fiber Ring Laser," *Journal of Lightwave Technology,* vol. 39, no. 10, pp. 3350-3357, 2021.

[9] D. Du *et al.*, "Ultrasensitive temperature sensor and mode converter based on a modal interferometer in a two-mode fiber," *Opt Express,* vol. 29, no. 20, pp. 32135-32148, Sep 27 2021.

[10] Z. Cao *et al.*, "Strain-insensitive and high temperature fiber sensor based on a Mach–Zehnder modal interferometer," *Optical Fiber Technology,* vol. 20, no. 1, pp. 24-27, 2014.

Proceedings of the 7th Optoelectronics Global Conference (OGC 2022)

The Micro-control Refractive Index Sensor of Dual-metal Antiresonance Optical Fiber

Boyao Li

School of Electronic Engineering and Intelligentization, Dongguan University of Technology
Dongguan, Guangdong, 523808, China
e-mail: liby@dgut.edu.cn

Tianrong Huang

School of Electronic Engineering and Intelligentization, Dongguan University of Technology
Dongguan, Guangdong, 523808, China
e-mail: 202041312112@qq.com

Abstract—With the development of artificial intelligence for complex environment monitoring technology, new sensors based on the combination of diversified customized optical fiber waveguides and functional materials have become an important solution to meet the diverse perceptions in complex environments. Then, for solving the concentration monitoring problem of multi-channel microfluidics, the bimetallic anti-resonant fiber structure is proposed in this paper, which realize the multi-band microfluidic sensor of bimetallic reverse resonant fiber. Theoretical results show that the sensor achieves the surface plasma resonance (SPR) coupling of the core mode and the metal film at multiple bands of 1450 nm, 1650 nm and 1700 nm, then uses this feature to achieve refractive index sensing in three bands. The corresponding results have potential applications in the field of multimetallic microstructure fiber fluid sensing in the air core, and provide new ideas for the design of multi-band SPR resonance sensors, which are expected to be applied in the field of biochemical monitoring.

Keywords-sensors; antiresonance optical fiber;multi-waveband sensing

I. INTRODUCTION

Due to its low transmission loss, long transmission distance and anti-interference properties, fiber optic sensors can not only transmit signals in a better and more stable and safe way in communications, but also in medicine to monitor liquids, temperatures and other biochemical information, fiber optics are playing its important value. In addition, microstructured optical fibers (MOFs), with their flexible design and excellent optical properties unmatched by traditional optical fibers, have greatly broken through the limitations of traditional optical fibers and brought about a profound change in fiber optics, providing new avenues for the development of optical fiber technology and its applications. In recent years, MOFs and their applications have become a hot topic at home and abroad, providing new opportunities for breakthroughs in optical communication, optical sensing, optical devices, biomedical science and other bottlenecks. Meanwhile, the presence of functional material modification, and the material modification of MOFs combines the excellent optical properties of the material with the flexible structural design of MOFs, providing a new way to enhance the performance of MOFs expand their application scope and develop new optical devices [1]. This is a new way to improve the performance of MOFs, expand

their applications and develop new optical devices. Among these methods, the surface plasmon resonance (SPR) effect of noble metals has a very high sensitivity response to the external world, which makes the application in fiber optic sensors more obvious [2]. The application of SPR in optical fiber sensors is more obvious due to the ultra-high sensitivity of precious metals to the external world.

In 1993, Jorgenson presented a new fiber optic SPR sensor by integrating it onto an optical fiber for the first time [3]. This brought about a milestone in fiber optic biochemical sensors and SPR detection technology. Conventional optical fiber SPR sensors require the processing of a swift field zone in the fiber and the deposition of a thin on the surface of the swift field zone to excite the plasma in the metal through the swift field to obtain the SPR detection signal. Singh et al. exploited this feature of large core diameter medical multimode fibers by depositing a Cu metal film and TiO_2 dielectric layer on the bare fiber core to produce SPR sensors [4]. Shrivastav et al. fabricated SPR sensors with molecularly imprinted films in large diameter bare fiber cores for the specific detection of chemical molecules [5,6]. The fiber side polishing technique is used to polish a plane on one side of the cylindrical surface of the fiber so that the swift field of the optical wave mode transmitted by the core can reach this plane, and this plane is ideal for planar film deposition. In China, Chen Yong et al. of the University of Science and Technology of China compared the performance of single-sided coated and cylindrically uniformly coated large-core diameter optical fiber SPR sensors and found that their sensitivity was almost identical, but the quality factor of the SPR resonance peaks of the single-sided coated sensors was somewhat worse [7]. However, most fibers have a glass cladding that cannot be easily removed. In order to allow the swift field to reach the glass-clad fiber surface, Coelho et al. produced a fiber SPR sensor with a sensitivity of 5100 nm/RIU by chemically etching the swift field region on the surface of a single-mode fiber using hydrofluoric acid [8].

Since conventional optical fibers have a fixed structure, are not suitable for fluid monitoring and are poorly embedded in multifunctional materials, the integration and miniaturization of devices with multifunctional measurement capabilities is one of the key research directions in the field of sensor applications. The foundation is laid for the integration of multifunctional optical fibers into micro structured optical fibers (MOFs).

978-1-6654-8699-6/22 $31.00 © 2022 IEEE

Unlike conventional optical fibers, MOFs are structurally flexible and have a variety of light-guiding mechanisms to meet a wide range of needs, and the desired optical properties can be obtained by adjusting the position of the optical hole, the cross-sectional structure and the size of the aperture. In addition, it can be filled with sensitive materials such as temperature and electric fields to achieve highly sensitive sensing of the corresponding physical quantities.

Some researchers have tried to combine optical fibers with certain metal modifications to adjust the polarizability of the devices based on the principle of surface plasmon resonance. 2010, Yu proposed a design for a surface plasmon resonance sensor based on photonic crystal fibers [9]. In 2013, Xue et al. investigated the polarization filtering properties of gold-plated and liquid-filled photonic crystal fibers using finite element methods [10]. In 2015, structures and the sensitivity of surface plasmonics of metals to achieve a polarization filter [11]. After a number of studies, it was found that although the polarization properties of the combined fibers improved, they were still weak for high-power transmission work. In 2016, Mousavi systematically investigated different approaches to introduce high birefringence and high polarization extinction ratios in hollow anti-resonant fibers [12]. In 2018, Yan proposed a new double-ring hollow-core ARF that enables single-polarization guidance [13]. Although the MOFs are structurally flexible, the light-guiding mechanism is complex and has yet to be developed for micron-scale fluid monitoring, and the SPRs are mostly monometallic in character and unsuitable for calibration.

Based on this ,this paper is based on a gold and silver bimetallic anti-resonant fiber microfluidic sensor, by placing gold and silver in the Cosmos region respectively, to achieve sensing in different wavelength bands, in addition, as the fiber itself is an antiresonance fiber, with the advantages of large scale, suitable for fluid transmission, simple light conduction mechanism, etc., very easy to integrate, the microfluidic integrated in a single fiber, through experimental simulation, to obtain a sensor in the 1.3um wavelength band to The microfluidics is integrated into a single optical fiber, and the experimental simulation results in a sensor that is sensitive in the 1.3 μm to 1.8 μm bands respectively, thus allowing the sensor to be used for multiple branching due to its multi-band sensitivity. The multifunctional sensor is simple in structure and stable in performance, and can be applied to the detection of physical quantities in many complex environments.

II. THEORETICAL STRUCTURE AND APPARATUS

Firstly, in order to achieve on-line monitoring of the refractive index of the fluid, a chamber is designed in which the fluid is injected through the left-hand opening at the top of the chamber and pumped out through the right-hand opening to form a liquid chamber so that the experimental fluid can be manipulated. The process is shown in Fig. 1. After the light wave has passed through the pigtail, the point light source is collimated and expanded through the objective lens, and the expanded light source is incident on the placed polarizer, then the polarizer is adjusted so that the required polarized light is incident on the focusing objective lens, which is focused through the objective lens into the designed bimetallic anti-resonant fiber, and then the signal is transmitted through the single mode fiber into the spectrometer, thus enabling the designed device to monitor the refractive index of fluid in practice. This enables the device to be used in practical fluid refractive index monitoring applications.

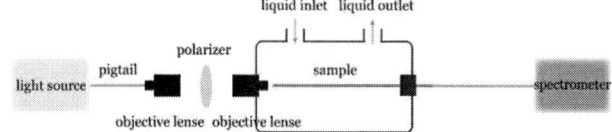

Figure 1. Device structure.

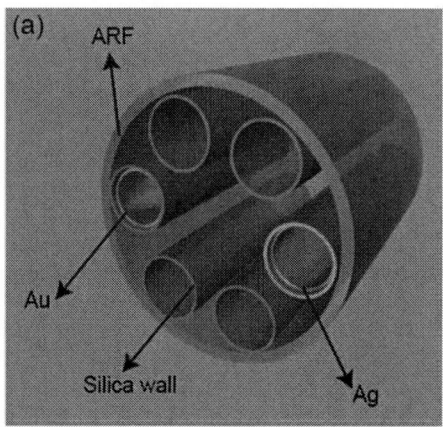

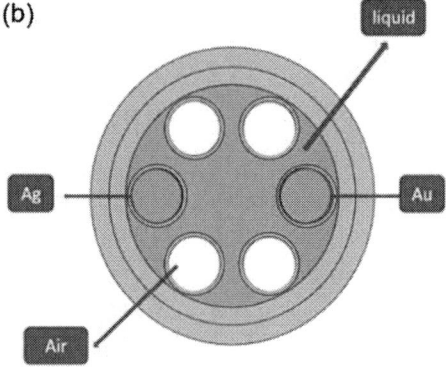

Figure 2. The design of the fiber optic sensor. (a) 3D schematic diagram. (b) Cross-section of ARF

Then, a corresponding bimetallic air-core anti-resonant fiber structure was further designed as shown in Fig. 2(a) is constructed in a configuration consisting of circular air holes arranged in a positive hexagonal lattice. The model is based on a silica glass substrate and the thickness of the perfectly matched layer is set to 5 μm The six-air holes with a radius of 8 μm The air holes are symmetrically distributed with an inner circle radius of 7 μm The inner circle is 7. The inner circles of the air holes in the horizontal direction are coated with silver and gold films from left to right. In addition, the upper and lower four air holes in the inner wall of the fiber are filled with air, while the fluid filling the rest of the inner wall is filled. The center of the fiber is a hollow core made of silica walls, which realizes guide light in low refractive index fiber core using anti resonance principle. Besides, the aperture is relatively simple, and then in the case of different metal wavelengths of gold and silver will produce different surface plasma resonance properties, very suitable for multi-wavelength sensing. The corresponding structure was modelled and analyzed using a common finite element simulation model analysis software. The two-dimensional structure of the corresponding model is shown in Fig. 2(b).

In the choice of spatial dimension choose two-dimensional, bimetallic materials choose gold and silver were plated in the inner membrane of the two air holes, in the physical field electromagnetic wave frequency domain to build photonic crystal fiber cross-section, according to the simulation process of the fiber structure parameters do not transform, the solid part of the fiber material is set to silicon dioxide, due to the existence of material dispersion, wavelength is different at the same time its refractive index refractive index with wavelength change.

In addition the dispersion equation for the quartz material in the model is described by Sellmeier as follows:

$$n^2(\lambda) = 1 + \sum_{j=1}^{m} \frac{B_j \lambda_j^2}{\lambda_j^2 - \lambda^2} \tag{1}$$

Where λ is the vacuum wavelength and B_j is the coefficient describing the different materials, available from the literature [14] found. The material dispersion equations for metallic gold and silver are given by the Drude-lorentz model [15]. In the analysis of the corresponding transmission characteristics, the limiting loss of the corresponding mode in the fiber is also of paramount importance, and the corresponding limiting loss can be calculated and analysed as follows.

$$\alpha(x,y) = 8.686 \times \frac{2\pi}{\lambda} im(n_{eff}) \times 10^4 \tag{2}$$

Where n_{eff} is the effective refraction of the material and is a complex number. $\alpha(x,y)$ corresponds to the limiting loss in the x-polarization direction and the y-polarization direction.

III. EXPERIMENTAL RESULTS AND NUMERICAL SIMULATIONS

In this section, FEA has been utilized to simulate and calculate the antiresonance fiber with the following results. According to Fig. 3(a) and (b), it can be visualized that the light in Fig. 3(a) represents the energy consumption of light in the x-polarized mode, with two loss peaks: while in Fig. 3(b) there is one loss peak in the y-polarized mode, so that

the change in refractive index of the liquid to be measured can be observed from the change in multiple peaks. When the refractive index of the environment in which the fiber is located changes, the sensitivity of the surface plasmon polaritons (SPP) mode changes and the conditions under which the coupling between the core film and the SPP film can take place are different. The peak value of the high loss is also different. Such a characteristic can be exploited in sensor devices.

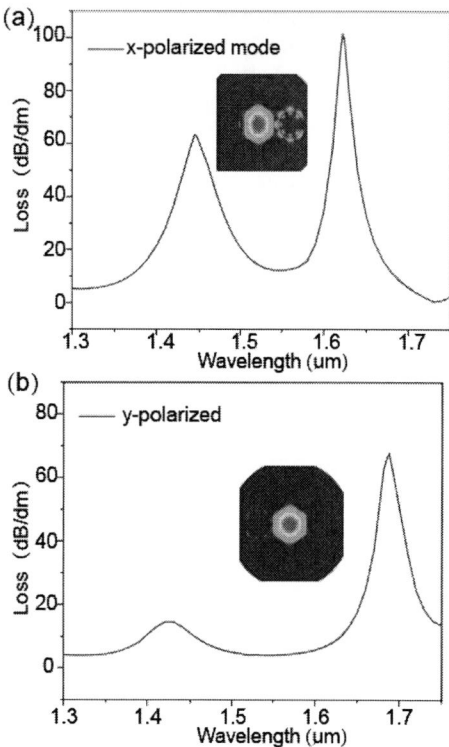

Figure 3. The confinement loss and power distribution of fundamental mode in core. (a) x-polarized mode (b) y-polarized mode.

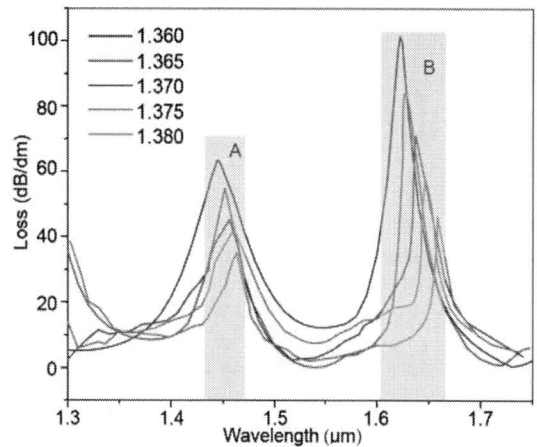

Figure 4. Resonance wavelength variation of the x-polarized core mode.

By further analysing the loss characteristics as the refractive index varies from 1.360 to 1.380, the corresponding results are shown in Fig. 4. It shows that the loss curve with wavelength varies for different refractive indices. the loss peak in the x-polarized mode moves in the direction of longer wavelengths depending on the increasing refractive index of the object we are studying. And it can be directly observed that there are two distinct loss peaks in the x-polarized mode, respectively in the A and B regions of the graph. And the two loss peaks are not the same. Besides, for the analysis of the simulation results in the y-polarized mode, it can be observed according to Fig. 5 below. Note of, the y-polarized mode that it can also be directly observed that there is a clear loss peak in the y-polarized mode in the A region. The motion of loss peak in the y-polarized mode is opposite to the x-polarized mode, and it will move in the direction of shorter wavelengths depending on the increase in the refractive index of the object we are studying. This reverse sensitivity behavior is very useful for the calibration of multisensor bands, and can avoid the systematic error of some instruments.

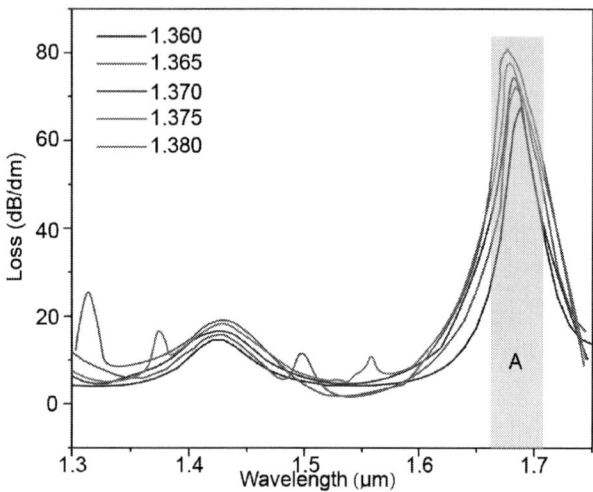

Figure 5. Resonance wavelength variation of the y-polarized core mode.

Then, fitting the two peaks in the x-polarized mode, A and B in Fig. 4, leads to Fig. 6 (a) and (b), which shows a sensitivity of 940 nm/RIU in the A region, and 1620 nm/RIU in the B region. Similarly, the changes of y-polarized mode can be also fitted in Fig. 6(c), which shows a sensitivity of -560 nm/RIU in the A region of Fig. 5. Through the different sensitivity of the orthogonal Hermite conjugate mode, the sensitivity calibration and diversified monitoring can be performed for different liquids in multi-directional and multi bandBy observing and analysing the fitted plots of region A and B in the x-polarized mode and the fitted plots of region A in the y-polarized mode, the linearity of the three fitted plots are 0.998, 0.977 and 0.933 respectively, so the linearity of the three fitted curves is good, and the sensitivity of all of them reaches 500 nm or more.

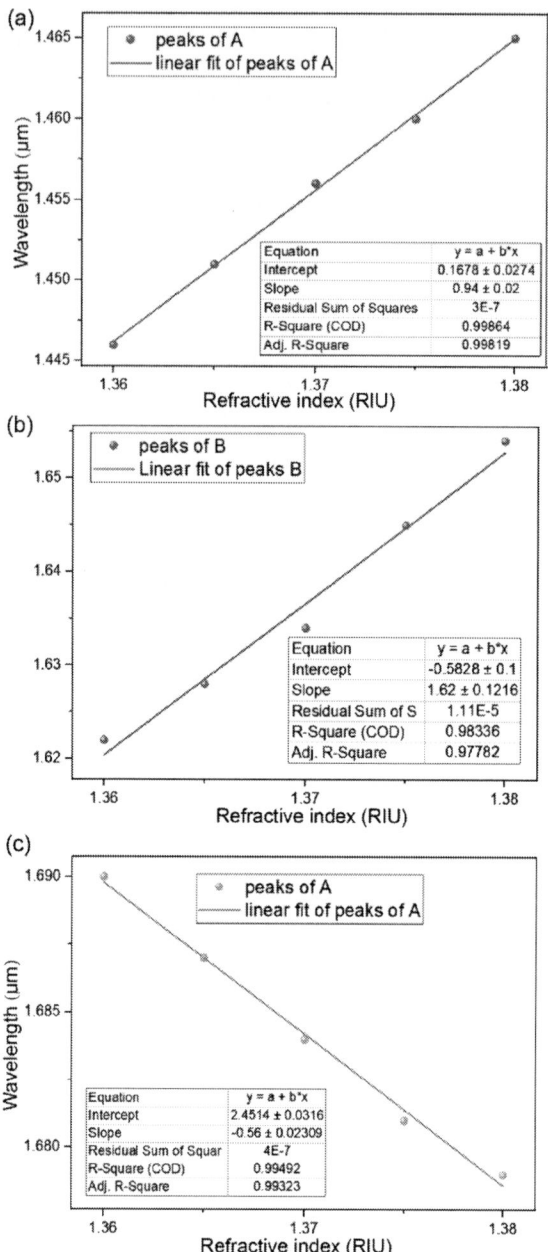

Figure 6. Sensing characteristics of the x-polarized and y-polarized core mode (a-b) Resonant wavelength variation fiting results of region A and B in x-polarized mode. (c) Resonant wavelength variation fiting results of region A in y-polarized mode.

IV. SUMMARY

This paper solves the problem of fixed structure of traditional optical fibers and low embedding with multifunctional materials through the flexibility of MOFs structure design combined with diversity of light guiding mechanism of noble metals. Utilizing gold and silver bimetals with MOFs, a bimetallic anti-resonant fiber multi-band microfluidic sensor is designed and simulated by finite element calculations. The final SPR response probes in multiple bands at 1450nm, 1650nm and 1700nm showed response linearities of 0.998, 0.977 and 0.993 for the three bands respectively, demonstrating the good linear response of the designed sensor. These excellent properties indicate that the use of the SPR properties of polymetallic and the waveguide flexibility of MOFs can achieve multi-band microfluidic sensing responses, which are potentially relevant in the corresponding functional sensing field.

FUNDING

Basic and Applied Basic Research Foundation of Guangdong Province (2022A1515110480).

REFERENCES

[1] B. Li, Y. Zhao, Y. Zhang, A. Y. Zhang, X. Li, J. Gu, S. Xi, G. Zhou, "Functionalized Micro Structured Optical Fibers and Devices for Sensing Applications: A Review," Journal of Lightwave Technology, vol. 39, no. 12, pp. 3812-3823, Jun. 2021.

[2] V. Semwal, B. D. Gupta, "Highly selective SPR based fiber optic sensor for the detection of hydrogen peroxide," Sensors and Actuators B: Chemical, vol. 329, pp. 129062, Feb. 2021.

[3] R. Jorgenson, S.S. Yee, " A fiber-optic chemical sensor based on surface plasmon resonance," Sensors and Actuators B: Chemical, vol. 12, pp. 213-220, Apr. 1993.

[4] S. Singh, S. K. Mishra, B. D. Gupta, "Sensitivity enhancement of a surface plasmon resonance based fibre optic refractive index sensor utilizing an additional layer of oxides," Sensors and Actuators A: Physical, vol. 193, pp. 136-140, Apr. 2013.

[5] A. M. Shrivastav, S. P. Usha, B. D. Gupta, "Fiber optic profenofos sensor based on surface plasmon resonance technique and molecular imprinting," Biosensors and Bioelectronics, vol. 79, pp. 150-157, May 2016.

[6] A. M. Shrivastav, B. D. Gupta, "SPR and Molecular Imprinting-Based Fiber-Optic Melamine Sensor With High Sensitivity and Low Limit of Detection," IEEE Journal of selected topics in quantum electronics, vol. 22, pp. 6900207, Jul. 2016.

[7] Y. Chen, "Biochemical sensing of SPR chips and research on fiber optic SPR sensing technology," Anhui: University of Science and Technology of China, 2012.

[8] L. Coelho L, J. M. M. de Almeida, J. L. Santos, R. A. S. Ferreira, P. S. Andre and D. Viegas, "Sensing Structure Based on Surface Plasmon Resonance in Chemically Etched Single Mode Optical Fibres," Plasmonics, 2015, 10: 319-327.

[9] X. Yu, Y. Zhang, S. S. Pan, P. Shum, M. Yan, Y. Leviatan, C. Li, "A selectively coated photonic crystal fiber based surface plasmon resonance sensor," J.Opt, vol. 12, pp. 015005, Dec. 2010.

[10] J. Xue, S. Li, Y. Xiao, W. Qin, X. Xin, X. Zhu, "Polarization filter characters of the gold-coated and the liquid filled photonic crystal fiber based on surface plasmon resonance," Optics express, pp. 13733-13740, May 2013.

[11] D. F. Santos, A. Guerreiro, J. M. Baptista, "SPR microstructured D-type optical fiber sensor configuration for refractive index measurement," IEEE Sensors Journal, vol.15, pp. 5472-5477, Oct. 2015.

[12] S. A. Mousavi, S. R. Sandoghchi, D. J. Richardson, F. Poletti, "Broadband high birefringence and polarizing hollow core antiresonant fibers," Optics express, vol. 24, pp. 22943-22958. Sep. 2016.

[13] S. B. Yan, S. Q. Lou, W. Zhang, Z. Lian, "Single-polarization single-mode double-ring hollow-core anti-resonant fiber," Optics express, 2018, vol. 26, pp. 31160-31171, Nov. 2018

[14] G. P. Agrawal, Nonlinear Fiber Optics. San Francisco, CA, USA: Academic, 2007.

[15] Vial A, A. Grimault, D. Macias, D. Barchiesi, and M. L. Chapelle Phys, "Improved analytical fit of gold dispersion: Application to the modeling of extinction spectra with a finite-difference time-domain method," Phys. Rev. B, vol. 71, pp. 085416, Feb. 2005.

Proceedings of the 7th Optoelectronics Global Conference (OGC 2022)

Non-Invasive Optical Fiber Sensing Vital Signs Monitoring Based on Envelope Extraction BCG Data Processing

Hanyu Zhao[1,2], Guo Zhu[1,2], Fei Liu[2], Xiaojun Liu[1,2], Jinhui Yuan[2], and Xian Zhou*[1,2]

[1] Shunde Graduate School of University of Science and Technology Beijing,
Foshan 528300, China

[2] School of Computer and Communication Engineering, University of Science and Technology Beijing,
Beijing 100083, China

*E-mail: zhouxian219@ustb.edu.cn

Abstract—An envelope-extraction-based ballistocardiography (BCG) data processing method is proposed with a non-invasive optical fiber sensor in this paper. The accuracy of heart rate calculation value can reach 1.28 ± 0.59 bpm (Mean Absolute Error ± Standard Deviation, MAE ± SD) for one minute of measured data. This designed system can identify human body status and analyze Heart Rate Variability (HRV).

Keywords-non-invasive optical fiber sensor; envelope-extraction; BCG data processing; HRV analysis

I. INTRODUCTION

In recent years, a number of people have suffered the sub-health states due to the increasing work pressure. A considerable study has grown up around the theme of long-term and accurate monitoring of vital signs and other health information in non-clinical and non-invasive state. Traditional electrocardiography (ECG) data acquisition and even wearable devices equipped with sign monitoring sensors can provide vital signs monitoring functions. However, all of them need to be in close contact with the skin. Compared with these systems, the non-invasive vital signs monitoring system based on optical fiber sensors has higher flexibility, portability, and wider application scenarios, which can meet the long-term effective monitoring requirement in a non-clinical state However, the matching data processing method generally requires manual calibration of feature points or pair matching of all feature points, resulting in high complexity, and the calculation accuracy needs to be improved.

In this paper, we propose and demonstrate an envelope extraction-based ballistocardiography (BCG) data processing for a non-invasive single-mode-multi-mode-single-mode (SMS) optical fiber sensor cushion, which does not need to match and identify each feature point of the BCG waveform, and can effectively reduce the influence of abnormal heart rate (HR) caused by heartbeat artifacts or environmental noise.

II. PRINCIPLE

In our work, the SMS optical fiber interference structure is applied for data acquisition. The fiber layer is located between a grid layer and a PVC layer, and is installed in a seat cushion, as shown in Figure 1. The grid layer helps to generate more deformation on the optical fiber to enhance its sensitivity by amplifying the tiny vibrational signals from the human body. When the beam from a light emitting diode

(LED) passes through the fusion point between single-mode and multi-mode fiber, it excites a large number of modes of light in multi-mode fiber, which will interfere at the fusion point between multi-mode and single-mode fiber.

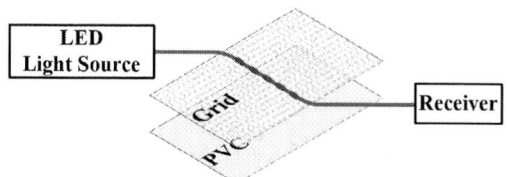

Figure 1. SMS interference structure.

Assuming that there are M order modes in the multi-mode fiber, the optical power detected at the section of the output fusion joint can be expressed as

$$P = \iint_S \left| \sum_{i=0}^{M} \vec{E}_i(x,y) e^{j\Delta\varphi i} \right|^2 ds$$

$$= \sum_{i=0}^{M} \iint_S \left| \vec{E}_i \right|^2 ds + 2 \sum_{i=0}^{M-1} \sum_{j=0}^{M} \vec{E}_i \cdot \vec{E}_j ds \cos(\Delta\varphi_i - \Delta\varphi_j) \quad (1)$$

where $\iint$ represent double integral, and E_i and $\Delta\varphi_i$ respectively represent the magnitude of the transverse electric field vector of the ith-order mode and the phase difference relative to the fundamental mode when the fundamental mode of the multimode fiber is used as a reference. Taking the first-order approximation of Equation (1), I_0 and I_1 are used to represent the fundamental mode light intensity and the first-order mode light intensity. Then, Equation (1) can be simplified as

$$p = \sum_{i=0}^{1} \iint_S \left| \vec{E}_i \right|^2 ds + 2 \iint_S \vec{E}_0 \cdot \vec{E}_1 ds \cos(\Delta\varphi_1 - 0) \quad (2)$$

$$= I_0 + I_1 + 2\sqrt{I_0 I_1} \cos(\Delta\varphi_1)$$

where $\Delta\varphi_1 = 2\pi(n_0 - n_1)L / \lambda$ (n_0 and n_1 represent the refractive index of the core fundamental mode and the cladding higher-order mode in the multi-mode fiber respectively, and L represents the length of the multi-mode fiber). During data collection, the user is told to keep at a resting state on the cushion. As heartbeat, respiration and other vital signs signals generate small vibration (denoted as

978-1-6654-8699-6/22 $31.00 © 2022 IEEE

$F(t)$) in the human body, leading to stress changes in the SMS structure. These stress changes will introduce phase variation between these two modes [7-8] as

$$\Delta\varphi_2 = \gamma F(t) \tag{3}$$

where γ is a constant between vibration and the phase vibration.

After it is received by a photodetector, digital signal processing (DSP) can be performed, as shown in Figure 2. It is known that the raw data includes vital sign signals of various frequencies, such as HR, respiration rate (RR), and noise signals. The normal RR of the human body is in the range of about 6 to 20 beats per minute (bpm), so the raw data can be bandpass filtered with a cutoff frequency of 0.1 Hz and 0.33 Hz. In this way, the respiration waveform can be obtained, and then the RR can be calculated.

During a normal cardiac cycle, the tiny vibrational signals produced by the heart contraction are more pronounced and easier to extract than those produced during diastolic. These feature points, mitral valve closure (MVC) and aortic valve opening (AVO) in the seismocardiogram (SCG), correspond to the H and J peaks in BCG, respectively. In the medical field, the frequency range of BCG waveforms is generally considered to be about nine times the frequency of the human heartbeat [5]. In this study, the HR was measured in the range of 50 bpm to 100 bpm, so the raw data should be band-pass filtered in the range of at least 7.47 Hz to 15 Hz to extract the BCG waveform containing obvious heartbeat information.

For the obtained BCG waveform, each cardiac cycle contains multiple heartbeat feature points, and the maximum value in each cardiac cycle is defined as J-peak. After a cardiac cycle, the heart has a short rest time, which significantly reduces the short-term energy of the BCG waveform in this period. The square and second-order differential processing of the bandpass-filtered signal can make each feature point more prominent, that is, emphasize

feature points. Then, the Extract Envelope process is performed, which can clearly divide each cardiac cycle preliminarily. Therefore, the highest peak of each cardiac cycle can be identified as the J-peak, which can be used as the feature point for calculating HR.

The next step is the process of HR Calculation. We chronologically arrange all the obtained J-peaks and denote them as J-peak-h (where h represents the hth cardiac cycle divided in the previous step), and define the time difference between the two groups of J-peaks as Interval-J. When the HR Calculation range is from 50bpm to 100bpm, 0.83s and 1.67s are used as the upper and lower limits of the threshold, which are represented by T-max and T-min, respectively.

When calculating the HR values of the hth and $(h+1)$th cardiac cycles, the Interval-J should be the time difference between J-peak-h and J-peak-$(h+1)$. If it is greater than T-max, this group of values needs to be discarded, and the calculation of the $(h+1)$th and $(h+2)$th cardiac cycles is performed. Otherwise, we continue to compare the Interval-J and T-min. If the Interval-J is smaller, the J-peak-$(h+1)$ needs to be discarded, and the time difference between J-peak-h and J-peak-$(h+2)$ will be used to update the value of Interval-J. On the contrary, if it is greater, the Interval-J can be divided by 60 to calculate the HR.

The obtained data contains several abnormal heart rate values, so it is necessary to further perform Outlier Removal processing. When the subject keeps quiet on the cushion for signal acquisition, it is less likely that there is a large difference between two cardiac cycles. We defined the threshold for identifying outliers as twice the standard deviation of all HR values. Each HR value is compared with its previous HR value, and if the absolute value of the difference is greater than the threshold, it is considered as an abnormal value. The abnormal value is replaced by the HR value of the previous cycle and filtered by the moving average of 20 cardiac cycles.

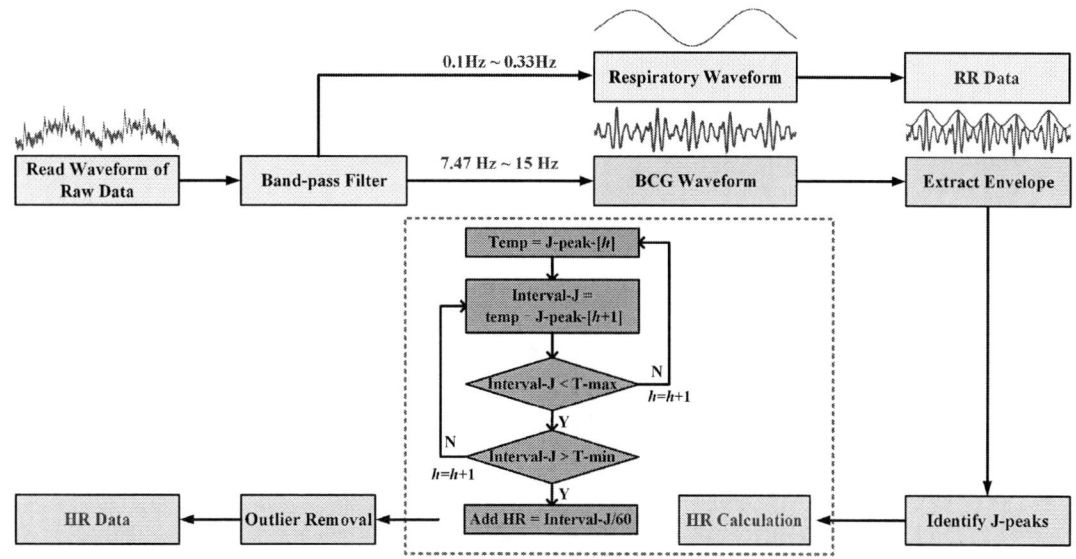

Figure 2. Envelope extraction BCG data processing.

Combining emphasizing feature points, extracting envelope peak, and moving average, this work doesn't need to match and identify each feature point in each cardiac cycle of BCG waveform. It reduces the complexity of calculation greatly, and mitigates the interference of J-peak to other peak identification effectively, so as to improve the accuracy of extracting HR feature points and calculating HR significantly.

III. EXPERIMENT RESULTS AND DISCUSSIONS

In this experiment, the SMS optical fiber sensing structure is used to collect the raw data of subjects' vital signs, and the medical IOCARE PM-900 ECG monitor is used to collect ECG signals simultaneously as the reference. Before data collection, the users will maintain a resting state or exercise violently for one minute so as to extract their vital sign data in the resting state or after exercise.

A. Resting-state Accuracy Analysis

The tested subjects are between 14 and 72 years old in good health state. Figure 3 shows their instantaneous heart rate values during the resting state by using our designed system and commercial medical ECG monitor, respectively. It can be seen that our HR calculation results show in good agreement with the ECG reference results. There are the same variation trends with small fluctuation deviation. Testing on different subjects, the data processing error results of the proposed processing method can reach 1.28 ± 0.59 bpm (Mean Absolute Error $\pm$ Standard Deviation, MAE $\pm$ SD) in one-minute-long data.

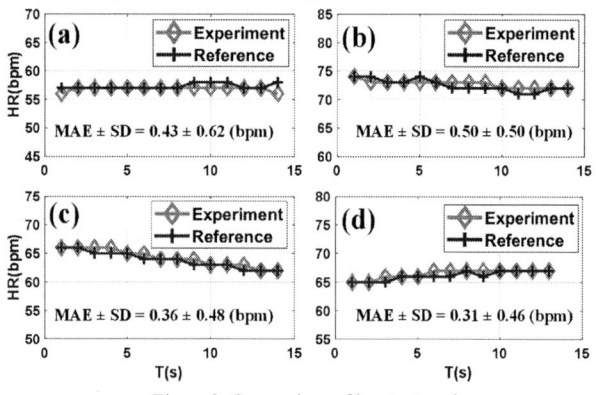

Figure 3. Comparison of heart rate values
(a) subject 1 (b) subject 2 (c) subject 3 (d) subject 4

B. Heart Rate Variability (HRV) Analysis

The R-R interval time series in the BCG signal can be used as the standard of HRV analysis after exercise [6]. The subjects were asked to take vigorous exercise for about one minute, and the cardiac cycle decreased. The HRV analysis data after exercise are shown in Figure 4. It can be seen that the subject's HR and RR increased significantly and the heartbeat amplitude increased significantly after exercise. As time went on, the subject's HR and amplitude gradually decreased.

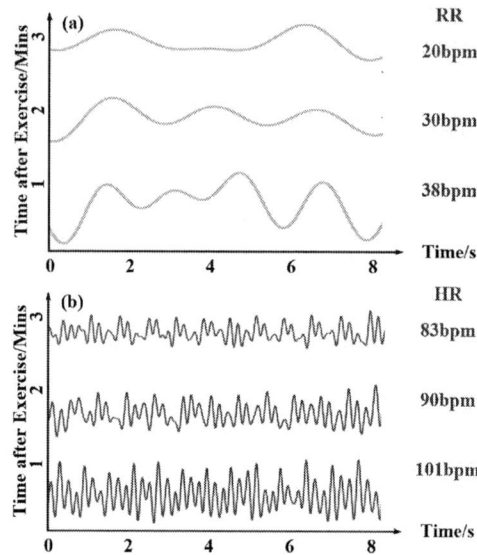

Figure 4. HRV analysis data after exercise
(a) RR value after exercise (b) HR value after exercise

C. Different state identification

In this study, we can not only calculate RR and HR but also identify the state of the user. As shown in Figure 5, we have studied three different states such as the "being used", "shaking", and "being unused". In Figure 5(a), the subject is working in a resting state, and an effective heartbeat waveform can be obtained by the above method. If the subject is still sitting on the cushion, but shaking the body or otherwise behaves. Since the amplitude and frequency of the vibration generated by the spontaneous movement of the human body are different from the heartbeat signal. Taking a subject who is shaking his legs on the cushion as another example (see Figure 5(b)), it can be easy to judge by monitoring the amplitude of the vibration signal. Figure 5(c) shows the signal in the "being unused" state. When no one is sitting on the cushion, the sensing optical fiber is not subject to vibration, bending, or other stress. By identifying different states of the human body, users with poor sitting postures can be reminded to adjust.

Figure 5. Different state identification
(a) Being used (b) Shaking (c) Being unused

IV. CONCLUSION

It can accurately divide the cardiac cycle without identifying all feature points to reduce the complexity of the algorithm and effectively avoid the influence of heartbeat artifacts to improve the accuracy of HR calculation. Non-invasive vital signs monitoring sensors are used to collect data of people in resting state, and the comparison with the ECG waveform data verifies the feasibility of the method. The system can effectively monitor the instantaneous HR of people working and resting on the cushion which the accuracy can reach 1.28±0.59 bpm (MAE±SD), and it also carries the function of HRV analysis and state recognition. This study can realize long-term monitoring in offices, homes, and other situations. It has the advantages of low-cost and high sensitivity and broad application prospects in realizing non-clinical health monitoring.

ACKNOWLEDGMENT

This work was supported by the Fundamental Research Funds for the Central Universities (Grant. No. FRF-BD-20-11A), the Scientific and Technological Innovation Foundation of Shunde Graduate School, USTB (Grant No. BK19AF005).

REFERENCES

[1] S. Han, W. Xu, S. You, B. Dong, F. Tan, C. Yu, W. Zhao, Y. Wang, "Investigation on Smart Cushion Based on SFS Structure and its Application in Physiological and Activity Monitoring", presented at Opto-Electronics and Communications Conference (OECC), Taipei, Taiwan, 2020, DOI:10.1109/OECC48412.2020.9273653.

[2] F. Tan, S. Chen, W. Lyu, Z. Liu, C. Yu, C. Lu, and H.Y. Tam, "Non-invasive human vital signs monitoring based on twin-core optical fiber sensors", Biomedical Optics Express, vol.10, no.11, pp. 5940-5952, 2019, DOI: 10.1364/BOE.10.005940.

[3] W. Lyu, W. Xu, F. Yang, S. Chen, F. Tan, and C. Yu, "Non-Invasive Measurement for Cardiac Variations Using a Fiber Optic Sensor",

IEEE Photonics Technology Letters, vol. 33, no. 18, pp. 990-993, 2021, DOI: 10.1109/LPT.2021.3078757.

[4] W. Chen, Y. Zhang, H. Yang, Y. Qiu, H. Li, Z. Chen, and C. Yu, "Non-Invasive Measurement of Vital Signs Based on Seven-Core Fiber Interferometer", IEEE Sensors Journal, vol. 21, no. 9, pp. 10703-10710, 2021, DOI: 10.1109/JSEN.2021.3061443.

[5] M. Wang, P. Liu, W. Gao, "Automatic Sleeping Posture Detection in Ballistocardiography", International Conference on Communications, Signal Processing, and Systems (CSPS), Springer, Singapore,2018, DOI: 10.1007/978-981-10-6571-2_216.

[6] Task Force of The European Society of Cardiology and The North American Society of Pacing and Electrophysiology, "Heart rate variability. Standards of measurement, physiological interpretation, and clinical use", Circulation, vol.93, no.5, pp.1043-1065,1996, DOI: 10.1161/01.CIR.93.5.1043

[7] P. Chen and C. You and P. Ding, "Event classification using improved salp swarm algorithm based probabilistic neural network in fiber-optic perimeter intrusion detection system", Optical Fiber Technology, vol.56, 2020, DOI: 10.1016/J.YOFTE.2020.102182.

[8] N. Irawati, A. Hatta, Y. Yhuwana, and Sekartedjo, "Heart Rate Monitoring Sensor Based on Singlemode-Multimode-Singlemode Fiber", Photonic Sensors, vol.10, pp.186-193, 2020, DOI : 10.1007/s13320-019-0572-7.

[9] A. H. Khandoker, H. F. Jelinek and M. Palaniswami, "Heart rate variability and complexity in people with diabetes associated cardiac autonomic neuropathy", Annual International Conference of the IEEE Engineering in Medicine and Biology Society (EMBC), Vancouver, BC, Canada, 2008, DOI: 10.1109/IEMBS.2008.4650261.

[10] L. B. Soldano and E. C. M. Pennings, "Optical multi-mode interference devices based on self-imaging: principles and applications", Journal of Lightwave Technology, vol. 13, no. 4, pp. 615-627, 1995, DOI: 10.1109/50.372474.

Proceedings of the 7th Optoelectronics Global Conference (OGC 2022)

Accurate Measurement of Large Strain under High-temperature Environment Based on Fiber Bragg Grating

Zhiyuan Wang, Jindong Wang*, Tao Zhu
Key Laboratory of Optoelectronic Technology & Systems (Ministry of Education)
Chongqing University
Chongqing, China
e-mail: tju_jdwang@163.com

Abstract—Strain measurement technology under high-temperature environment has been a hot and difficult research issue in the field of measurement. On the one hand, conventional resistive strain gauges are susceptible to electromagnetic interference at high temperature. And on the other hand, common fiber sensors will be invalid under high-temperature environment, and may fall off under large strain conditions. In this paper, a precision measurement scheme that combining plasma surface treatment and metal oxide adhesive based on fiber Bragg grating (FBG) of large strain under high temperature environment is proposed, where three types of protection for the grating area of a FBG sensor were established, and a new plasma surface treatment method is theoretical analyzed and experimental studied. Suitable adhesive is also carefully selected according to the characteristics of the sample to be measured. After optimizing the mechanical transfer effect with the proposed method, effective coupling between the surface of the sample to be measured and the fiber grating sensor is realized, and large strain measurement up to 1200με under 1000°C environment is experimentally achieved.

Keywords- fiber Bragg grating; high temperature; large strain measurement; Fiber sensing; Precision measurement

I. INTRODUCTION

Strain measurement is one of the key aspects of structural health monitoring (SHM), which is of great significance to assess the state of structures and to secure minimum safety conditions Strain measurement includes room temperature strain and high temperature strain measurement [1]-[4]. For dynamical systems under high temperature conditions, high temperature strain measurement is not only an important support in the evaluation of aerospace engine′s reliability, but also a visual basis for structural optimization and refinement design. The measurement of large strain under high-temperature environment [5]-[6] has always been a huge problem in the development of aero-engines, The mechanical properties of materials and structures under high-temperature service environments are significantly different from those in normal temperature environments, high temperature will degrade the mechanical properties of materials and structures, which will reduce the safety and reliability of structures.

Traditional high-temperature strain measurement technology generally uses resistance strain gauges, it is made of metallic materials that are easily oxidized under high-temperature environment, which seriously affects the measurement accuracy. In addition, in the respect of fix mode of strain gauges under high-temperature environment, there are some problems. The commonly used organic adhesive is unable to work properly under high temperatures above 500 °C; Inorganic adhesive is composed of ceramic material, which can bear high temperature above 1000 °C, but it has a small thermal expansion coefficient and prone to thermal mismatch with the substrate, which eventually result in fall off of the sensor and failure of the strain measurement [7]-[9].

As an innovative technology of sensing, Fiber Bragg grating (FBG) [10]-[12] sensor carry the signal by the light and transfer signal by the fiber. Compared with the traditional electrical sensors such as resistance strain gages, fiber Bragg grating has the characteristics of small size, light weight, low loss, high sensitivity, and resistance to electromagnetic interference, corrosion resistance and multiplexing, etc... However, there are still several key technologies that must be solved in order to effectively use fiber optic sensing technology for strain monitoring of aerospace engines under high-temperature environment. Not only does the FBG itself need to be able to bear large strain under high-temperature environment, but also have to package and mount the fiber optic sensor on the actual structure. Therefore, the world's leading space power are developing FBG strain sensing technology to break through the key technology of aerospace engine strain measurement [13]-[21].

In this paper, a precision measurement scheme that combining plasma surface treatment and metal oxide adhesive based on fiber Bragg grating (FBG) of large strain under high temperature environment is proposed, where three types of protection for the grating area of a FBG sensor were established, and a new plasma surface treatment method is theoretical analyzed and experimental studied. Suitable adhesive is also carefully selected according to the characteristics of the sample to be measured. After optimizing the mechanical transfer effect with the proposed method, effective coupling between the surface of the sample to be measured and the fiber grating sensor is realized, and

978-1-6654-8699-6/22 $31.00 © 2022 IEEE

large strain measurement up to 1200με under 1000°C environment is experimentally achieved.

II. EXPERIMENTAL SET UP

A. Experimental System

The experimental system diagram is shown in Fig. 1, The encapsulated FBG sensor is mounted on the surface of a certain model of engine after plasma surface treatment with high-temperature adhesive, and the other end of the FBG sensor is connected to the MOI fiber grating interrogator through a fixed platform and finally to the PC. During the experiment, once the engine is started, the strain signals from its surface is transmitted to the FBG sensor. The FBG transforms the strain signals into the shift of the central wavelength and transmits it to the interrogator for demodulation and processing.

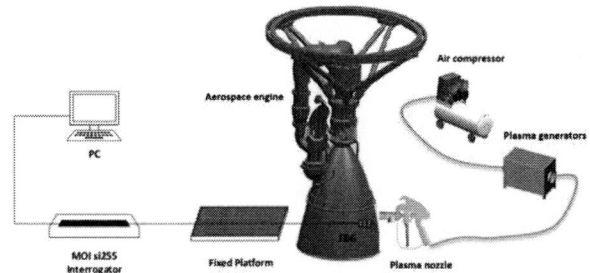

Figure 1.   Experimental system diagram.

B. Surface Treatment and Adhesive

1) Surface treatment: In order to optimize the effective coupling efficiency between the fiber grating sensor and the surface of to-be-tested sample, a targeted surface treatment is performed on the surface in this paper.

Plasma surface treatment is an excellent surface modification method, which is widely used in surface modification of various materials. Since all substances will become plasma state at an extreme high temperature, so the plasma state is called the "fourth state" of substance. Plasma contains a large number of active particles, such as high energy electrons, ions, free radicals, excited gas atoms, molecules and photons. Plasma surface treatment causes collision, scattering, excitation, heterogeneous, defects, crystallization and amorphization by injecting particles or gases into the material surface. These particles interact with the material surface such as etching and cleaning, oxidation, activation and polymerization, so as to achieve the treatment effect of changing the surface properties of the material.

On the one hand, plasma can remove organic pollutants and impurities on the surface, remove dust particles and static electricity. When the plasma interacts with the surface of the object to be cleaned, on the one hand, as shown in Fig. 2(a), plasma or plasma-activated chemically active substances are used to react chemically with the material surface dirt, such as the oxidation reaction between the reactive oxygen species in the plasma and the organic matter on the surface of the material. The organic pollutants on the

surface of the material are decomposed into carbon dioxide and water. On the other hand, the physical effects such as the bombardment of pollutants by high-energy particles of plasma are used. For example, the surface pollutants of the object are cleaned by active argon plasma (as shown in Fig. 2(b)), and the volatile pollutants are discharged by vacuum pump due to the bombardment.

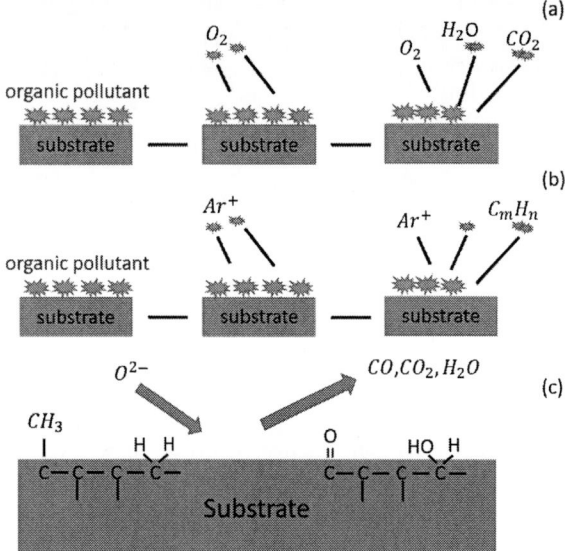

Figure 2.   Schematic diagram of plasma surface treatment. (a)(b) Plasma removes impurities from the surface. (c) Plasma activates the surface.

On the other hand, plasma can introduce new functional groups on the material surface and form highly active chemical bonds with the treated surface. As shown in Fig. 2(c), the highly active chemical bonds are more likely to react with other substances and form stable chemical bonds, thus achieving the purpose of improving the adhesion paste effect, as shown by the increase of surface tension after plasma treatment.

After the plasma treatment, the impregnation properties and adhesion of the surface are improved, Therefore, the next step of coating adhesive should be carried out immediately after the completion of surface treatment, so that the surface and the adhesive were closely integrated and the mechanical transmission were maintained, Thus, the precision measurement of large strain under extreme high temperature environment can be realized. Compared with other dry processes such as radiation, laser, electron beam and corona treatment, the unique feature of plasma is that the depth of plasma surface treatment only involves a very thin layer of the substrate surface. According to the observation results of electron spectroscopy for chemical analysis (ESCA) and scanning electron microscope (SEM) for chemical analysis, it can be inferred that it is generally within the range of tens to thousands of angstrom(nanoscale) from the surface, so that the interface physical properties can be significantly improved without any influence on the material phase.

2) Selection of Adhesives: In this paper, three types of

adhesives are used, namely SL high-temperature adhesive, *Ausbond* high-temperature adhesive and RESBOND high-temperature adhesive, the parameters of the three adhesives are compared in the following Table I.

TABLE I. PARAMETERS AND PROPERTIES OF DIFFERENT ADHESIVES

Adhesive	SL	AUSBOND	RESBOND
Main components	Silicate	Aluminosilicate	Zirconia
Maximum Temp	1300°C	1200°C	2200°C
Thermal expansion	/	17.8×10⁻⁶/°C	7.4×10⁻⁶/°C
Compressive Strength	≥6 MPa	20.8MPa	41.3MPa
Flexural Strength	≥2 MPa	30 MPa	20.7 MPa
Cure	40°C×6h	R.T.×24h+ 150°C×2h	R.T.×24h

The ideal high-temperature adhesive is characterized by good fluidity before curing and no cracks and other defects in the high-temperature environment after curing; the Thermal expansion should match that of the sample to be tested, and the compressive strength and flexural strength should be large; the curing method should be simple.

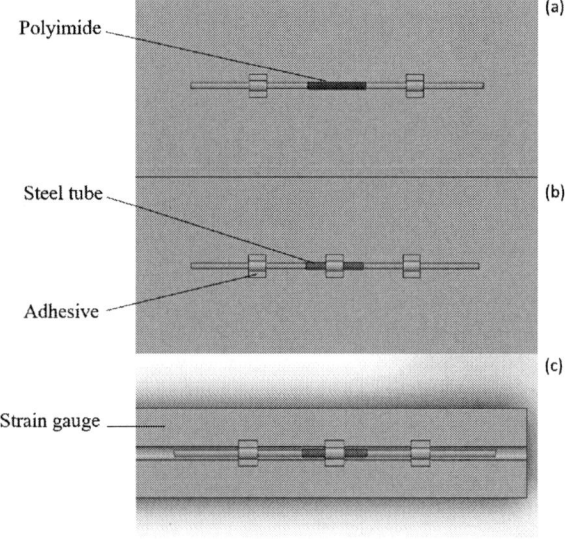

Figure 3. Three types of grating protection. (a) Coated polyimide (b) Grating suspended inside the tube (c) Mounting on strain gauges.

In this paper, the chosen adhesive RESBOND bonding technique is discussed in order to determine the appropriate bonding and curing method. The bonding method should follow the following criteria: The adhesive, as the intermediate between the sample and the FBG, should be in good contact with the sample to ensure the bond strength and avoid falling off; The adhesive needs to be applied evenly to avoid damage to the fiber grating under high temperature; The bonding process should be simple to minimize potential damage. In the process of strain sensing, there is a multi-stage strain transfer process: the to be tested sample

→substrate of sensor → adhesive layer → FBG, and each stage causes strain loss, which reduces the strain transfer efficiency of FBG and makes the strain felt by FBG inaccurate with the real strain of the measured structure. Therefore, it is necessary to compare the strain efficiency of the three adhesives. Since different adhesives have different modulus of elasticity, the relationship between the strain transfer efficiency and the modulus of elasticity of the adhesive layer can be calculated as shown in the Fig. 3.

It can be seen from the Fig. 3 that as the Elastic Modulus of the adhesive increases, the strain transfer efficiency also increases. After the Elastic Modulus is more than 30 GPa, the growth of strain transfer efficiency tends to slow down and gradually approaches the limit. The strain transfer efficiency of RESBOND is the highest because the modulus of elasticity of SL is about 30 GPa, that of Osborn is about 45 GPa, and that of RESBOND is 210 GPa. The main component of RESBOND is zirconium oxide, which has a fracture toughness of 5~6MPa √ m. The greater the fracture toughness, the stronger the resistance to brittle fracture. The geometrical parameters of the hand-applied production adhesive are not easily controlled, resulting in human interference in the stability and reliability of the test data, so the adhesive shape should be standardized as much as possible.

C. Protection of Sensors

Since FBG is thin diameter, fragile and easily broken, it is difficult to use directly. Therefore, the FBG is generally protected by encapsulation. In this paper, three types of encapsulations are proposed to protect the grating, as shown in the following Fig. 4.

The first is a small section of steel tube (slightly longer than the length of the grating) with high-temperature adhesive glued to the surface of the to-be-tested sample, then the FBG is passed through so that the grating is suspended inside the tube, and the two ends of the FBG outside the grating are fixed to the surface with adhesive, as shown in Fig. 4(b). The second one is to fix the tube to the strain gauge with a thin groove by adhesive, and then pass the FBG through the tube so that the grating is suspended inside the tube, while the fiber is fixed to the strain gauge by adhesive on both sides of the tube as shown in Fig. 4(c). The third one is to apply the polyimide coated FBG directly to the surface with adhesive, as shown in Fig. 4(b). Considering that in the first two encapsulation methods, the strain of the measured material is first transferred to the metal encapsulation shell, then to the adhesive, and finally to the FBG, and the multilayer strain transfer link increases the possibility of strain loss along the strain transfer, the third method simplifies the multilayer strain transfer structure.

After curing, it was found that the first two methods have the possibility of falling off by vibration, but the third method avoid this since the bare fiber is directly pasted and the adhesive layer is thinly applied.

Encapsulation protection is also required for non-grating segments of FBG, this paper uses nylon casing and armored hoses to provide double protection. Nylon casing provides flexible protection for the fiber and prevents the fiber from being damaged by irregular edge cutting when passing through the armored hose; The armored hose further provides rigid protection, which can make the fiber work normally and stably under the harsh environment of high temperature and large strain.

III. EXPERIMENTAL RESULTS

The following groups of controlled experiments were conducted, and the experimental system is shown in Fig.1(b). The grating is encapsulated by physical protection, and the FBG is installed on the surface of the sample to be tested with adhesive.

There are the following two groups of sensors 1 and 2, set up two control experiments, both experiments use the same sensor and keep the installation position in the same place, the first is under room temperature environment, sensor 1 with high-temperature adhesive mounted on the surface without plasma surface treatment, sensor 2 with high-temperature adhesive mounted on the plasma treated surface.

The second group is under high temperature environment, sensor 1 with high-temperature adhesive mounted on the surface without plasma surface treatment, sensor 2 was mounted on the plasma treated surface with high temperature adhesive. After the sensors were mounted, excitation was applied to them, and the response curves of the sensors to the excitation in both experiments are shown in Fig. 4.

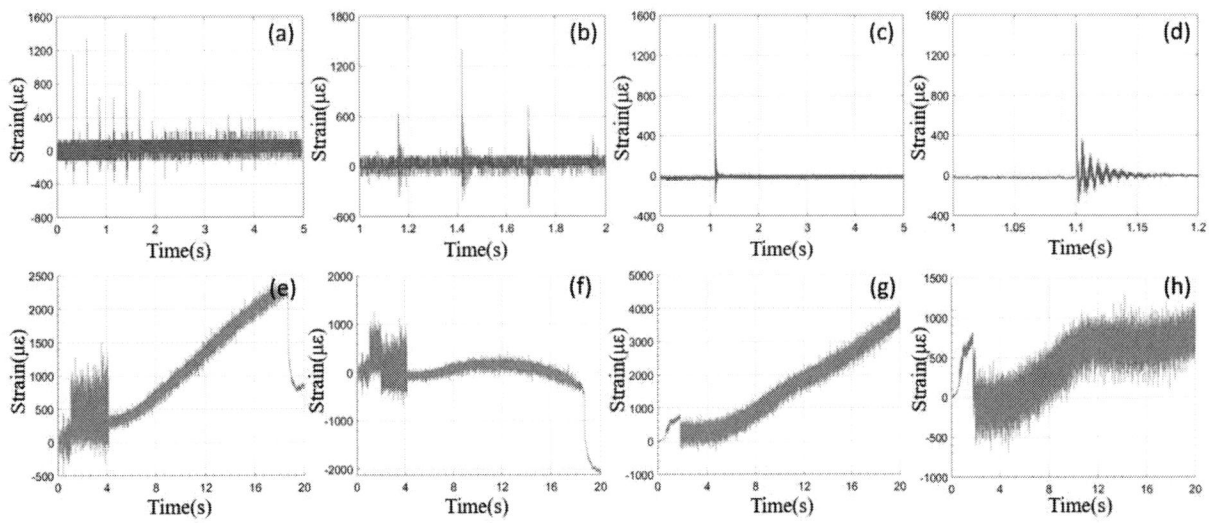

Figure 4. Experimental curves of FBG sensors under room temperature and high temperature. (a, b) sensor 1 bonding with high-temperature adhesive without plasma surface treatment under room temperature. (c, d) sensor 2 bonding with high-temperature adhesive with plasma surface treatment under room temperature. (e, f) sensor 1 bonding with high-temperature adhesive without plasma surface treatment under high temperature. (g, h) sensor 2 bonding with high-temperature adhesive with plasma surface treatment under high temperature.

The following experiments are conducted under room temperature environment. As the Fig. 4(a) shows, sensor 1 sensed the strain signal throughout the whole time and obtained the maximum strain of 1079με at 1.42 s, Fig. 4(b) shows the time period during the experiment when the sensor responds to the maximum strain excitation, the sensor did not fall off and performed well throughout the whole time. From Fig. 4(c), it can be seen that the sensor 2 senses the strain signal throughout the whole time and obtains the maximum strain of 1515με at 1.1 s, Fig. 4(d) shows the time period during the experiment when the sensor responds to the maximum strain excitation, the sensor does not fall off and performs well throughout the whole time. It can be seen that whether the plasma surface treatment is applied under room temperature has little effect on the experimental results, and the sensors do not fall off.

The following experiments are conducted under high temperature environment. Thermocouples are installed near these two groups of sensors for measuring temperature data under high temperature environment for temperature compensation. The experimental results are shown in Fig.

4(e)-(h). From Fig. 4 (e) (f), it can be seen that the sensor 1 started to respond to weak signals at 1.06s, strong signals at 1.81s, a certain degree of off only to respond to weak signals at 4.09s, and falling completely at 14.93s. From Fig. 4 (f), it can be seen that the maximum strain of the group of sensors before falling is 1274με at 2.08s after the temperature compensation. From the Fig. 4(g)(h), it can be seen that the group of sensors starts to respond to the signal at 2.3s, and the sensors do not fall off and respond well to the signal throughout the experiment. And Fig. 4 (h) can see the maximum response after temperature compensation at 11.6s, which is 1280με. It can be seen that the sensors mounted on the surface without plasma surface treatment under high temperature will have some degree of falling when subjected to large strains due to the high temperature, and will completely fall off as the temperature increases. The sensors mounted on the plasma-treated surface performed well when subjected to large strains and did not falling.

Comparing the experimental results of the two groups of sensors, it can be seen that the plasma treatment improves the mechanical transfer efficiency between the surface of the sample to be measured and the adhesive, making the effective coupling between the sensor and the surface and achieving the accurate measurement of large strains of at least 1200με under high temperature environment.

IV. CONCLUSION AND OUTLOOK

In this paper, aiming at the problem of strain measurement of FBG under high-temperature environment, we proposed the encapsulation protection method and plasma surface treatment scheme for FBG sensors, analyzed and selected the suitable adhesive according to the surface characteristics of the sample to be measured. The plasma surface treatment improves the impregnation properties and adhesion of the surface, and achieves a strong and effective coupling between the surface of the sample and the adhesive; the selected zirconium dioxide adhesive has good adhesion performance under high temperature, stable and does not fall off; Finally, the precision measurement of large strain at least 1200με under high temperature is achieved. We now outline several practical and structural refinements to our approach that could be addressed in the future.

One consideration is the further protection for FBG sensors. In previous work we have shown that three types of encapsulation protection can protect the FBG and make it work properly under high temperature environment. In the future, we will design an encapsulated structure with thermal insulation, which can be used to protect the FBG and make it more convenient and efficient for strain testing under high temperature environments.

Another practical issue is fixing the FBG sensors onto the surface. In this work, the sensors were held simply by adhesive. To increase the mechanical robustness, other techniques to fix the FBG sensor to the surface could be considered, such as flame spraying as has been reported.

REFERENCES

[1] Zhang Han, Zhaohui Du, Zuowei Fang, Shibing Wang, and Xuefeng Chen. "Sparse Decomposition Based Aero-Engine's Bearing Fault Diagnosis." Journal of Mechanical Engineering 51, no. 1 (2015): 97-105.

[2] Shi X. "Tehl Analysis of Aero-Engine Mainshaft Roller Bearing Based on Quasi-Dynamics." Journal of Mechanical Engineering 52, no. 3 (2016): 86.

[3] Wang Y. S., X. M. Dong, W. Liu, M. G. Huang, L. I. Xin, and H. X. Cui. "Research on Developments of High Temperature Testing Technology for Aero-Engine." Measurement & Control Technology (2017).

[4] Yang X. "Data Acquisition and Process System for Temperature Field of Some Aeroengine." Computer Automated Measurement & Control (2003).

[5] Bian Qiang, Constantin Bauer, Andrea Stadler, Markus Lindner, Martin Jakobi, Wolfram Volk. "In-Situ High Temperature and Large Strain Monitoring During a Copper Casting Process Based on

Regenerated Fiber Bragg Grating Sensors." Journal of Lightwave Technology 39, no. 20 (2021): 6660-69.

[6] Guo Kuikui, Jun He, Laipeng Shao, Gaixia Xu, and Yiping Wang. "Simultaneous Measurement of Strain and Temperature by a Sawtooth Stressor-Assisted Highly Birefringent Fiber Bragg Grating." Journal of Lightwave Technology 38, no. 7 (2020): 2060-66.

[7] Shiuh-Chuan, HER, and TSAI Chang-Yu. "Strain Measurement of Fiber Optic Sensor Surface Bonding on Host Material." Transactions of Nonferrous Metals Society of China 19 (2009): s143-s49.

[8] Tahir BA, J Ali Saktioto, M Fadhali, RA Rahman, and A Ahmed. "A Study of Fbg Sensor and Electrical Strain Gauge for Strain Measurements." Journal of optoelectronics and advanced materials 10, no. 10 (2008): 2564-68.

[9] Zhao Hai-tao, Quan-bao Wang, Ye Qiu, Ji-an Chen, Yue-ying Wang, and Zhen-min Fan. "Strain Transfer of Surface-Bonded Fiber Bragg Grating Sensors for Airship Envelope Structural Health Monitoring." Journal of Zhejiang University SCIENCE A 13, no. 7 (2012): 538-45.

[10] Rao Yun-Jiang. "In-Fibre Bragg Grating Sensors." Measurement science and technology 8, no. 4 (1997): 355.

[11] Kersey, Alan D, Michael A Davis, Heather J Patrick, Michel LeBlanc, KP Koo. "Fiber Grating Sensors." Journal of Lightwave Technology 15, no. 8 (1997): 1442-63.

[12] Kashyap, Raman. Fiber Bragg Gratings: Academic press, 2009.

[13] Ferreira, Marta S, Paulo Roriz, Jörg Bierlich, Jens Kobelke, Katrin Wondraczek. "Fabry-Perot Cavity Based on Silica Tube for Strain Sensing at High Temperatures." Optics Express 23, no. 12 (2015): 16063-70.

[14] Habisreuther, Tobias, T Elsmann, A Graf, and MA Schmidt. "High-Temperature Strain Sensing Using Sapphire Fibers with Inscribed First-Order Bragg Gratings." IEEE Photonics Journal 8, no. 3 (2016): 1-8.

[15] Petrie Christian M, Niyanth Sridharan, Adam Hehr, Mark Norfolk, and John Sheridan. "High-Temperature Strain Monitoring of Stainless Steel Using Fiber Optics Embedded in Ultrasonically Consolidated Nickel Layers." Smart Materials and Structures 28, no. 8 (2019): 085041.

[16] Havermann, Dirk. "Study on Fibre Optic Sensors Embedded into Metallic Structures by Selective Laser Melting." Heriot-Watt University, 2015.

[17] Goossens, Sidney, Ben De Pauw, Thomas Geernaert. "Aerospace-Grade Surface Mounted Optical Fibre Strain Sensor for Structural Health Monitoring on Composite Structures Evaluated against in-Flight Conditions." Smart Materials and Structures 28, no. 6 (2019): 065008.

[18] Piazza Anthony, Lance W Richards, and Larry D Hudson. High-Temperature Strain Sensing for Aerospace Applications. WRSGC Summer Test and Measurements Conference, 2008.

[19] Meng S, C Du, W Xie, S Huo, L Jiao, H Jin, and L Song. "Application of High-Temperature Optical Fiber Sensor in Temperature and Strain Testing of Hot Structure." J. Sol. Rocket Technol 36 (2013): 701-05.

[20] Zeng P, Y. Wang, F. Chen, Y. Sun, and Z. You. "Analysis of the Effect of Adhesives on Strain Transfer for Surface Bonded Polyimide Fiber Bragg Grating." Chinese Journal of Sensors and Actuators 32, no. 1 (2019): 43-49.

[21] Zhang, Wei, Weimin Chen, Yuejie Shu, Jun Wu, and Xiaohua Lei. "Degradation of Sensing Properties of Fiber Bragg Grating Strain Sensors in Fatigue Process of Bonding Layers." Optical Engineering 53, no. 4 (2014): 046102.

Proceedings of the 7th Optoelectronics Global Conference (OGC 2022)

Design of a Hollow-core Microstructured Optical Fiber with Low Loss and High Polarization-maintaining

Aoyan Zhang
Department of Electronic and Electrical Engineering
Southern University of Science and Technology
Shenzhen, China
e-mail: 2019022157@m.scnu.edu.cn

Zhipeng Deng
School of Information and Optoelectronic Science and Engineering
South China Normal University
Guangzhou, China
e-mail: 2019022104@m.scnu.edu.cn

Jialong Li
Department of Electronic and Electrical Engineering
Southern University of Science and Technology
Shenzhen, China
e-mail: 12131038@mail.sustech.edu.cn

Guiyao Zhou
School of Information and Optoelectronic Science and Engineering
South China Normal University
Guangzhou, China
e-mail: gyzhou@scnu.edu.cn

Perry Ping Shum*
Department of Electronic and Electrical Engineering
Southern University of Science and Technology
Shenzhen, China
e-mail: shenp@sustech.edu.cn

Abstract—**In this paper, a hollow-core microstructured optical fiber is proposed. By adding several rounded hexagonal air-hole arrays to the cladding of the hollow-core polarization-maintaining fiber, the requirements of low loss and high polarization-maintaining are achieved. In the wavelength ranges of 1.540 μm - 1.585 μm and 1.609 μm - 1.653 μm, the confinement loss is less than 0.1 dB/km, and the birefringence is higher than 5×10^{-5}. Such a fiber performance heralds new opportunities for hollow-core anti-resonant fibers in practical applications.**

Keywords-hollow-core microstructured optical fiber; hollow-core anti-resonant fiber; polarization-maintaining

I. INTRODUCTION

As a new type of optical fiber, hollow-core microstructured optical fiber [1,2] has the advantages of low nonlinearity, low transmission delay, and high damage threshold because its transmission energy is mainly concentrated in the air. It has incomparable advantages [3] in optical communication, optical sensing, and optoelectronic devices. With the rapid development of aerospace and communication fields, higher requirements are put forward for the transmission bandwidth and polarization-maintaining characteristics of optical fiber [4].

The use of the advantages of hollow-core microstructured fiber to achieve wider conduction band and higher polarization-maintaining characteristics is a hot topic in research [5]. Because the structure and light-guiding mechanism of hollow-core microstructured fiber are different

from solid core fiber, the design concept of bandwidth and polarization-maintaining of solid core fibers can not be used in the hollow-core fiber [6].

In this paper, we proposed a hollow-core microstructured optical fiber with low loss and high polarization-maintaining (HPBG-HCARF). Based on the light-guiding mechanism of hollow-core fiber, the structure is designed through theoretical simulation.

II. STRUCRURE DESIGN OF HPBG -HCARF

Fig. 1 shows the design model of the HPBG-HCARF. The six capillaries are embedded in the air core of the hollow-core photonic bandgap fiber. The thickness of two capillaries in the vertical direction is not consistent with that of the other four capillaries to introduce high birefringence in the HPBG-HCARF.

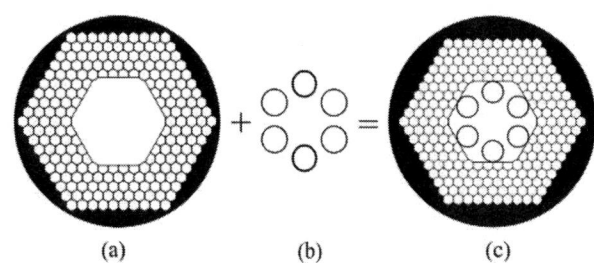

Figure 1. Design model of the HPBG-HCARF. (a) The 61-hole hollow-core photonic bandgap fiber; (b) Six capillaries; (c) HPBG-HCARF.

978-1-6654-8699-6/22 $31.00 © 2022 IEEE

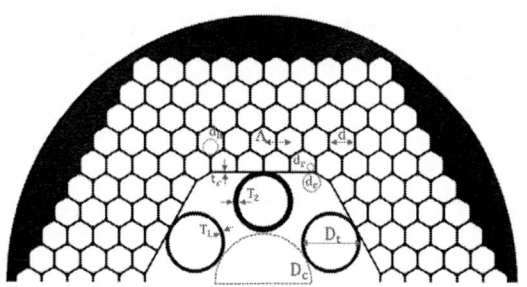

Figure 2. Structural parameters of the HPBG-HCARF.

Fig. 2 shows the structural parameters of the HPBG-HCARF. The diameter of the air core is D_c, the outer diameter of six capillaries is D_t, the two different thicknesses of capillaries are T_1 and T_2, respectively. The hole spacing of rounded hexagonal air hole is Λ, the diameter of hole is d, the diameter of rounded hexagonal air hole is d_h, the number of air holes arranged is N, the quartz thickness between the inner and outer cladding is t_c, and the diameters of rounded angle on two sides are d_r and d_c, respectively.

In order to achieve low loss transmission in the range of 1300 nm to 1600 nm, the bandgap structure parameters are set to the following fixed values: Λ=4.7 μm, d=0.98Λ, d_h=0.44Λ, d_c=0.94Λ, d_r=0.2Λ, t_c=0.5(Λ-d). The diameter of the six capillaries is fixed proportional to the size of the air core, D_t=0.6D_c. By combining the structure parameters of bandgap and deriving the geometric structure relationship, D_c and D_t are approximately 17.365 μm and 10.419 μm, respectively.

III. NUMERICAL SIMULATION AND ANALYSIS

A. The Number of Bandgap Layers

For traditional photonic bandgap optical fibers, the number of layers of periodic structure plays a crucial role in confinement loss (CL). The more layers of air holes, the lower the confinement loss. Fig. 3 shows the structure of optical fibers with different bandgap layers. Fig. 4 shows the confinement loss spectrum of different structures in Fig. 3.

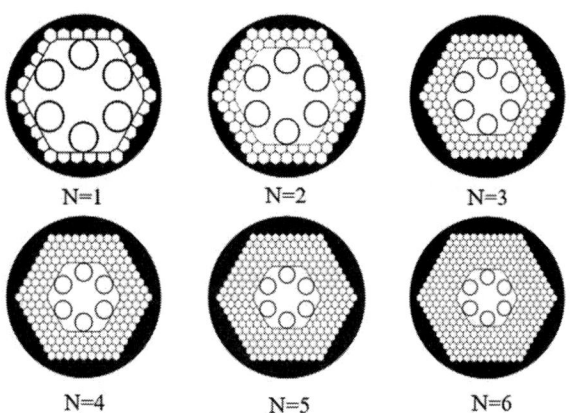

Figure 3. Structure diagram of HPBG-HCARF with different bandgap layers.

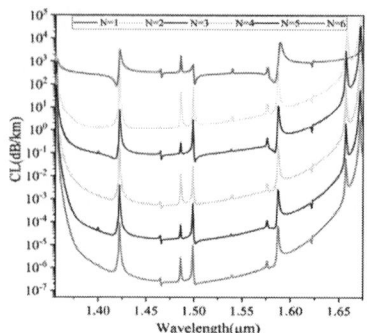

Figure 4. Confinement loss spectrum of different number of bandgap layers.

When the number of bandgap layers reaches 5 or more, the confinement loss in the range of 1.355 μm to 1.65 μm is less than 1 dB/km. Although simply increasing the number of bandgap layers can effectively reduce the confinement loss, too many layers will make the size of optical fiber larger. Therefore, to consider the power limitation ability to light, N=5 is selected as the best parameter of this structure.

B. The Thicknesses of Six Capillaries

The most appropriate values of capillary thickness T_1 and T_2 are explored according to the principle of the conventional resonant band in the polarization-maintaining hollow-core anti-resonant fiber. Firstly, set T_2=T_1 to determine the area of low confinement loss. The first-order anti-resonant thickness is 0.372 μm calculated from the anti-resonant condition, which corresponds to the low loss region of Fig. 5. Therefore, the thickness of T_1 is selected as 0.372 μm.

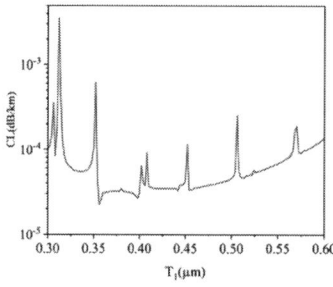

Figure 5. Confinement loss spectrum for different T_1 of the HPBG-HCARF.

Then, T_2 is gradually changed to be in the first-order anti-resonant thickness region and achieve high polarization-maintaining. In order to narrow the range of thickness variation, according to the resonant formula, the first-order resonant thickness is about 0.744 μm. Therefore, increasing the calculation range from 0.650 μm to 0.750 μm can ensure that the thickness in both directions is within the first-order anti-resonant thickness area. Fig. 6(a) represents the change of limiting loss in different polarization directions with T_2. Fig. 6(b) represents the change of birefringence with T_2, and the gray dotted line represents the reference line with birefringence of 5×10^{-5}.

978-1-6654-8699-6/22 $31.00 © 2022 IEEE

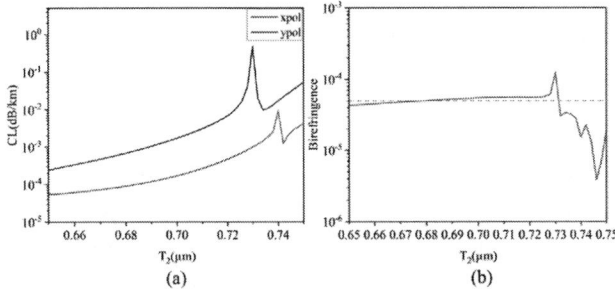

Figure 6. The influence of different T_2 on the HPBG-HCARF. (a) Confinement loss; (b) birefringence.

As shown in Fig. 6, with the increase of T_2, the confinement loss increases gradually, while the birefringence increases gradually and then decreases. This is mainly because of the anti-resonant principle. When the value of T_2 increases towards the first-order resonant region, the fundamental mode will leak along this direction, resulting in increased loss and changes in birefringence. When the size of T_2 is between 0.680 μm and 0.710 μm, it shows a low confinement loss and high birefringence greater than 5×10^{-5}. Therefore, the thickness of T_2 is selected as 0.710 μm.

C. Confinement Loss and Polarization-maintaining Characteristics of HPBG-HCARF

Fig. 7(a) is the optimized structure of HPBG-HCARF with N=5, T_1=0.372 μm, T_2=0.710 μm. In order to compare the confinement loss and polarization-maintaining characteristics of HPBG-HCARF, the inner capillaries in Fig. 7(a) are extracted as the optical fiber structure (SR-HCARF) in Figure 7(b). The two fibers have the same internal structural parameters.

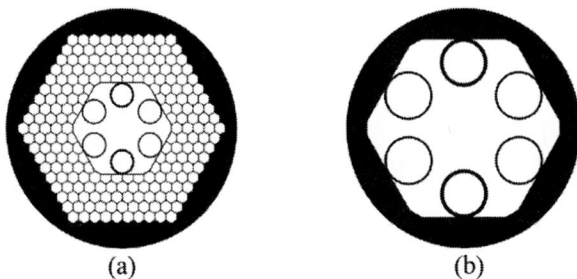

Figure 7. (a) Optimized HPBG-HCARF; (b)SR-HCARF.

The confinement loss spectrum of the two optical fibers are shown in Fig. 8. It can be seen that the confinement loss of HPBG-HCARF is much lower than that of SR-HCARF, about seven orders of magnitude. This shows that the air holes in the bandgap layer can significantly reduce the loss. Although the existence of surface modes makes the confinement loss spectrum of HPBG-HCARF have small peaks, in the range of 1.540 μm - 1.585 μm (Band1) and 1.609 μm - 1.653 μm (Band2), the confinement loss spectrum of HPBG-HCARF changes slowly and the influence of surface mode can be almost ignored.

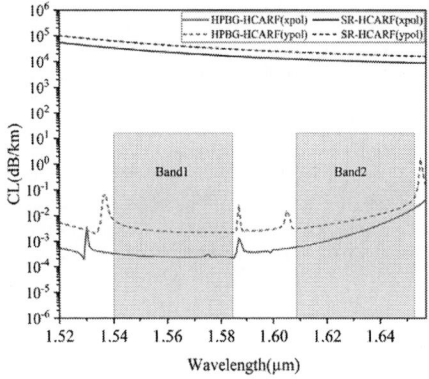

Figure 8. Confinement loss spectrum of the two optical fibers.

The birefringence values of the two optical fibers are shown in Fig. 9. Due to the coupling between the fundamental mode and surface mode of HPBG-HCARF, the birefringence value at some wavelengths deviates greatly from SR-HCARF. The birefringence values of the two fibers are basically the same, indicating that the realization of birefringence in HPBG-HCARF depends on the thickness difference of the capillaries. Birefringence is greater than 5×10^{-5} in Band1 and Band2.

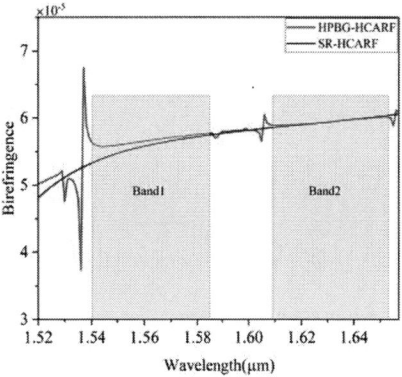

Figure 9. Birefringence of the two optical fibers.

IV. CONCLUSION

A hollow-core microstructured optical fiber with low loss and high polarization-maintaining is presented theoretically. The structure parameters of the optical fiber are optimized, including the number of bandgap layer N, the thickness of the inner capillary T_1 and T_2. The results of calculating the confinement loss and birefringence of the optimized optical fiber show that in the wavelength ranges of 1.540 μm - 1.585 μm and 1.609 μm - 1.653 μm, the confinement loss is less than 0.1 dB/km, and the birefringence is higher than 5×10^{-5}.

ACKNOWLEDGMENT

The work of this paper is conducted with support from the Foundation of the National Natural Science Foundation of China (61935007).

REFERENCES

[1] F. Poletti, N. V. Wheeler, M. N. Petrovich, *et al.*, "Towards high-capacity fibre-optic communications at the speed of light in vacuum," Nature Photonics, vol. 7, March 2013, pp. 279-284, doi:10.1038/nphoton.2013.45.

[2] J. C. Knight, J. Broeng, T. A. Birks, *et al.*, "Photonic band gap guidance in optical fibers," Science, vol. 282, November 1998, pp. 1476-1478, doi:10.1126/science.282.5393.1476.

[3] H. Sakr, Y. Hong, T. D. Bradley, *et al.*, "Interband short reach data transmission in ultrawide bandwidth hollow core fiber," Journal of Lightwave Technology, vol. 38, September 2020, pp. 159-165, doi: 10.1364/JLT.38.000159.

[4] W. Ding, Y. Y. Wang, "Hybrid transmission bands and large birefringence in hollow-core anti-resonant fibers," Optics Express, vol. 23, August 2015, pp. 21165-21174, doi:10.1364/OE.23.021165.

[5] J. M. Fini, J. W. Nicholson, B. Mangan, *et al.*, "Polarization maintaining single-mode low-loss hollow-core fibres," Nature Communications, vol. 5, October 2014, doi:10.1038/ncomms6085.

[6] S. Yerolatsitis, R. Shurvinton, P. Song, *et al.*, "Birefringent Anti-Resonant Hollow-Core Fiber," Journal of Lightwave Technology, Vol. 38, June 2020, pp. 5157-5162, doi: 10.1109/JLT.2020.3000706.

Proceedings of the 7th Optoelectronics Global Conference (OGC 2022)

Interference Fading Suppression for Multi-frequency Φ-OTDR

Yu Wang [1,2], Junhong Wang [2], Bin Liang [1],

1. Shanxi Transportation Technology Research &
Development Co., Ltd
Taiyuan, China
e-mail: wangyu@tyut.edu.cn,
wangjunhong1226@link.tyut.edu.cn, liangbin@sxjt.net

Yan Li [2], Qing Bai [1,2], Baoquan Jin [1,2*]

2. Key Laboratory of Advanced Transducers and
Intelligent Control Systems (Ministry of
Education and Shanxi Province)
Taiyuan University of Technology
Taiyuan, China
e-mail: liyan0931@link.tyut.edu.cn,
baiqing@tyut.edu.cn, jinbaoquan@tyut.edu.cn

Abstract—In phase-sensitive optical time domain reflectometer (Φ-OTDR), due to interference fading, the intensity of Rayleigh backward scattering (RBS) signals may be close to zero at the fading position and submerged in noise, resulting in the abnormal demodulated phase of vibration signal. In this paper, a multi-frequency Φ-OTDR is proposed for the interference fading suppression. Three acousto-optical modulators (AOMs) are used to generate multi-frequency probe pulses. Data alignment processing is carried to ensure the consistency of multi-frequency beat signals. Rotated-vector-sum method is adopted to aggregate multi-frequency signals. Experimental results show that with the help of data alignment processing, the fading probability of aligned curve can be reduced to 1.90%, and the SNR of positioning curve can reach 16.03dB.

Keywords-distributed optical fiber sensing; Φ-OTDR; interference fading suppression; data alignment

I. INTRODUCTION

As a typical distributed optical fiber vibration sensing technology, phase-sensitive optical time domain reflectometer (Φ-OTDR) uses Rayleigh backward scattering (RBS) signals to detect and locate vibration signals, which has been widely concerned and studied [1]. However, interference fading exists in the Φ-OTDR system [2], so the intensity of RBS signals may be close to zero at the fading position and submerged in noise, resulting in the distortion of the reconstructed vibration signal. Interference fading noise has been the bottleneck of Φ-OTDR system in practical engineering applications.

In recent years, many researchers have studied various interference fading suppression theories and schemes [3], [4]. As a representative method, multi-frequency detection method is to use probe pulses with different frequencies to obtain mutually independent RBS signals, and then through signal aggregation processing to reduce the fading effect of the received signal. In 2017, the research group of Shanghai Jiao Tong University proposed to effectively eliminate interference fading by combining the detection signals of multiple degrees of freedom [5]. In 2018, this research group used light intensity modulator to generate positive and negative frequency components of linear frequency modulation detection pulses to suppress fading noise [6]. In 2018, the research group of Schlumberger Fiber Optics

Technology Center used fiber ring to produce multiple frequency pulsed light to suppress the interference fading noise [7]. In 2019, the research group of Chongqing University used the electro-optic modulator to modulate multi-frequency nonlinear detection pulses, and to reduce the influence of interference fading [8]. In 2019, the research group of Nanjing University connected three acousto-optic modulators in parallel, significantly reducing the distortion caused by fading [9]. In 2019, the research group of University of Electronic Science and Technology of China effectively eliminated interference fading by spectrum extraction by using the sideband feature information of the rectangular pulse spectrum [10]. In 2022, the research group of Southwest Jiaotong University utilize the rotated-vector-average method to synthesize the decomposed signals, the demodulated phase noise is greatly suppressed [11]. In 2022, the research group of Taiyuan University of Technology proposed an active frequency transformation method, which can reduce the fading probability by means of multi-frequency detection [12].

Therefore, multi-frequency detection method has been widely concerned to suppress interference fading in Φ-OTDR. Considering the complexity and the cost, multi-frequency detection with multiple acousto-optical modulators (AOMs) may be the effective choice. However, the delay of different modulated pulse may lead to misaligned RBS signals and invalidity of fading suppression. In this paper, in order to improve the reliability of interference fading suppression, multi-frequency detection with three AOMs in parallel is adopted in Φ-OTDR. The consistency of RBS signals is then ensured through data alignment processing.

II. SYSTEM STRUCTURE AND PRINCIPLES

A. Multi-frequency Detection Structure

The multi-frequency Φ-OTDR structure based on multiple AOMs is shown in Fig. 1. Continuous light emitted by narrow linewidth laser is divided into probe light (99%) and local light (1%) by the 99:1 coupler. The probe light passes through the 1×3 coupler into AOM1, AOM2 and AOM3, which can be modulated into multi-frequency pulsed light. The pulsed light passes through the 3×1 coupler into erbium-doped fiber amplifier (EDFA) for power amplification, and then is filtered by dense wavelength division multiplexer (DWDM). The multi-frequency pulsed light after amplification and noise

978-1-6654-8699-6/22 $31.00 © 2022 IEEE   137

filtering is incident into the sensing fiber through the circulator (CIR). An optical isolator (ISO) is connected to the end of the sensing fiber to avoid Fresnel reflection. The optical beat frequency between the RBS light generated within the sensing fiber and the local light will appear at the 2×2 coupler. The beat signal is converted into electrical signal by the balanced photodetector (BPD), collected by the data acquisition (DAQ) card and then uploaded to the upper computer (PC) for signal processing. In order to ensure the consistency of sensing signals, the driving signals of AOM1, AOM2 and AOM3 must be synchronized by the any waveform generator (AWG), and the DAQ card must be synchronized to trigger the synchronous data acquisition.

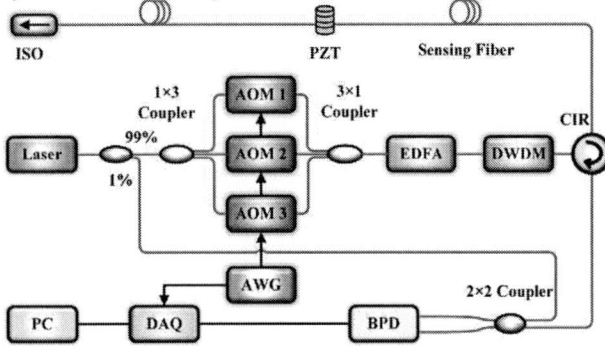

Figure 1. Multi-frequency detection structure

In the experiment, the total length of sensing fiber is 4.2km, which is composed of two sensing fibers of 4km and 200m connected to the piezoelectric transducer (PZT) device. The PZT device is driven by an electronic sine signal with a frequency of 200Hz and an amplitude of 2Vpp, in order to generate vibration signals. The AOMs modulate the continuous light to pulsed light with a pulse width of 200ns and a repetition rate of 8kHz. The frequency shift of AOM1 is 80MHz, the frequency shift of AOM2 is 150MHz, and the frequency shift of AOM3 is 200MHz. The sampling rate of DAQ card is set to 500MHz.

B. Signal Aggregation by Rotated-vector-sum Method

In the process of data processing, multi-frequency signal aggregation is the key step of multi-frequency detection to achieve fading suppression. The main methods of signal aggregation are direct summation method and rotated-vector-sum method [5]. Beat signal is a vector signal, and the phase angle difference between vectors is the decisive factor to determine the sum of vectors. The direct summation method is the result of direct summation of two signals. When the phase angle difference between two vectors is less than 90°, the vector strength will be enhanced by the summation. When the phase angle difference between two vectors exceeds 90°, the sum of two vectors may be smaller than the single vector. Therefore, the direct summation of beat signals strength may weaken its own strength.

Rotated-vector-sum method refers to rotating the vectors to be summed in one direction and then summing them, as shown in Fig. 2. The conjugate of each vector is firstly calculated and normalized, and then the original vector is rotated to zero phase by multiplying the normalized conjugate

of each vector. When all vectors are rotated to zero phase, the signal strength can be maximized by summation calculation. For any vector signal, this method can achieve the maximum strength of any vector summation. So this paper uses the rotated-vector-sum method to aggregate multi-frequency beat signals, and then achieves the purpose of interference fading suppression.

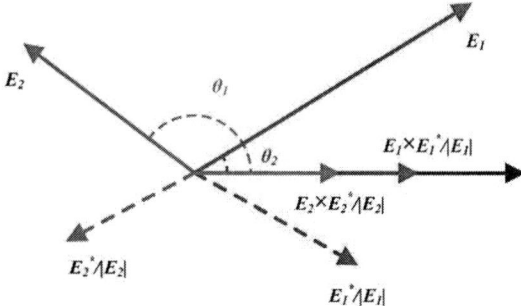

Figure 2. Schematic diagram of rotated-vector-sum method

C. Analysis of Misaligned Multi-frequency Pulsed Light

For AOMs with different frequency shifts, their modulation performance may be different. The key performance parameters of AOM include pulse extinction ratio, rise time and insertion loss. According to the manufacture values of AOM, the insertion losses of three AOMs are about 2dB. Their extinction ratios are 39.8dB, 48.46dB and 62.25dB, respectively. In order to obtain the actual modulated pulsed signal, the oscilloscope is used to measure modulation performance, which can be plotted in Fig. 3. The trigger pulse is plotted by black curve. The rise time of pulsed light modulated by AOM1 (plotted by blue curve) is about 23.4ns. The rise time of pulsed light modulated by AOM2 (plotted by purple curve) is about 23.7ns. The rise time of pulsed light modulated by AOM3 (plotted by orange curve) is about 9ns.

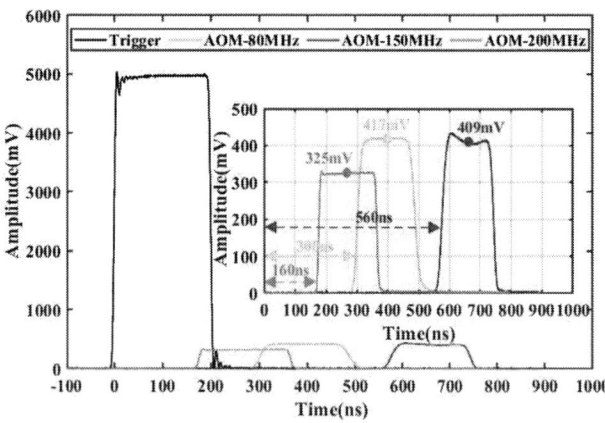

Figure 3. Performance of pulsed light modulated by AOMs

According to the measurement of modulated pulsed signal, the rise time of AOM3 is smaller than that of the other two AOMs, the response time to trigger signal is about 160ns, but the amplitude of 325mV after modulation is lower than those of the other two with 417mV and 409mV. Since the

978-1-6654-8699-6/22 $31.00 © 2022 IEEE

pulsed light modulated by AOM is generally amplified by EDFA, the amplitude of pulsed light may have little influence on the experimental results after amplification. It should be noted that the response time of pulsed signal modulated by AOM1 and AOM2 are about 300ns and 560ns. Therefore, the misaligned multi-frequency pulsed light can be clearly obtained because of different response time. In the multi-frequency detection system based on the parallel AOMs, the response time of AOM mainly affects the system performance, and the delay of modulation pulses may cause the delay of beat signals to different degrees. Therefore, it is necessary to align the beat signal before demodulation to ensure the consistency of signal.

D. Principle of Data Alignment Processing

Different AOMs are excited by the same trigger signal of AWG to generate pulsed signal. Because of the delay between modulation pulses generated by AOMs, the beat signals collected at the same time correspond to different positions on the sensing fiber. In order to ensure the consistency of beat signals, one beat signal with the response time of τ_0 is selected as the reference for data alignment processing of beat signals. According to the pulse width, the delay distance can be calculated as:

$$\delta = \frac{c \times (\tau_i - \tau_0)}{2\mu_f} \qquad (1)$$

where δ is the delay distance caused by different response time of two AOMs, c is the speed of light in vacuum, τ_i is the response time of another probe pulse, and μ_f is the refractive index of fiber. According to the delay distance obtained, two of three beat signals can be aligned to obtain the aggregation signals at the same position of sensing fiber. This data alignment processing may be carried out several times to obtain the optimized alignment results.

III. Experimental Results and Discussions

A. Multi-frequency Beat Signals

Time domain and frequency domain diagrams of beat signals based on three AOMs are obtained as shown in Fig. 4. Fig. 4 (a) and (b) show the collected original beat signal, including three frequencies of 80MHz, 150MHz and 200MHz, which are consistent with the frequency shifts of AOMs used in the system. Similarly, bandpass filtering is firstly carried out to obtain three beat signals as shown in Fig. 4 (c) and (d). The blue curve plots the 80MHz beat signal, the purple curve plots the 150MHz beat signal, and the orange curve corresponds to the beat signal with the frequency of 200MHz. Based on the minimum response time of 200MHz beat signal, the time delays of 80MHz beat signal and 150MHz beat signal are about 140ns and 400ns respectively, corresponding to the delay distance of 14m and 40m. In order to ensure the consistency of beat signals, data alignment processing is performed on original signals before demodulation, as shown in Fig. 4 (e) and (f). Fig. 4 (f) clearly highlights the consistency of aligned beat signals at the beginning section of sensing fiber.

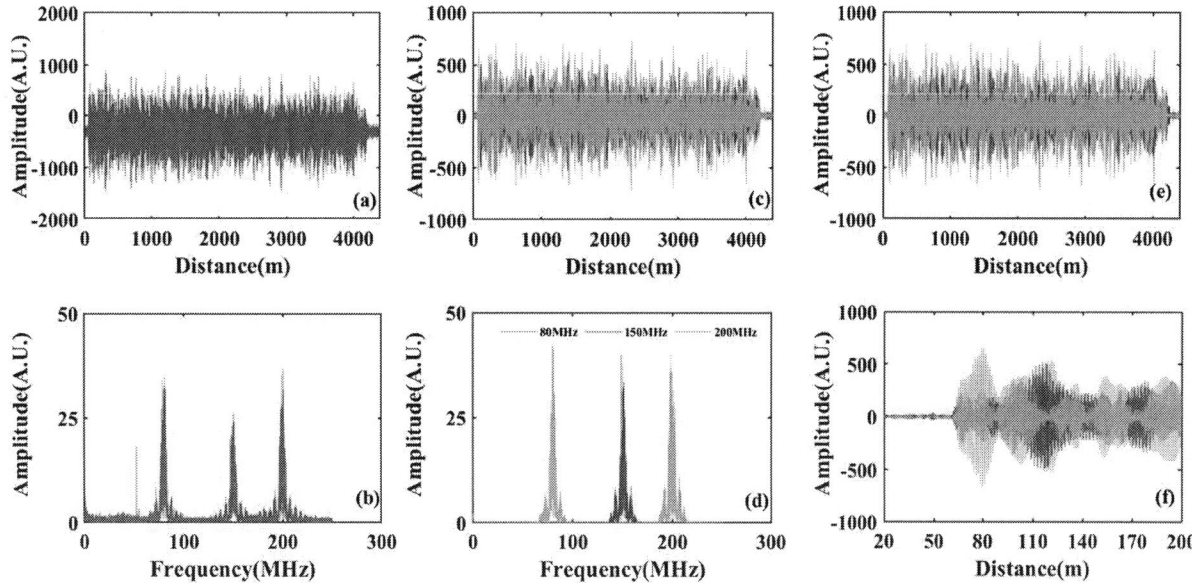

Figure 4. Time domain and frequency domain diagrams of multi-frequency beat signals

B. Amplitude Signals and Demodulated Phase Signals

Amplitude signals can be obtained with the help of in-phase and quadrature (IQ) demodulation processing of beat signals [12]. The superposition of 50 normalized original amplitude signals after aggregation is shown in Fig. 5(a). The distance-time waterfall diagram without data alignment is shown in Fig. 5(b). The region of vibration signals located by the amplitude signal without data alignment is significantly widened between 4020m to 4070m. The corresponding amplitude signals with data alignment are shown in Fig. 5 (c)

978-1-6654-8699-6/22 $31.00 © 2022 IEEE   139

and (d). Similarly, the region of vibration signals with data alignment is between 4015m to 4035m, which is consistent with the pulse width of 200ns. In addition, according to data statistics, the fading probability of misaligned and aligned amplitude signals is 2.08% and 1.90%, respectively. Therefore, if multiple AOMs are used to achieve fading suppression, the collected data need to be aligned.

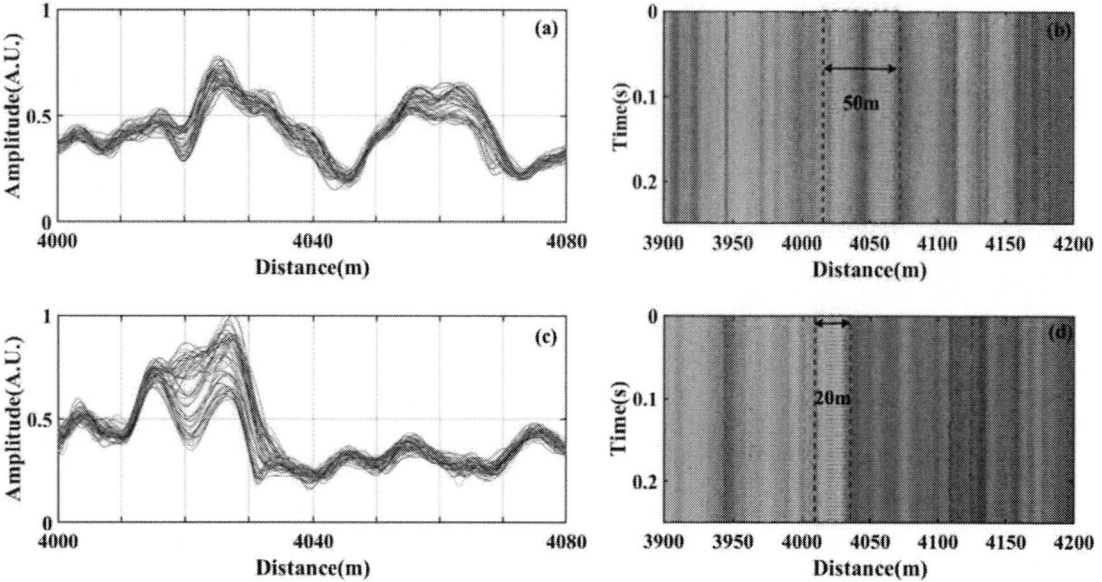

Figure 5. Amplitude signals and distance-time waterfall diagrams without and with data alignment

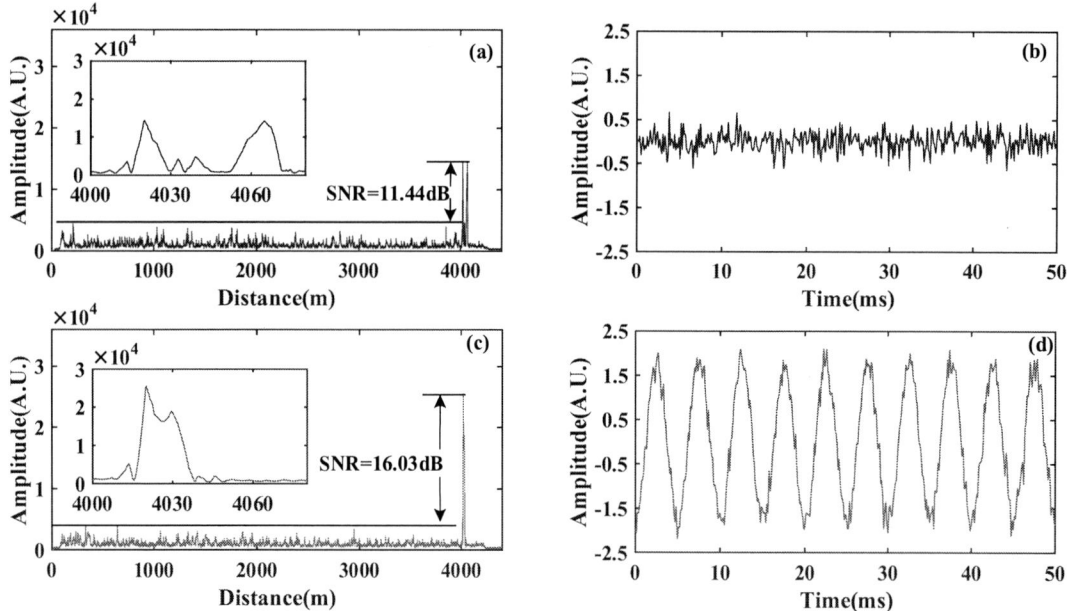

Figure 6. Positioning curves and demodulated phase signals

Positioning curve and demodulated phase signal of vibration signal obtained without data alignment are shown in Fig. 6 (a) and (b). It can be seen that the vibration positioning is affected by the response time of AOM, leading to multiple false peaks in the positioning curve. It may result in positioning error of vibration signal, and the SNR of positioning curve is only 11.44dB. Due to the broadening of vibration region and the existence of positioning error, the whole vibration region may be not covered by differential region of phase demodulation, and thus the vibration signal can not be demodulated, shown as Fig. 6 (b). Positioning curve and demodulated phase signal of vibration signal

978-1-6654-8699-6/22 $31.00 © 2022 IEEE   140

obtained with data alignment processing are shown in Fig. 6 (c) and (d). It can be seen that only one peak value in the positioning curve can be obtained, and the SNR can reach 16.03dB. In addition, the vibration signal can be effectively demodulated.

IV. CONCLUSION

In this paper, an interference fading suppression method based on three AOMs is proposed. By using three AOMs in parallel to modulate the continuous light emitted by laser, multi-frequency detection signals are generated. The beat signals generated in the sensing fiber are aligned, aggregated, and demodulated to obtain the amplitude and phase information of the vibration signal. Experimental results show that the interference fading suppression method based on the parallel AOMs and the data alignment processing can achieve a fading probability of 1.90% and a positioning SNR of 16.03dB.

ACKNOWLEDGMENT

This work was supported in part by the National Natural Science Foundation of China under Grants 62175175, and in part by the Fundamental Research Program of Shanxi Province under Grants 202103021222010.

REFERENCES

[1] M. Zabihi, Y. S. Chen, T. Zhou, J. X. Liu, Y. Y. Shan, Z. Meng, et al., "Continuous Fading Suppression Method for Φ-OTDR Systems Using Optimum Tracking Over Multiple Probe Frequencies," *Journal of Lightwave Technology*, vol. 37, no. 14, pp. 3602-3610, July, 2019.

[2] Z. Y. Zhao, H. Wu, J. H. Hu, K. Zhu, Y. L. Dang, Y. X. Yan, et al., "Interference fading suppression in φ-OTDR using space-division multiplexed probes," *Optics Express*, Vol. 29, Issue 10, pp. 15452-15462, 2021.

[3] K. X. Cui, F. Liu, K. R. Wang, X. J. Liu, J. H. Yuan, B. B. Yan, et al., "Interference-Fading-Suppressed Pulse-Coding Φ-OTDR Using Spectrum Extraction and Rotated-Vector-Sum Method," *IEEE Photonics Journal*, vol. 13, no. 6, pp. 1-6, Dec. 2021.

[4] X. Wang, B. Lu, Z. Y. Wang, H. R. Zheng, J. J. Liang, L. C. Li, et al., "Interference-Fading-Free Φ-OTDR Based on Differential Phase Shift Pulsing Technology," *IEEE Photonics Technology Letters*, vol. 31, no. 1, pp. 39-42, Jan., 2019.

[5] D. Chen, Q. W. Liu, Z. Y. He, "Phase-detection distributed fiber-optic vibration sensor without fading-noise based on time-gated digital OFDR," *Optics Express*, vol. 25, no. 7, pp. 8315-8325, 2017.

[6] D. Chen, Q. W. Liu, Z. Y. He, "High-fidelity distributed fiber-optic acoustic sensor with fading noise suppressed and sub-meter spatial resolution," *Optics Express*, vol. 26, no. 13, pp. 16138-16146, 2018.

[7] A. H. Hartog, L. B. Liokumovich, N. A. Ushakov, O.I. Kotov, T. Dean, T. Cuny, et al., "The use of multi-frequency acquisition to significantly improve the quality of fibre-optic-distributed vibration sensing," *Geophysical Prospecting*, vol. 66, no. Suppl S1, pp. 192-202, 2018.

[8] J. D. Zhang, H. T. Wu, H. Zheng, , J. S. Huang, G. L. Yin, T. Zhu, et al., "80 km Fading Free Phase-Sensitive Reflectometry Based on Multi-Carrier NLFM Pulse Without Distributed Amplification," *Journal of Lightwave Technology*, vol. 37, no. 18, pp. 4748-4754, 2019.

[9] M. Zabihi, Y. S. Chen, T. Zhou, J. X. Liu, Y. Y. Shan, Z. Meng, et al., "Continuous Fading Suppression Method for Φ-OTDR Systems Using Optimum Tracking Over Multiple Probe Frequencies," *Journal of Lightwave Technology*, vol. 37, no. 14, pp. 3602-3610, 2019.

[10] Y. Wu, Z. N. Wang, J. Xiong, J. L. Jiang, S. T. Lin, Y. X. Chen, "Interference Fading Elimination With Single Rectangular Pulse in Phi-OTDR," *Journal of Lightwave Technology*, vol. 37, no. 13, pp. 3381-3387, 2019.

[11] H. Qian, B. Luo, H. J. He, Y. Zhou, C. Hu, X. H. Zou, et al. "Fading-Free Φ-OTDR With Multi-Frequency Decomposition," *IEEE Sensors Journal*, vol. 22, no. 3, pp. 2160-2166, 2022.

[12] Y. Wang, Y. Li, L. Xiao, B. Liang, X. Liu, Q. Bai, B. Q. Jin, "Interference Fading Suppression Using Active Frequency Transformation Method With Auxiliary Interferometer Feedback," *Journal of Lightwave Technology*, vol. 40, no. 3, pp. 872-879, 2022.

Proceedings of the 7th Optoelectronics Global Conference (OGC 2022)

Taper Optical Fiber for Distributed Light-driven Soft Robots

Minghui Niu, Ziyan Zhao, Jiayuan Min, Jie Hu, Huanhuan Liu*, Dan Luo*, Liyang Shao, and Perry Ping Shum*

Department of Electronic and Electrical Engineering,
Southern University of Science and Technology, China
e-mails: shenp@sustech.edu.cn; luod@sustech.edu.cn; liuhh@sustech.edu.cn

Abstract—**We have proposed and demonstrated that the taper fiber-enabled motion of soft robots. A thin film with a high refractive index is encapsulated on the top of the taper fiber as well as soft robot. By controlling power of light along the taper fiber, the soft robot can realize a grasping action. Our results may contribute to the effort of exploring distributed light field to drive soft robots.**

Keywords- Taper fiber; Soft robots; Distributed light field

I. INTRODUCTION

Optical fiber exhibiting merits of wide frequency band, low loss, capability of working in long distance and anti-interference ability, has broad application in the fields of optical communication and optical sensing [1-3]. Recently, exploring optical fiber in robots has attracted much attention because it enables robots with additional functions such as smart sensing or motion control.

In particular, soft robot is of great interest because of natural flexibility and ability to imitate soft tissue, which shows great potential in human-computer interaction, medicine, detection and *etc.* [4]. Currently, one of the most widely used soft robots is photo-controlled soft robot. However, the current photo-controlled braking of soft robot using free-space light source requires bulky components [5], and suffers from the interference of external environment and difficulty to remote control.

As an excellent optical waveguide, optical fiber can accurately transmit the light to the position where the light energy is needed. For example, in 2020, Zmyślony *et al.* grew liquid crystal elastomers on optical fiber optic cross-sections and mimic insect mouthpieces [6]. In 2022, Xiao *et al.* embedded tapered fiber with a tapered waist width of 700 nm into the soft robot. Light from the end facet of tapered fiber is used to drive the motion of soft robot and it achieves the bending of the soft robot over 270 degrees [7]. However, light from the taper fiber end experiences diffraction and is attenuated quickly, which induces a limited working distance [8-11]. Therefore, a distributed light source from tapered fiber to drive soft robots is worth of investigation.

In this work, we propose a method of encapsulated taper fiber and soft robot with high refractive index material, which can drive soft robots to bend by a distributed energy field. By controlling power of light along the taper fiber, the soft robot can realize a grasping action. As a proof of concept, the obtained results may contribute to the effort of exploring optical fiber at the onset of soft robots.

II. DESIGN AND FABRICATION OF TAPER FIBER

Our proposed tapered-fiber scheme for driving soft robots is shown in Fig. 1. The tapered fiber is embedded in the soft robots. A high refractive index packaging material is proposed to extract the light energy bound in the fiber cladding. The mode energy in the cladding can enter the high refractive index material and eventually be absorbed by the soft robot, which is conducive to the bending of the soft robot.

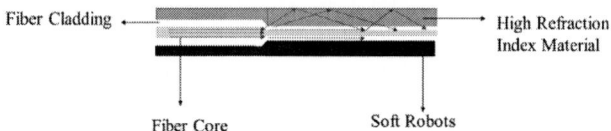

Figure 1. Schematic diagram of extracting light energy from high refractive index materials

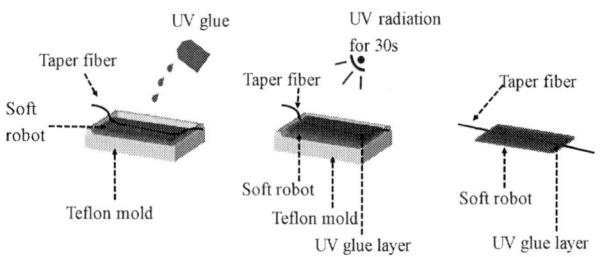

Figure 2. Fabrication process of taper fiber driven soft robot

The fabrication process of taper fiber driven soft robot is shown in Fig. 2. in the experiment, the soft robots used are liquid crystal elastomer doped with a photo-thermal conversion material carbon black which are photo-thermal soft robot [5,12]. After pulling the optical fiber into taper fiber by using hydrogen and oxygen flame machine [13], we package the liquid crystal elastomer and taper fiber by UV glue adhesive film with a thickness of 1 mm and refraction index of 1.8 in a Teflon mold which is well design for making certain thickness UV glue layer. We put the soft robots into the Teflon mold and put a taper fiber with taper waist around 12-μm on the soft robot and drop the UV glue on the soft robots to form a film, and irradiate it with UV for 30s, then we take the soft robot out of mold. The chosen 12-μm taper waist is due to that the taper fiber is not frigile and also has a large energy leakage. The length of taper region is 5 mm, and the embedded position is in the middle position of the soft robot.

978-1-6654-8699-6/22 $31.00 © 2022 IEEE

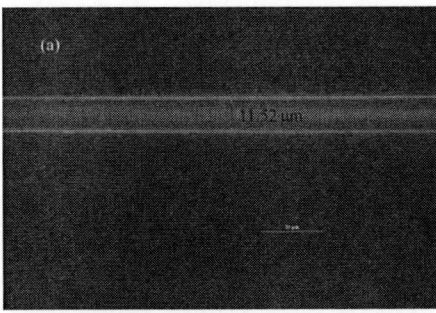

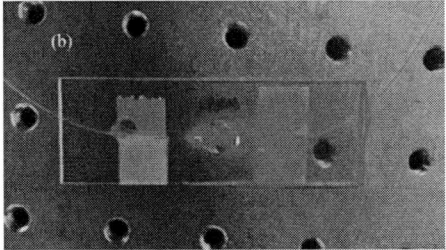

Figure 3. (a) SEM for the taper fiber used in experiment (b) characteristic of taper fiber covered by UV glue

Figure 3(a) shows the microscopic image of taper fiber with diameter around 12 μm. Figure 3(b) shows that light from a 650 nm laser is launched into a taper fiber coated by a drop of UV glue. We can clear see that the light leak from the taper region and form a divergent beam.

III. APPLICATION OF TAPER FIBER FOR DRIVING SOFT ROBOT

Figures 4 and 5 shows our experimental setup for the application of taper fiber to drive soft robots. The laser source is 976 nm, which is connected to the optical fiber of the soft robots through an isolator. The output power of the laser is 350 mW. The soft robot is based on liquid crystal elastomer, with a length of 5 cm, a width of 1 cm and a thickness of 1 mm. As is shown in Fig. 6, the middle part of the soft robot begins to twist, the curved area is the area where the taper fiber is embedded which is proved distributed energy area. And in this case, the soft robot can realize a grasping action. When the laser turns off, the soft robot returns to the origin status.

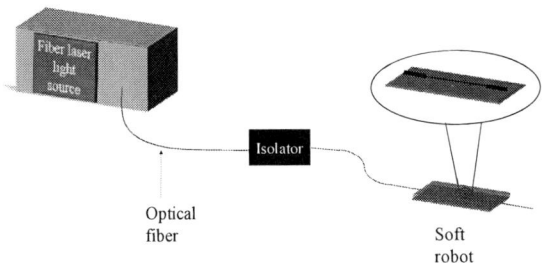

Figure 4. Optical path for driving soft robot

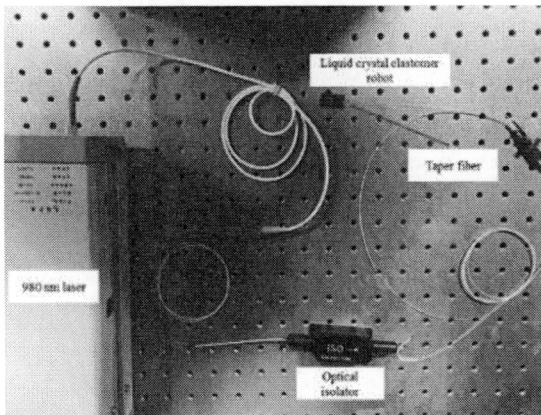

Figure 5. Experimental setup for taper fiber driven soft robot

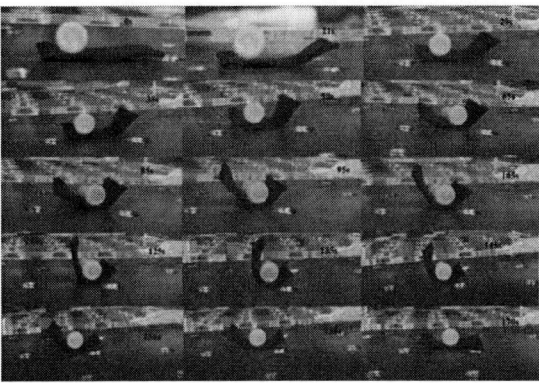

Figure 6. Taper fiber driven soft robot bending diagram

IV. CONCLUSION

In this paper, we have proposed and demonstrated that the taper fiber-enabled motion of soft robots. And we use high refraction index material-UV glue to package the taper fiber and soft robot, which avoid using nanometer level taper fiber instead of taper fiber with about 12 μm taper waist strengthening the stability and reducing the difficult of manufacturing taper fiber. Contrast to the single light spot driven soft robots, the taper fiber with distributed energy to drive soft robots in this paper have the following advantages: (1) the taper fiber driven soft robot can twist from the middle, causing bending at both sides, which enables taper fiber to realize the grasping action; (2) the taper fiber can achieve a relatively long distance for light-matter interaction, which can be applied in large size soft robots. Our results may contribute to the effort of exploring distributed light field to drive soft robots.

ACKNOWLEDGMENT

This project is supported from Stable Support Program for Higher Education Institutions from Shenzhen Science, Technology Innovation Commission (20200925162216001); Special Funds for the Major Fields of Colleges and Universities by the Department of Education of Guangdong Province (2021ZDZX1023); Guangdong Basic and Applied

Basic Research Foundation (2021B1515120013); Open Fund of State Key Laboratory of Information Photonics and Optical Communications (Beijing University of Posts and Telecommunications, No. IPOC2020A002), Natural Science Foundation of Guangdong Province (No. 2022A1515011434). The Open Projects Foundation of State Key Laboratory of Optical Fiber and Cable Manufacture Technology (No. SKLD2105).

REFERENCES

[1] T. O. Charrett, S. W. James, and R. P. Tatam, "Optical Fibre Laser Velocimetry: A Review," Measurement Science and Technology, vol. 23, no. 3, p. 032001, 2012.

[2] I. García, J. Zubia, G. Durana, et al. "Optical fiber sensors for aircraft structural health monitoring," Sensors, vol. 15, no. 7, pp. 15494-15519, 2015.

[3] D. J. Richardson, J. M. Fini, and L. E. Nelson, "Space-division multiplexing in optical fibres," Nature Photonics, vol. 7, no. 5, pp. 354-362, 2013.

[4] M. Pilz da Cunha, M. G. Debije, and A. P. Schenning, "Bioinspired light-driven soft robots based on liquid crystal polymers," Chemical Society Reviews, vol. 49, no. 18, pp. 6568-6578, 2020.

[5] S. Li, H. Bai, Z. Liu, X. Zhang, C. Huang, L. W. Wiesner, M. Silberstein, and R. F. Shepherd, "Digital Light Processing of liquid crystal elastomers for self-sensing artificial muscles," Science Advances, vol. 7, no. 30, 2021.

[6] M. Zmyślony, K. Dradrach, J. Haberko, et al. "Optical pliers: Micrometer-scale, light-driven tools grown on optical fibers," Advanced Materials, vol. 32, no. 33, p. 2002779, 2020.

[7] J. Xiao, T. Zhou, N. Yao, et al. "Optical fibre taper-enabled waveguide photoactuators," Nature Communications, vol. 13, no. 1, 2022.

[8] T. A. Birks, W. J. Wadsworth, and P. S. Russell, "Supercontinuum generation in tapered fibers," Optics Letters, vol. 25, no. 19, p. 1415, 2000.

[9] L. Tong, R. R. Gattass, J. B. Ashcom, et al. "Subwavelength-diameter silica wires for low-loss Optical Wave Guiding," Nature, vol. 426, no. 6968, pp. 816-819, 2003.

[10] J. D. Love, W. M. Henry, W. J. Stewart, et al. "Tapered single-mode fibres and devices. part 1: Adiabaticity criteria," IEE Proceedings J Optoelectronics, vol. 138, no. 5, p. 343, 1991.

[11] F. Orucevic, V. Lefèvre-Seguin, and J. Hare, "Transmittance and near-field characterization of sub-wavelength tapered optical fibers," Optics Express, vol. 15, no. 21, p. 13624, 2007.

[12] Q. He, Z. Wang, Y. Wang, A. Minori, M. T. Tolley, and S. Cai, "Electrically controlled liquid crystal elastomer-based soft tubular actuator with multimodal actuation," Science Advances, vol. 5, no. 10, 2019.

[13] X. Wu and L. Tong, "Optical microfibers and nanofibers," Nanophotonics, vol. 2, no. 5-6, pp. 407-428, 2013.

Proceedings of the 7th Optoelectronics Global Conference (OGC 2022)

Comparison of the Simulation Algorithms for Nonlinear Pulse Propagation in Multimode Fibers

Jiayu Lu, Lili Kong, Xiaosheng Xiao*

State Key Laboratory of Information Photonics and Optical Communications, School of Electronic Engineering

Beijing University of Posts and Telecommunications

Beijing, China

*e-mail: xsxiao@bupt.edu.cn

Abstract—**Nonlinear dynamics of pulse propagation in single-mode fiber have been investigated extensively. Recently, with the wide use of multimode fiber in optical technologies such as fiber lasers, imaging, optical fiber communication systems etc., multimode fiber has gradually received attention. Therefore, the investigation of nonlinear pulse propagation in multimode fibers becomes more and more important. However, the numerical simulation of nonlinear pulse propagation in multimode fibers is a thorny problem. In this Presentation, we compare three simulation algorithms for the nonlinear propagation in multimode fibers. The first algorithm is Massively Parallel Algorithm, based on the Generalized Multimode Nonlinear Schrodinger equation (GMMNLSE); the second is, three-dimensional (3D) algorithm, based on the traditional Generalized Nonlinear Schrodinger Equation (GNLSE), which describes the propagation of whole 3D optical field in multimode fibers; and the third is Radial Coordinate (RC) algorithm, based on the RC GNLSE which is simplified from the GNLSE by the Hankel Transform. These three algorithms are compared from the aspects of operation speed, numerical error and applicable scenarios, and their advantages and disadvantages are analyzed. This investigation will provide some guidelines and suggestions for the numerical investigations of nonlinear pulse propagation in multimode fibers, including the fields of optical fiber communications with multimode fibers, nonlinear fiber optics, spatiotemporal mode-locking in multimode cavities, etc.**

Keywords-Numerical simulation; nonlinear pulse propagation; multimode fiber

I. INTRODUCTION

In the field of optical fiber communication, with the popularity of multimode optical fiber, people use spatial mode multiplexing to greatly improve the communication capacity. Compared with single-mode fiber, multimode fiber has multiple spatial modes, so there are more abundant nonlinear space-time effects. In the field of fiber lasers, traditional mode locking means the phase locking of multiple longitudinal modes of a single transverse mode in single-mode fiber. In 2017, spatiotemporal mode locking, i.e., mode locking multiple transverse and longitudinal modes simultaneously, demonstrated in multimode fiber cavities by Logan et al [1]. In multimode fiber, the evolution of pulsed light is (3+1) dimensional ((3+1) D). When the dispersion

effect is balanced with the nonlinear effect in the multimode fiber cavity, the phase locking of transverse and longitudinal modes can be achieved at the same time. In addition, there are many studies on the nonlinear pulse propagation in passive multimode fibers, such as Kerr beam self-cleaning effect [2], soliton self-frequency shift [3], soliton bound state [4] and so on.

The purpose of this Presentation is to compare three algorithms based on generalized multimode nonlinear Schrodinger equation (GMMNLSE), generalized nonlinear Schrodinger equation (GNLSE) and radial coordinate (RC) GNLSE, respectively, and to provide some suggestions for using these tools. We first briefly introduce the models and basic principles of the three algorithms in Section II. Then we compare these three algorithms by calculating some special scenarios in Section III. In Section IV, we briefly summarize the results.

II. MODELS AND ALGORITHMS

A. GMMNLSE

The GMMNLSE [5] could be simplified as [6]:

$$\frac{\partial A_p}{\partial z} = i\left(\beta_0^{(p)} - \Re\left[\beta_0^{(0)}\right]\right)A_p - \left(\beta_1^{(p)} - \Re\left[\beta_1^{(0)}\right]\right)\frac{\partial A_p}{\partial t} + i\sum_{n\geq 2}\frac{\beta_n^{(p)}}{n!}\left(i\frac{\partial}{\partial t}\right)^n A_p \quad (1)$$

$$+i\frac{n_2\omega_0}{c}\left(1+\frac{i}{\omega_0}\frac{\partial}{\partial t}\right)\sum_{l,m,n}\left\{(1-f_R)S_{plmn}^K A_l A_m A_n^* + f_R S_{plmn}^R A_l\left[h*\left(A_m A_n^*\right)\right]\right\}$$

where A_p is the electric field envelope of the *p-th* spatial mode, $\beta_n^{(p)}$ is the *n-th* order dispersion coefficient of the *p-th* mode, n_2 is the nonlinear coefficient (we set $n_2 = 2.3\times10^{-20}\, m^2/w$ in our simulations), ω_0 is the central angular frequency, f_R is the contribution of Raman effect to nonlinearity (we set $f_R = 0$ in our simulations), S_{plmn}^K and S_{plmn}^R are the modal overlap factors of Kerr term and Raman phase respectively (*p,l,m,n* refers to the different spatial modes in the multimode fiber), and h is the Raman response function, z is the distance along the multimode fiber.

In principle, the GMMNLSE is based on the mode-decomposition of the 3D light in multimode fibers. To solve the set of equations (1) quickly, Massively Parallel

978-1-6654-8699-6/22 $31.00 © 2022 IEEE   145

Algorithm (MPA) was proposed to calculate it [6,7].

In this way, one can decompose the space into discrete steps along the z direction, and calculate the integrand function at each z_i point in parallel, thus the operation speed can be greatly improved. Two step sizes $\triangle_z$ and L are used to calculate the integrand. The relationship between them is described by $L = m\triangle_z$ and m refers to the parallelization extent, i.e., the integrand will be calculated at m points ($z_1, z_2 \ldots z_m$) simultaneously.

B. 3D GNLSE

Instead of decomposing the light into spatial modes, the nonlinear propagation of 3D pulse in waveguide can be described by the traditional GNLSE [8,9]:

$$\frac{\partial A(x,y,z,\omega)}{\partial z} = \frac{i}{2\beta_{eff}(\omega)}\nabla_T^2 A + \hat{D}_\omega(\omega)A + i\frac{\beta_{eff}(\omega)}{2}\left(\left(\frac{n(x,y,\omega)}{n_{eff}(\omega)}\right)^2 - 1\right)A$$
$$+ \mathcal{F}_t\left\{i\frac{n_2\omega_0}{c}\left(1 + \frac{i}{\omega_0}\frac{\partial}{\partial t}\right)\left\{(1-f_R)|A|^2 A + f_R A\left[h_R*|A|^2\right]\right\}\right\}$$
$$(2)$$

where $\beta_{eff}(\omega)$ is the effective propagation constant, $\nabla_T^2 = \partial_x^2 + \partial_y^2$, $\hat{D}_\omega = \sum_{m=2}^{3} i^{m+1}\frac{\beta_m}{m!}\partial_t^m$ is the dispersion operator using Taylor approximation, $n(x,y,\omega)$ is the spatial refractive index, $n_{eff}(\omega)$ is the effective refractive index, $\mathcal{F}_t$ refers to the Fourier transform in time domain, h_R is the Raman response of the multimode fiber. One can use a spectral split-step algorithm to calculate the GNLSE [9].

C. RC GNLSE

The RC GNLSE is based on an assumption that the 3D field is radial symmetry:

$$\frac{\partial A(r,z,\omega)}{\partial z} = \frac{i}{2\beta_{eff}(\omega)}\nabla_r^2 A + \hat{D}_\omega(\omega)A + i\frac{\beta_{eff}(\omega)}{2}\left(\left(\frac{n(r,\omega)}{n_{eff}(\omega)}\right)^2 - 1\right)A$$
$$+ F_t\left\{i\frac{n_2\omega_0}{c}\left(1 + \frac{i}{\omega_0}\frac{\partial}{\partial t}\right)\left\{(1-f_R)|A|^2 A + f_R A\left[h*|A|^2\right]\right\}\right\}$$
$$(3)$$

The RC algorithm is similar to the 3D algorithm except that the Hankel transform is applied to convert $A(f_x, f_y, \omega)$ into $A(\rho, \omega)$ [10]:

$$\rho = \sqrt{f_x^2 + f_y^2} \quad f_x = \rho cos\phi \quad f_y = \rho cos\phi \quad \phi = arctan\frac{f_y}{f_x}$$
$$(4)$$

$$A(\rho) = 2\pi\int_0^\infty A(r)J_0(2\pi\rho r)rdr$$
$$(5)$$

where J_0 represents the 0-order Bessel function.

The rest of the RC algorithm is the same as the 3D algorithm (using the split-step reckoning method).

III. COMPARISONS OF THE ALGORITHMS

The computer we use for simulation is a laptop with i5-

12500H CPU, RTX3050 GPU and 16G RAM. Though better computer may improve the performances of these three algorithms, the conclusion of the comparison among them is still valid. In all the considered examples, we do not consider the loss in the graded-index multimode fiber (radius of 25μm), and uniformly set the length of the fiber to 0.5 m. Numerically solving found that about ~100 (scalar) spatial modes can be supported by this multimode fiber.

A. Calculation Speed

For all these three algorithms, we use a Gaussian pulse with an initial energy of 6nJ as the input, and evenly distribute the energy to the first 6 modes. For 3D (RC) GNLSE, these 6 modes are combined into a 3D (2D) light field, serving as the input field. First of all, we compare the calculation speed of these three algorithms. Here, we uniformly set the linear step size of the three algorithms to 20μm.

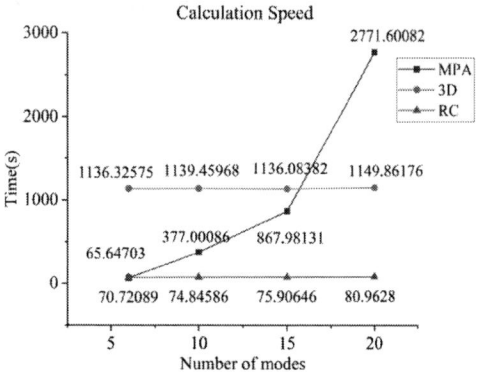

Figure 1. Comparison of the calculation speed of the three algorithms, with the increase of the number of modes considered in MPA.

Fig. 1 shows that the RC algorithm has the superiority in the operation speed since the light considered by this algorithm is (2+1) D. However, we have to remember that the RC algorithm is based on the assumption of the field in radial symmetry which means only LP_{0n} modes can be considered in the algorithm. As a result, this algorithm is limited to simulate some special optical systems. The MPA algorithm is very fast for a relatively small number of modes. While as the number of modes increases, the algorithm gets much slower for that the mode-resolved method becomes inefficient. The 3D algorithm may be the slowest (for the small number of modes) while it will not be affected by the number of modes and is applicable to more general situations.

B. Numerical Error

Next, we compare the numerical errors of the algorithms. Considering the particularity of applicable scenarios of RC, we firstly compare the 3D and MPA algorithms. The input is the same as Part A.

Taking the step length as a variable, we get the total energy loss (due to the numerical error) under different linear step sizes. In MPA algorithm, we calculated the total energy of the six modes considered (due to the mode-decomposition characteristics of GMMNLSE, the energy will not get into

978-1-6654-8699-6/22 $31.00 © 2022 IEEE

other modes), while in 3D algorithm we calculated the total energy of the 3D light field in Fig. 2. We can see from Fig. 2 that the loss of 3D algorithm is the smallest when the step size is 20μm. This is contrary to the intuition that the smaller the step size leads to the higher accuracy. It may result from the truncation error of the numerical simulations, or from the large discretization separation of the 3D field due to the limited computer memory. We therefore suggest that the linear step size should be set to 20μm in the simulation, and the nonlinear step size is an integral multiple of the linear step size, but usually no more than 100μm. Benefited by the mode-decomposition method, the energy of MPA is automatically conserved [9]. As a result, there is almost no energy loss in the simulation of MPA.

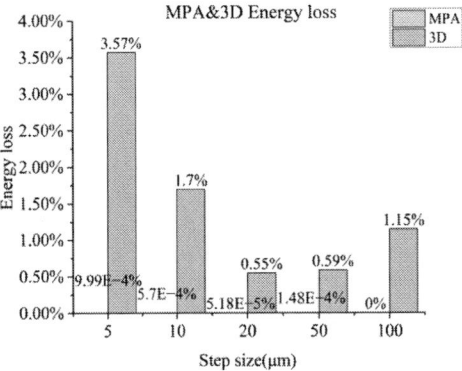

Figure 2. The numerical errors of MPA and 3D algorithms. Note that the error of MPA algorithm is very small and the values are labelled in the figure.

The output beam profiles of MPA and 3D algorithms are compared in Fig. 3. As shown in Fig. 3, there is a slight difference between the beam profiles calculated by MPA and 3D algorithms. The difference results from two aspects, one is the numerical errors of the algorithms, and the other will be discussed in the following.

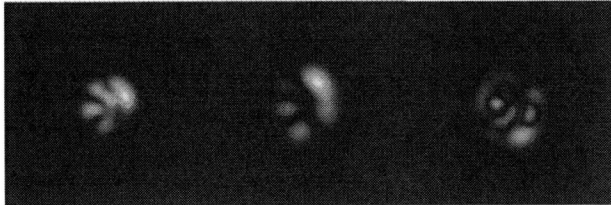

Figure 3. Difference (right) between the beam profiles calculated by MPA (left) and 3D (middle) algorithms.

Then we decomposed the 3D light field into the first 6 modes and calculated the total energy of these 6 modes. We still present the results in the form of energy loss. Fig. 4 shows that the total energy of the first 6 modes is less than that of the 3D light field.

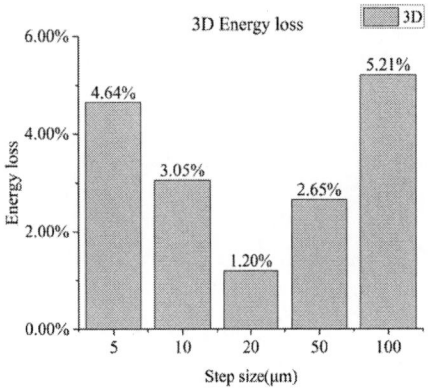

Figure 4. Difference between energy of the first 6 modes and total energy in 3D algorithm

In order to show that the energy may transfer from the input modes to higher-order spatial modes, i.e., higher spatial mode could be generated from the input modes by intermodal nonlinearity (e.g., intermodal four-wave mixing), we consider more modes when decomposing the output 3D field and calculate the loss for the 3D algorithm. Fig. 5 indicates that with more modes considered, the loss is further reduced and tends to be stable. This could be explained by that 3D algorithm can include high-order modes though the input field is the summation of the first 6 modes. These higher-order modes also get some energy due to nonlinearity even if we did not allocate energy to them at first. This means 3D algorithm is more general than MPA algorithm. We believe that taking more high-order modes into consideration can decrease more energy loss for the 3D algorithm.

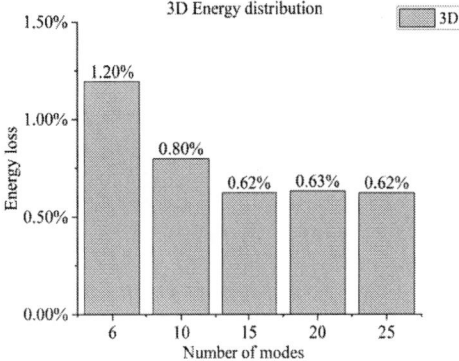

Figure 5. The total energy of 3D algorithm, versus the number of the modes considered in the decomposing process

In order to compare the 3D and RC algorithms, here we choose $LP_{01}, LP_{02}, LP_{03}$ as the loaded modes. Fig. 4 shows that the loss increases with the step size in RC algorithm which is in accordance with our common sense. The reason may be that the end correction of the Hankel transformation is taken in the RC algorithm [11] which eliminates the truncation error. Same as 3D algorithm, RC algorithm includes higher-order modes, so there is also loss of the total energy of the three modes. Compared with 3D algorithm, a smaller step size can achieve higher accuracy in RC algorithm, so we suggest step size should be set as small as possible (though small step size will increase the calculation time).

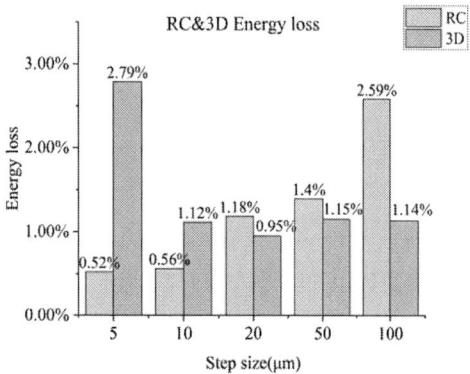

Figure 6. Total energy of the first 3 modes in RC&3D algorithm, versus the step size

IV. CONCLUSION

In summary, the advantages of MPA algorithm lie in its computational speed and high applicability, while the disadvantage is that when considering a large number of modes, the speed will decrease significantly or even slower than 3D algorithm. We suggest that the mode number used in MPA algorithm should not exceed 20. In addition, MPA algorithm makes the assumption that the energy will not transfer to the unconsidered mode. If the energy transfers to the unconsidered mode by nonlinear optical effects (e.g., intermodal four-wave mixing or Raman effect among the spatial modes), the result will be inaccurate. The advantages of 3D algorithm are its wide applicability and good scalability. Its running speed will not be affected by the number of modes, and the results are more general. Its disadvantages are also obvious. It requires a relatively large amount of calculation. Because the 3D light needs to be sampled in x, y and t, the actual operation is likely to exceed the running memory of the current computer, resulting in the inability to improve the resolution in these three dimensions and finally the numerical error may occur. MPA and RC algorithms only sample in two dimensions, so it is easy to increase the number of sampling points in any dimension without exceeding the memory of computer. RC algorithm is the most special of the three algorithms. Limited

by Hankel transform, it can only be used in systems where radial symmetric modes are dominant. However, it has the computational speed of MPA algorithm and the good scalability of 3D algorithm. In the future, more investigation of these algorithms will be conducted.

ACKNOWLEDGMENT

We would like to thank Dr. Logan G. Wright for the help of the simulation code. This work is supported by the Fundamental Research Funds for the Central Universities (BUPT 2021RC08), and the State Key Laboratory of IPOC (BUPT), P. R. China (No. IPOC2020ZT02).

REFERENCES

[1] Wright, L. G., Christodoulides, D. N., & Wise, F. W. (2017). Spatiotemporal mode-locking in multimode fiber lasers. *Science*, *358*(6359), 94-97.

[2] Krupa, K., Tonello, A., Shalaby, B. M., Fabert, M., Barthélémy, A., Millot, G., ... & Couderc, V. (2017). Spatial beam self-cleaning in multimode fibres. *Nature Photonics*, *11*(4), 237-241.

[3] Antikainen, A., Rishøj, L., Tai, B., Ramachandran, S., & Agrawal, G. P. (2019). Fate of a soliton in a high order spatial mode of a multimode fiber. *Physical review letters*, *122*(2), 023901.

[4] Ding, Y., Xiao, X., Wang, P., & Yang, C. (2019). Multiple-soliton in spatiotemporal mode-locked multimode fiber lasers. *Optics express*, *27*(8), 11435-11446.

[5] Poletti, F., & Horak, P. (2008). Description of ultrashort pulse propagation in multimode optical fibers. *JOSA B*, *25*(10), 1645-1654.

[6] Wright, L. G., Ziegler, Z. M., Lushnikov, P. M., Zhu, Z., Eftekhar, M. A., Christodoulides, D. N., & Wise, F. W. (2017). Multimode nonlinear fiber optics: massively parallel numerical solver, tutorial, and outlook. *IEEE Journal of Selected Topics in Quantum Electronics*, *24*(3), 1-16.

[7] Korotkevich, A. O., & Lushnikov, P. M. (2011). Proof-of-concept implementation of the massively parallel algorithm for simulation of dispersion-managed WDM optical fiber systems. *Optics letters*, *36*(10), 1851-1853.

[8] Wright, L. G., Renninger, W. H., Christodoulides, D. N., & Wise, F. W. (2015). Spatiotemporal dynamics of multimode optical solitons. *Optics express*, *23*(3), 3492-3506.

[9] Ziegler, Z. *Numerical tools for optical pulse propagation in multimode fiber* (Thesis (Cornell University 2017)).

[10] Siegman, A. E. (1977). Quasi fast Hankel transform. *Optics letters*, *1*(1), 13-15.

[11] Agrawal, G. P., & Lax, M. (1981). End correction in the quasi-fast Hankel transform for optical-propagation problems. *Optics letters*, *6*(4), 171-173.

Proceedings of the 7th Optoelectronics Global Conference (OGC 2022)

Hollow Core Bragg Fiber-based Gas Pressure Sensor Using Parallel Fabry-Perot Interferometers

Zongru Yang, Weihao Yuan, Changyuan Yu*

Photonics Research Centre, Department of Electronic and Information Engineering,
The Hong Kong Polytechnic University
Hong Kong, China
*changyuan.yu@polyu.edu.hk

Abstract—An ultra-high sensitivity parallel-connected Fabry-Perot interferometers (FPIs) pressure sensor based on hollow core Bragg fiber (HCBF) and harmonic Vernier effect is proposed and demonstrated. One FPI (FPI-1) acts as the sensing unit while the other FPI (FPI-2) is used as the reference unit to generate the Vernier effect. The FPI-1 was prepared by fusion splicing a section of HCBF between a single-mode fiber (SMF) and a hollow silica tube (HST), and the FPI-2 was fabricated by sandwiching a piece of HCBF between two SMFs. Two FPIs with very different free spectral ranges (FSR) in the fringe pattern were connected to the 3-dB coupler parallelly, which realizes the harmonic Vernier effect and ensures the stability of the interference fringe. Both measurements of the air pressure in the range of 0-0.24 MPa and the temperature in the range of 25-55 °C were conducted using the dual FPIs sensor. Experimental results exhibited that high sensitivity of 124.35 nm/MPa with excellent linearity of 0.9967 was achieved by the sensing probe. Moreover, the calculated temperature crosstalk was as low as ~0.072 kPa/°C. The proposed sensor can be a promising candidate for real-time and high-precision gas pressure monitoring.

Keywords-harmonic Vernier effect; hollow core Bragg fiber; pressure measurement; Fabry-Perot interferometer sensor

I. INTRODUCTION

Optical fiber gas pressure sensors have been widely used in various sensing areas. Several kinds of structures for gas pressure measurement have been developed based on long-period fiber gratings (LPFGs) [1], fiber Bragg gratings (FBGs) [2], anti-resonant reflecting optical waveguides (ARROWs) [3], Mach-Zehnder interferometers (MZIs) [4], and Fabry-Perot interferometers (FPIs) [5]. Among them, considering robustness and convenient manufacturing, opened FPIs are extensively employed recently [6]. Opened FPI sensors estimate the gas pressure via observing the change of the refractive index (RI) or the cavity length of the FPI.

Opened FPIs take advantage of the gas circulating between the intrinsic FP cavity and the external environment, which indicates their durability and fast response. The RI of gas in the air cavity will equally change with outside, therefore the linearity and wide sensing range can be guaranteed. There are primary two methods to develop opened FPIs. One

approach is based on the femtosecond laser drilling technique to create a micro-channel in the side of the fiber [7]. By using this method, this type of sensor possesses high endurance and well fringe visibility. Nonetheless, the fabrication devices are generally pricey and complex, and the debris produced from laser drilling usually disturbs the quality of the interference pattern. Another scheme can be realized by cascading a hollow silica tube (HST) [8] to the FPI. The air hole of HST could serve as a gas channel while the facet of HST functions as a reflector of FPI. Due to the easy operation, flexible combination, and low cost, this structure is certainly competitive in gas pressure measurement. However, a theoretical analysis proves that the sensitivity of opened FPI gas sensors is limited below 5 nm/MPa [9], which restricts their practical application expecting high sensitivity. Accordingly, maintaining a large measuring range while ensuring relatively high sensitivity is the major motivation for optical fiber gas pressure sensing.

To improve the sensitivity of opened structure gas pressure sensors, FPI sensors based on the Vernier effect have been extensively studied [10–12]. Similar to the Vernier caliper, the optical Vernier effect utilizes a reference interferometer as a fixed scale and a sensing interferometer as a sliding part. The overlap between two interference signals with a slight free spectral range (FSR) difference would generate a superimposed spectrum, which contains a large envelope that provides a wavelength shift magnification. By detecting the response of the extracted spectrum envelope, the sensing capabilities can be amplified by an order of magnitude compared to that of the individual sensing interferometer, leading to a new generation of high-resolution fiber optic devices.

In this paper, we introduced harmonics to the optical Vernier effect and proposed an ultra-high sensitivity fiber gas pressure sensor with two separated FPIs. The harmonic Vernier effect is an extended concept of the fundamental Vernier effect [13]. Theoretically, the fundamental case occurs when the difference between the two optical path lengths (OPLs) is tiny, while the harmonic Vernier effect is that the OPL of the reference unit pluses multiple integers ($i, i \geq 0$, i being the harmonic order) of the OPL of the sensing unit. Compared to the fundamental Vernier effect, the

978-1-6654-8699-6/22 $31.00 © 2022 IEEE

harmonic case has further promoted the sensing ability, where the magnification factor is in proportion to the harmonic order. For the proposed sensor, these FPIs are based on the SMF-HCBF-HST (opened FPI) structure, and the SMF-HCBF-SMF (closed FPI) structure, respectively. As an air cavity of FPI, the HCBF has the advantage of low transmission loss. An inner diameter of the HST much smaller than that of the HCBF provides the mirror reflection and the gas channel. The opened FPI as the sensing element and the closed FPI as the reference part are connected by a 3-dB coupler in the parallel configuration. Two FPIs with very different OPLs form the first-order optical harmonic Vernier effect. In the experiment, the proposed sensor demonstrated a tremendous gas pressure sensitivity of 124.35 nm/MPa ranging from 0 to 0.24 MPa with high linearity of 0.9967, which is 30.5 times larger than the individual sensing interferometer. Moreover, a fairly low-temperature crosstalk of 0.072 kPa/°C was realized benefiting from the isolation of two interferometers.

II. FABRICATION AND PRINCIPLE

A. Sensor Fabrication

Fig. 1 (a, b) illustrates the schematic diagram of the proposed gas pressure sensor, which is constructed by two FPIs in a parallel configuration. The commercial fusion splicer (FURUKAWA, FITEL S178A) was used for splicing. To ensure the ideal mechanical strength of the splicing joint and non-deformation of the air core, the arc power, arc duration, and repetition time were optimized. The sensing unit denoted as FPI-1 was fabricated by fusion splicing a section of the homemade HCBF, with a length of about 0.96 mm and air core diameter of 32 μm, between the SMF (Corning, SMF28) and the HST (Polymicro Technologies, TSP005150), as shown in Fig. 1 (a). The gas diffuses from the external environment to the air cavity of HCBF through the hollow core of HST, in the meanwhile, the head face of HST serves as a reflector. There is an offset between the size of the gas inlet channel and the reflection ratio, hence an HST inner diameter of 5 μm was chosen to ensure sufficient mirror reflection and fast response. It is noted that the last end face of HST must be polished to avoid redundant reflection. As depicted in Fig. 1 (b), the reference unit named FPI-2 was formed by sandwiching a piece of HCBF with a length of around 1.86 mm between two SMFs. The gas tightness of the reference unit guarantees that its interference pattern can avert being disturbed by the outside environment.

The sketch map of the cross-section of homemade HCBF is presented in Fig. 1 (c). The HCBF is formed by an air core surrounded by a four-bilayers annular rings structure with commutative high-low RIs distribution. Unlike the hollow core capillary (HCC), which is usually used as the air cavity of the FPI, the homemade HCBF based on the ARROW mechanism has a comparatively low transmission loss that can offer a well-defined interference as a result. The detailed parameter and fabrication process can be referred to [14].

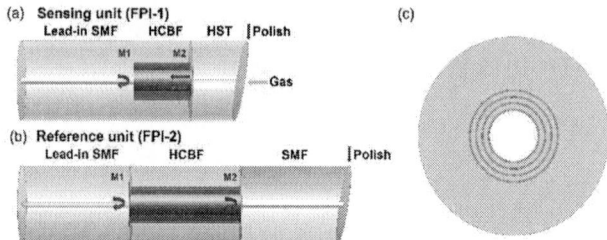

Figure 1. Schematic diagram of the proposed sensor: (a) sensing unit; and (b) reference unit; and (c) cross-section of the HCBF.

B. Harmonic Vernier Effect

The essential condition of the fundamental Vernier effect is that OPLs of both FPIs should be similar but not equal, nevertheless, the application of the harmonic Vernier effect provides the possibility to fabricate two interferometers with great distinction in the OPL. Harmonics of the Vernier effect relies on the OPL of reference FPI being increased by i ($i \geq 0$, i being an integer) multiples of the OPL of the sensing FPI ($OPL_2 = n_2 L_2 + i n_1 L_1$), where i is the harmonic order. The FSR of the reference FPI becomes [13]:

$$FSR_2^i = \left| \frac{\lambda^2}{2(n_2 L_2 + i n_1 L_1)} \right|. \tag{1}$$

It is worth noting that the case $i = 0$ signifies the fundamental Vernier effect. And the harmonics introduced re-forms the envelope intensity with the same frequency and FSR, which can be verified by the FSR formula of harmonic Vernier spectrum as follow:

$$FSR_e^i = \left| \frac{FSR_1 \cdot FSR_2^i}{FSR_1 - (i+1)FSR_2^i} \right|. \tag{2}$$

The magnification (M) factor depends on the concept of the internal envelope [13], which is related to the order of harmonics. In consequence, the M factor as a function of i is defined as:

$$M^i = \left| \frac{(i+1)FSR_2^i}{FSR_1 - (i+1)FSR_2^i} \right| = (i+1)M, \tag{3}$$

where M is the magnification factor of the fundamental Vernier effect [12]. From Eq. (2) and (3), it is apparent that the sensitivity can be increased by $(i+1)$ times with the same FSR of envelope compared to that of the fundamental case [12]. But there is a trade-off between the sensitivity and the visibility since the increasing harmonic order will result in the evanescence of envelope peaks [13].

Practically, the length of in-line FPIs is commonly at a sub-millimeter scale to acquire relatively large sensitivity.

However, the guarantee of accuracy is challenging using the current slicing process. Thus, the harmonic Vernier effect can be implemented for a higher fabrication tolerance. For the case of the proposed sensor, the introduction of the first order harmonics is capable of enhancing the sensitivity over the fundamental Vernier effect by double.

C. Sensing Principle

For gas pressure sensing, the crucial parameter of the proposed structure is the OPL of the cavity, that is, changes in both RI (n) and cavity length (L) could induce wavelength shift. However, the change of cavity length of the in-line FPI directedly induced by pressure is unfeasible, the sensitivity is primarily contributed by variation of the RI as a result. According to the Edlen equation [7], n is the function of the temperature and the pressure:

$$n_{air} = 1 + \frac{2.8793 \times 10^{-9}}{1 + 0.00367 \times T} \cdot P, \quad (4)$$

where T stands for temperature in degrees Celsius (°C), and P denotes the gas pressure in Pascal (Pa). Therefore, the pressure sensitivity of the sensing FPI can be simply derived from $\lambda_m = 4nL/(2m + 1)$ and Eq. (4) as:

$$
\begin{aligned}
S_P &= \frac{\partial \lambda_m}{\partial P} \\
&= \frac{4L}{2m+1} \cdot \frac{\partial n_{air}}{\partial P} \\
&= \lambda_m \frac{\partial n_{air}}{\partial P} \cdot \frac{1}{n_{air}} \\
&= \lambda_m \frac{2.8793 \times 10^{-9}}{1 + 0.00367 \times T} \cdot \frac{1}{n_{air}},
\end{aligned} \quad (5)
$$

where λ_m is the center wavelength of the m order (m being an integer beginning from 1) interference dip for the sensing unit. Typically, the sensitivity of the sensing FPI at a temperature of 25 °C and wavelength of 1550 nm is 4.09 nm/MPa, which indicates that the pressure sensitivity of the in-line FPI is nearly constant and is irrelevant to the cavity length. Since the variation of temperature has a tiny influence on the RI of air, the sensitivity of temperature is mainly from the thermal expansion effect induced cavity length change

For the proposed sensor in parallel configuration based on the first harmonic Vernier effect, the pressure sensitivity can be obtained by multiple Eq. (3) and Eq. (5):

$$S_p^1 = M^1 S_P, \quad (6)$$

where M^1 is the magnification factor for the first harmonics of the Vernier effect.

III. EXPERIMENTAL RESULT AND DISCUSSION

A. Gas Pressure Experiment

The schematic diagram of the experimental setup for gas pressure measurement using parallel structured FPIs based on the harmonic Vernier effect and the HCBF is demonstrated in Fig. 2. Both the sensing unit and reference unit were parallelly connected to the output arms of the 3-dB coupler. The incident light launching from the amplified spontaneous emission (ASE; MC Fiber Optics) with a spectral range of 1528–1603 nm was guided into the coupler and was divided equally into two parts. One part propagated to the sensing unit while another one was transmitted to the reference unit. Then the reflection beams overlapped at the coupler and were detected by the optical spectrum analyzer (OSA; YOKOGAWA, AQ6374) with a resolution of 0.05 nm. The sensing unit was sealed into the chamber of the pressure test bench (MY Instrument Technology Co., LTD., YFY-XK600) with ultraviolet (UV) glue. A high-accuracy pressure gauge (MY Instrument Technology Co., LTD., GJ100-M05) was installed on the bench to measure the pressure with a precision of 0.0001 MPa.

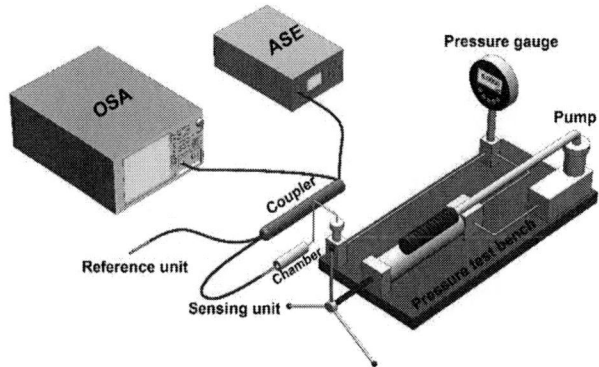

Figure 2. Schematic diagram of the experimental setup.

Gas pressure experiments were repeated with pressure increasing and decreasing from 0 to 0.24 MPa with a step of 0.02 MPa at room temperature (~ 23.5 °C). Reflection spectra with pressure increasing and decreasing are shown in Fig. 3 (a) and (b), respectively. Four envelope peaks in the range from 1528 to 1603 nm can be detected. The peak pointed by a red arrow shown in Fig. 3 (a) at the wavelength of about 1542 nm was arranged to be the traced peak, beginning from the gas pressure of 0 MPa. As pressure rises, the envelope peak reveals a steady and remarkable redshift. As shown in Fig. 4, the pressure sensitivity in the process of pressure increasing is measured to be 121.87 nm/MPa, with the R square value of 0.9959. Inversely, the blueshift of the Vernier envelope occurs when the pressure reduces, and the calculated slope and linearity are 124.35 nm/MPa and 0.9967, respectively. The M factor is calculated to be 30.5, which is in line with the theoretical result $M^1 = 30$ according to Eq. (3). The pressure sensitivity of the proposed sensor is 1.56 times higher than [15] and 2.71 times higher than [16]. Both pressures increasing and decreasing exhibits ultra-high sensitivity and great linearity, and the similarity of results from the two groups signifies the high repeatability.

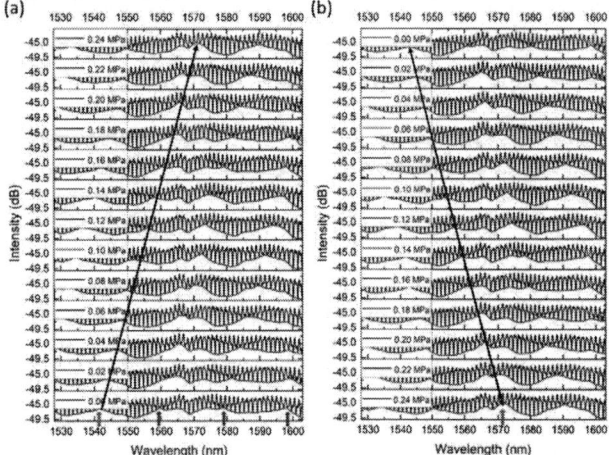

Figure 3. Wavelength shift of the envelope: (a) pressure increasing; and (b) temperature decreasing.

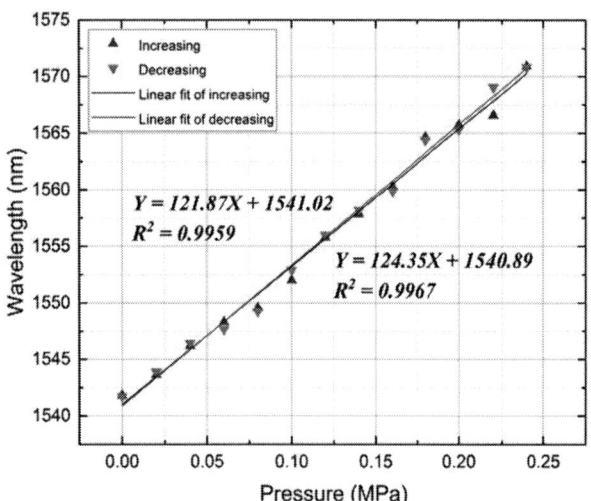

Figure 4. Linear fit of the pressure sensing.

B. Temperature Experiment

For temperature sensing, the HCBF-based harmonic Vernier sensor was located in a temperature chamber, and the temperature was set varying from 25 to 55 °C with a step of 10 °C. Fig. 5 (a) shows the reflection spectra of a single envelope, which specify the blue-shift trend of the envelope peak with temperature increasing. The linear response of peak wavelengths with the change of temperature was also collected. Fig. 5 (b) reveals that there has been an insignificant decrease in peak wavelength when the temperature rises. The fitting slope is calculated to be -8.8 pm/°C with a linearity of 0.9954. Compared to the wavelength shift of the envelope peak caused by pressure change, that induced by temperature is negligible. The temperature crosstalk is calculated to be ~0.072 kPa/°C, which means that the proposed HCBF-based sensor can realize temperature-insensitive pressure sensing.

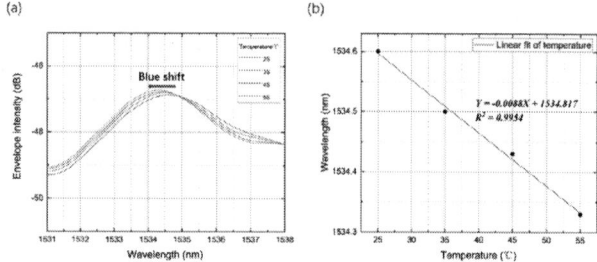

Figure 5. (a) Response of the proposed sensor to temperature. (b) Linear fit of the temperature sensing.

IV. CONCLUSION

In conclusion, a harmonic Vernier effect-based parallel-connected FPI sensor using HCBF. is proposed for pressure measurement. According to the envelope peak shift, the measured pressure sensitivity of the proposed sensor is 124.35 nm/MPa in the range of 0-0.24 MPa, with a linearity of 0.9967, and the sensitivity of temperature varying from 25 to 55 °C is -8.8 pm/°C, with linearity of 0.9954. And the magnification factor reaches as high as 30.5. The ultra-high sensitivity, low temperature crosstalk, compact size, cost-effective production, and simplicity will make the proposed sensor continuously progress in the sensing area.

ACKNOWLEDGMENT

The project is funded by Shenzhen-HK-Macao Science and Technology Plan C SGDX2020110309520303.

REFERENCES

[1] X. Zhong, Y. Wang, C. Liao, S. Liu, J. Tang, and Q. Wang, "Temperature-insensitivity gas pressure sensor based on inflated long period fiber grating inscribed in photonic crystal fiber," *Optics letters*, vol. 40, no. 8, pp. 1791–1794, 2015, doi: 10.1364/OL.40.001791.

[2] Q. Zhang, N. Liu, T. Fink, H. Li, W. Peng, and M. Han, "Fiber-Optic Pressure Sensor Based on π-Phase-Shifted Fiber Bragg Grating on Side-Hole Fiber," *IEEE Photon. Technol. Lett.*, vol. 24, no. 17, pp. 1519–1522, 2012, doi: 10.1109/LPT.2012.2207715.

[3] M. Hou *et al.*, "Antiresonant reflecting guidance mechanism in hollow-core fiber for gas pressure sensing," *Optics express*, vol. 24, no. 24, pp. 27890–27898, 2016, doi: 10.1364/OE.24.027890.

[4] H. Lin, F. Liu, H. Guo, A. Zhou, and Y. Dai, "Ultra-highly sensitive gas pressure sensor based on dual side-hole fiber interferometers with Vernier effect," *Optics express*, vol. 26, no. 22, pp. 28763–28772, 2018, doi: 10.1364/OE.26.028763.

[5] M. Quan, J. Tian, and Y. Yao, "Ultra-high sensitivity Fabry-Perot interferometer gas refractive index fiber sensor based on photonic crystal fiber and Vernier effect," *Optics letters*, vol. 40, no. 21, pp. 4891–4894, 2015, doi: 10.1364/OL.40.004891.

[6] Z. Zhang, Y. Wang, M. Zhou, J. He, C. Liao, and Y. Wang, "Recent advance in hollow-core fiber high-temperature and high-pressure sensing technology [Invited]," *Chin. Opt. Lett.*, vol. 19, no. 7, p. 70601, 2021, doi: 10.3788/COL202119.070601.

[7] Z. Li *et al.*, "High-Sensitivity Gas Pressure Fabry–Perot Fiber Probe With Micro-Channel Based on Vernier Effect," *J. Lightwave Technol.*, vol. 37, no. 14, pp. 3444–3451, 2019, doi: 10.1109/JLT.2019.2917062.

[8] Z. Zhang *et al.*, "Diaphragm-free gas-pressure sensor probe based on hollow-core photonic bandgap fiber," *Optics letters*, vol. 43, no. 13, pp. 3017–3020, 2018, doi: 10.1364/OL.43.003017.

[9] G. Z. Xiao, A. Adnet, Z. Zhang, Z. Lu, and C. P. Grover, "Fiber-optic Fabry-Perot interferometric gas-pressure sensors embedded in pressure fittings," *Microw. Opt. Technol. Lett.*, vol. 42, no. 6, pp. 486–489, 2004, doi: 10.1002/mop.20345.

[10] A. D. Gomes, H. Bartelt, and O. Frazão, "Optical Vernier Effect: Recent Advances and Developments," *Laser & Photonics Reviews*, vol. 15, no. 7, p. 2000588, 2021, doi: 10.1002/lpor.202000588.

[11] Y. Liu, X. Li, Y.-n. Zhang, and Y. Zhao, "Fiber-optic sensors based on Vernier effect," *Measurement*, vol. 167, p. 108451, 2021, doi: 10.1016/j.measurement.2020.108451.

[12] Y. Chen, L. Zhao, S. Hao, and J. Tang, "Advanced Fiber Sensors Based on the Vernier Effect," *Sensors (Basel, Switzerland)*, vol. 22, no. 7, 2022, doi: 10.3390/s22072694.

[13] A. D. Gomes *et al.,* "Optical Harmonic Vernier Effect: A New Tool for High Performance Interferometric Fibre Sensors," *Sensors (Basel, Switzerland)*, vol. 19, no. 24, 2019, doi: 10.3390/s19245431.

[14] Z. Yang, W. Yuan, and C. Yu, "Hollow Core Bragg Fiber-Based Sensor for Simultaneous Measurement of Curvature and Temperature,"

[15] X. Yang *et al.,* "Simplified highly-sensitive gas pressure sensor based on harmonic Vernier effect," *Optics & Laser Technology*, vol. 140, p. 107007, 2021, doi: 10.1016/j.optlastec.2021.107007.*Sensors (Basel, Switzerland)*, vol. 21, no. 23, 2021, doi: 10.3390/s21237956.

[16] X. Song *et al.,* "High Sensitivity Fiber Gas Pressure Sensor with Two Separated Fabry–Pérot Interferometers Based on the Vernier Effect," *Photonics*, vol. 9, no. 1, p. 31, 2022, doi: 10.3390/photonics9010031.

Proceedings of the 7th Optoelectronics Global Conference (OGC 2022)

Self-adjusting Light Source Based on a Dual-Function GaN Light-Emitting Diode

Yumeng Luo, Jiahao Yin, and Kwai Hei Li*
School of Microelectronics
Southern University of Science and Technology
Shenzhen 518055, China
*Corresponding author: khli@sustech.edu.cn

Abstract—**GaN light-emitting diodes (LED) play a vital role in modern lighting technology, and the further development of smart lighting systems capable of automatically adjusting the brightness has received extensive attention. Herein, we present a simple and elegant approach based on a single GaN LED that can self-adjust the output intensity in response to the changes in ambient intensity. The GaN LED with InGaN/GaN multi-quantum wells can operate in both luminescence and photodetection modes, and its electrical and optical performances are thoroughly investigated. Driven by a microcontroller board under pulse-width modulation, the device acts as a detector to provide photocurrent signals that reflect the ambient light intensity at the off state, and provides the desired intensity level at the on state. This work also exhibits a proof-of-concept demonstration of real-time stabilization of blue and white light irradiances at target areas despite large variations in ambient irradiance. The proposed novel self-adjusting scheme based on a dual-function LED chip without the need for external photosensors can be an alternative approach for smart lighting applications.**

Keywords - **GaN, light-emitting diode, light source, pulse-wdith modulation.**

I. INTRODUCTION

With the tremendous progress of growth and microfabrication techniques, GaN-based light-emitting diodes (LED) have been considered to be the core element of light sources, owing to their superior advantages in high efficiency, long lifespan, small footprint, and low carbon emission [1] [2] [3] [4], compared with traditional incandescent bulbs and fluorescent tubes. According to a statistics report,lighting continuously accounts for about 19 % of the global electricity consumption in recent decades [5] [6]. Although LED devices have been proved to be an effective means for saving energy, further reduction in energy consumption in lighting remains an important task to alleviate the energy crisis. Recently, various smart lighting schemes have been proposed to automatically adjust the brightness according to environmental changes and human activities [7], which often requires the implementation of sensing units to recognize these changes. According to the sensed reference signal, the dimming of LED light sources can be readily controlled through amplitude modulation (AM) and pulse-width modulation (PWM). [8]

Among the possible intelligent lighting strategies, real-time automatic brightness adjustment based on variations in ambient light can be widely used for a variety of indoor and outdoor lighting applications [9] [10]. However, attaching photodetectors generally involves multiple optical and mechanical components, which inevitably reduce the efficiency, robustness, and compactness of the lighting system. Moreover, the complex integration of light sources with external sensing elements may be affected by unexpected environmental factors, such as shock or vibration, thus weakening the stability of the system.

To address the above problems, a simple and elegant approach is proposed in this work by using a GaN LED with InGaN/GaN multi-quantum wells (MQWs) as both emitter and detector. Under PWM driving, the MQWs in the device will detect the ambient light to generate the electron-hole pairs in the off state and the photocurrent can act as a reference signal for indicating ambient light intensity. During the on state, light emits from MQW due to the radiative combination and the output intensity is optimized by adjusting the duty cycle of the PWM signal, as illustrated in Figure 1. The luminescence and light detection properties of the LED device are studied and its auto-adjusting response to ambient light change is investigated to verify the performance of the proposed approach.

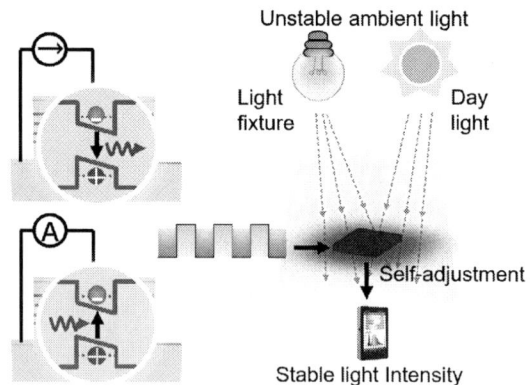

Figure. 1. Schematic diagram showing the working principle of self-adjusting light source based on dual-function GaN LED.

II. DEVICES AND FABRICATION

The process starts with metal-organic chemical vapor deposition (MOCVD) of an epilayer structure composed of undoped GaN, Si-doped n-GaN, InGaN/GaN multi-quantum wells (MQW), and Mg-doped p-GaN, on a 4-inch sapphire

978-1-6654-8699-6/22 $31.00 © 2022 IEEE

substrate. After epitaxial growth, a series of standard wafer-scale processes including photolithography, etching, and metal evaporation are applied to fabricate the LED device. The sapphire wafer is lapped and polished to a thickness of 150 μm, and the processed wafer is cut into 780×780 μm^2 dies via laser micromachining. The LED die is attached and wire-bonded to a printed circuit board (PCB). The PCB is designed with a drilled hole slightly smaller than the die to allow the light to propagate through, as shown in Figure 1(b). Figure 1(c) shows the operating device with blue emission. To obtain the white light emission, the surface of the LED is coated with yellow phosphor to achieve the mixing of blue and yellow light with chromaticity coordinates (0.3, 0.32), as shown in Figure 1(d).

redshift of 1.0 nm is obtained due to a bandgap shrinkage effect.

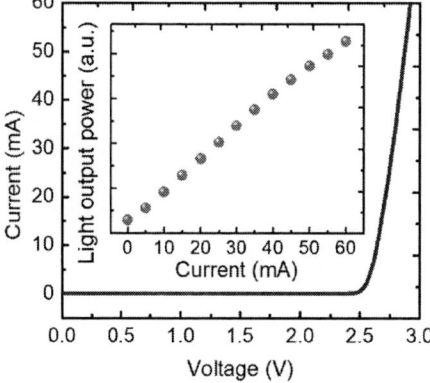

Figure 3. I-V characteristic of the device. The inset shows the L-I plot of the device.

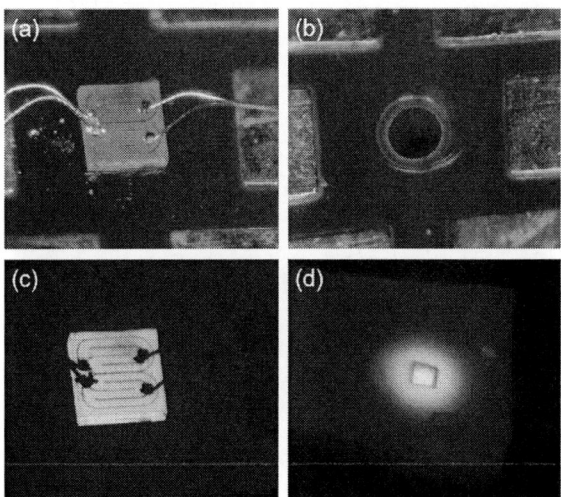

Figure 2. Optical images of (a) packaged device and (b) PCB. (c) Microphotograph of the working device. (d) Optical image of the blue LED with yellow phosphor.

III. RESULTS AND DISCUSSIONS

The electrical and optical properties of the LED are studied. Plotted in Figure 3, the I-V curve of the device shows a nonlinear characteristic, in which the forward voltage of the device at 20 mA is approximately 2.7 V. The inset of Figure 3 indicates a positive linear correlation between light output power and injection current. The electroluminescence (EL) spectra of the LED under different injection currents are measured and shown in Figure 4(a). Due to the quantum-confined stark effect (QCSE) in the InGaN/GaN MQW [11] [12], slight changes in the emission profiles with increasing current injection are observed. To investigate this, the peak wavelengths and spectral widths extracted from Figure 4(a) are plotted in Figure 4(b). The emission peak exhibits a blueshift of 0.3 nm as the current rises from 10 to 30 mA, attributed to band filling effects together with carrier-induced screening of the built-in electric field in quantum wells. When the current is further increased to 70 mA, the self-heating effect dominates and introduces a spectral broadening of 5.7 nm, and a spectral

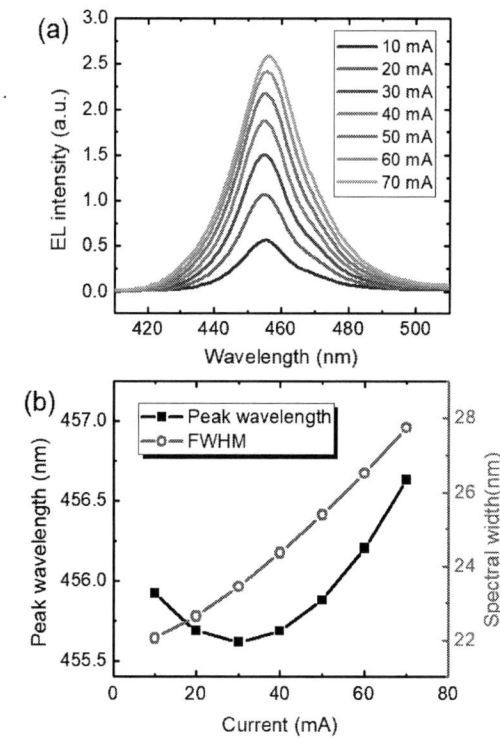

Figure 4. (a) Room-temperature EL spectra of the LED under different current. (b) Plot of peak emission wavelengths and spectral widths as functions of driving current.

In addition to emitting light, the GaN LED can also work oppositely as a detector. For the InGaN-based MQW capable of generating blue light, the LED can partially respond to half of its emission spectrum with higher photon energies [13]. In other words, the blue LED chip adopted in this work can detect the blue-violet emission or blue-violet spectral region of broadband white light. To investigate the light-detecting ability, another LED acts as an ambient light source and the device

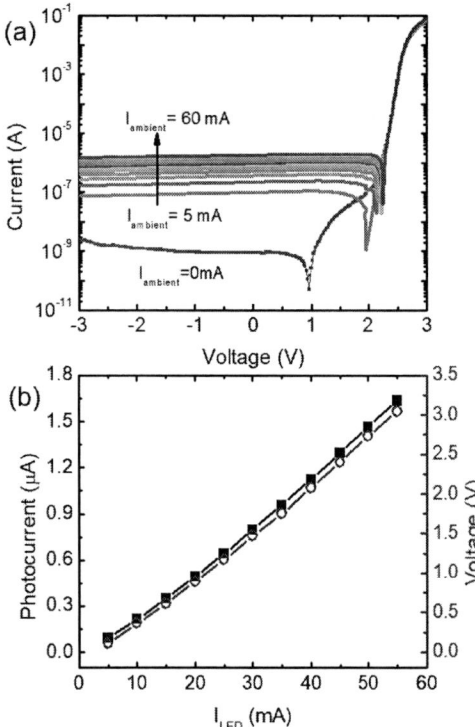

Figure 5. (a) I–V characteristics of the device measured under dark and illumination conditions with the external LED operating in the current range of 5-60mA. (b) Photocurrent measured under irradiance of the external LED operating at different currents. The voltage is converted from the photocurrent through the TIA.

responding to this ambient light is investigated and plotted in Figure 5(a). Under dark conditions, the photocurrents at reverse bias voltages are at a low level of $10^{-8} \sim 10^{-9}$A. While the ambient light source is operating at 5 mA, the photocurrents increase dramatically to 10^{-7}A, revealing the device is sensitive to the incoming light. Moreover, the device can provide a linear response to the ambient light intensity, as evident in Figure 5(b).

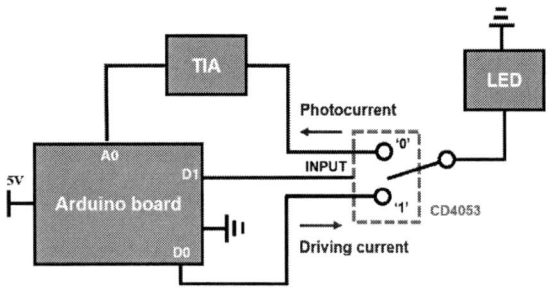

Figure 6. Circuit diagram of the self-adjusting light source system

After the emission and detection capabilities of the GaN LED are determined, the GaN LED is driven by the PWM driving technique with a frequency of 125 Hz by connecting it with an Arduino programmable circuit board. To avoid crosstalk between the input and output signals, a digitally con-

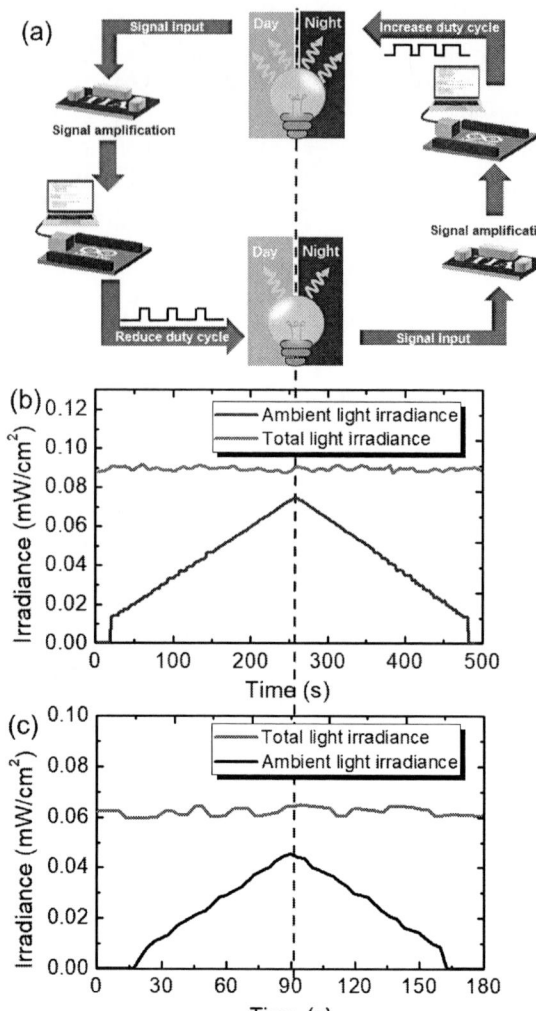

Figure 7. (a) Schematic diagram of self-adjusting schemce for increasing and decreasing ambient irradiance. Meausred irradiance of ambient irradiance and total irradiance under (b) blue and (c) white emission operations.

trolled analog switch (CD4053) is inserted, as shown in Figure 6. During the off state of the PWM signal, the unbiased LED functions as a detector and generates photocurrents according to ambient light intensity. To enable the microprocessor to read the signal, a transimpedance amplifier (TIA) is employed to convert the photocurrent into voltage while maintaining its linear response characteristics, as shown in Fig. 5(b). After analyzing the reference signal, the duty cycle of the driving PWM signal will be adjusted accordingly, as depicted in Figure 7(a). In the case of intense ambient light, the duty cycle of the on-state will be shortened, thereby lowering the LED output intensity. On the other hand, when ambient light is weak, the duty cycle will be increased.

To verify the performance of the proposed self-adjusting scheme, the irradiance of the target area is measured by a commercial optical power meter (OHSP-3501, Hopocolor). Without operating the GaN LED, the measured irradiance from the ambient light source varies widely, from 0 mW/cm^2 to

0.074 mW/cm^2, as shown in Figure 7(b). When the dual-function GaN LED is involved, the total light irradiance originating from the ambient light and LED emission can be stabilized at around 0.09 mW/cm^2 with a fluctuation of less than 5%, suggesting that the device can provide a real-time response to changes in ambient light intensity. Under white emission operation, it can be seen from Figure 7(c) that the total irradiance is around 0.062 mW/cm^2 with a fluctuation close to 9%. Notably, the detecting efficiency and light intensity of the LED is reduced since part of the blue light is converted into yellow emission. Moreover, the yellow emission has small photon energy which is unable to be detected by the blue MQWs. In other words, the coverage of the yellow phosphor weakens the light emission and light detection of the device. Although there exist fluctuations in both sets of results, the irradiance at the target area reaches a certain degree of stability. Moreover, this work aims to provide a proof-of-concept demonstration of self-adjusting intensity based on a single LED chip without relying on external photosensors. The light intensity and detection capability can be readily enhanced by scaling up the sizes or the number of devices.

IV. Conclusion

In this work, a self-adjusting light source based on a GaN LED operating in PWM mode has been demonstrated. During the off and on state of PWM signal, the LEDs can detect the intensity of ambient light and emit an optimized amount of emission in the off and on states, respectively. With the aid of microprocessor, the device can respond in real-time to ambient light intensity and stabilize the irradiation at the target area when operating with blue and white light. The proposed device without external photosensor can be an innovative approach adopted in modern smart lighting applications.

Acknowledgment

College Students' Innovative Entrepreneurial Training Plan Program under Grant 2020X65; Special Funds for the Cultivation of Guangdong College Students' Scientific and Technological Innovation ("Climbing Program" Special Funds.) under Grant pdjh2021c0070.

References

[1] S. Nakamura, T. Mukai and M. Senoh, "Candela-class high-brightness InGaN/AlGaN double-heterostructure blue-light-emitting diodes", Appl. Phys. Lett., vol. 64, pp. 1687-1689, 1994.

[2] S. Nakamura and M. R. Krames, "History of gallium–nitride-based light-emitting diodes for illumination", Proc. IEEE, vol. 101, no. 10, pp. 2211-2220, Oct. 2013.

[3] M. H. Crawford, "LEDs for solid-state lighting: Performance challenges and recent advances", IEEE J. Sel. Topics Quantum Electron., vol. 15, no. 4, pp. 1028-1040, Jul./Aug. 2009.

[4] S. P. DenBaars et al., "Development of gallium-nitride-based light-emitting diodes (LEDs) and laser diodes for energy-efficient lighting and displays", Acta Materialia, vol. 61, pp. 945-951, Feb. 2013.

[5] P. J. G. Carreira, P. J. C. Esteves, C. A. R. Samora and A. T. De Almeida, "Efficient and adaptive led public lighting integrated in évora smart grid", 22nd International Conference and Exhibition on Electricity Distribution (CIRED 2013), pp. 1-4, 2013.

[6] A. L. Langner, L. C. Siebert, A. R. Aoki, E.K. Yamakawa, R. J. Riella, L.C. da Rosa, D. Ribera Neto, "A pilot project of street lighting Telemanagement in a Smart Grid environment", Innovative Smart Grid Technologies Latin America (ISGT LATAM) 2015 IEEE PES, pp. 304-309, 2015.

[7] Yansong Liang, Zhouding Jia, Huanzhong Yao and Jing Chen, "Housing intelligent lighting control strategy research", 2014 IEEE 3rd International Conference on Cloud Computing and Intelligence Systems, pp. 728-731, 2014.

[8] Chinchero, H.F., Alonso, J.M. and Ortiz T, H, "LED lighting systems for smart buildings: a review". IET Smart Cities, 2: 126-134, 2020.

[9] Yusaku Fujii, Noriaki Yoshiura, Akihiro Takita, and Naoya Ohta, "Smart street light system with energy saving function based on the sensor network", Proceedings of the fourth international conference on Future energy systems (e-Energy '13), pp. 271–272, 2013.

[10] Marc Füchtenhans, Eric H. Grosse and Christoph H. Glock, "Smart lighting systems: state-of-the-art and potential applications in warehouse order picking", International Journal of Production Research, 59:12, 3817-3839, 2021.

[11] T. Takeuchi et al., "Determination of piezoelectric fields in strained GaInN quantum wells using the quantum-confined Stark effect", Appl. Phys. Lett., vol. 73, pp. 1691-1693, Sep. 1998.

[12] T. Tetsuya et al., "Quantum-confined Stark effect due to piezoelectric fields in GaInN strained quantum wells", Jpn. J. Appl. Phys., vol. 36, pp. L382-L385, Apr. 1997.

[13] K. H. Li, H. Lu, W. Y. Fu, Y. F. Cheung and H. W. Choi, "Intensity-Stabilized LEDs With Monolithically Integrated Photodetectors,", IEEE Transactions on Industrial Electronics, vol. 66, no. 9, pp. 7426-7432, Sept. 2019.

978-1-6654-8699-6/22 $31.00 © 2022 IEEE

Proceedings of the 7th Optoelectronics Global Conference (OGC 2022)

Modeling and Analysis of Zinc Diffusion Effect within InP-Based Mach-Zehnder Modulators

Ruoyun Yao[1*], Wanshu Xiong[1], Zhangwan Peng[1], Yiti Xiong[2], Chaodan Chi[2], Chen Ji[1]

[1] College of Information Science and Electronic Engineering, Zhejiang University, Hangzhou, China
[2] Research Institute of Intelligent Networks, Zhejiang Lab, Hangzhou, China
*Corresponding author.
e-mail: yaoruoyun@zju.edu.cn

Abstract—We investigate the influence of Zinc diffusion on the modulation efficiency and optical insertion loss of InP-based MZI modulators. By analyzing the electric-optical field distribution and free-carrier absorption loss of three types of Zinc diffusion profiles, the mechanisms between Zn diffusion and modulation efficiency and optical insertion loss inside MZI modulators are systematically studied. We show that InP-based MZI modulator modulation efficiency is limited by electro-optic field distribution overlap and optical mode interacts with p-type dopant increases device insertion loss when Zn diffuses during MOCVD epitaxial process.

Keywords-MZI modulator; Zinc diffusion; modulation efficiency; loss; MOCVD.

I. INTRODUCTION

Optical fiber data communication technology has experienced dramatic bandwidth increases over the past two decades, driven by the exponentially increasing internet traffic and hyper scale data center I/O bandwidth demands, which in turn requires higher speed modulator technology to supply the necessary optical data transmission bandwidth. The current mainstream component technology for 100G, 200G and 400G optical transceivers (Tx/Rx) are based on discrete InP directly modulated laser (DML) and electro-absorption modulated laser (EML) chips [1-3]. In meeting the next generation high speed optical interconnect technology scaling demands, photonic integrated circuit (PIC) technology stands out for its compact size, lower packaging cost and reduced power consumption, comparing to discrete component solutions. Silicon photonics MZI modulator integrated Tx fabricated by silicon on insulator (SOI) process is not able to achieve monolithic integration with III-V laser source, which will result in extra cost on heterogeneous integration and packaging [4]. Furthermore, comparing with SiP modulators working on plasma dispersion effect, the electric field induced electro-optic effect of InP modulators, such as quantum confined stark effect (QCSE), is 10 times stronger, demonstrating a greater superiority of III-V material for high speed modulation applications [5].

The InP-based MZI modulator fast electro-optic response is attributed to the rapid electric field effect. It has been shown that strong electric field intensity at active region will significantly improve the electro-optic modulation efficiency of InP-based MZI modulators [6]. A standard InP-based MZI

modulator structurally consists of a p-type InP cladding layer, an n-type InP cladding layer, an undoped multi-quantum well active region, and two undoped spacer layers on the lower and upper side of the MQW region separately. The state of art InP-based MZI modulator with over twenty quantum wells contributes to a thick intrinsic region (generally more than 0.6 μm), indicating weaker electric field intensity at a given reverse bias voltage. In order to strengthen the electric field intensity in the MQW region, neither thick spacer layer nor high doping concentration in the active region is allowed. While the former case can be avoided by careful design, the latter situation is determined by manually material doping control during epitaxy process. Zinc and Silicon are commonly used as p-type and n-type dopant in Metalorganic Chemical Vapor Deposition (MOCVD) epitaxial process for InP-based optoelectronic devices [7]. While Si dopant is stationary after growth, Zn doping has a strong tendency to diffuse toward layers of lower concentration during high temperature epitaxy. Consequently, the p-type doping profile can strongly deviate from design, affecting overall electric field distribution and degrading device performance. Zn diffusion process is highly dependent on MOCVD growth conditions, thus minimizing Zn diffusion effect is one of the main challenges in MOCVD grown InP epitaxial material.

In this paper, we systematically analyzed the effect of Zn diffusion on an InP-based MZI modulator design, which to the best of our knowledge has not been reported in literature. Based on device performance simulations we showed for the first time that Zn diffusion from the p-type doped InP layer by $\sim 10^{17}$ cm^{-3} can strongly reduce the electric-optical modulation efficiency and increase free-carrier absorption, degrading the MZI modulator half-wave voltage and insertion loss. From electric-optical field distribution overlap analysis, the relationship between Zn diffusion and modulation efficiency, which ultimately affects half-wave voltage in the MZI modulators, is analyzed in detail. We believe our work identified for the first time the critical role Zn diffusion control plays in MOCVD grown InP-based MZI modulator device performance, and provides a systematic approach for optimizing InP-based MZI modulator design.

II. DEVICE STRUCTURE

Fig. 1 shows the schematic layer structure of a standard deep etched InP-based MZI modulator structure with 2 μm width ridge, grown on semi-insulating (SI) InP substrate,

978-1-6654-8699-6/22 $31.00 © 2022 IEEE

including the n-type doping InP layer (1.7 μm), lower undoped InP layer (0.196 μm), multi-quantum well (MQW) layer (0.408 μm), upper undoped InP layer (0.196 μm), p-type doping InP layer (1.3 μm) and heavily p-type doped InGaAs contact layer (0.2 μm). The undoped InP layer serves as a spacer to confine the optical mode in the MQW active region and prevent mode leakage into p- or n-doped layers, while the MQW layer serves as the active region to realize high speed electro-optic effect at applied electric field and dominants the MZI modulator electro-optic modulation response. The unintentional doped MQW layer and spacer layers will be completely depleted during operation, and the built-in E field will rapidly change the refractive index of optical mode in the waveguide, resulting in phase shift in the drive arm of the MZI and ultimately intensity modulation of the MZI modulator output.

In order to reduce the MZI modulator switching voltage, the electro-optic response efficiency is maximized by carefully selecting the quantum well structures, while the quantum well confinement factor and the electro-optic field overlap are maximized by utilizing a deep-etched waveguide profile. Different strength of the electric field generated in the p-i-n diode structure by the applied reverse voltage will directly influence the electro-optic effects and cause different levels of required driving voltage accordingly. Zinc diffusion during MOCVD growth will modify the E field distribution and carrier concentration in the spacer and MQW active layers, impacting the electro-optic efficiency, optical insertion loss, and the overall MZI modulator performance.

III. MODELING AND SIMULATION

We simulated different levels of Zinc diffusion effect within a MOCVD grown MZI modulator epitaxy structure, using commercial optical simulators APSYS and Lumerical. Fig. 2 illustrates the Quantum-Confined Stark Effect of a 12 nm InGaAsP quantum well. Considering a single quantum well placed in an electric field perpendicular to the quantum well layer, the so-called Quantum-Confined Stark effect forces a shift of the band gap towards lower energies and a decrease of the absorption coefficient for energies above the band gap. As shown in Fig. 2(a) and (b), the electric field pulls the electrons and holes towards opposite sides of the layer, resulting in an overall net reduction to the energy of the electron-hole pair and a corresponding shift of the exciton absorption [8]. Fig. 2(c) shows the absorption spectra of the

QW at 10 kV/cm to 34 kV/cm applied electric field simulated with APSYS, where heavy-hole and light-hole absorption peaks are displayed, indicating a phenomenon of absorption peak red-shift, which will in turn cause refractive index change due to Kramers-Kronig relations [9]. The QCSE is determined by quantum well structure and electric field intensity simultaneously. The MQW structure of our model is optimized by maximizing Δn and minimizing Δκ of the refractive index under certain applied electric field, indicating the optimal modulation efficiency in an ideal working condition, while E field intensity reduction will cause a degradation in MZI modulator electro-optic efficiency, as we can see from more quantitative Lumerical simulation results below.

Fig. 3(a) shows the baseline MZI modulator structure in our study, assuming the Zn dopant does not diffuse and the doping profile is unchanged from our optimized design for maximum modulation efficiency. In the baseline design, an n-type doped InP cladding layer (1×10^{18} cm^{-3}) with 1.7μm thickness is grown on a semi-insulating InP substrate, serving as an n-contact layer. An intrinsic region with total thickness of 800nm is divided into two 196 nm thickness undoped InP spacer layers and a 408 nm thickness undoped MQW layer placing in the middle, assuming an ideal natural n-type background doping concentration of 1×10^{15} cm^{-3}. A 1.3 μm light p-type doped InP cladding layer (5×10^{17} cm^{-3}) is grown on the top of the intrinsic region, following with a 0.2 μm heavy p-type doped InGaAs contact layer (1×10^{19} cm^{-3}) serving as a p-contact layer.

Fig. 3(b) and (c) show the same MZI modulator epitaxy design with two doping profiles representing different levels of Zn diffusion effect within the upper spacer layer and MQW layer of the intrinsic region. In Fig. 3(b), the upper spacer layer and the MQW layer are light p-type doped (5×10^{16} cm^{-3}) with Zinc. Fig. 3(c) represents a more severe Zn diffusion case, with the upper spacer layer and MQW layer p-type doped at a higher concentration (1×10^{17} cm^{-3}).

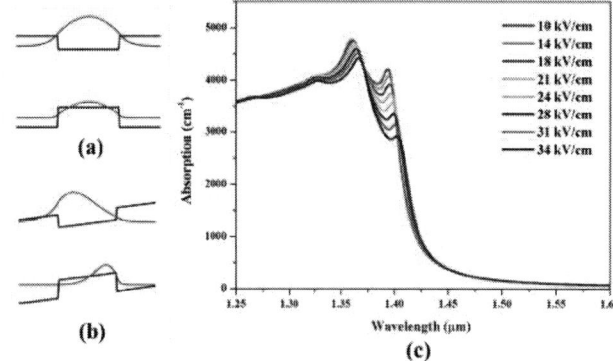

Figure 2. Illustration of the Quantum-Confined Stark Effect of a 12 nm InGaAsP quantum well: band edge and wavefunctions are plotted without (a) and with (b) applied electric field; (c) absorption spectra of the QW at 10 kV/cm to 34 kV/cm applied electric field are simulated with APSYS.

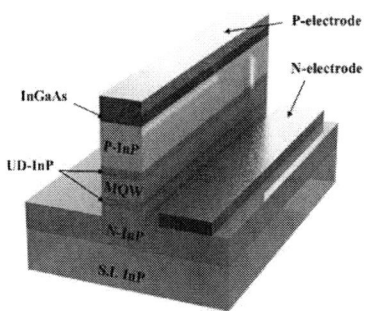

Figure 1. Schematic of an MZI modulator waveguide structure.

(a)	Type-A		(b)	Type-B		(c)	Type-C	
InGaAs; 0.2µm; p:1×10^{19}cm^{-3}			InGaAs; 0.2µm; p:1×10^{19}cm^{-3}			InGaAs; 0.2µm; p:1×10^{19}cm^{-3}		
InP; 1.3µm; p:5×10^{17}cm^{-3}			InP; 1.3µm; p:5×10^{17}cm^{-3}			InP; 1.3µm; p:5×10^{17}cm^{-3}		
InP; 0.196µm; n:1×10^{15}cm^{-3}			InP; 0.196µm; p:5×10^{16}cm^{-3}			InP; 0.196µm; p:1×10^{17}cm^{-3}		
InGaAsP MQW; 0.408µm; n:1×10^{15}cm^{-3}			InGaAsP MQW; 0.408µm; p:5×10^{16}cm^{-3}			InGaAsP MQW; 0.408µm; p:1×10^{17}cm^{-3}		
InP; 0.196µm; n:1×10^{15}cm^{-3}			InP; 0.196µm; n:1×10^{15}cm^{-3}			InP; 0.196µm; n:1×10^{15}cm^{-3}		
InP; 1.7µm; n:1×10^{18}cm^{-3}			InP; 1.7µm; n:1×10^{18}cm^{-3}			InP; 1.7µm; n:1×10^{18}cm^{-3}		
Semi-insulating InP substrate			Semi-insulating InP substrate			Semi-insulating InP substrate		

Figure 3. MZI modulator epi designs with three different Zinc doping profiles. Layers with doping variations shown in bold frame.

A. Zinc Diffusion Effect on Electric-optical Field Distribution

For an InP-based MZI modulator, high electro-optic field distribution overlap is important to achieving optimal electro-optic efficiency because it indicates strong electric field and light interaction, which will cause refractive index change as a result of QCSE. To demonstrate the impact on electric field due to Zinc diffusion effect, we performed electric field distribution simulation for the three proposed doping profiles from Fig. 3 using Lumerical. The simulation results are shown in Fig. 4 for an external reverse bias voltage of 2V, demonstrating reduced electric field intensity in active region with increasing Zn diffusion concentration. As shown in Fig. 4(a), the electric field is stronger in the vicinity of the p-contact and decreases towards the n-contact due to the n-type nature of background doping. In the none Zn diffusion case, the n-type nature background doping leads to the formation of the built-in field at the 'p-type-to-intrinsic' junction and causes the built-in field reaches further towards the p-doping layer. As Zn diffuses into the intrinsic region, the n-type background dopant neutralizes with the p-type Zn dopant until p-type doping dominants the intrinsic region doping concentration. In Type-B model, as Zn diffusion concentration reaches 5×10^{16} cm^{-3}, the electric field intensity in the upper spacer layer turns to zero and the E field in the lower spacer layer increases. In Type-C, when Zn diffusion concentration reaches 1×10^{17} cm^{-3}, the electric field intensity in the entire upper spacer layer and partial MQW layer become zero and the electric field intensity in the lower spacer layer turns to the highest.

As for optical power distribution, which is situated in the middle of the MQW region, remains unchanged with different Zn diffusion concentration. According to our simulation, about 41% optical power is confined within the quantum well layers in the simulated deep etched waveguide structure.

B. Zinc Diffusion Effect on Modulaton Efficiency

The modulation efficiency, defined as $d\Phi/dV \cdot L$ is a key parameter for evaluating performance of an electro-optic modulator, which directly determines the drive voltage and the device size required for the operational wavelength ideally drop to zero due to interference. The modulation efficiency of an InP-based MZI modulator can be calculated as a product of three factors: quantum well confinement factor, which describes the optical power ratio confined in the quantum well layers; QCSE factor, which refers to the refractive index change due to QCSE at applied electric field; and electro-optic field overlap, which represents the efficiency of applied voltage interacts with optical mode within the active region. As we apply the same MQW and waveguide structure design in our simulation models, it is reasonable to assume that the QCSE factor and the quantum well confinement factor keep constants in the same working conditions. Hence, the electro-optic efficiency is directly determined by electric field intensity and optical mode overlap within the active layer. We calculate the electro-optic field distribution overlap integration Γ_{EO} according to equation (1).

$$\Gamma_{EO} = \frac{\iint_{MQW} E_{ele}(x,y) \cdot |E_{opt}(x,y)|^2 \, dxdy}{\iint_{-\infty}^{+\infty} |E_{opt}(x,y)|^2 \, dxdy} \tag{1}$$

The overlap integral is calculated in the active MQW region denoted by 'MQW' in the integral for both x and y direction. The overlap integral is normalized by the total power confined in the optical mode.

Fig. 5 shows the electro-optic modulation efficiency for our MZI modulator design, varying with Zn diffusion concentration in the upper spacer layer and MQW layer from 0 to 1×10^{17} cm^{-3}. Our calculation result demonstrates that

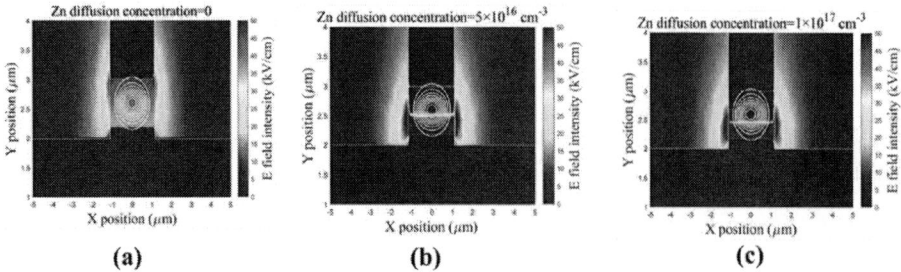

Figure 4. Electric field (colored) and optical power (white lines) distributions of an MZI modulator under different Zn diffusion conditions.

978-1-6654-8699-6/22 $31.00 ©2022 IEEE

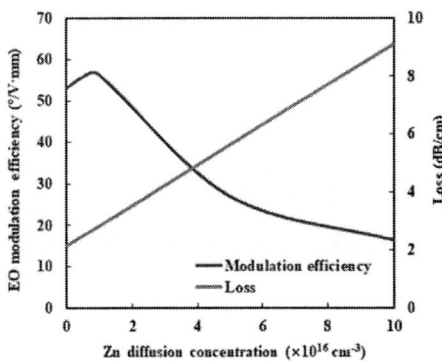

Figure 5. Simulated electric-optical modulation efficiency and free-carrier absorption loss with varying degrees of Zn diffusion effect.

at low Zn diffusion concentration ($<1 \times 10^{16}$ cm^{-3}), EO modulation efficiency increases with higher Zn diffusion level, which can be explained by a net reduction of p- and n-type dopant in the intrinsic region. As Zn concentration keeps increase, modulation efficiency starts to drop rapidly, which comes from the dramatic reduction of E field intensity in the MQW layer. As shown in Fig. 5, the modulation efficiency drops from 53 °/V·mm to 16 °/V·mm when Zn diffusion concentration in the upper spacer layer and MQW active layer reaches 1×10^{17} cm^{-3}.

C. Zinc Diffusion Effect on Optical Insertion Loss

The insertion loss is one of the critical parameters of an MZI modulator of achieving the optimal performance. When Zinc diffuses into the intrinsic region of an MZI modulator, the free-carrier concentration in the active layer is about to increase, which will result in extra free-carrier absorption loss. The relationship between free-carrier absorption loss and p- and n-type carrier concentration is described as follows

$$\text{Loss} = \iint \Gamma_{opt} \cdot (\alpha_n N_n + \alpha_p N_p) \mathrm{dxdy} \qquad (2)$$

where Γ_{opt} is the confinement factor of a certain layer where free-carrier exists, α_n and α_p are absorption coefficients of n- and p-type carriers accordingly, and N_n and N_p are n- and p-type carrier concentrations. It is acknowledged that α_p is generally ten times larger than α_n, thus make preventing optical mode reaching p-type doped regions more critical to achieving low absorption loss. When Zn diffuses into the intrinsic region, the p-type carrier absorption loss increases and results in a degradation in device insertion loss. A quantitative simulation result is plotted in Fig. 5, which demonstrates an approximately linear increase in free-carrier absorption loss with growing Zn diffusion concentration from 0 to 1×10^{17} cm^{-3}. It is observed that the insertion loss will increase ~7 dB/cm when Zn diffusion concentration in the upper spacer layer and MQW layer reaches 1×10^{17} cm^{-3}.

IV. Analysis and Discussion

According to our simulation results shown above, the electro-optical modulation efficiency and optical insertion

loss of MZI modulator structures with Zn diffusion effect degrade to some degree, comparing with Type-A baseline design. Further simulations on half-wave voltage of MZI modulator with Zn diffusion effect are performed to demonstrate MZI modulator performance degradation more distinctly.

Fig. 6 shows the simulated transmission function of Type-A, B and C MZI modulator structures, assuming the same device length of 4 mm. For Type-A model, the half-wave voltage (V_π) is 1V, while for Type-B and C the V_π increases to 2V and 3V accordingly, which is inversely proportional to the corresponding modulation efficiency. In real world applications, low $V_\pi \cdot L$ modulator design is one of the main technical challenges that people seek to overcome. Our work demonstrates that Zn diffusion effect within InP-based MZI modulators will degrade modulation efficiency and result in higher $V_\pi \cdot L$, rising the electrical driver demands and increasing consumptions.

Overall our simulation results show that electro-optic field distribution plays an important role in determining InP-based MZI modulator characteristics. Electric field intensity reduction in MQW region created by Zn diffusion, especially when it reaches 1×10^{17} cm^{-3}, can be highly detrimental to MZI modulator modulation efficiency and will increase half-wave voltage. Moreover, Zn diffusion effect induced free-carrier absorption will increase device insertion loss and degrade modulator performance.

V. Conclusion

In conclusion, we have investigated the influence of Zinc diffusion effect on the performance of InP-based MZI modulator design. Effects of different doping profiles representing varying levels of Zinc diffusion during MOCVD epitaxial growth are analyzed in detail using commercial simulation software. The modulation efficiency and free-carrier absorption loss have been quantified for a waveguide with core layer situated in the center of the intrinsic region. We showed that Zinc diffusion into the MQW depletion region can reduce the electro-optic efficiency and increase optical insertion loss. Our work demonstrated that carefully controlling Zinc diffusion effect during MOCVD growth is critical for maintaining InP-based MZI modulator device performance.

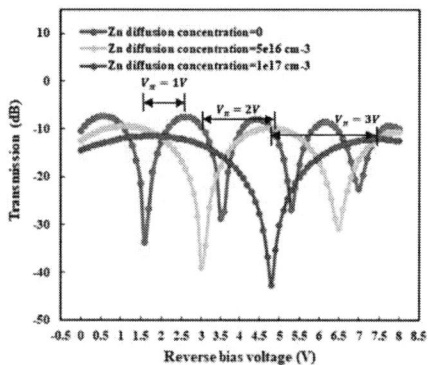

Figure 6. Simulated transmission function of MZI modulator at different levels of Zn diffusion effect.

ACKNOWLEDGMENT

This work was supported in part by the National key research & development (R&D) plan 2020YFB1805701, the National key research & development (R&D) plan 2019YFB2203801, the Zhejiang Lab grant. 2020LC0AD01/001, the Zhejiang Lab grant. 2020LC0AD02/001, and the National Natural Science Foundation of China under Grants 61974132.

REFERENCES

[1] S. Arafin and L. A. Coldren, "Advanced InP photonic integrated circuits for communication and sensing," IEEE Journal of Selected Topics in Quantum Electronics, vol. 24, no. 1, pp. 1-12, 2017.

[2] Zhou, X., R. Urata , and H Liu. "Beyond 1Tb/s Datacenter Interconnect Technology: Challenges and Solutions (Invited)." Optical Fiber Communication Conference 2019.

[3] R. Urata, H. Liu, X. Zhou, and A. Vahdat, "Datacenter Interconnect and Networking: from Evolution to Holistic Revolution," in Optical Fiber Communication Conference 2017.

[4] Doerr, C. R. , et al. "Single-Chip Silicon Photonics 100-Gb/s Coherent Transceiver." Optical Fiber Communication Conference IEEE, 2014.

[5] D. A. Miller et al., "Band-edge electroabsorption in quantum well structures: The quantum-confined Stark effect," Physical Review Letters, vol. 53, no. 22, p. 2173, 1984.

[6] Chuang, S. L., N. Peyghambarian ,and S. Koch. "Physics of Optoelectronic Devices." Physics Today 49.7(1996):62-62.

[7] D. Zhou et al., "The effect of zinc diffusion on extinction ratio of MQW electroabsorption modulator integrated with DFB laser," in Semiconductor Lasers and Applications VI, Nov. 2014, vol. 9267, p. 926714, doi: 10.1117/12.2073556.

[8] H. Klein, "Integrated InP Mach-Zehnder Modulators for 100 Gbit/s Ethernet Applications using QPSK Modulation," 2010.

[9] Lucarini, V. , et al. "Kramers-Kronig Relations in Optical Materials Research." Springer Berlin Heidelberg, 2005.

Proceedings of the 7th Optoelectronics Global Conference (OGC 2022)

Advanced Getter Solutions for Gas Contaminants Absorption in Optoelectronic Devices

Giovanni Zafarana, Enea Rizzi, Luca Mauri, Alessio Corazza, Marco Moraja

SAES Getters S.p.A.
Lainate, Italy
e-mail: marco_moraja@saes-group.com

Abstract—**Sealed optoelectronic devices may experience performance issues, or failures, due to outgassing and release of gas species like H_2O, H_2, and VOCs during the operating lifetime. Hermetic packaging is the standard solution for sealing opto-electronic devices in order to protect them from external atmosphere or harsh environments, ensuring higher reliability and longer life. Unfortunately, hermeticity is not effective to prevent the release of gases from materials that are inside the sealed device. Accumulation of H_2, H_2O or VOCs could significantly affect the devices, leading to poor performances and drift over the lifetime. An effective way to fix this issue is to integrate getter materials. The engineered absorbing materials developed by SAES can be integrated into the packages to selectively absorb gases and preserve the device performances.**

Keywords: optoelectronic device, getter, harmful contaminants, outgassing

I. INTRODUCTION

Hermetically sealed packages are used in many fields like micro-electronics, optoelectronics, telecom, bio-medical and other sectors. Hermeticity is commonly adopted to protect the device from external atmosphere and from aggressive environments.

Inside opto and photonic devices, gaseous contaminants may be responsible of device malfunctioning, performance degradation or even failure over lifetime. The main gaseous species that can be harmful for hermetic sealed devices are moisture, hydrogen and volatile organic compounds (VOCs). It is also worth noting that moisture may cause problems also in semi-hermetic devices.

In optical transceiver modules hydrogen and moisture are the main gas contaminants to be removed. Critical levels of these gases are 1000 ppmv for H_2 and 5000 ppmv for H_2O.

There are several problems induced by the presence of these gases. For instance, H_2 is responsible of electrical performance degradation, it can diffuse through metal layers, causing shifts in currents and trans-conductance [1-3], it can reacts with surface oxides inside hermetic packages promoting the formation of moisture [4].

Moisture can be responsible of electrical shorting or corrosion of solder joints. Photodetectors components are affected by water from dark-current increase [5].

In laser diodes-based devices, like transmitters, the gases that are considered harmful are water and VOCs. The main

issues are related to signal attenuation because of gas condensation.

In order to manage such potential issues related to gases evolution inside the hermetic package, one interesting and well-proved solution is to integrate a getter material that can absorb the gaseous contaminants.

SAES has been developing for many years engineered sorbing materials, combined with special polymeric matrixes and with solventless formulation, easy to be dispensed and cured on packaging components.

The paper presents the latest results achieved in the functional performances of getter materials, developed by SAES and belonging to the ZeDry™ family [6], being ZeDry™ the trade-mark of the getters family, of which the products named ZeDry/H_2 and ZeDry/VOC, specifically engineered for sorption of H_2O+H_2 and $H_2O+VOCs$, respectively, are the two main representatives.

II. DISPENSABLE GETTERS FEATURES

Dimensions of optoelectronic devices are typically small and the technological trend is always to get smaller and smaller footprint and volumes. Consequently, the available space for integration of components and materials is always limited.

In order to accomplish to this trend, the engineered getters materials have been developed in the form of thick film that can be directly applied to the device lid, or eventually to a subcomponent of the package.

Integration can be effectively achieved by dispensing the getter materials on lids and patterning the shape according to specific geometrical requirements, see Figure 1.

Figure 1.   Getter coatings on lids: ZeDry/VOC (top) and ZeDry/H_2 (bottom).

978-1-6654-8699-6/22 $31.00 © 2022 IEEE

After dispensation, a thermal treatment is necessary to cure and consolidate the material.

The active getter material that reacts with the gas to be removed (H_2, VOC or H_2O) is dispersed into a polymeric matrix without the need of using solvents.

The solventless formulation ensures that no additional, undesirable gases are generated and released inside the device because of the getter formulation.

Moreover, the getter composition is compatible with the typical temperatures of the hermetic sealing processes, like laser and seam welding or soldering with eutectic materials, i.e. AuSn.

Based on the specific device, the getter can be selected in order to interact with the main gases of interest.

Tailored materials have been developed and patented for absorption of moisture and hydrogen, or moisture and VOCs. Depending on the gas species, the interaction can be reversible or irreversible: for H_2 the sorption is irreversible, while for VOCs and H_2O the adsorption occurs in a reversible way.

The active materials have been designed to guarantee very low moisture concentration in the devices, even well below 1,000 ppm; in addition their high gettering action allows to achieve even lower H_2 or VOC levels, thus creating conditions suitable to improve the reliability of optical transceivers and of other optoelectronic devices.

III. GETTER ACTIVATION EFFICIENCY

In order to interact properly with gases, the getter must be previously activated by heating the material in vacuum, nitrogen or dry air conditions, immediately before sealing the package.

In particular, the water absorbing properties of the getters have been investigated by changing activation parameters, like temperature and vacuum level, simulating different fabrication conditions of the real devices.

Temperature and vacuum parameters affect in different ways the level of activation of the materials.

The goal of these measurements was to demonstrate the efficiency and regeneration level of the getter material, in terms of water desorption, when treated in different vacuum and temperature conditions. The water adsorption-desorption capacity and the overall performances were investigated using a digital microbalance coupled to a temperature controlled vapor analyzer.

A study of water desorption in isothermal conditions was set up to verify the achievable re-activation level of SAES products, belonging to the ZeDry™ getter family, especially in terms of recovered water sorption capacity. Tests were carried out on the ZeDry/H_2 product, but results are valid also for the ZeDry/VOC getter. The materials were deposited as composite paste coatings on metallic substrates with thickness of 0.2 mm and they were cured at 170 °C for 1 hr.

A specific analysis protocol was used for these tests, starting from the saturation with water vapor of the sample before activating the material in different ways. We report in Table I the steps of the protocol analysis:

TABLE I.   TEST CONDITIONS

PROCESS STEPS		
Step	Operation	Temperature (°C)
1	Water adsorption	RT
2	Vacuum activation	RT
3	Water re-adsorption	RT
4	Vacuum re-activation	50
5	Water re-adsorption	RT
6	Vacuum re-activation	100

The steps related to water desorption (getter activation) were performed with two different set of vacuum pumps: a scroll pump with a low and medium vacuum range (10^{-2}-10^{-3} mbar) and a turbo pump with high vacuum range (10^{-5}-10^{-6} mbar).

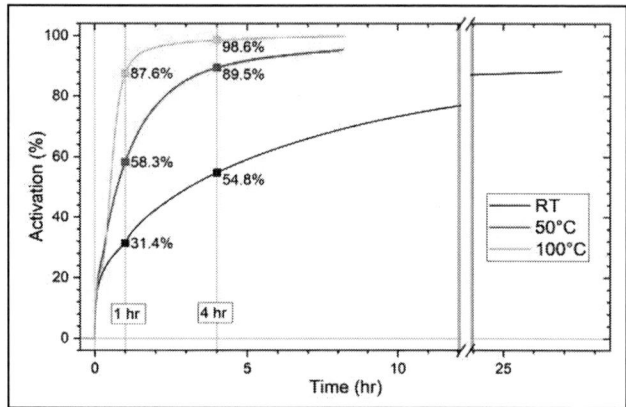

Figure 2.   Activation levels of ZeDry™ product achieved for H_2O after heating at RT, 50°C and 100°C under pumping with turbo pump (high vacuum range).

In Figure 2 it is reported the result of the getter activation efficiency related to high vacuum condition (10^{-5}-10^{-6} mbar) and different heating temperatures. After complete saturation of the material exposed to water vapor (15 mbar partial pressure), the subsequent step consists in the activation of the material at room temperature (RT) with a vacuum pump (turbo pump, in Figure 2) that forces the water desorption from the sample. During this step the water that was physically sorbed is slowly released and the signal (corresponding to the recovered activation level) needs a quite long time to stabilize (black line). Once the sample signal is stabilized under these conditions, water is re-admitted at room temperature and the getter is saturated again (step 3).

Subsequently, the vacuum pumping is re-started, but desorption is performed raising the temperature to 50 °C (red line) (step 4). The last steps, 5 and 6, are another repetition of re-adsorption of water and getter activation in high vacuum at 100 °C (green line). Increasing the activation temperature, the physico-chemical desorption of water from

978-1-6654-8699-6/22 $31.00 © 2022 IEEE

the material is greatly accelerated, taking place in a much shorter time.

Measurements at low vacuum conditions were also performed on ZeDry™ getters, in order to compare the results with those obtained at high vacuum settings. All the results are compared in Table II.

TABLE II. GETTER ACTIVATION EFFICIENCY IN LOW AND HIGH VACUUM CONDITIONS

	Low range vacuum		High range vacuum	
T (°C)	Activation Efficiency (%)	Activation Time (hrs)	Activation Efficiency (%)	Activation Time (hrs)
RT	75	6.5	86	4.5
50	90	8.5	93	6
100	100	26	100	21

The data from Table II suggest operating always at temperatures higher than 100 °C in order to activate properly the getter and maximizing its functionality. At high temperature (100 °C), regardless of vacuum conditions, the regeneration provides the same results in terms of removed water and activation level, but the corresponding time is different; for example to achieve an high activation level (≥97%) at 100°C 5 hours are needed in low vacuum conditions, 3 hours in high vacuum.

On the other hand, lower temperature and low-range vacuum conditions lead to decreased activation efficiency, reaching the values of 75.6% at RT and 90.5% at 50 °C. When activation is carried out in high vacuum conditions, the activation efficiency is better: 86% at RT and 93.6% at 50 °C. It is worth noting that activation of the material at relatively low temperatures, like 50 °C, is enough to get 90% of the getter performances.

IV. WATER ABSORPTION EFFICIENCY AT DIFFERENT TEMPERATURES

Another important feature that has been investigated is the characteristic of water absorption at different temperatures, from room temperature up to more extreme demanding conditions, like 120 °C.

Some demanding high temperature operation conditions for opto-electronic devices, suggested to set up tests in order to assess the performance of water uptake of the getter in a warm environment.

In Figure 3 we show results of moisture retention measured on ZeDry™ getters at constant partial pressure of water (15 mbar) when increasing temperature from RT, to 50°C, 80°C, 100°C and 120°C, considering the last one as the maximum temperature that opto-electronic devices can reach in drastic operating conditions.

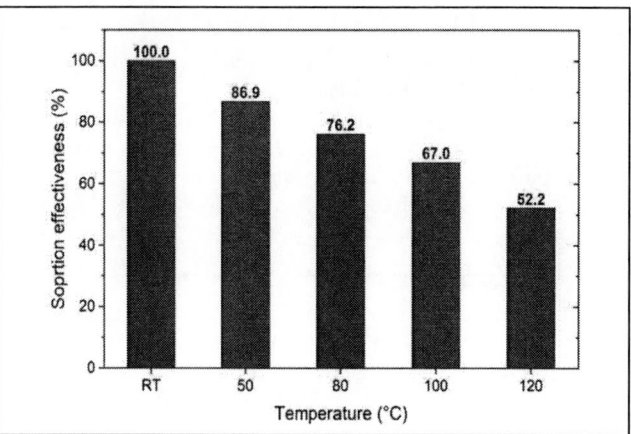

Figure 3. Water uptake performances of ZeDry™ at different temperatures.

As we can see in Figure 3, after activation and moisture sorption at RT we set the water uptake at room temperature at 100%. By heating the getter, we observe a weight decrease due to partial water desorption from the sample.

The increase in temperature negatively affects the absorption properties with about 50% reduction in functionality at high operating temperatures.

Despite this, the getter still provides more than half (52.1 %) of the full efficiency at 120°C, guaranteeing good performances even in the most tough working conditions of the device.

V. GETTER FOR H_2O AND H_2: ZEDRY/H_2

The specific sorption performances of the ZeDry/H_2 getter material are reported in the following Figure 4.

The left red graph shows the sorption curve for hydrogen carried out at room temperature.

Hydrogen uptake signal was followed in a continuous way for 24 hours. Most of the H_2 quantity is absorbed after 6 hours of exposure.

The right blue graph highlights the water uptake at room temperature of the ZeDry/H_2 getter. The kinetics in this case is even faster, since most of the absorption of water takes place within 1.5 hours.

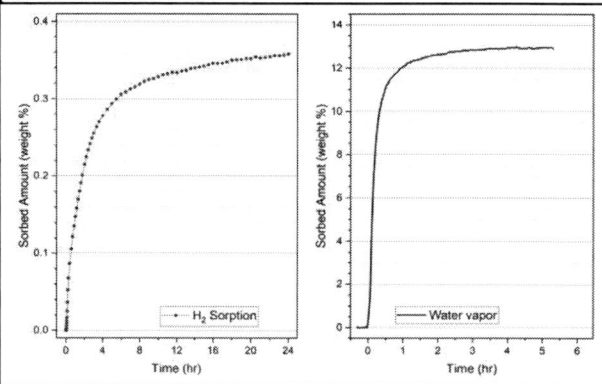

Figure 4. Hydrogen (left) and water (right) uptake performances of ZeDry/H_2 getter at RT.

VI. Getter for H_2O and VOC: ZeDry/VOC

The ZeDry/VOC getter has been developed to capture water and VOCs contaminants.

In Figure 5 the results of H_2O and VOCs adsorption tests are reported in order to highlight the sorbed quantities and adsorption kinetics of these species.

The sorption curve of Methyl Ethyl Ketone (MEK) is reported as an example for VOC typical compound.

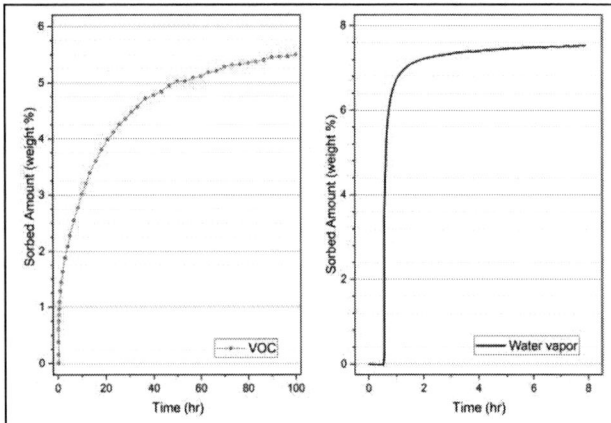

Figure 5. VOCs (left) and water (right) uptake performances of ZeDry/VOC getter at RT.

The VOC adsorption test was carried out with prolonged exposure (100 hours), until product saturation was almost achieved. The total amount of organic adsorption was 5.45% by weight of getter paste.

As shown by the water absorption curve, instead, the moisture sorption on ZeDry/VOC has the same behavior of that measured on the ZeDry/H_2 getter. Most of the H_2O uptake takes place in 1.5 hours.

VII. Conclusions

In this paper, we have presented the characteristics and performances of new getter products used in hermetically sealed packages of optoelectronic devices. SAES developed engineered absorber materials, with solvent-free formulation, belonging to the ZeDry™ family suitable to remove significant amounts of water even under tough temperature conditions; they can be easily activated at moderate temperatures under different vacuum conditions.

Furthermore, the developed ZeDry/H_2 and ZeDry/VOC products exhibit also excellent performances, in terms of hydrogen and VOCs sorption, respectively, ensuring the possibility to remove all the detrimental contaminants inside optoelectronic devices. These getters have been designed to keep very low water concentration in the devices, even well below 1,000 ppm and to achieve hydrogen or VOC levels much lower than 500 ppm; the effective gettering action assures improved device reliability, guaranteeing proper operation and performances along their entire lifetime.

Acknowledgements

We would like to thank Dr. Stefano Zilio and his research team for the measurements and characterizations performed in the R&D laboratories of SAES, supporting this work with data and scientific assistance.

References

[1] W.O. Camp, R. Lasater, V. Genova, R. Hume, "Hydrogen effects on reliability of GaAs MMICs," Proceedings of 11th GaAs Integrated Circuit Symposium, pp. 203-206, 1989, DOI: 10.1109/GAAS.1989.69326

[2] Ronald L. Pease, Dale G. Platteter, Gary W. Dunham, John E. Seiler, Philippe C. Adell, Hugh J. Barnaby, Jie Chen, "The Effects of Hydrogen in Hermetically Sealed Packages on the Total Dose and Dose Rate Response of Bipolar Linear Circuits", IEEE Transactions on Nuclear Science, Vol. 54, No. 6, pp. 2168-2173, 2007, DOI: 10.1109/TNS.2007.907870

[3] R.R. Blanchard; J.A. del Alamo; A.C. Calveras, "Hydrogen-induced changes in the breakdown voltage of InP HEMTs2", IEEE Transactions on Device and Materials Reliability, Vol. 5, No. 2, June 2005, DOI: 10.1109/TDMR.2005.846825

[4] M. Albarghouti,N. El Dahdah, G. Perosevic, S. Jain, J.-M. Papillon, and S. Fernandez, "Moisture and Hydrogen Release in Optoelectronic Hermetic Packages", Journal of Microelectronics and Electronic Packaging, 2014, vol. 11, pp. 75-79, DOI: 10.4071/imaps.403

[5] P. Somani, D.P. Amalnerkar, and S. Radhakrishnan, "Effect of moisture (in solid polymer electrolyte) on the photosensitivity of conducting polypyrrole sensitized by prussian blue in solid-state photocells" Synthetic Metals, Vol. 110, No. 3, pp. 181-187, 2000, DOI: 10.1016/S0379-6779(99)00279-9

[6] P. Vacca, M. Riva, "High-Capacity Dispensable Getter", EP3883683

Proceedings of the 7th Optoelectronics Global Conference (OGC 2022)

Modeling and Analysis of High-Speed Modified Uni-Travelling-Carrier Photodiodes Under High Optical Power Injection

Zhangwan Peng[1*], Wanshu Xiong[1], Ruoyun Yao[1], Chaodan Chi[2], Yiti Xiong[2], Chen Ji[1]

[1]College of Information Science and Electronic Engineering, Zhejiang University, Hangzhou, China
[2]Research Institute of Intelligent Networks, Zhejiang Lab, Hangzhou, China
*Corresponding author.
e-mail: pengzhangwan@zju.edu.cn

Abstract—**We simulated InGaAs/InP Modified Uni-Traveling-Carrier Photodiodes (MUTC-PDs) with different reverse bias voltages under high optical power injection, using a commercial device level simulator Apsys. By optimizing the absorber layer structure and the operating voltage, this work achieved an MUTC-PD design with an opto-electric responsivity of 0.16A/W and over 300GHz 3-dB bandwidth under 10 mW/μm² optical injection density.**

Keywords-MUTC-PD; terahertz; high optical power injection

I. INTRODUCTION

Terahertz (THz) technology has attracted more and more interests in recent years due to its various applications, such as broadband wireless communications [1,2], spectroscopic of materials [3] and imaging for security [4]. The uni-traveling photodiode (UTC-PD) is better suited for generating continuous Terahertz wave than the photoconductive switch which is a representative device operation with sharp optical input pulses [5]. The UTC-PD has been used widely in the photomixing process as a key component for it is easier to be built in terms of system configuration [6]. To alleviate the tradeoff between the bandwidth and efficiency of UTC-PDs, MUTC-PDs structure with hybrid absorber are proposed [7,8]. Bandwidth and responsivity are the most important indicators of PD performance. However, the space charge effect is the main limiting factor for achieving higher bandwidth while maintaining a relatively sufficient responsivity under high optical injection conditions. High density photocarriers cannot be swept out of the depletion layer timely and heap up heavily in the transition layer, which in turn causes electric field collapse, thus degrading the transit time [9].

In this paper, high speed MUTC-PD under high optical power injection conditions are achieved by optimizing depletion region structure and increasing reverse bias voltage. The transit-time-bandwidth of the MUTC-PD with 20-μm²-area reaches up to over 300GHz under 10mW/μm² optical density, and the responsivity is 0.16A/W.

II. DEVICE STRUCTURE

The epitaxial layers are fabricated on a semi-insulating InP substrate and grown by low-pressure MOCVD technology, including the lightly doped n-type InP collector,

p-type InGaAs absorber, and InGaAsP block layers. The inverse bias voltage can apply to the coplanar waveguide (CPW) electrodes which are fabricated by magnetron sputtering technology. The structure of the UTC-PD that has a bottom illumination design is shown in Fig. 1. To reduce the junction capacitance, the absorption area was designed to be 20 μm².

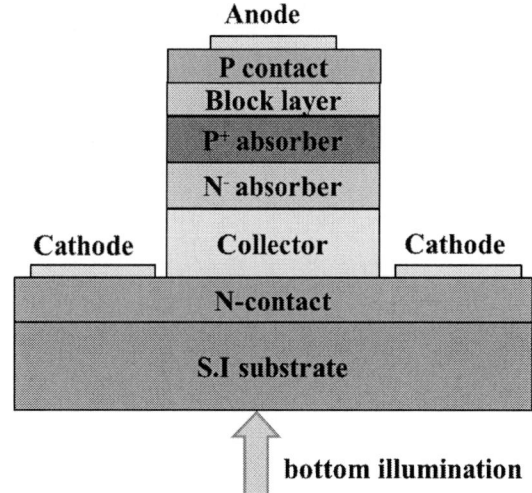

Figure 1. Schematic diagram of a UTC-PD structure

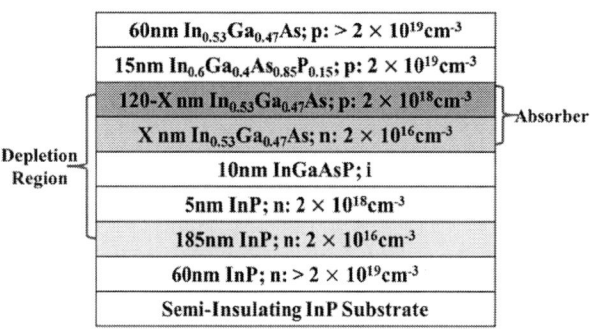

Figure 2. Epitaxy structure of a MUTC-PD

Fig. 2 shows a detailed epitaxy structure of a MUTCPD. The collection layer is slightly n-doped to provide a fixed positive charge background, thus can compensate for the

978-1-6654-8699-6/22 $31.00 © 2022 IEEE

suppression of space charge under high optical power levels. And it serves as a barrier for hole diffusion toward the cathode due to its larger bandgap.

A 5nm cliff layer is inserted to increase the electrical field of the interface. Above the cliff layer, a 10nm undoped InGaAsP quaternary transition layer is grown to form a grading bandgap transition to "smooth" out the heterojunction interface.

The overall thickness of the InGaAs absorber layer is 120nm in order to ensure the adequate light absorption. It contains two parts: the X nm heavy doping p-type (2e18 cm^{-3}) layer will not be depleted at the operating voltage, while the light (120-X) nm doping n-type (2e16 cm^{-3}) layer is the depleted absorber, generating fast drift electrons. The highly p-doped InGaAsP block layer acts as a barrier against the diffusion of electrons towards the anode.

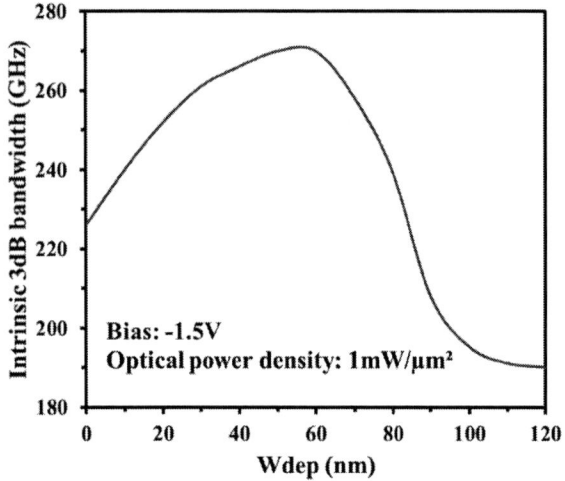

Figure 3. The corresponding transit-time-bandwidth

Fig. 3 displays that the intrinsic 3-dB bandwidth of different device structure. In order to acquire the maximum intrinsic 3-dB bandwidth, the depleted absorber thickness X is varied from zero to 120nm. In the beginning, the bandwidth is 227GHz when the thickness of depleted absorber is zero. Then the bandwidth adds up to its maximum with the thickness around 60nm as the electron diffusion time is dramatic decreased with the increasing of X. Finally, instead of being collected during the dielectric relaxation time, holes in absorption layer begin to drift with saturation velocity leading to a longer transit time. The bandwidth reduces to 190GHz as transport delay of holes turns into the main limiting factor. Consequently, the optimized depleted absorber thickness X is 60nm.

Fig. 4 shows the simulated intrinsic 3-dB bandwidth under the bias voltage of 1.5V. The RC effect limitation is not under our consideration because of its small absorption area.

Therefore, it is estimated that our MUTC-PD displays a transit-limited-bandwidth as large as 270, 248, and 187GHz under a reverse bias voltage of 1.5V at 1, 5, 10mW/µm^2, respectively. The space charge effect is the main reason that

the bandwidth decreases with the increasing of optical power injection, which will be discussed in the next section.

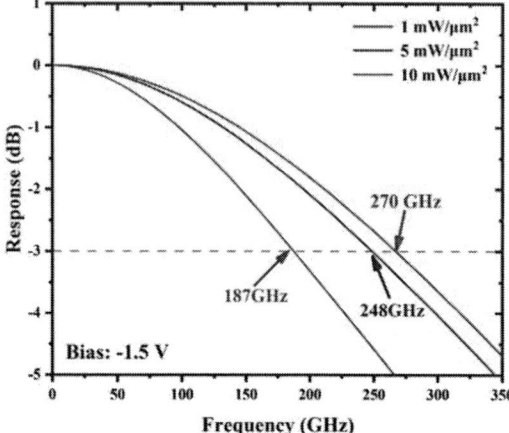

Figure 4. Simulated relative frequency response under different optical density

III. ANALYSIS AND DISCUSSION

Fig. 5 shows the distribution of the carrier concentration when the input optical power density is 1, 5, and 10 mW/µm^2, respectively. Photocarriers are generated rapidly and cannot be swept out in time with the optical input power density increasing in InGaAs absorber layers, low level photo-electrons accumulation is found in the depleted n-InP collector region, strong photo-electrons are heaping up in the narrow InGaAsP transition layer while photogenerated holes accumulation can be observed in the depleted absorber layer.

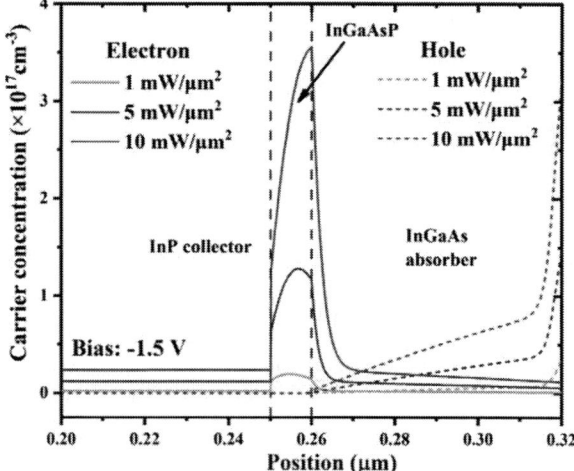

Figure 5. Simulated carrier distribution under different optical density (solid: electron; dash: hole)

IV. OPTIMIZATION AND RESULTS

It is obvious that the severe space charge effect at high optical level lead to the bandwidth decreasing rapidly. To solve this bad phenomenon, a method to increase the E-field in the depletion region by increasing the reverse bias voltage is introduced.

A. Carrier Concentration under Different Reverse Bias Voltage

Fig. 6(a), (b) and (c) shows the electron and hole concentration under different reverse bias voltage. Fig. 6 (a) displays that electron accumulation in transition layer is little under low optical level. Electron concentration increases first and then reduces at the operating voltage of 1.5V, but gradually goes up in a relatively low value at the voltage of 3.0V, the same trend at the voltage of 5.0V and 10.0V. The space charge effect caused by the accumulation of photogenic carriers becomes severe under high optical level, from the Fig. 6(b), it can be seen that when the reverse bias voltage is up to 5V, the electron accumulation presents a linear growth until a fixed value ($0.6e17cm^{-3}$). Although the voltage is increasing further, the accumulation of electrons at the boundary between the transition layer and the collector layer remains at this value, resulting in bandwidth saturation. Fig. 6(c) shows that the value of the operating voltage required to cause the same phenomenon above-mentioned increases further as the optical power density increasing.

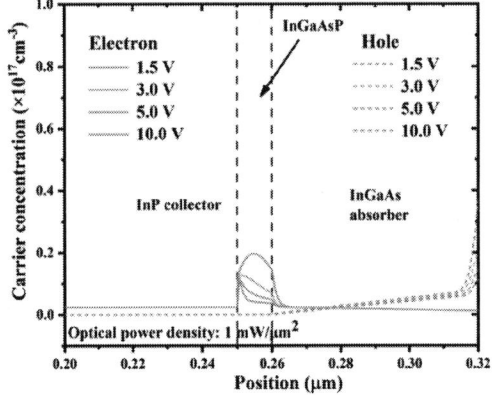

Figure 6. (a) Carrier concentration distribution under 1mW/μm² optical power density

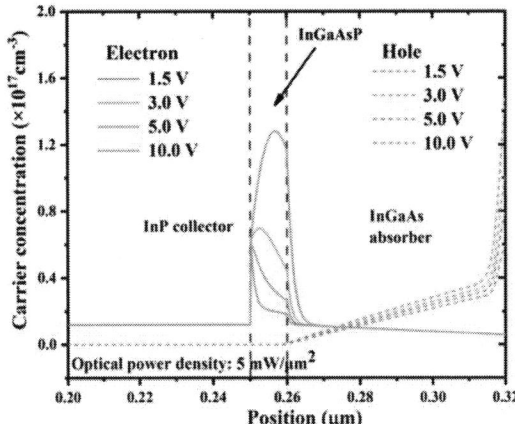

Figure 6. (b) Carrier concentration distribution under 5mW/μm² optical power density

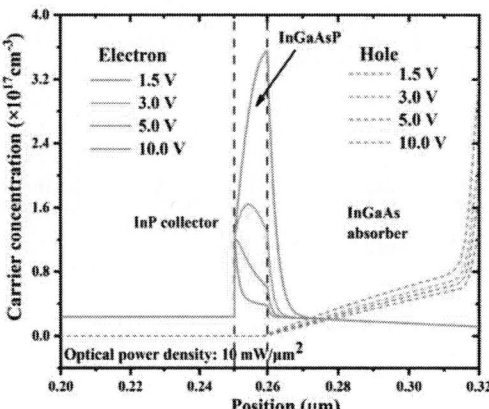

Figure 6. (c) Carrier concentration distribution under 10mW/μm² optical power density

B. Relative Frequency Response under Different Reverse Bias Voltage

Fig. 7 shows the intrinsic 3-dB bandwidth varies with the reverse bias voltage at different optical power densities. The bandwidth first increases with the rise of reverse bias voltage and then tends to saturation. Due to the rise of reverse bias voltage, the accumulation of electrons in the depletion layer will be reduced, resulting in the increment of bandwidth. For the sake of device safety, it is better to achieve a faster response at a relatively low bias voltage as much as possible. The figure shows that a MUTCPD acquired a 3-dB bandwidth over 300GHz under the reverse bias voltage of 3V, 5V, 8V when the input optical power density is set to 1, 5, and 10 mW/μm², respectively.

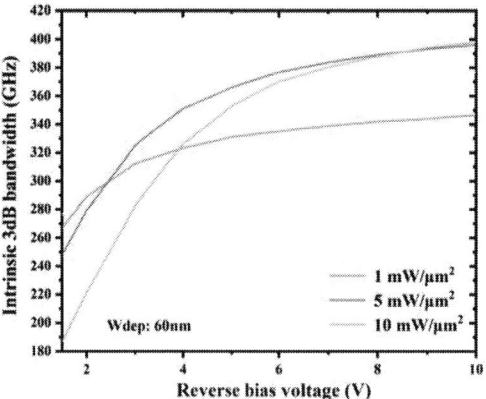

Figure 7. Simulated intrinsic 3dB bandwidth under different reverse bias voltage

V. CONCLUSION

We have investigated the space charge effect on the high-speed performance of Modified UTC-PD under high input optical power. The depleted absorber region in the epitaxy structure and the operating voltage has been carefully analyzed using commercial simulator software to achieve the maximum intrinsic speed response. Our work demonstrated that increasing the reverse bias voltage can accelerate the drift of electrons in the depletion region which will significantly increase UTC-PD frequency bandwidth. As a result, our proposed MUTC-PD, under 1, 5, 10mW/μm^2, exhibits a 3-dB bandwidth over 300GHz at the 3V, 5V, 8V respectively with theoretical responsivity of 0.16A/W at 1.55μm wavelength. The simulation work can be utilized for the fabrication and measurement of high-speed modified uni-traveling-carrier photodiodes.

ACKNOWLEDGMENT

This work was supported in part by the National key research & development (R&D) plan 2020YFB1805701, the National key research & development (R&D) plan 2019YFB2203801, the Zhejiang Lab grant. 2020LC0AD01 /001, the Zhejiang Lab grant. 2020LC0AD02/001, and the National Natural Science Foundation of China under Grants 61974132.

REFERENCES

[1] Koenig S, Lopez-Diaz D, Antes J, et al, "Wireless sub-THz communication system with high data rate," Nature photonics, 2013, 7(12): 977-981.

[2] Shi J W, Huang C B, and Pan C L, "Millimeter-wave photonic wireless links for very high data rate communication," NPG Asia Materials, 2011, 3(4): 41-48.

[3] Nagatsuma T, Ito H, and Ishibashi T, "High-power RF photodiodes and their applications," Laser & Photonics Reviews, 2009, 3(1-2): 123-137.

[4] Dülme S, Grzeslo M, Morgan J, et al, "300 GHz photonic self-mixing imaging-system with vertical illuminated triple-transit-region photodiode terahertz emitters," 2019 International Topical Meeting on Microwave Photonics (MWP). IEEE, 2019: 1-4.

[5] C. C. Renaud, M. Natrella, C. Graham, J. Seddon, F. Van Dijk and A. J. Seeds, "Antenna Integrated THz Uni-Traveling Carrier Photodiodes," in IEEE Journal of Selected Topics in Quantum Electronics, vol. 24, no. 2, pp. 1-11, March-April 2018, Art no. 8500111, doi: 10.1109/JSTQE.2017.2725444.

[6] Ishibashi T, Muramoto Y, Yoshimatsu T, et al, "Unitraveling-carrier photodiodes for terahertz applications," IEEE Journal of Selected Topics in Quantum Electronics, 2014, 20(6): 79-88.

[7] Wakatsuki, T. Furuta, Y. Muramoto, T. Yoshimatsu and H. Ito, "High-power and broadband sub-terahertz wave generation using a J-band photomixer module with rectangular-waveguide output port," 2008 33rd International Conference on Infrared, Millimeter and Terahertz Waves, 2008, pp. 1-2, doi: 10.1109/ICIMW.2008.4665566.

[8] T. Nagatsuma et al., "Giga-bit wireless link using 300–400 GHz bands," 2009 International Topical Meeting on Microwave Photonics, 2009, pp. 1-4.

[9] J. Zhang et al., "Analysis of Carrier Transportation in High Power Uni-Traveling-Carrier Photodiodes based on Self-Consistent Monte Carlo Model," 2020 Asia Communications and Photonics Conference (ACP) and International Conference on Information Photonics and Optical Communications (IPOC), 2020, pp. 1-3.

[10] N. Shimizu, N. Watanabe, T. Furuta, and T. Ishibashi, "Improved Response of Uni-Traveling-Carrier Photodiodes by Carrier Injection," Japanese Journal of Applied Physics, vol. 37, no. Part 1, No. 3B, pp. 1424-1426,19.

Proceedings of the 7th Optoelectronics Global Conference (OGC 2022)

Research and Application System Design of Intelligent Inspection of Multispectral Segment Optoelectronic Devices Based on 5G

Tielin Lu
Instrumentation Technology and Economy Institute,
Beijing, China
lutielin@126.com

Lei Yue
Siemens Industry Software (Beijing) Co., Ltd.,
Beijing, China
yuelein@126.com

Bowen Lu
Beijing Jiaotong University,
Beijing, China
827418257@qq.com

Xiawei Feng
Instrumentation Technology and Economy Institute,
Beijing, China
fxw@tc124.com

Hongqi Han
Institute of Scientific and Technical Information of China,
Beijing, China
bithhq@163.com

Abstract—Massive real-time image test results put forward higher requirements for network transmission and intelligent technology. Intelligent detection technology using 5G network has become one of the most important research directions of multi-spectral optoelectronic devices. However, they are still confused about the architecture of the 5G network and how to design the system with unmanned control for different user needs. At present, it is urgent to solve how to overcome the instability of 5G optoelectronic device network structure through multi-spectral segments, achieve high-quality transmission of results, and obtain intelligent optimal results. This paper summarizes the configuration and mode of 5G network, application system of multispectral optoelectronic devices. This paper also analyzes the intelligent detection system based on 5G scheme. Finally, we have developed a case of multi-spectral segment photoelectric detection architecture based on 5G communication structure. We have validated the application scenario of intelligent detection in the design of transmission line detection and the feasibility purpose of intelligent detection in smart grid.

Keywords-5G network; intelligent inspection; communication structure; smart grid; information systems

I. INTRODUCTION

This paper takes the inspect process of multispectral segment optoelectronic devices as the background. The background of this paper is the detection process of multispectral optoelectronic devices. There are many auxiliary devices in the operation of the power grid system and each of them needs to be monitored [1].One of the main differences between fifth-generation mobile communications (5G) and previous generations of cellular networks is 5G pay more attention to machine communication and Internet of things (IoT).The capabilities of 5G thus extend far beyond mobile broadband with ever increasing data rates. Therefore, with the increasing data rate,5G's capabilities go far beyond mobile broadband In particular, 5G supports communications with unprecedented reliability and extremely low latency, as well as massive IoT

connections[2-4].These features pave the way for many new use cases and applications in many different verticals, including the automotive, healthcare, agriculture, energy, and manufacturing sectors. Especially in the field of grid inspection, 5G and multispectral segment optoelectronics may have an impact on the efficiency of smart grids, as relevant building blocks such as wireless connectivity, edge analysis, optoelectronic device applications or network communications will find their future direction in smart grids.

With the rapid development of electric power inspection business, the demand of unmanned inspection management is increasing, and the scope of inspection is expanding day by day. Photoelectric multi-device combination, daily inspection and alarm monitoring need to be real-time and synchronous. In order to achieve multi-level comprehensive management and coordination, it is urgent to establish a communication system for multi-photoelectric device scheduling. At present, the communication of broadband multispectral imaging detectors is mainly stored at the local end, and the equipment and mission terminals cannot achieve point-to-point communication or networking communication [5-8]; some use satellite communication and network communication [9-14].

In this paper, we provide an overview of 5G's basic potential for network communication, and outline relevant communication architecture and key edge analysis with cloud technology. Furthermore, we introduce some of the main communication structure of data model and certain major challenges that have not yet been resolved. Finally, intelligent inspection of multispectral segment optoelectronic devices system is presented in more detail as an important initiative to ensure that 5G for the smart grid domain.

II. SYSTEM INTRODUCTION AND KEY FUNCTION

The rationale for the development of 5G network was not only to expand the broadband capabilities of communicate network, but also to provide advanced wireless connectivity for a wide variety of vertical industries, which can be easily used in power grid. In order to obtain this, 5G network

978-1-6654-8699-6/22 $31.00 © 2022 IEEE   171

supports three essential types of communication: enhanced mobile broadband (eMBB), massive machine-type communication (mMTC), and ultra-reliable low-latency communications (URLLC)[15].

eMBB provides extremely high data rates (of up to several Gb/s) and offers enhanced coverage, well beyond that of 4G. mMTC is designed to provide wide-area coverage and deep indoor penetration for hundreds of thousands of IoT devices per square kilometer. In addition, mMTC is designed to provide ubiquitous connectivity with low software and hardware requirements from the devices, and will support battery-saving low-energy operation. URLLC can facilitate highly critical applications with very demanding requirements in terms of end-to-end (E2E) latency (down to the millisecond level), reliability and availability.

The current power inspection system has many limitations, such as high communication costs when all 5G communication methods are used. In order to ensure the communication stability of mobile devices and the high rental fees for 5G communication bandwidth, each multi-spectral optoelectronic device is equipped with a 5G communication module, which greatly increases the cost. Moreover, as a power grid maintenance department, it only needs to repair and maintain the abnormal inspection results, and the requirements for multi-level user management and regional management are complex. The traffic fee is also very high. Using the imaging results of multi-spectral optoelectronic devices, after the edge image preprocessing, when combined with 5G communication, the subordinates through the 5G network can avoid the problems of limited network scale and inflexible networking, and can identify the specific location information. Since the power grid is composed of cross-regional and multi-departmental decision-making management, remote interaction is deployed in the decision-making management department, and the management department communicates with various regions through 5G. When inspection faults occur or are about to occur, real-time updates are made through the 5G network in time, match the feedback instructions of the fault knowledge base, and guide the maintenance personnel to carry out the emergency control and operation maintenance of the real-time power grid.

The intelligent inspection system is divided into four parts: multispectral optoelectronic device acquisition layer, edge analysis layer, 5G network transmission layer, and intelligent maintenance application layer. The edge analysis and remote management center and different maintenance departments use 5G networks to achieve wide-area interconnection, and internally use broadband multimedia clusters to establish a central network with local inspection as the core. On this basis, the system application of multi-modal information such as voice, video, data (text, picture, location information) can be further realized [16-17]. The construction of the actual inspection system can rent the base stations, group scheduling gateways, edge node control units, multi-spectral photoelectric detection devices, and equipment such as operation and maintenance applications or real screens. In order to further prevent interference and security

in the inspection communication process, reliability and electromagnetic compatibility tests should be carried out during the communication process, and security measures such as firewalls and user permissions should be set up. This also includes image inspection results characteristics (for example user image data quality) and procedures for 5G network registration, identification, safety and security, etc.

III. 5G Network and Inspection Key Technology

In this parts, we would discuss the network transmission to set up 5G network and also the key technology of inspection system process of multispectral optoelectronic device. With the edge data analysis technology, it needs to be tested to ensure the multi-modal information of the implemented service are fulfilled. Additionally, the architecture should focus the optoelectronic devices performance since it may impact the application results from the application of detection data.

A. 5G Network Model

For the 5G network, there are three parts of the 5G model. As the Fig.1, the 5G network consist with the social grid system, 5G operators and grid operators. The common network share the same cellular network physical architecture. For the grid operators logic isolation, which is realized by network slicing and other technologies.

The social grid system can be safer and more stable inspection connection with the service through the internal network of the grid slice network system, and internal connection is realized on the UAV or vehicle-mounted system. The data detected by multi-spectral optoelectronic devices can be stored in the form of wired or private network to realize local processing.

The 5G operators can use the core admin of the 5G network communication and determination, such as the operating capability, charging admin and give the position of the inspection location information. The 5G operators can be used by the local telecom provider or the power grid operators separate private network, which can be public network and optional connections.

The grid operators collect the relevant data sent by the 5G operator, which is transmitted to different users in stages. For the management department, it mainly establishes functions such as predictive maintenance knowledge base, alarm, logical operation and maintenance, result classification and identification, and sends the processing results to different users through the network in time. The object of the inspection is to track the location found in the inspection in time for the inspection site, maintenance users to deal with necessary minor repairs in a timely manner, and managers to grasp the maintenance progress in a timely manner.

B. Inspection System's Work Process

In the application of inspection system, the business model refers to the data type, bearing protocol, data volume, transmission interval and other business models transmitted from the multispectral optoelectronic device. The end of the output image data to the processing unit of the received data,

978-1-6654-8699-6/22 $31.00 © 2022 IEEE

which are used to assist the analysis of different system architectures. The process to achieve the image as the Fig. 2, which shows the multispectral optoelectronic device working process in the inspection system.

Multi-spectral optoelectronic detection devices are mainly divided into UV-visible detectors and infrared detectors. UV-visible detectors are integrated devices that can simultaneously collect real-time operating images and arc discharge phenomena of transmission and distribution circuits. Infrared detectors are separate thermal imaging devices. It is used to identify heating and abnormal temperature changes on site. Infrared detection devices need to perform a temperature correction process on the identification target. Therefore, infrared and visible/ultraviolet photoelectric devices are generally used in combination and mounted on drones or inspection vehicles.

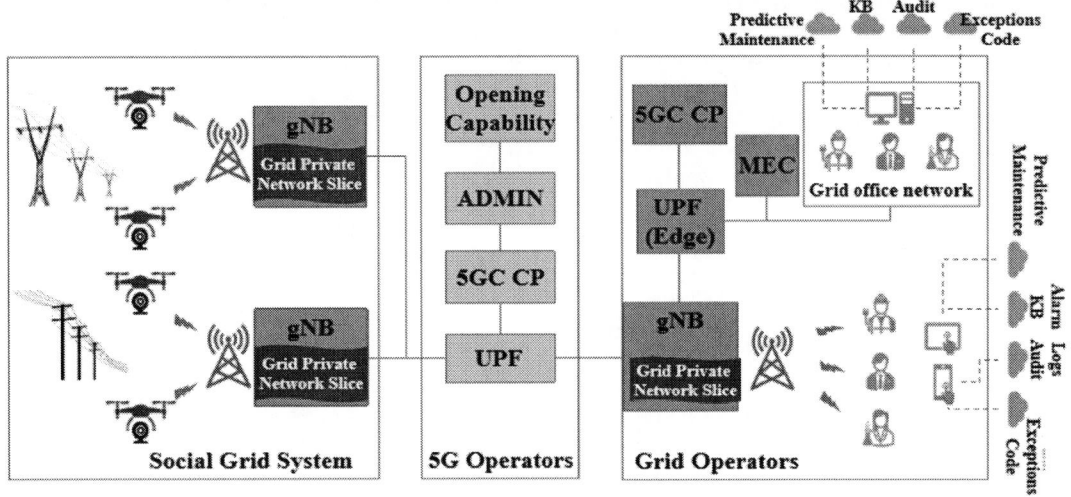

Figure 1. 5G network of the grid system

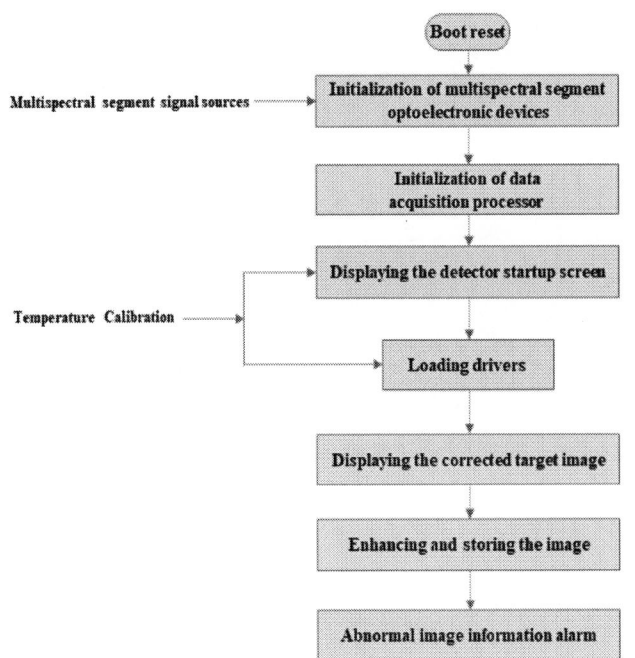

Figure 2. The process of inspection working system

After the detectors are turned on at the same time, the infrared temperature control module first needs to adjust the temperature so that the temperature of the photo detector is stable. During this process, the current will oscillate erratically. It is necessary to initialize each detector and initialize the memory. Connect with the storage unit and link to the client, at the same time calibrate the working temperature, start to correct the working temperature of the infrared detector, collect real-time image information at the same time, and calibrate the uniformity of visible light and infrared images to complete the image correction output, Generally, image correction and system correction tests should be set up. Finally, after the corrected image is completed, the video output display will start. If necessary, the image should be enhanced. The processed image should be stored and compared locally, and abnormal image result data can be found in time.

C. Edge Data Analysis Technology

Edge data analysis technology [18] solves the real-time demand of multi-modal information such as voice, video, data (text, picture, location information). Edge data analysis technology uses near the edge of the inspection detectors, satisfies the requirement of fast response. The current inspection scenarios need to quickly identify malfunction or abnormality position exist in the power grid process, detect the thermal and discharge. It improves the efficiency of detection and needs strong image processing speed. We consider deployment of UPF server in the drones or inspection vehicles. With deployment of intelligent image recognition algorithm in the server, the optoelectronic devices, 5G network, edge recognition analysis technology can complete power line of quickly identify and monitor the grid process and equipment.

By the multispectral detection results and 5G communication technology, it can significantly reduce the missed and false detection. In the case of complex texture image classification and background interference, it realizes real-time synchronization of detection result data. The edge computing realizes efficient and rapid iteration closed-loop of data model and continuously improves the accuracy of field model. Through the big data software system and historical quality data association, which initialize the detection sample database, through depth learning training to collect images data. By manual confirm, we can identify the grid failure and abnormal state, which is used to maintain the equipment iteratively and improve the adaptability and robustness of the detection.

IV. SYSTEM DESIGN AND APPLICATION FUNCTION

To illustrate the architecture that we have developed and how it can be applied to a inspection system, we describe a

5G network model for the inspection application scenario for power grid. The end detection terminal has a variety of service interfaces, including high-definition video, scheduling voice, multi-type image processing, positioning data, etc. In progress of the power grid lines, each of power-related running activities shall be monitored and consequently the set of activities to be recorded in multi-spectral optoelectronic detector, which is realized the efficient image data transmission between recognition and processing system, 5G network, AI vision cloud platform and grid management system.

A. System Architecture

For the 5G communication structure of intelligent system, we builds a five-systems together as the whole architecture. The architecture is shown as Fig.3, it contains end point detector system, edge analysis platform,5G communicate network, AI cloud platform and applications of the monitoring model with the cloud applications.

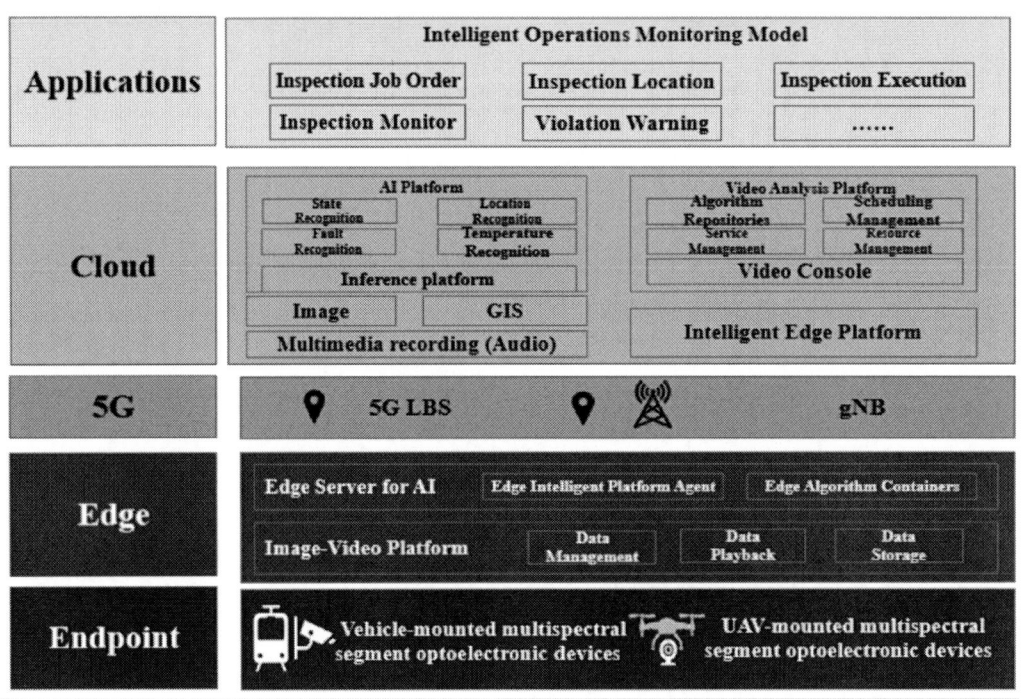

Fig. 3. 5G with industrial vision detection system architecture

These system can be considered within the scope of real-time communication and non-real-time mode. The image recognition management system can set up the determined training model and analysis the acquisition of images for inference prediction, and feedback results to the realization of related power grid equipment maintain functions. The AI cloud platform can use the computer video analysis model, algorithm base and sample image, combined with deep

learning AI algorithm, to complete the functions of data pre-processing, data annotation and training to generate real-time remote or predictive maintenance model.

B. System Functions

The main functions of intelligent inspection include the ability to access other networks (such as the intranet of the power system and the public Internet) to ensure that the inspection system can directly connect to other networks. The cloud platform can realize point-to-point video

---

This work has been supported in part by the National Key Research and Development Program of China (2020YFB2009304)

conferencing, video surveillance, voice and text communication for multi-level users at the edge, management department, and maintenance department. In principle, the multi-spectral inspection equipment only realizes communication through the edge to ensure the independence of the result data. The 5G network can adopt the central networking method. The base station can manage multiple edge inspection terminals at the same time. The collection results of all inspection terminals will be automatically integrated into one video and sent back to the cloud platform through rights management when necessary. At the same time, there should be different inspection edges that can realize 5G communication, so as to ensure that the inspection data in the area can be synchronized and shared in time, avoid the occurrence of repeated inspections, and facilitate the interconnection and unity of inspection results.

V. Conclusion and Directions for Further Research

In this research an approach of intelligent inspection with 5G network, the miss inspection can be greatly reduced in the case of complex real-time image classification and background interference, the inspect process can be accurately controlled, and the real-time synchronization of the inspection results can be realized synchronously. The intelligent inspection system can be iteratively closed loop efficiently and quickly to improve the accuracy of the smart grid. It can be easily made that multispectral segment optoelectronic devices combination edge data analysis, and 5G communicate network will become principal criteria to be considered for future inspection design and development for varies industries not only in grid scenario. We are setting up a 5G and inspection test system for applications based on our design architecture, and evaluated it by the grid scenario.

In the implementation scheme, the network can realize the image acquisition and upload the image data through 5G network by the edge data analysis platform to complete the video detection. In order to reduce the end cost and reduce the difficulty of deployment, 5G modules can be integrated into the equipment to flexibly adapt to various grid monitoring application scenarios processing. As the high-definition detector has 5G communication ability, direct access 5G network and also the high-definition detector does not have 5G communication ability, uses the edge premises equipment to access 5G network. There are several methods how this research could be continued in the future. Under the new situations of smart data model settled in the edge information platform, we identified capability gaps requiring further study effort in order to obtain the results by real-time work for the scenario. In particular, it would set a bridge that mutual understanding between the 5G communicate technology and inspection application together be facilitated further. This is one of the reasons for establishment of the intelligent inspection system, which has the goal of ensuring that using 5G and intelligent means ultimately becomes a great use case.

Acknowledgmen

All the things should thank to Pro.Chunxi Wang, Pro.Xiaojing liu and Dr.Yujia Shang, who encourage me to do the research and give me hands. This work has been supported in part by the National Key Research and Development Program of China (2020YFB2009304).

References

[1] S. Zhang et al., "Analysis of intelligent inspection program for UAV grid based on AI," 2020 IEEE International Conference on High Voltage Engineering and Application (ICHVE), 2020, pp. 1-4, doi: 10.1109/ICHVE49031.2020.9279634.

[2] L. Fan, J. Li, Y. Pan, S. Wang, C. Yan and D. Yao, "Research and Application of Smart Grid Early Warning Decision Platform Based on Big Data Analysis," 2019 4th International Conference on Intelligent Green Building and Smart Grid (IGBSG), 2019, pp. 645-648, doi: 10.1109/IGBSG.2019.8886291.

[3] M. Xu, "Research on 5G UAV Smart Grid Inspection Technology Based on Digital Twin," 2021 IEEE 7th International Conference on Cloud Computing and Intelligent Systems (CCIS), 2021, pp. 367-370, doi: 10.1109/CCIS53392.2021.9754652.

[4] Y. Fengchun, H. Ninghui and W. Xutao, "Research on Power Grid Inspection Path Based on Edge Computing," 2020 International Conference on Urban Engineering and Management Science (ICUEMS), 2020, pp. 231-234, doi: 10.1109/ICUEMS50872.2020.00058.

[5] T. Xiao-Xu, M. Bao-Kun, G. Li-Yuan, C. Jin-Pei and Y. Xi-Ming, "An Intelligent Inspection Robot of Power Distribution Network Based on Image Automatic Recognition System," 2018 China International Conference on Electricity Distribution (CICED), 2018, pp. 97-100, doi: 10.1109/CICED.2018.8592477.

[6] L. Ruchao, P. Chigang, Z. Feng, W. Yong and L. Pingyuan, "EMC analysis of power grid intelligent inspection equipment based on finite element method," 2017 EPTC Power Transmission and Transformation Technology Conference, 2017, pp. 1-6, doi: 10.1049/cp.2017.0558.

[7] L. Jiezhong, L. Yong and Z. Linjiang, "Design and Research of Intelligent Inspection Management System for Distribution Network (CICED 2020)," 2021 China International Conference on Electricity Distribution (CICED), 2021, pp. 27-32, doi: 10.1109/CICED50259.2021.9556721.

[8] R. Liu, X. Hai, S. Du, L. Zeng, J. Bai and J. Liu, "Application of 5G network slicing technology in smart grid," 2021 IEEE 2nd International Conference on Big Data, Artificial Intelligence and Internet of Things Engineering (ICBAIE), 2021, pp. 740-743, doi: 10.1109/ICBAIE52039.2021.9389979.

[9] B. Priya and J. Malhotra, "An Intelligent User-RAT association for 5G enabled Smart Grid," 2020 IEEE International Conference on Computing, Power and Communication Technologies (GUCON), 2020, pp. 300-304, doi: 10.1109/GUCON48875.2020.9231264.

[10] M. Forcan, M. Maksimović, J. Forcan and S. Jokić, "5G and Cloudification to Enhance Real-Time Electricity Consumption Measuring in Smart Grid," 2020 28th Telecommunications Forum (TELFOR), 2020, pp. 1-4, doi: 10.1109/TELFOR51502.2020.9306518.

[11] ISO/IEC 9646-1:1994, "Information technology - Open Systems Interconnection – Conformance testing methodology and framework - Part 1: General concepts".

[12] Zhang, Xiaoling , et al. "Adaptive Frame Rate Optimization Based on Particle Swarm and Neural Network for Industrial Video Stream." 2019 24th IEEE International Conference on Emerging Technologies and Factory Automation (ETFA) IEEE, 2019.

[13] Lu, Tielin , et al. "Recent research in data description of the measurement property resource on common data dictionary." Young Scientists Forum Young Scientists Forum 2017, 2018.

[14] Lu, Tielin , et al. "Research on industrial robot reducer state monitoring cloud platform ."Proceedings Volume 11565, AOPC 2020: Display Technology; Photonic MEMS, THz MEMS, and Metamaterials; and AI in Optics and Photonics; 115650E 2020.

[15] T. Lu, Z. Fan, Y. Lei, Y. Shang and C. Wang, "The Edge Computing Cloud Architecture Based on 5G Network for Industrial Vision Detection," 2021 IEEE 6th International Conference on Big Data Analytics (ICBDA), 2021, pp. 328-332, doi: 10.1109/ICBDA51983.2021.

[16] Z. Zhou, W. Wang, F. Dong, J. Jiang and M. Kavehrad, "Application of Optical Wireless Communications for the 5G enforced Smart Grids," 2020 IEEE 4th Conference on Energy Internet and Energy System Integration (EI2), 2020, pp. 756-760, doi: 10.1109/EI250167.2020.9347307.

[17] V. Casella, M. Franzini and M. E. Gorrini, "Crop marks detection through optical and multispectral imagery acquired by UAV," 2018 Metrology for Archaeology and Cultural Heritage (MetroArchaeo), 2018, pp. 173-177, doi: 10.1109/MetroArchaeo43810.2018.13615.

[18] Ali, M. , et al. "Edge Enhanced Deep Learning System for Large-scale Video Stream Analytics." 2018 IEEE 2nd International Conference on Fog and Edge Computing (ICFEC) IEEE Computer Society, 2018.

Proceedings of the 7th Optoelectronics Global Conference (OGC 2022)

Portable Microscopic Phase Retrieval System Using the Transport of Intensity Equation on Android Platform

Yu Chen, Hong Cheng[*], Zhengguang Tian, Xunting Yang, Fen Zhang, Wei Li

School of Electronic and Information Engineering

Anhui University

Hefei, China

e-mail: chenghong@ahu.edu.cn

Abstract—The microscope used in the traditional phase retrieval method based on the transport of intensity equation (TIE) is not portable enough. In this paper, a microscopic phase retrieval software and hardware system with good portability, high precision and strong adaptability is designed under the Android system. The intensity images are captured by the system which uses a smart phone with a portable microscope, and controls the lifting and lowering of the stage through a high-precision motor. Then the self-designed Android application software calculate the phase of the obtained image by using TIE method. Finally, the effectiveness of the system is verified by simulation experiments and experimental measurements.

Keywords-phase retrieval; transport of intensity equation; android system; portable microscope

I. INTRODUCTION

The monochromatic coherent light wave field can be completely described by a two-dimensional complex amplitude function, which amplitude part [1] is the phase, and the square of the amplitude is the intensity. Most of the information carried in an object is stored in the phase [2], but the human eye and other detectors can only identify changes in amplitude but not in phase, resulting in inability to distinguish parts of an object with different thicknesses or refractive indices. How to use the intensity information to obtain the lost phase is an especially important problem in the field of phase retrieval.

TIE is a quantitative non-interference phase retrieval method derived by Teague [3] in 1983 under the paraxial approximation condition. The equation describes the quantitative relationship between the intensity differential component in the direction of the optical axis and the phase of the light wave in the plane perpendicular to the optical axis. The phase retrieval technique based on TIE is widely used in the field of optical measurement [4], X-ray [5], electron microscope [6], computer imaging [7] and so on.

The axial intensity differential in TIE can be approximated by an intensity difference component in the plane perpendicular to the optical axis. Therefore, it is usually necessary to collect in-focus, under-focus and over-focus images for calculation in experiments [8]. Hong Cheng [9] of Anhui University used a binocular microscope based on registration restoration for phase retrieval. This method is

only suitable for laboratory conditions, requiring high microscopy equipment, which are not easy to carry. Bogoch [10] proposed a mobile phone microscope, but it was based on the traditional microscopes, which are ineffective for colorless objects. Meng [11] proposed a smartphone-based hand-held quantitative phase microscope system, which used a smartphone to capture in-focus, under-focus, and over-focus images for phase retrieval. But the system required manual focus to capture the image, which lacked precision.

In view of the above problems, this paper proposes a microscopic phase retrieval system based on TIE under the Android system: First, a smartphone is used in conjunction with a portable microscope system, and a single-chip microcomputer is used to control a high-precision motor to lift the stage to capture in-focus, under-focus, and over-focus images. The captured images are then phase-retrieved by using a self-developed Android application. Finally display and save the retrieval results inside the application. The system has good portability, high cost-performance and high accuracy.

II. THE TRANSPORT OF INTENSITY EQUATION

A. Definition of the Transport of Intensity Equation

By deriving the wave equation under the condition of paraxial approximation, the TIE in the monochromatic coherent light field can be obtained. This equation represents the relationship between the amount of change in light intensity in the direction of the optical axis and the phase of the light wave in the plane perpendicular to the optical axis:

$$-k\frac{\partial I(x,y,z)}{\partial z} = \nabla_\perp \left[I(x,y,z)\nabla_\perp \varphi(x,y,z) \right] \quad (1)$$

Where $I(x,y,z)$ represents the light intensity at the focus in the vertical direction of the optical axis; $\nabla_\perp$ represents 2D gradient operator; z represents transmission direction; k represents a wave number, and $k = 2\pi/\lambda$, where λ is the wavelength of light; $\varphi(x,y,z)$ represents the phase distribution of the object. $\dfrac{\partial I(x,y,z)}{\partial z}$ is the axial strength differential, which can be approximated by the finite difference method, and the results are shown in Fig. 1.

978-1-6654-8699-6/22 $31.00 © 2022 IEEE

Let $I(x,y,z-\Delta z)$ and $I(x,y,z+\Delta z)$ be the under-focus and over-focus intensities of the defocus distance Δz, respectively, then the relationship between the axial intensity differential and the over-focus and under-focus strengths is:

$$\frac{\partial I(x,y,z)}{\partial z} \approx \frac{I(x,y,z+\Delta z)-I(x,y,z-\Delta z)}{2\Delta z} \quad (2)$$

Substituting Equation (2) into Equation (1), the TIE in the monochromatic coherent light field is obtained:

$$-k\frac{I(x,y,z+\Delta z)-I(x,y,z-\Delta z)}{2\Delta z} = \nabla_\perp\left[I(x,y,z)\nabla_\perp\varphi(x,y,z)\right] \quad (3)$$

When the light field to be measured is a quasi-monochromatic light field, the light wave field approximately only contains a single light frequency. Using the Wigner distribution function [12] $W(x,u)$, the generalized TIE of the spatially partially coherent light field can be obtained:

$$\frac{\partial I(x)}{\partial x} = -\lambda\nabla\cdot\int uW(x,u)\mathrm{d}u \quad (4)$$

In the formula: u is the spatial frequency coordinate corresponding to x. When a monochromatic light wave is completely coherent in space, the relationship between the first-order conditional spatial frequency moment of its Wigner function and the lateral gradient of the phase in the case of partial coherence:

$$\frac{\int uW(x,u)\mathrm{d}u}{\int W(x,u)\mathrm{d}u} = \frac{1}{2\pi}\nabla\phi_{\text{out}}(x,y,z) \quad (5)$$

In a partially coherent light field, $\phi_{\text{out}}(x,y,z)$ is a generalized phase, which is the superposition of the phase $\phi_{\text{in}}(x,y,z)$ of the illuminating light and the phase $\phi(x,y,z)$ of the object:

$$\phi_{\text{out}}(x,y,z) = \phi_{in}(x,y,z)+\phi(x,y,z) \quad (6)$$

When the primary light source satisfies the condition of optical axis symmetry, that is, $\phi_{\text{in}}(x,y,z)=0$, the phase of the partially coherent light is consistent with the traditional phase, and the corresponding the TIE is exactly the same as that of the fully coherent illumination. Therefore, the unknown quantity of phase can be obtained by solving equation (3). The classic solution methods include fast Fourier transform method [13], Green's function method [14], multigrid method [15], and the most commonly used is fast Fourier transform method, which has the characteristics of simplicity and high efficiency.

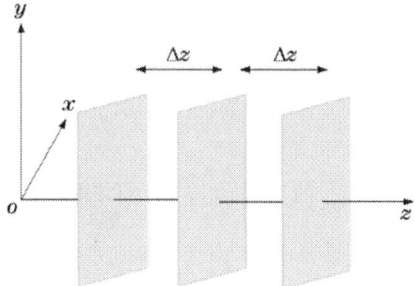

Figure 1. Schematic diagram of light intensity change

B. Fourier Solution of the Transport of Intensity Equation

Assuming that the intensity of the light wave field is uniform, that is, the intensity I_0, an auxiliary function $\nabla\phi(x,y,z)=I(x,y,z)\nabla_\perp\varphi(x,y,z)$ is introduced, and the transport of intensity equation (1) can be expressed by two Poisson equations:

$$-k\frac{1}{I_0}\frac{\partial I(x,y,z)}{\partial z} = \nabla^2\varphi(x,y,z) \quad (7)$$

$$\nabla\cdot\left(\frac{\nabla\phi(x,y,z)}{I_0}\right) = \nabla^2\varphi(x,y,z) \quad (8)$$

When the object is a pure phase object, the phase $\varphi(x,y,z)$ can be obtained by solving the above two equations.

Combining formulas (2) and (8), the object phase is obtained by two Fourier solutions:

$$\varphi(x,y,z_0) \approx \frac{k}{I_0}\mathrm{J}^{-1}\frac{\mathrm{J}\left[I(x,y,z_0+\Delta z)-I(x,y,z_0-\Delta z)\right]}{4\pi^2\left(f_x^2+f_y^2\right)\Delta z} \quad (9)$$

Where, J and J^{-1} represent the forward and inverse Fourier transforms, respectively, f_x and f_y represent the frequency domain coordinates.

III. MICROSCOPIC PHASE RETRIEVAL SYSTEM UNDER ANDROID SYSTEMEM

The system consists of a hardware microscopic imaging part and a software phase retrieval part. The two cooperate with each other and can accurately perform phase retrieval. First, install the mobile phone on the portable microscopic imaging system, and use the high-precision motor to move the stage up and down to adjust the focal length. Then use the phone camera to capture in-focus, under-focus, and over-focus images. Finally, the self-designed Android application is used to process the intensity image, and the high-precision phase retrieval result is obtained.

A. The Hardware Design of Microscopic Phase Retrieval System

Fig.2(a) shows the hardware design of the microscopic phase retrieval system, which consists of the following parts: (1) Control component (2) Platform drive component (3) Light source component (4) Microscope component (5) Mobile phone support component. The optical path of the microscope imaging system module designed in this paper is shown in Fig.2(b): The light emitted by the LED point light source is firstly diffused into a uniform and bright beam by the diffuser plate (P1), and then filtered by the filter (P2) to obtain monochromatic coherent light to illuminate the sample. The objective lens (L1) and eyepiece (L2) constitute a 4f imaging system to ensure the equivalence of the object plane and the image plane, and finally obtain the required image through the mobile phone camera. Fig.2(c) is the picture of hardware.

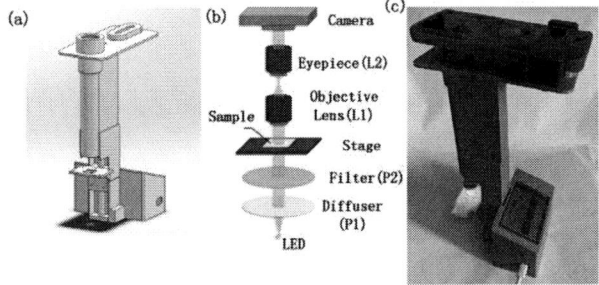

Figure 2. Microscopic imaging equipment. (a) 3D rendering; (b) optical path diagram; (c)hardware

1) Control component

We used a single-chip microcomputer (manufacturer: STMicroelectronics, model: STM32F407ZGT6) as the controller (as shown in Fig.3), which interacts with the user through a touch screen. The user only needs to control the stepper motor platform through the touch screen to make it reach the focus position, and then input the defocus distance Δz in the controller system interface, and the control system will automatically send out control commands to drive the stage to move to the corresponding position. The power supply of the controller adopts the current mainstream PD power supply protocol and supports 9-20V wide voltage input. Any power adapter that can use the PD protocol can be used in this system.

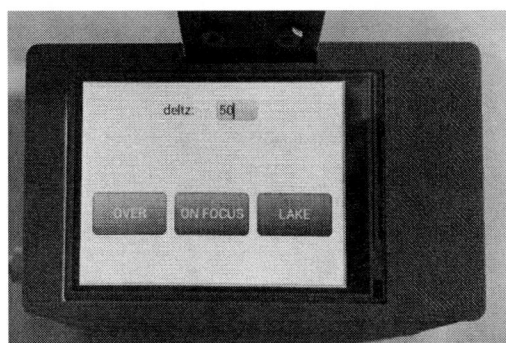

Figure 3. controller

2) Platform drive component

In the process of phase retrieval, in addition to collect the in-focus image at the focus position, it is also necessary to collect the under-focus and over-focus images with a distance of $\pm \Delta z$ from the focus plane. The accuracy of the distance Δz from the focal plane at the time of under-focus and over-focus image acquisition has a direct impact on the error margin of the retrieval result. Therefore, in order to accurately measure the moving distance of the focusing plane of the object, we use a high-precision stepping motor (28*30 size, step angle 1.8°), with a high-performance stepping motor driver chip (manufacturer: TRINAMIC, model: TMC2209) and precision screw structure (specification T6*1-50mm). The theoretical control precision of high-precision stepper motor can reach 0.16um.

3) Light source component

In this system, we used LEDs (shown in Fig.4) as the light source to provide partially coherent light. The light source adopts a replaceable modular design to meet the situation that the light source of different wavelengths may need to be replaced when shooting different objects to ensure the accuracy of the retrieval results. The LED light-emitting components are individually soldered on a PCB, connected to the control part through re-pluggable connectors (PH2.0-2P), and connected to the stage on which the measured object is placed through M3 screws. A filter slot is designed on the stage, and the user can use the filter to obtain a single wavelength of light, which can further improve the retrieval accuracy.

Figure 4. LED light source

4) Microscope component

The microscope component part of the system uses a 3D-printed housing to assemble the individual components. In addition, the standard interface specified by the national standard (GB/T 22055.2-2008, GB-T 22132-2008) is adopted at the interface, which means that users can choose different objective and eyepiece groups according to different materials to be observed. In the experiment, in order to ensure that the edge part of the field of view captured by the smartphone does not have image field curvature, a flat-field achromatic objective lens (MRN70040, 4×) and a low chromatic aberration plan eyepiece (WF10X, 10×) are used to guarantee the imaging accuracy of the edge part.

5) Mobile phone support component

Considering the complexity of the mobile phone model, the camera positions and mobile phone sizes of different mobile phones are quite different. The mobile phone support component still adopts a modular design, and only needs to modify the frame and other positions on the existing template to adapt to different types of mobile phones.

B. The Software Design of Microscopic Phase Retrieval System

We designed an Android phase retrieval program (as shown in Fig.5), which cooperated with the above-mentioned microscopic phase imaging equipment to build a portable and high-precision microscopic phase retrieval system under the Android system. This system uses the mobile phone as the medium for data acquisition, and uses the Java language to write software in the programming environment of Android Studio. The flow chart of the software is shown in Fig.6.

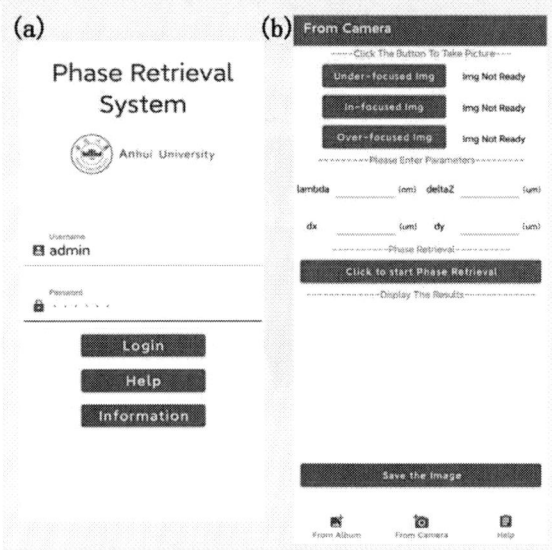

Figure 5.   Phase retrieval Android app. (a) login interface; (b) phase retrieval interface

The user can call the camera by clicking the buttons of "Under-focused Img", " In-focused Img" and " Over-focused Img" in the program's "Use camera to take pictures for phase retrieval" interface (as shown in Fig.5(b)). With the portable microscope imaging equipment matched with the software, the in-focus image, under-focus image and over-focus image required for phase retrieval can be captured. And the user can also directly select the corresponding image from the album for phase retrieval on the interface of "Select photos from album for phase retrieval". Then, the user enters the parameters required for the calculation: lambda (wavelength of the shooting light source), deltaZ (changeable distance of the focal length), dx (sampling period in the x direction) and dy (sampling period in the y direction). After all the pictures and parameters are input, click the "Click to start Phase Retrieval " button to start the phase retrieval, and the retrieval result will be displayed on this interface. Finally,

users can click the "Save the Image" button to save the retrieval photo to the album.

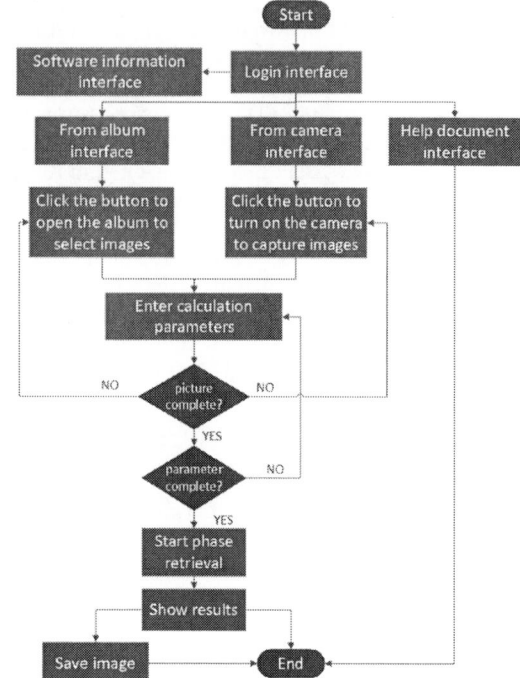

Figure 6.   The flow chart of the software

IV.   EXPERIMENT AND ANALYSIS

A.   Simulation Experiment

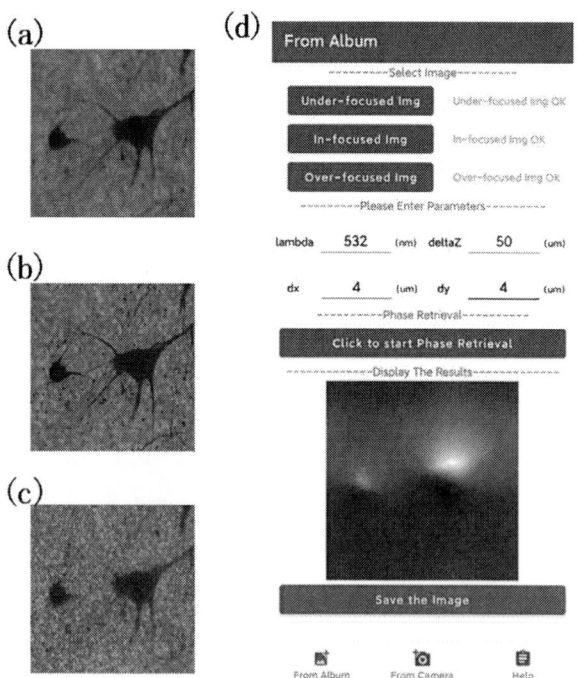

Figure 7.   Phase retrieval of neuronal slices. (a)under-focus image;(b) in-focus image;(c) over-focus image;(d) retrieval result

The first set of experiments tests the phase retrieval effect of the system software. The simulation experiment assumes that the nerve cell slice shown in Fig.7(b) is the corresponding in-focus image, the light wavelength is set to 532nm, the defocus distance is set to 50um, and the under-focus image and over-focus image obtained by angular spectrum propagation are shown in Fig.7(a) and Fig.7(c). The results obtained using the system software are shown in Fig.7(d).

B. Experimental Measurements

In the experimental measurements, the filter with the center wavelength of 532nm is used, and the defocus distance is set to 50um to conduct experiments on the test object. The test object is a lens array made of drops of silicone oil with a refractive index of 1.579. The filling material of the template is polydimethylsiloxane with a refractive index of 1.403, and the maximum thickness of each lens is about 1.15mm. First install the filter on the portable microscope, and then install the object to be measured on the stage. The defocus distance was then set in the controller, and the in-focus, under-focus and over-focus images were taken using the microphase retrieval program. Finally, the phase retrieval can be performed by inputting the experimental parameters in the software.

There is a certain relationship between the physical thickness of the object to be measured and the obtained phase $\varphi(x, y)$ and refractive index [16]:

$$L(x, y) = \frac{\lambda}{n_o - n_m} \frac{\varphi(x, y)}{2\pi}$$

(10)

n_m is the refractive index of the medium around the test object, and n_o is the refractive index of the test object. The specific depth value of the micro-lens array can be calculated according to formula (10).

We measured a single micro-lens, and Fig.8 shows the measurement results. Fig.8(a) is the under-focus image, Fig.8(b) is the in-focus image and Fig.8(c) is the over-focus image. The phase retrieval results of the above three images are shown in Fig.8(d). And Fig.8(e) is the depth value along the line in Fig.8(d), the depth value measured by the proposed method is about 1.21mm, and the relative error between it and the real height is 5.2%.

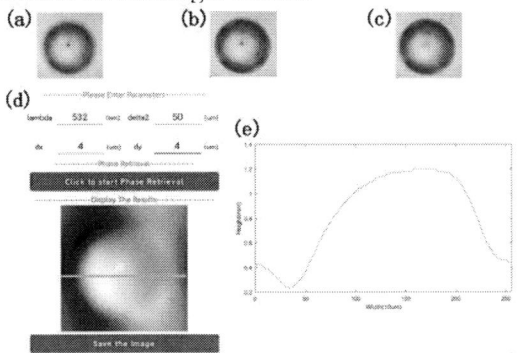

Figure 8. Micro-lens phase retrieval. (a)under-focus image;(b) in-focus image;(c) over-focus image; (d) retrieval result;(e) the retrieval phase of the dashed part in (d)

V. CONCLUSIONS

In this paper, a microscopic phase retrieval system under Android is designed. The system uses a smart phone with a 3D printed portable microscope system and uses a single-chip microcomputer to control a high-precision motor to raise and lower the stage to capture in-focus, under-focus, and over-focus images. Then use the self-developed Android application to perform phase retrieval on the captured images, and finally display and save the retrieval results inside the application. And simulation experiments and experimental measurements are given to verify that the system has good portability, high cost-performance and high accuracy.

ACKNOWLEDGEMENT

The research was supported by the Natural Science Project of Anhui Higher Education Institutions of China (No. KJ2020ZD02, KJ2019ZD04), Nature Science Foundation of Anhui Province, China (No. 2008085MF209).

REFERENCES

[1] Cowley J M. Diffraction physics[M]. Elsevier, 1995.J. Clerk Maxwell, A Treatise on Electricity and Magnetism, 3rd ed., vol. 2. Oxford: Clarendon, 1892, pp.68–73.

[2] Zuo C, Li J, Sun J, et al. Transport of intensity equation: a tutorial[J]. Optics and Lasers in Engineering, 2020, 135: 106187, doi:10.1016/j.optlaseng.2020.106187.

[3] Teague M R. Deterministic phase retrieval: a Green's function solution[J]. JOSA, 1983, 73(11): 1434-1441, doi:10.1364/JOSA.73.001434.

[4] Gong Q, Wei Q, Xu J, et al. Digital field of view correction combined dual-view transport of intensity equation method for real-time quantitative imaging[J]. Optical Engineering, 2018, 57(6): 063102, doi:10.1117/1.OE.57.6.063102.

[5] Thompson D A, Nesterets Y I, Pavlov K M, et al. Fast three-dimensional phase retrieval in propagation-based X-ray tomography[J]. Journal of synchrotron radiation, 2019, 26(3): 825-838, doi:10.1107/S1600577519002133.

[6] Omar E Z, El-Bakary M A. An immersion microscopy method for determining the optical anisotropy in fibres using transport intensity equation technique[J]. Applied Physics B, 2019, 125(8): 1-12, doi:10.1007/s00340-019-7268-y.

[7] Zuo C, Chen Q, Tian L, et al. Transport of intensity phase retrieval and computational imaging for partially coherent fields: The phase space perspective[J]. Optics and Lasers in Engineering, 2015, 71: 20-32, doi:10.1016/j.optlaseng.2015.03.006.

[8] Lu Z, Qijian T, Dingnan D, et al. Field-of-View Correction for Dual-Camera Dynamic Phase Imaging Based on Transport of Intensity Equation[J]. Chinese Journal of Lasers, 46(8): 0804005, doi: 10.3788/CJL201946.0804005.

[9] Hong C, Qiyang Z, Chuan S, et al. Dual-Camera Phase Retrieval Based on Registration Restoration[J]. Acta Optica Sinica,41(12):1210002, doi:10.3788/AOS202141.1210002.

[10] Bogoch I I, Koydemir H C, Tseng D, et al. Evaluation of a mobile phone-based microscope for screening of Schistosoma haematobium infection in rural Ghana[J]. The American journal of tropical medicine and hygiene, 2017, 96(6): 1468, doi:10.4269/ajtmh.16-0912.

[11] Meng X, Huang H, Yan K, et al. Smartphone based hand-held quantitative phase microscope using the transport of intensity equation method[J]. Lab on a Chip, 2017, 17(1): 104-109, doi:10.1039/C6LC01321J.

[12] Xingyuan L, Chengliang Z, Yangjian C. Research Progress on Methods and Applications for Phase Reconstruction Under Partially

978-1-6654-8699-6/22 $31.00 © 2022 IEEE

Coherent Illumination[J]. Chinese Journal of Lasers, 2020, 47(5): 0500016, doi: 10.3788/CJL202047.0500016.

[13] Allen L J, Oxley M P. Phase retrieval from series of images obtained by defocus variation[J]. Optics communications, 2001,199(1-4):65-75, doi:10.1016/S0030-4018(01)01556-5.

[14] Woods S C, Greenaway A H. Wave-front sensing by use of a Green's function solution to the intensity transport equation[J]. JOSAA,2003,20(3):508-512, doi:10.1364/JOSAA.20.000508.

[15] Xue B, Zheng S, Jiang Z. Phase Retrieval Using Transport of Intensity Equation Solved by Full Multigrid Method[J]. Acta Optica Sinica, 2009, 29(6): 1514.

[16] Cheng H, Lv Q, Wei S, et al. Rapid phase retrieval using SLM based on transport of intensity equation[J]. Infrared and Laser Engineering,2018,47(7):0722003, doi:10.3788/irla201847.0722003.

Proceedings of the 7th Optoelectronics Global Conference (OGC 2022)

Visible and Infrared Luminescence and Applications of Er-doped AlN Thin Films

Zhiyuan Wang[a], Feihong Zhang[a], Sergii Golovynskyi[a], Zhenhua Sun[a], Baikui Li[a], Honglei Wu[a,*]

[a]Key Laboratory of Optoelectronic Devices and Systems of Ministry of Education and Guangdong Province, College of Physics and Optoelectronic Engineering, Shenzhen University, 518060, Shenzhen, P.R. China

*Corresponding author.

E-mail addresses: hlwu@szu.edu.cn

Abstract— **An Er^{3+}-doped AlN film has been prepared by radio frequency magnetron sputtering and its photoluminescence (PL) characteristics were studied. It shows high PL efficiency in a wide range of the visible (540 and 560 nm) and infrared (817, 864, 980 and 1534 nm) ranges. In the transition of 4f levels of Er^{3+}, the PL from $^2H_{11/2} \rightarrow {}^4I_{15/2}$ and $^4S_{3/2} \rightarrow {}^4I_{15/2}$ shows obvious temperature dependence, which exhibits a functional relationship between their PL intensity ratio I$_{540nm}$/I$_{560nm}$ and temperature over 100-550 K. It has a relatively high sensitivity (little above 0.01 K^{-1}) among the currently researched temperature sensing materials and can be used as a new contactless temperature sensor in a harsh environment. In the field of infrared luminescence, 864 and 1534 nm are the low loss transmission window in silica fibers and suitable for communication systems.**

Keywords-AlN; Erbium; Visible and infrared; Temperature sensor; Communication wavelength

I. INTRODUCTION

AlN is the great advance semiconductor with a wide bandgap of 6.2 eV and excellent photoelectric performances[1,2]. Erbium ions (Er^{3+}) have been widely studied because of their excellent luminescence properties[3-5]. In recent years, Er^{3+} has been widely studied in contactless optical temperature sensing based on different materials[6-8]. AlN has very stable physical properties, which is very suitable as Er doped matrix to prepare optical sensor.

In this paper, the luminescence characteristics of AlN:Er^{3+} are studied, which shows a very wide luminous range (from 535 to 1540 nm). As a temperature sensing, the ratio of 540 nm (I_{540nm}) to 560 nm luminous intensity (I_{560nm}) as a function of temperature was determined, and the temperature detection sensitivity is calculated. In addition to its stable detection at high temperature, its sensitivity is several times higher than that of other reported Er-doped materials. The results show that Er^{3+}-doped AlN film has great potential in the application of optical temperature sensing. Moreover, infrared luminescence in 816 and 1534nm make AlN:Er have the potential to be used as a communication materials.

II. PREPARATION AND CHARACTERIZATION OF THIN FILMS

AlN:Er^{3+} films were prepared on a sapphire (c-plane) substrate by radio frequency magnetron sputtering. The target used an Al:Er^{3+} alloy (99.9 %) with a concentration of Er of 2 at%. The deposition pressure in a chamber was 0.3 Pa and the deposition temperature was 350 °C. The gas atmosphere in the chamber was nitrogen (99.999%) and argon (99.999%), with the flow velocities of 10 SCCM and 20 SCCM, respectively. The

sputtering time of the film was 20 min and the power was 150 W.

Table I and Fig. 1 show the EDS composition characterization of the AlN:Er^{3+} film. The ratio of Al to N is close to 1:1, and the concentration of Er is about 0.9 at%, the oxygen impurity may come from the AlN oxidation.

TABLE I.   COMPOSITION CHARACTERIZATION OF THIN FILMS BY EDS

Element	Weight%	Atomic%
Al	56.02	45.28
N	32.70	49.17
O	3.68	4.85
Er	6.59	0.90

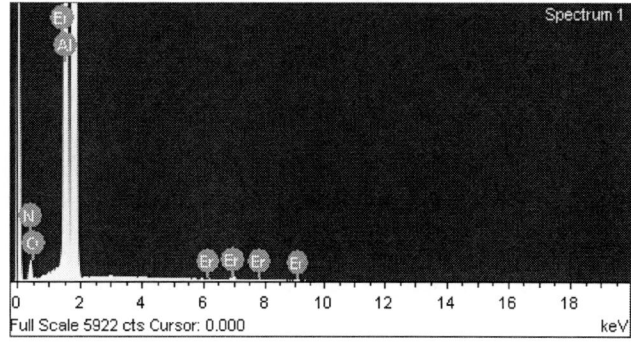

Figure 1.   EDS spectrum of the AlN:Er^{3+} thin film.

Fig. 2a shows the thickness of the AlN:Er^{3+} film characterized by scanning electron microscopy (SEM). The thickness of the film is about 430 nm. It can be calculated that the deposition rate is about 21 nm/min. As for the crystalline structure of the film, X-ray diffraction spectroscopy (XRD) can be used to characterize the lattice orientation. As shown in Fig. 2b, the film may not be a single crystal film, and there is a multiple orientation growth (mainly 002 and 101 orientations). The characterization results of composition and structure have proved the effective doping of Er^{3+} in AlN thin films.

978-1-6654-8699-6/22 $31.00 © 2022 IEEE

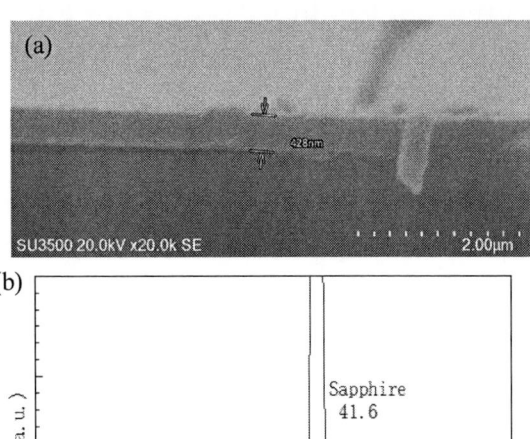

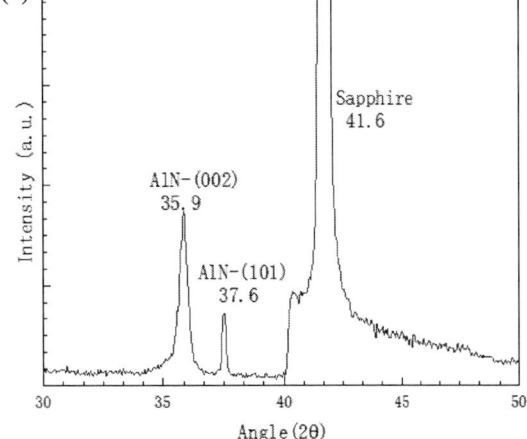

Figure 2. (a) SEM characterization and (b) XRD pattern of the film.

III. LUMINESCENCE CHARACTERIZATION AND RESULT DISCUSSION

A. Photoluminescence from the Vis to NIR Ranges

The room-temperature PL spectrum of the AlN:Er^{3+} film has been measured under 532 nm excitation. In the visible range (Fig. 3a), two luminescence bunch of peak from $^2H_{11/2} \rightarrow ^4I_{15/2}$ and $^4S_{3/2} \rightarrow ^4I_{15/2}$ (540 and 560 nm) are observed. In the NIR range (Fig. 3b,c), the bunches are observed at 816, 869, 985 and 1534 nm from the transitions of $^2H_{11/2} \rightarrow ^4I_{13/2}$, $^4S_{3/2} \rightarrow ^4I_{13/2}$, $^4I_{11/2} \rightarrow ^4I_{15/2}$ and $^4I_{13/2} \rightarrow ^4I_{15/2}$, respectively. Fig. 3d shows the luminescence transition of the Er-4f level. Because of a high excitation efficiency, apart of a strong green emission Er^{3+} has a rather strong PL of NIR bands. Among them, 869 and 1534 nm are the low loss transmission window in silica fibers and suitable for communication systems. It can be seen from the figure that the PL of 816/864 nm is from the excited state $^2H_{11/2}/^4S_{3/2}$ to the same lower state $^4I_{13/2}$. When the number of photogenerated carriers on the $^2H_{11/2}/^4S_{3/2}$ reaches saturation, some carriers will transfer to the intermediate states $^4I_{11/2}$ and $^4I_{13/2}$. There they recombine to the ground state $^4I_{15/2}$, resulting in a far NIR emission at 985 and 1534 nm. This 4f level system is similar to the laser medium, in which the luminescence efficiency of the NIR band depends on the efficiency of the sample being excited. The more photogenerated carriers excited to $^2H_{11/2}/^4S_{3/2}$, the stronger emission of the NIR luminescence band.

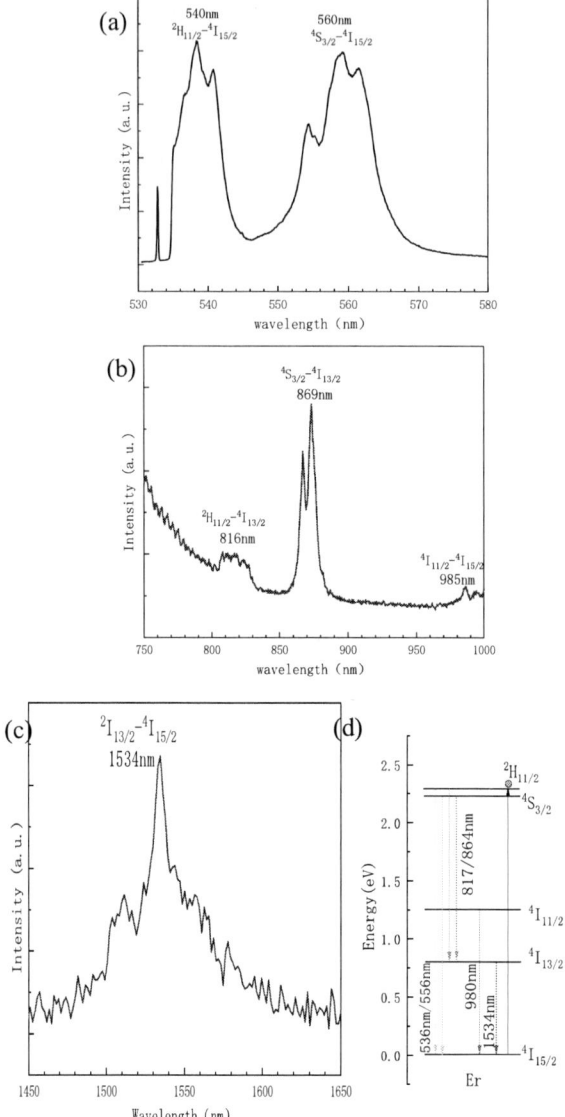

Figure 3. PL spectra of the AlN:Er^{3+} film excited by 532 nm laser, measured in the (a) Vis and (b&c) NIR ranges. (d) Luminescence mechanism of Er^{3+} 4f level.

B. Temperature Sensing Effect

Fig. 4a shows the change of the Er^{3+} characteristic luminescence intensity with temperature. When rising temperature, the PL intensity at 540 nm increases and, on the contrary, that at 560 nm decreases. This occurs because the thermal excitation of electrons of lower level $^4S_{3/2}$ to the higher level $^2H_{11/2}$ with rising temperature.

In the transition of the RE-4f levels, the ratio of the transition intensity of two adjacent energy levels is generally proportional to temperature[9]. In the changing temperature, the fluorescence intensity ratio (FIR) I_{540nm}/I_{560nm} of AlN:Er^{3+} has been plotted and fitted (Fig. 4b). The tendency of the black spot indicates that the FIR value increases with the temperature. The red line is a polynomial fitting result of FIR versus temperature, and the fitting result is shown in Eq. (1).

$$FIR = 0.7057 - 0.0108 \cdot T + 5.3 \cdot 10^{-5} \cdot T^2 - 4.1 \cdot 10^{-8} \cdot T^3 \quad (1)$$

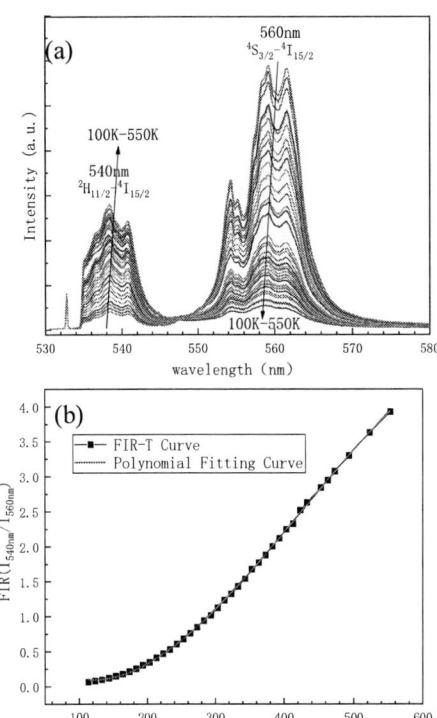

Figure 4. (a) Temperature-dependent PL of the green spectrum range of the AlN:Er^{3+} film excited by 532 nm laser. (b) Fluorescence intensity ratio of two green PL bands.

FIR has been fitted perfectly in the temperature change process from 100 to 550 K. Moreover, the absolute detection sensitivity S_a at high temperatures (above 500 K) reaches 0.012T^{-1}, which is several times higher than that of temperature sensing materials studied in recent years, shown in Tab. II. On the other hand, AlN has very high temperature stability, which can be stably used in environments of relatively high temperatures.

TABLE II. THE MAXIMUM SENSITIVITY OF DIFFERENT ER^{3+} DOPED MATERIALS BASED ON FIR TECHNIQUE IN RECENT REPORTS

Material	Max S_a (10^{-3} K^{-1})	Temperature range (K)	Ref.	Year
AlN:Er	>12	100-550	This work	2022
Er/Yb:Ba$_3$Lu$_2$Zn$_5$O$_{11}$	1.7	303-573	[10]	2021
Er/Yb:La$_{9.31}$Si$_{6.24}$O$_{26}$	3.65	283-523	[11]	2022
Er:(AlGa)$_2$O$_3$	5.6	77-400	[6]	2021
Er/Yb:CaF$_2$	2.24	298-548	[7]	2022
Er/Yb:Ba$_5$Y$_8$Zn$_4$O$_{21}$	3.9	296-563	[12]	2021
Yb/Er:NaGdF$_4$	3.65	303-343	[13]	2019
Er/Yb:Na$_3$Y(VO$_4$)$_2$	6.7	303-603	[14]	2021

IV. CONCLUSION

An Er^{3+}-doped AlN film has been prepared by radio frequency magnetron sputtering and its photoluminescence (PL) characteristics were studied. It shows high PL efficiency in a wide range of the visible (540 and 560 nm) and infrared (817, 864, 980 and 1534 nm) ranges. In the transition of 4f levels of Er^{3+}, the PL from $^2H_{11/2} \rightarrow {}^4I_{15/2}$ and $^4S_{3/2} \rightarrow {}^4I_{15/2}$ shows obvious temperature dependence, which exhibits a functional relationship between their PL intensity ratio I$_{540nm}$/I$_{560nm}$ and temperature over 100-550 K. It has a relatively high sensitivity of 0.012 K^{-1} among the currently researched temperature sensing materials. So, it can be used as a new contactless temperature sensor in a harsh environment. As a temperature sensor, it has high sensitivity and temperature stability. Therefore, the results confirm that AlN:Er^{3+} is a multifunctional material, which can be used not only for visible/NIR light-emitting devices but also for low loss communication systems and sensors.

ACKNOWLEDGMENT

This work was supported in part by National Natural Science Foundation of China (61974094); Key Research and Development Project of Guangdong Province (2020B010169003); Science and Technology Innovation Commission of Shenzhen (JCYJ20200109105413475).

REFERENCES

[1] Y. H. Hao, X. Wang, and Y. H. An, "A Ga$_2$O$_3$/AlN heterojunction for self-powered solar-blind photodetection with high photo-to-dark current ratio and fast response speed," Phys Scr, vol. 96, no. 12, 2021.

[2] S. Q. Lu, X. J. Jiang, Y. Z. Wang, K. Huang, N. Gao, D. J. Cai, Y. H. Zhou, C. C. Yang, J. Y. Kang, and R. Zhang, "Enhancing deep-UV emission at 234 nm by introducing a truncated pyramid AlN/GaN nanostructure with fine-tuned multiple facets," Nanoscale, vol. 14, no. 3, pp. 653-662, 2022.

[3] L. P. Fang, A. Y. Yin, S. F. Zhu, J. J. Ding, L. Chen, D. X. Zhang, Z. Pu, and T. W. Liu, "On the potential of Er-doped AlN film as luminescence sensing layer for multilayer Al/AlN coating health monitoring," J. Alloys Compd., vol. 727, pp. 735-743, 2017.

[4] K. Pavani, J. S. Kumar, K. Srikanth, M. J. Soares, E. Pereira, A. J. Neves, and M. P. F. Graca, "Highly efficient upconversion of Er^{3+} in Yb^{3+} codoped non-cytotoxic strontium lanthanum aluminate phosphor for low temperature sensors," Sci. Rep., vol. 7, 2017.

[5] X. Wang, X. P. Li, H. Q. Yu, S. Xu, J. S. Sun, L. H. Cheng, X. Z. Zhang, J. S. Zhang, Y. Z. Cao, and B. J. Chen, "Effects of Bi^{3+} on down-up-conversion luminescence, temperature sensing and optical transition properties of Bi^{3+}/Er^{3+} co-doped YNbO$_4$ phosphors," J. Rare Earths, vol. 40, no. 3, pp. 381-389, 2022.

[6] G. F. Deng, K. Saito, T. Tanaka, and Q. X. Guo, "Improvement of sensing sensitivity based on green emissions from Er-doped (AlGa)$_2$O$_3$ films," J. Lumin., vol. 232, 2021.

[7] M. J. Li, L. B. Su, X. Y. Chen, Q. Wu, and B. Zhang, "Effect of Yb^{3+} concentration on Er^{3+} doped CaF$_2$ single crystal for temperature sensor applications," Opt. Commun., vol. 520, 2022.

[8] W. H. Lin, F. Zhao, L. Y. Shao, M. I. Vai, P. P. Shum, and S. M. Sun, "Temperature Sensor Based on Er-Doped Cascaded-Peanut Taper Structure In-Line Interferometer in Fiber Ring Laser," IEEE Sens. J., vol. 21, no. 19, pp. 21594-21599, 2021.

[9] M. D. Dramicanin, "Trends in luminescence thermometry," J. Appl. Phys., vol. 128, no. 4, 2020.

[10] J. Chen, J. J. Guo, Y. H. Chen, X. S. Peng, G. A. Ashraf, and H. Guo, "Up-conversion properties of Ba$_3$Lu$_2$Zn$_5$O$_{11}$: Yb^{3+}, Er^{3+} phosphors for optical thermometer based on FIR technique," J. Lumin., vol. 238, 2021.

978-1-6654-8699-6/22 $31.00 © 2022 IEEE

[11] J. Zhang, S. S. An, Y. N. Zhang, and Y. M. Zhang, "Optical temperature-sensing properties based on upconversion luminescence of $La_{9.31}Si_{6.24}O_{26}$:Er^{3+}, Yb^{3+} with different strategies," Spectrochim. Acta A Mol. Biomol. Spectrosc., vol. 265, 2022.

[12] J. Chen, W. N. Zhang, S. F. Cui, X. S. Peng, F. F. Hu, R. F. Wei, H. Guo, and D. X. Huang, "Up-conversion luminescence properties and temperature sensing performances of $Ba_5Y_8Zn_4O_{21}$: Yb^{3+}, Er^{3+} phosphors," J. Alloys Compd., vol. 875, 2021.

[13] J. M. Wang, H. Lin, Y. Cheng, X. S. Cui, Y. Gao, Z. L. Ji, J. Xu, and Y. S. Wang, "A novel high-sensitive upconversion thermometry strategy: Utilizing synergistic effect of dual-wavelength lasers excitation to manipulate electron thermal distribution," Sens. Actuators B Chem., vol. 278, pp. 165-171, 2019.

[14] Q. Xiao, X. Y. Dong, X. M. Yin, H. Wang, H. Zhong, B. Dong, and X. X. Luo, "Dual-color up-conversion luminescence and temperature sensing of novel $Na_3Y(VO_4)_2$: Yb^{3+}, Er^{3+} phosphor under multi-wavelength excitation," Mater. Res. Bull., vol. 141, 2021.

Proceedings of the 7th Optoelectronics Global Conference (OGC 2022)

Wave-front Coding Technology to Extend Depth of Field in Remote Sensing Optical System

1st Qixiang Gao
Chang Guang Satellite Technology
Co., Ltd,
Changchun 130102, China
gqx17@mails.tsinghua.edu.cn

2nd Yuanhang Wang
Changchun Institute of Optics, Fine
Mechanics and Physics, Chinese
Academy of Sciences,
Changchun 130102, China
wyh4896@163.com

3rd Xing Zhong
Chang Guang Satellite Technology
Co., Ltd,
Changchun 130102, China
ciomper@163.com

4th Yu Li
Chang Guang Satellite Technology
Co., Ltd,
Changchun 130102, China
hitliyu06@163.com

Abstract—Wave-front coding technology refers to placing a phase mask at the aperture stop of a traditional optical system, modulating the wavefront and phase of the incident light field at the pupil, and finally using the digital image processing method to decode to realize computational imaging. Depth of Field (DOF) is a significant issue, the wavefront of the optical system is modulated by inserting a phase plate, so that optical modulation transfer function (MTF) and point spread function (PSF) are not sensitive to the changes of the object distances within a large depth of field range. Finally, image restoration is achieved by decoding, and clear images with a large depth of field are obtained. Compared with previous work, we innovatively proposed a high-order polynomial phase plate design, considering manufacturing difficulty as an optimization factor, and finally realized 200m-100km ultra-large depth-of-field imaging in a remote sensing optical system without a focusing mechanism.

Keywords—wave-front coding; phase plate design; extend depth of field

I. INTRODUCTION

Among traditional optical satellites, mechanical focusing mechanisms such as motor drive and hydraulic drive are mostly used, which not only have low precision but also increase the volume, mass of the optical system, and satellite cost. In addition, the focusing mechanism cannot meet the requirements for an accurate and fast response when the target passes quickly from far to near in the field of view [1]-[2].Therefore, high-quality images of such targets cannot be captured. Since Dowski and Cathey first proposed the cubic phase plate for depth of field extension and completed the basic theoretical analysis in 1995[3]. Researchers have proposed many improved phase plate types, such as asymmetrical, sine, logarithm, etc[4]-[9], which are all theoretical analyses, without considering practical application scenarios. The key of the phase plate design is to make the MTF and PSF of the optical system insensitive to changes from different object distances. There are two methods for evaluating the performance of depth-of-field extension wave-

front coding: one is imaging evaluation criteria based on Fisher information theory, and the other one is a method based on Strehl ratio standard deviation. In the second method, the Strehl ratio can be replaced by MTF or PSF, all of which have the same effect. Inspired by the phase plate technology, we add a phase plate in the remote sensing optical system to achieve a large depth of field imaging of 200m-100km. In this paper, a high-order polynomial phase plate design is implemented, and considering manufacturing difficulty, the peak to valley(PV) of the phase plate surface shape is used as part of the optimization function in the process of optimizing the phase plate parameters.

II. TRADITIONAL OPTICAL SYSTEM

As mentioned above, although various types of phase plates have been proposed, there are still many limitations in the practical application. For example, the cubic phase plate is hard to detect and assemble, and the exponential phase plate has a steep surface shape which is difficult to manufacture. Considering the actual processing and assembly, we finally chose to adopt the phase plate in the form of a multi-order odd polynomial for design.

First of all, an available optical system is employed with the following basic parameters as in Table I.

TABLE I. BASIC PARAMETERS

Focal length	Field	Aperture	Wavelength
160mm	5°×5°	45mm	450nm-800nm

The initial optical structure diagram is shown in Fig. 1, and the detector's pixel is $3.2\,\mu m$, Nyquist frequency is 156 cycles/mm, MTF in different object distances are illustrated in Fig. 2. For the traditional optical system, it satisfies clear imaging at a distance of 200m or 1000m or 100km, but cannot be applied to other distances.

978-1-6654-8699-6/22 $31.00 © 2022 IEEE

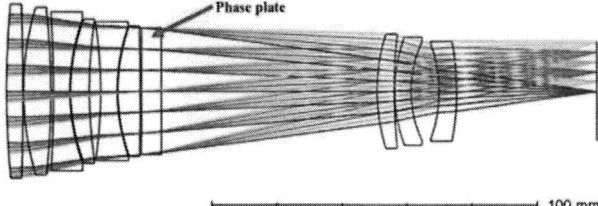

Figure 1. Optical system layout.

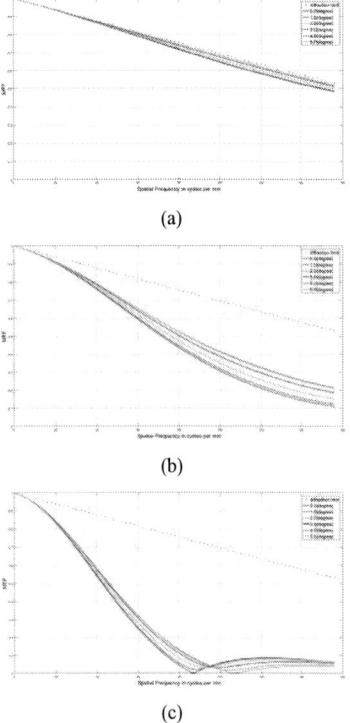

(a)

(b)

(c)

Figure 2. MTF of the optical system focused on 200m at different object distances. (a)200m, (b)1000m, (c)100km.

III. PHASE PLATE DESIGN

Although focus in different object distances can be satisfied by the focusing mechanism, fast target capture cannot be achieved. Through the analysis of the defocus, we set 1000m clear imaging as the initial structure. Considering the central part of the phase plate is close to the plane, in order to avoid the generation of ghost image, we choose to insert the phase plate radius of 18.5mm at the stop position. The phase plate is expressed as

$$f(x, y) = \alpha(x^3 + y^3) + \beta(x^5 + y^5) + \gamma(x^7 + y^7) \quad (1)$$

where x and y are the normalized pupil coordinates, α, β, γ are the parameters of the phase plate.

Next, the parameters of the phase plate need to be optimized. Given a set of phase plate parameters, we first consider the consistency of MTF, i.e., the standard deviation of MTF under different distances and different fields of view. We set the minimum MTF threshold larger than 0.1 at the

Nyquist frequency. Meanwhile, take it that the surface steepness of the phase plate should not be too high into account by setting PV of the surface shape to 7 μm(r=18.5mm). For 200m-100km large depth of field imaging, 200m, 1000m, and 1000km are set as distances, and 0 degree, 1 degree, 2 degree, 3 degree, 4 degree, and 5 degree are set as the fields. Each group of phase plate parameters is taken as input, under the condition that MTF>0.1 at Nyquist frequency and surface PV< 7 μm, the best phase plate parameters are found with the smallest standard deviation of MTF at different distances and different fields. In the optimization process, considering that the particle swarm optimization algorithm (PSO)[10]-[11] has a fast convergence speed, few parameters, and the algorithm is simple to implement, PSO algorithm is employed to optimize the parameters of the phase plate consequently, the flow chart of PSO algorithm is shown in Fig. 3.

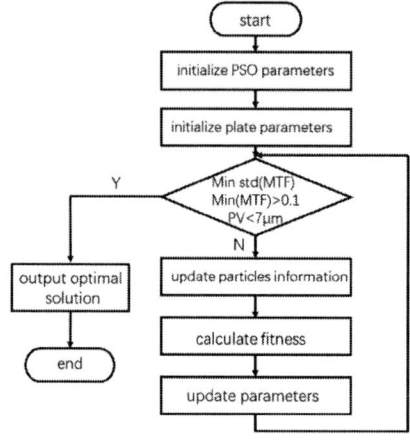

Figure 3. Optimization process

Finally, the phase plate parameters α, β, γ are determined as 0.55, -0.009, -0.016, respectively. The surface shape is illustrated in Fig. 4.

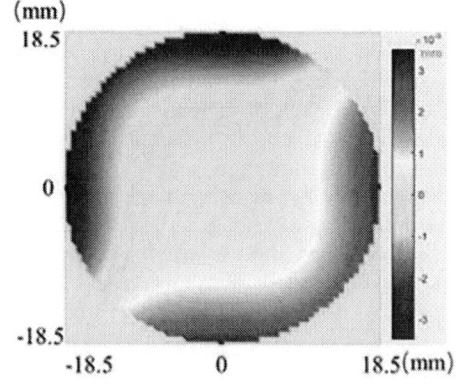

Figure 4. Designed phase plate surface sag

The system has high consistency in MTF and PSF under different distances and fields as shown in Fig. 5.

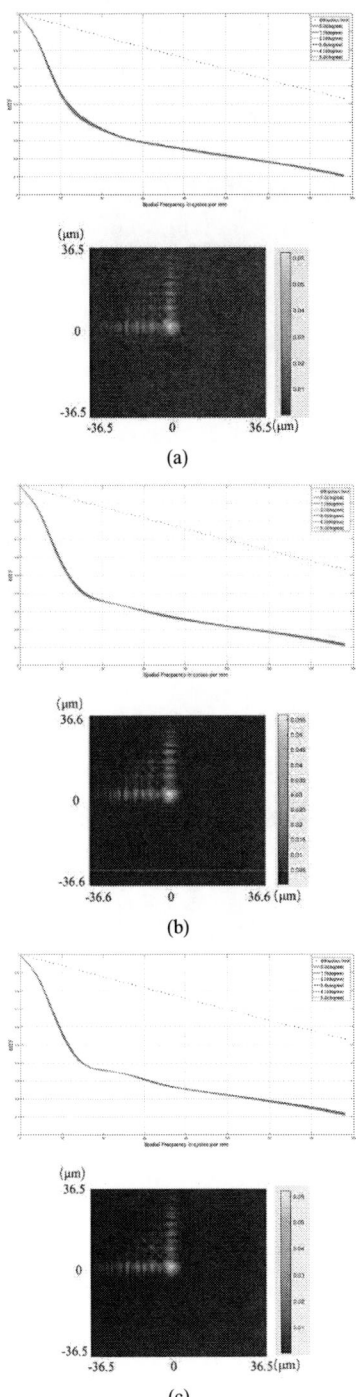

Figure 5.   MTF and PSF of the optical system focused on 1000m at different object distances after inserting the designed phase plate. (a)200m, (b)1000m, (c)100km.

For comparison, we optimized a cubic phase plate for the initial optical system under the traditional constraints, and the results are shown in Fig. 6 and Fig. 7.

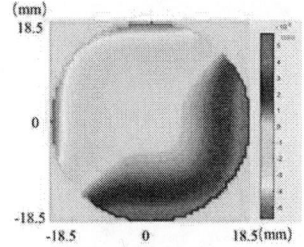

Figure 6.   Cubic phase plate surface sag

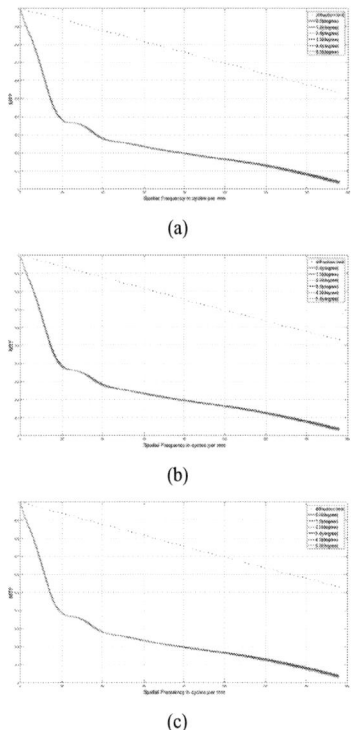

Figure 7.   MTF of optical system focused on 1000m after inserting cubic phase plate at different object distances. (a) 200m, (b)1000m, (c) 100km.

After comparison, there are without obvious differences in MTF consistency, restricting MTF>0.1 at Nyquist frequency, which improves the quality of the image. At the same time, the designed phase plate surface PV is only 7 μm, less than the cubic phase plate PV 11μm, easier to manufacture.

IV.   IMAGE RESTORATION

We extract PSF at different distances, perform convolutional simulation imaging, and restore the image to verify the effect of phase plate technology on the depth of field extension. The results are shown in Fig.8, compared with traditional optical system imaging, it can be seen that although the image encoded by the phase plate is blurred at each observation position, image blurring under different observation distances and different fields have similarity, and after the image restoration technology[11]-[12], the images at the three object distance positions 200m, 1000m, 100km at different depth of field are clear.

978-1-6654-8699-6/22 $31.00 © 2022 IEEE

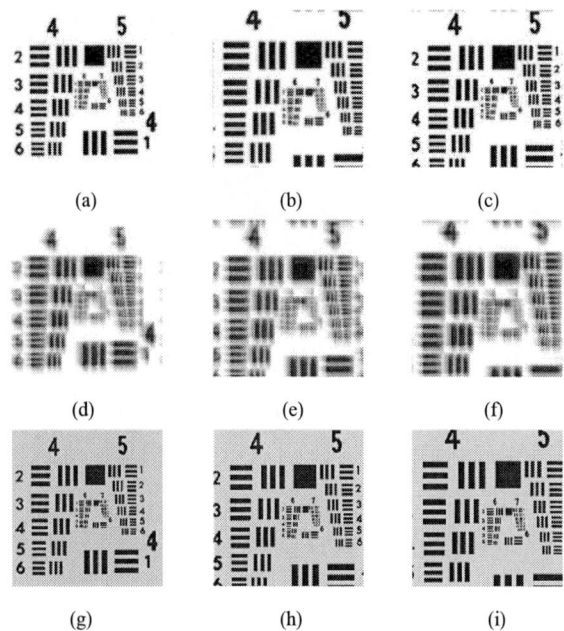

Figure 8. Traditional optical system imaging focused at 1000m for different object distances: (a) 200m.(b) 1000m.(c) 100km; After inserting designed phase plate: (d) 200m.(e) 1000m; (f) 100km; After image restoration: (g) 200m.(g) 1000m.(i) 100km.

V. CONCLUSION

By inserting a high-order even-order polynomial phase plate under the constraint of MTF at Nyquist frequency, surface PV, compared with the traditional cubic phase plate, without reducing the depth of field extension range and imaging quality, we considered the feasibility of manufacturing and assembly in practical applications, and finally completed the design of the phase plate. The remote sensing optical system in this paper has achieved large depth-of-field imaging of 200m-100km.

REFERENCES

[1] B. Witherspoon, L. Huff, M. Jacoby, and P. Mammini, "Results of environmental testing of the focus and alignment mechanism of the near-infrared camera on the James Webb Space Telescope," Proc. of SPIE., vol. 7439, pp. 74391F-1–13, Sep. 2009.

[2] Pollock, and Randy. "System for establishing best focus for the Orbiting Carbon Observatory instrument," Opt. Eng., vol. 48(7), pp. 073605-1–9, Jul. 2009.

[3] E. R. Dowski, and W. T. Cathey, "Extended depth of field through wave-front coding." App. Opt., vol. 34, no. 11, pp. 1859–1866, Apr. 1995.

[4] S. S. Sherif, W. T. Cathey, and E. R. Dowski, "Phase plate to extend the depth of field of incoherent hybrid imaging systems," App. Opt., vol. 43, no. 13, pp. 2709–2721, May. 2004.

[5] A. Castro and J. O. Castañeda, "Asymmetric phase masks for extended depth of field," App. Opt., vol. 43, no. 17, pp. 3474–3479, Jun. 2004.

[6] A. Castro and J. O. Castañeda, "High focal depth with fractional-power wave fronts," Opt. Lett., vol. 29, no. 6, pp. 560–562, Mar. 2004.

[7] Q. Yang, L. Liu, and J. Sun, "Optimized phase pupil masks for extended depth of field," Opt. Commun., vol. 272, pp. 56–66, Nov. 2006.

[8] A. Castro, J. O. Castañeda, and A. W. Lohmann, "Bow-tie effect: differential operator," App. Opt., vol. 45, no.30, pp. 7878–7884, Oct. 2004.

[9] Z. Hui, L. Qi, and H. Feng, "Optimized sinusoidal phase mask to extend the depth of field of an incoherent imaging system," Opt. Lett., vol. 35, no. 2, pp. 267–269, Jan. 2010.

[10] J. Kennedy and R. Eberhart, "Particle swarm optimization,"Proc. Of IEEE Int. Conf. Neural Networks., vol. 4, pp. 1942–1948, Dec. 1995.

[11] X. Li, X. L. Lia, K. Wang, and Y. Li, "A Multi-Objective Particle Swarm Optimization Algorithm Based on Enhanced Selection," IEEE. Access, vol. 7, pp. 168091–168103, Nov. 2019.

[12] S. Qi, J. Jia, and A. Agarwala, "High-quality motion deblurring from a single image," ACM Trans. Graph., vol. 27, no.3, pp.1–10, Aug. 2008.

[13] S. Zhuo, and T. Sim, "Defocus map estimation from a single image," Pattern Recognit, vol. 44, pp. 1852–1858, Mar. 2011.

Proceedings of the 7th Optoelectronics Global Conference (OGC 2022)

Influence of Target Layout on the Accuracy of Monocular Vision 6DOF Spatial Pose Measurement

LIU Xingtan[1], HUA Baocheng[1], LIU Qihai[1], DENG Loulou [1], LIU Jing[1], TAO Liqing[2], WU Xiuyu[2], LEI Kaiyu[1]
[1]Beijing Institute of Control Engineering, China Academy of Space Technology, Beijing 100190, China
[2]Hangzhou Center of China Academy of Space Technology, Hangzhou 310024, China
xingtanliu@163.com

Abstract—**The method of pose measurement using computer monocular vision is widely used in the field of six-degree-of-freedom(6DOF) pose measurement in industry, aerospace and other fields. By analyzing the factors that affect the accuracy of monocular pose measurement, the research on the target layout of cooperative positioning is carried out. First, a sub-millimeter monocular spatial positioning simulation system is built, and the key components of the target layout are defined as Target Scale(TS) and Elevation Difference(ED); then, as the input, the 6DOF measurement accuracy of the monocular spatial positioning system under different conditions is obtained. The results show that the three-axis position accuracy is better than 0.5mm, and is affected by the positive correlation between the TS and the ED, and the axial variation is more significant. The three-axis attitude accuracy is better than 0.03° (1.8 arc seconds), which is greatly positive affected by the TS, and the axial pose is more robust than azimuth pose.**

Keywords-monocular vision; spatial pose measurement; accuracy analysis; target layout

I. INTRODUCTION

Vision measurement is an advanced system that uses high-density, low-noise and low-distortion cameras, a high-speed real-time image acquisition system, a dedicated image hardware processing system and a high-performance computer to effectively process images. Vision measurement is widely used in aerospace rendezvous and docking, industrial intelligent precision measurement, medical surgery positioning and other fields.

There are many classification methods of vision measurement, which can be divided into monocular vision measurement, binocular vision measurement and multi-eye vision measurement according to the number of vision sensors used. Among them, monocular vision measurement refers to using only one camera to shoot a single cooperative target image for 6DOF pose measurement, while binocular vision is based on the principle of parallax, which requires two cameras to obtain surrounding scenes from different angles at different times. In contrast, the monocular principle only needs one camera, which has the advantages of simple structure and simply equipped camera calibration, and also avoids the shortcomings of the binocular stereo vision principle[1], such as small effective field of view, difficult to match stereo baseline. Besides, due to the complexity of the binocular stereo system, the accuracy is not easy to guarantee.

For monocular measurement, under the premise that non-coplanar cooperative feature targets can be constructed, its characteristics include but are not limited to being able to observe a long distance[2] and a large field of view, avoiding

the possibility of restricted occlusion of the full field of view for binocular observation. The camera design is simple, and there are no special constraints on installation similar to the long baseline and calibration of the binocular system; the small amount of data occupies less computing resources, and the 6DOF pose can be calculated in real time; in addition, the system has advantages of high robustness and low cost. When the target is far away from the camera, the effect of monocular vision in pose measurement is better than that of binocular vision, and the attitude angle error is obviously better than that of monocular, because binocular does not introduce the actual model of the target[3].

The main content of this paper is to use the monocular vision method to measure the high-precision pose and attitude of the space cooperative target, and to deeply analyze the influence of the target layout on the accuracy of the monocular pose measurement.

II. SPACE POSITIONING SIMULATION SYSTEM

A. Principle of monocular vision

Monocular vision systems use only one vision sensor. The monocular vision system loses depth information due to the projection from the three-dimensional objective world to the N-dimensional image during the imaging process. Therefore, specific targets need to be constructed to restore depth.

Nevertheless, the monocular vision system is simple in structure, mature in algorithm and of less computationally intensive. At the same time, monocular vision is the basis of other types of vision systems, such as binocular stereo vision are realized by adding other means and measures on the basis of monocular vision system.

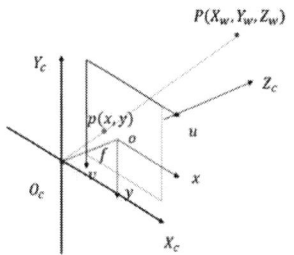

Figure 1. Coordinate systems involved in monocular vision.

The coordinate systems involved in monocular vision include: world coordinate system, camera coordinate system, image coordinate system and pixel coordinate system.

Through coordinate system transformation, the real point P in the world coordinate system is projected to the image coordinate system P. It is necessary to pre-calibrate the camera's internal parameters (focal length, resolution, pixel size, distortion parameters) accurately, extract the image plane coordinates (u,v) accurately, and design a reasonable target layout and pre-measure its relative spatial position relationship[4].

The depth information of the spatial feature point P in the camera coordinate system is obtained by formula (1):

$$
\begin{bmatrix} u_i \\ v_i \\ 1 \end{bmatrix} = \begin{bmatrix} f_u & 0 & u_0 & 0 \\ 0 & f_v & v_0 & 0 \\ 0 & 0 & 1 & 0 \end{bmatrix} = \begin{bmatrix} X_{ci} \\ Y_{ci} \\ Z_{ci} \\ 1 \end{bmatrix}. \tag{1}
$$

The specific process of the algorithm is as follows: first, establish the rigid body transformation relationship between the world coordinate system, the camera, and the target (cooperative feature point); then, establish the feature point and image point equation system according to the spatial geometric transformation and the camera model; finally, solve the feature point in the camera. The coordinate value under the coordinate system is obtained, and the relative pose between the target and the characteristic coordinate system is finally obtained according to the rigid body transformation relationship.

B. Monocular vision measurement system composition

The composition of the monocular vision 6DOF spatial pose measurement system is as follows:

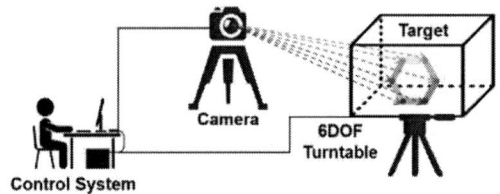

Figure 2. 6DOF spatial pose measurement system.

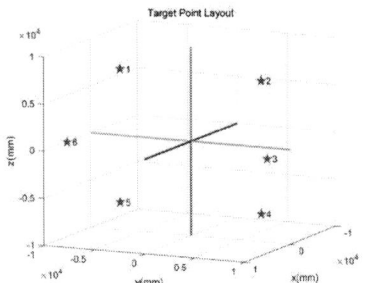

Figure 3. Target layout in 3D space

The measurement system is mainly composed of three functional subsystems: a computational control system, a camera imaging system and a 6DOF cooperative target turntable system. The 6DOF turntable is operated by the computational control system, and the true value of the pose change is provided by the turntable system. At the same time, the camera is controlled to image the target under different working conditions, and the imaging data is synchronously transmitted to the computational control system. Combined with the accurately calibrated camera internal parameters, the accurately extracted target image surface coordinates and the precisely measured relative position relationship of the target, the 6DOF results can be obtained through the monocular visual space pose algorithm, and the relevant results can be converted into the turntable pose. The same coordinate system is used to complete the comparison and accuracy evaluation.

C. Monocular space positioning simulation system

By simulating the composition of the monocular imaging system, the monocular spatial positioning simulation system is built, and set the following input parameters:

TABLE I.   INPUT PARAMETERS OF MONOCULAR SPATIAL POSITIONING SIMULATION SYSTEM

Input parameters	Numerical value
Focus	16.128mm
Pixel Size	0.00369mm
Resolution	1698.884118pix, 1348.957267pix
Pixel extraction accuracy	0.1pix
Target Scale	2000.0mm
Elevation Difference	500.0mm

Among them, random offsets in the X and Y directions are applied to the camera, as shown in the figure below, to test the robustness of the system.

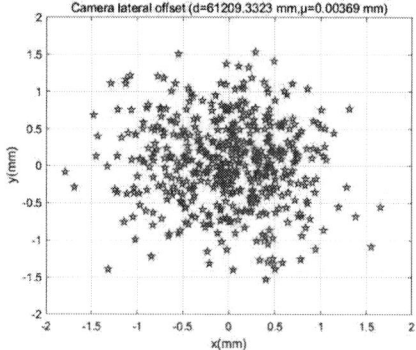

Figure 4. Camera Offset.

In the calculation control terminal, the output three-axis position accuracy is obtained: 0.053597mm, 0.059750mm, 0.397904mm; the output three-axis attitude accuracy: 0.005508°, 0.005849°, 0.003912°. In this way, the monocular pose measurement function can be verified, thereby obtaining a sub-millimeter monocular spatial positioning simulation system.

978-1-6654-8699-6/22 $31.00 © 2022 IEEE

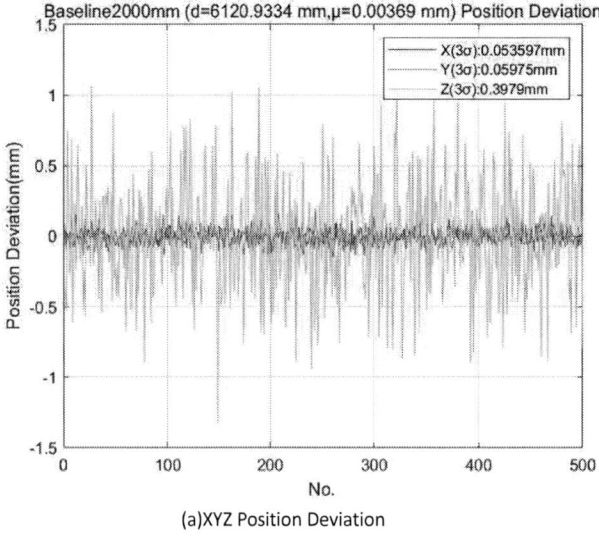

(a)XYZ Position Deviation

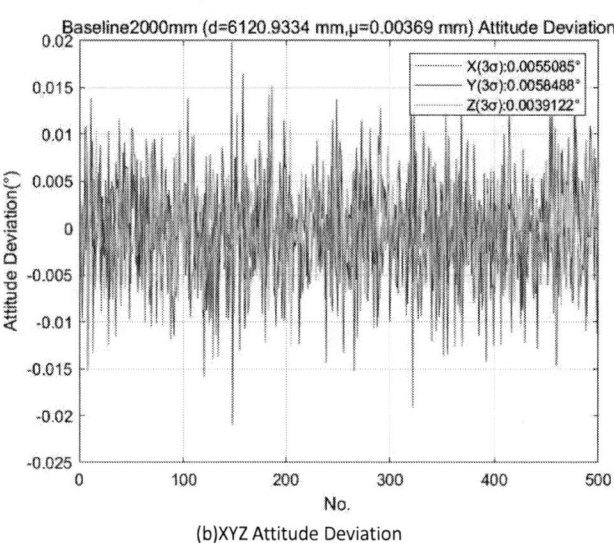

(b)XYZ Attitude Deviation

Figure 5. 6DOF Spatial Pose Measurement Result.

D. Analysis of Influencing Factors of Accuracy

From the monocular principle and system composition, analyze the influencing factors on the 6DOF pose measurement results of the monocular measurement system, including the following aspects[5]:

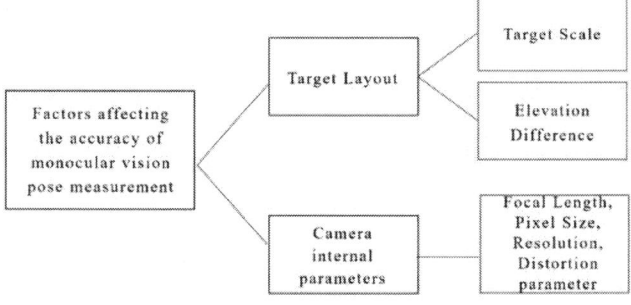

Figure 6.   Influencing factors of monocular measurement accuracy.

Among them, the camera internal parameters follow the system settings, and at the same time consider the system stability, it is generally not suitable to perform too much optimization, so choose to analyze and optimize the target layout. Since the target layout can be decomposed into Target Scale(TS) and Elevation Difference(ED), TS is the maximum envelope circumcircle radius of the target array, and ED is the maximum target drop parallel to the incident direction of the camera light. The two factors do not affect each other, so they can be designed independently of each other. Therefore, quantitative analysis of the TS and ED is carried out below.

III.   ANALYSIS OF THE INFLUENCE OF TARGET LAYOUT ON MEASUREMENT ACCURACY

A. Target Scale(TS) and Elevation Difference(ED)

The decoupling analysis of the scale factor and the axial height difference is carried out, and it is used as the input to the sub-millimeter monocular spatial positioning simulation system built in this paper, and the 6DOF measurement accuracy of the monocular spatial positioning system under different conditions is obtained.

By setting the target scale factor range of 50~1000mm and the axial height difference of 50~1000mm, keep the input parameters Focus, Pixel Size, Resolution, Pixel extraction accuracy, Target Scale, and Elevation Difference consistent with the simulation system. The output results are the spatial position (figures (a), (c), (e)) and poses (figures (b), (d), (f)) under different target scales and elevation differences.

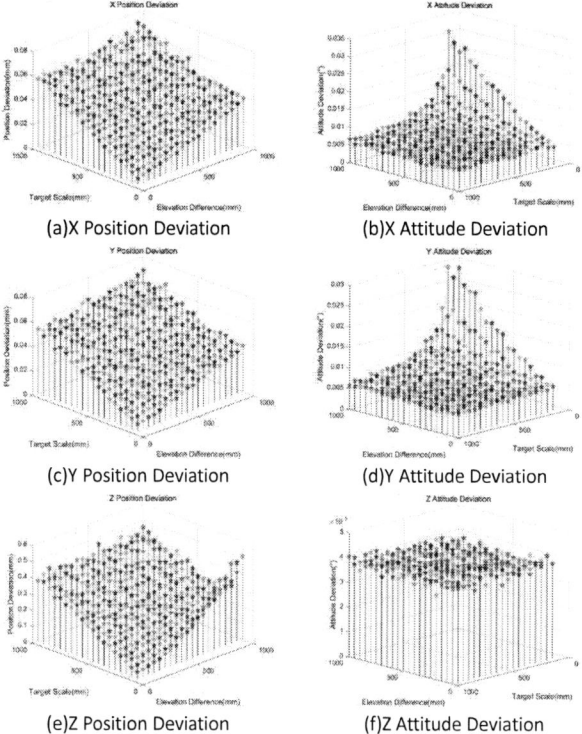

(a)X Position Deviation   (b)X Attitude Deviation

(c)Y Position Deviation   (d)Y Attitude Deviation

(e)Z Position Deviation   (f)Z Attitude Deviation

Figure 7.   6DOF Spatial Pose Measurement Result.

B. Summary

The results show that, on the one hand, the pose results in the X and Y directions (azimuth) are basically the same, but the accuracy in the Z direction (axial) is much larger than the azimuth accuracy due to the limitation of the monocular principle. On the other hand, the position accuracy of monocular measurement increases with the increase of the scale factor and increases with the increase of the axial height difference; the attitude accuracy decreases with the increase of the scale factor, but the change with the axial height difference is not obvious.

IV. CONCLUSION

By analyzing the influencing factors of monocular pose measurement accuracy, a set of sub-millimeter monocular spatial positioning simulation system was built, and the key factors in the composition of the target layout were clarified: scale factor and axial height difference. And input the monocular spatial positioning system to obtain the monocular 6DOF pose measurement accuracy under different target layout characteristics.

It is concluded that the monocular system built in this paper has high measurement accuracy, the position accuracy is less than 0.5mm, and the attitude accuracy is less than 0.03 ° (also 1.8 arc seconds). Among them, the three-axis position accuracy is positively affected by the TS and the ED, and the axial change is more significant; the three-axis attitude accuracy is more positive affected by the TS. where the axial attitude is more robust than the azimuth attitude change. In the follow-up design of a monocular high-precision measurement system, the target layout can be specially designed according to this standard, such as aerospace rendezvous and docking, industrial automation measurement, medical surgery navigation and other fields, which are widely used.

ACKNOWLEDGMENT

To express my heartfelt thanks to my colleagues, engineers: Gong Dezhu, Liu Lu, and the precision measurement teams members: Zou Yue, Guo Jian, Zhang Lihua, Zhang Chenglong, Wei Xinpeng, for their supports and assistance in my testing process. At the same time, thanks to all the experts who provided valuable opinions on the writing of the thesis.

REFERENCES

[1] ZHU Feng, HE Lei, HAO Yingming. Contrast of Calculated Accuracy between Monocular Vision and Binocular Vision without Modeling ［J］. Chinese Journal of Scientific Instrument, 2007, 28 (S):165.

[2] ZHAO C H, GAO W W, LIU L, et al. A Vision Guidance Sensor for Shenzhou-8 Spacecraft Autonomous Rendezvous and Docking ［J］. Aerospace Control and Application, 2011, 37(6): 6-13.

[3] LIU Q H, GONG D Z, HUA B C, et al. New Generation Camera-Type Rendezvous and Docking Sensor ［J］. Aerospace Control and Application, 2018, 44(2): 56-61.

[4] Xu WeiGao. Key Technology Research on Pose Measurement Based on Monocular Vision[D]. Xian, University of Chinese Academy of Sciences, 2008.

[5] HAO R J, WANG Z Y, LI Y R, Error Analysis Method for Monocular Vision Pose Measurement System ［J］. Journal of Applied Optics, 2019, 40(1): 79-85.

Proceedings of the 7th Optoelectronics Global Conference (OGC 2022)

Design of LED Array Control Module for Optical Camera Communication

Han Liu,[1]
School of Computer & Communication Engineering
University of Science and Technology Beijing
Beijing, China
e-mail: ustblh2018@163.com

JianPing Wang,[2] HuiMin Lu,[3]
School of Computer & Communication Engineering
University of Science and Technology Beijing
Beijing, China

Abstract—**Visible light communication technology is a communication system that uses light-emitting diodes as the emission source and electronic devices equipped with cameras or photodiodes as the receiving end. Optical camera communication, also known as optical imaging communication, belongs to a special kind of visible light communication, using image sensors as receivers. In this paper, an LED array control module for optical camera communication is designed, which realizes the data generation and signal transmission functions of the transmitter of the optical imaging communication system, and performs signal output and data transmission tests in indoor and outdoor environments.**

Keywords-component; Visible Light Communication, LED Array, Optical Imaging Communications

I. INTRODUCTION

As an optical wireless communication technology, visible light communication is not limited by spectrum resources, and it is a route to expand wireless communication solutions. Visible light communication uses optical signals in the visible light region, and mainly uses LEDs as transmitter light sources. The advantages of LED in lighting are reflected in the aspects of energy efficiency, life and safety, and LED lamps are energy-saving and environmentally friendly, have a long service life, and will not cause overheating problems as a cold light source. The most important thing is that LED, as the luminous source of visible light communication, has the advantages of low radiation, energy saving and environmental protection, and good confidentiality in the field of communication. Gas stations, aircraft interiors and other places have requirements for the power of electromagnetic signals, and visible light communication has application scenarios in these places.

Traditional visible light communication uses photodiodes (PDs) as receivers. With the popularization of digital cameras, optical camera communication, also known as visible light imaging communication, is gradually emerging. It can use smartphone cameras and ordinary digital cameras as receivers. While inheriting the advantages of visible light communication, it can directly replace the receiving end of the communication system without changing the transmitting end.

Optical imaging communication system is a kind of visible light communication that is easier and more convenient to use than traditional visible light communication. Traditional visible light communication is based on PD to receive signals, mainly using PIN and PIN array, while optical imaging communication system uses more popular lens group and image sensor array. As a receiver for receiving signals, this receiver can be implemented using the cameras of modern smart devices. Because of the different receiving methods, optical imaging communication can more easily separate the wavelength of the carrier optical signal, and the signal processing at the receiving end can also be implemented under the Android or iOS system.

The transmitters of visible light communication and optical imaging communication are basically the same. Both transmitters use LEDs to transmit signals after encoding and modulating the signals. The principle of the two receivers is also different. The PIN receiver of visible light communication directly converts the optical signal into an electrical signal after receiving the signal. The electrical signal is demodulated and decoded after the amplifier circuit, and then the data is output. After receiving the image sensor receiver of optical imaging communication, it needs to output the signal through the pixel sweep analog-to-digital converter (ADC), and then perform the work of demodulation and decoding. In contrast, optical imaging communication has a larger operating frequency bandwidth and Scalable range, lower signal-to-noise ratio and decoding complexity.

II. FUNDAMENTALS OF OPTICAL IMAGING COMMUNICATION

Optical imaging communication has developed rapidly, and the related technologies involved are also changing with each passing day. This chapter mainly introduces the principle of LED light radiation, key technologies of optical imaging communication system and the application of MIMO technology in optical imaging communication.

A. Principle of LED light radiation

LED chip is composed of semiconductors, including P-type semiconductors and N-type semiconductors, and a P-N junction is formed between the P-type semiconductors and the N-type semiconductors. In a static state, the chip as a whole is neutral to the outside world. When a positive external voltage is applied, the holes in the P region and the electrons in the N region move to the P-N junction. Within the range of a few microns from the P-N junction to the P region, the high-energy holes and the high-energy electrons

978-1-6654-8699-6/22 $31.00 © 2022 IEEE

combine to generate photons. This process is self-luminous radiation. Phenomenon, in this process, electrical energy is converted into light energy, and the higher the released energy, the shorter the light wave.

$$E = \frac{nhc}{\lambda} \qquad (1)$$

In the above formula, E is the total photon energy, n is the total number of photons, h is Planck's constant, c is the speed of light, and λ is the wavelength.

In optical imaging communication, luminous flux and luminous intensity are the main transmitter parameters. The luminous flux of LED refers to the radiant power perceived by the standard human eye per unit time, symbol lm. As the current increases, the LED luminous flux increases. The relationship between luminous flux and radiant flux is as follows.

$$\Phi = K_m \int V(\lambda) \psi_{e\lambda} d\lambda \qquad (2)$$

Luminous intensity is referred to as luminosity, the unit is cd, which characterizes the strength of a light-emitting device in a specified direction.

$$I = \frac{d\varphi}{d\Omega} \qquad (3)$$

B. Key Technologies of Optical Imaging Communication System

A complete visible light communication/optical imaging communication system is shown in Figure 1. The system needs to complete two photoelectric conversions. The input binary information bits at the transmitting end are encoded and modulated, and the electrical signal to the optical signal is realized by the LED drive circuit. The optical signal is transmitted in the channel of free space, and the receiving end uses a photodiode or imaging sensor to convert the optical signal into an electrical signal, and then demodulates and decodes the signal, and finally obtains the transmitted data and outputs the signal.

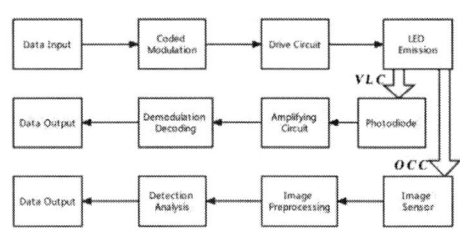

Figure 1.　VLC and OCC communication system structure diagram.

Optical imaging communication can use a variety of modulation methods, the more common are the following three methods.

- OOK modulation, also known as Amplitude shift keying (ASK). Binary OOK modulation is binary amplitude keying, also known as 2ASK. The anti-noise performance of this modulation method is not strong, but the modulation method is simple to implement and is the most basic modulation method. 2ASK uses the signal amplitude as a variable. There

are only two types of carrier amplitudes. The information is also expressed through these two amplitudes. The two amplitude states correspond to "0" or "1" respectively, and the unipolar NRZ controls the on or off of the signal sine wave.

- The principle of Pulse Position Modulation (PPM) is to group the original data, divide the time occupied by the data of k bits of information into time slots, and then use the pulse signal of one of the time slots to indicate that it belongs to A phase modulation.

- Color-shift keying modulation (Color-shift keying, CSK), there is a kind of RGB-LED in white LED, which is composed of LEDs of three primary colors (red light, blue light, green light). There is a filter in the image sensor of the optical camera at the receiving end, which can separate the three colors of red, blue and green according to frequency, and the three separated colors can be regarded as the signals from three different LED lights. The LED lights transmit information independently, detect the intensities of the three types of light and demodulate and decode them respectively, which can triple the communication rate.

C. MIMO Technology and Its Application in OCC System

Multiple-Input Multiple-Output technology, also known as MIMO technology, has great advantages in the field of wireless communication. Data is sent after being divided, which reduces the data length of each transmitted signal and increases the transmission distance of the communication system. At the same time, this kind of technology increases the communication rate without occupying additional spectrum.

The main idea of MIMO technology is that multiple signals are sent in parallel in the transmitter, and the receiver receives in parallel and then performs technical processing on the received signals to obtain the original data transmitted by the transmitter. In the practical application of OOC, in order to have both lighting and communication functions, multiple LEDs are usually used for lighting and communication. These multiple LEDs can be regarded as LED arrays. Multiple LEDs work at the same time, and the luminosity is stronger, which is conducive to improving outdoor Optical imaging communication system communication distance.

In the OCC system, the transmitter processes the data to be transmitted, and divides the serial original data into multiple parallel sub-data. Using spatial multiplexing technology, each sub-data is transmitted through LED lights. At the receiving end, using receivers with not less than the number of sub-data copies to receive, obtain multiple parallel sub-data. After receiving these parallel sub-data by single or multiple image sensors at the receiving end of the optical imaging communication system, they can receive the sub-data according to each LED light According to the channel characteristics of the receiver, the signals sent by different LED lights are separated, and the parallel sub-data stream is restored to a serial data stream according to the set sequence. The LED array composed of multiple LED lights is used, and the occupied bandwidth remains unchanged. case, the communication capacity of the system is increased.

III. DESIGN AND IMPLEMENTATION OF LED ARRAY TRANSMITTER AT TRANSMITTER OF OPTICAL IMAGING COMMUNICATION SYSTEM

A. Design of LED Array Coding for Optical Imaging Communication System

At the transmitter, the host computer inputs a group of 8-bit or 16-bit bit data, and sends it to the FPGA development board in hexadecimal through UART. After serial-to-parallel conversion on the development board, a group of 8-bit or 16-bit stream will be processed as parallel 8 or 16 bit data. At this time, these "0" and "1" information bits are assigned to the designated FPGA output I/O port through Manchester encoding or (7, 4) Hamming code, and through OOK modulation, the LED array is driven to send optical signals in a high and low level. through the MIMO channel of visible light.

It is captured as an image at the receiving end, the obtained image is processed, and the data at the sending end is recovered after decoding. Each LED light of the LED array forms an independent communication link with the receiver. As long as the boundary is determined and the positive direction of the array is set, the receiver can identify and locate each LED light in the pixel plane of the array through image processing technology. relative position to ensure the normal operation of the entire communication link.

The modulation method of OOK is relatively simple to implement in the optical imaging communication system. When the high level is transmitted to each LED driving circuit in the LED array, the LED light is on, and the data is "1" at this time. When the driving circuit of the LED light is low power Usually, the LED light is off, and the data at this time is "0". When a series of data streams are continuously input into the driving circuit of the LED array, the LEDs will continuously switch on and off according to the input data to send optical signals.

Manchester coding is to add a transition to the middle of each bit data, so it is also called split-phase code. The advantage of this is that each bit of data transmission will undergo a transition, from "0" to "1" or Jumping from "1" to 0, this coding method can realize the self-synchronization of the transmitting end and the receiving end, because the interval of each jump is one clock cycle or half a clock cycle, which can be judged the actual clock signal from the received code pattern.

Using Manchester encoding, if the hardware device is not damaged, there can only be two consecutive "0" or "1" at most. Although the detection function is provided, it cannot correct the erroneous data bits. Therefore, the (7,4) Hamming code is used for improvement. (7,4) Hamming code belongs to linear block code, the code length is 7 bits, the information bit is 4 bits, and the supervision bit is 3 bits. This encoding method can detect one bit of the code through the relationship between the supervision bit and the information bit. mistake.

$$G = \begin{Bmatrix} g_0 \\ g_1 \\ g_2 \\ g_3 \end{Bmatrix} = \begin{bmatrix} 1 & 1 & 0 & 1 & 0 & 0 & 0 \\ 0 & 1 & 1 & 0 & 1 & 0 & 0 \\ 1 & 1 & 1 & 0 & 0 & 1 & 0 \\ 1 & 0 & 1 & 0 & 0 & 0 & 1 \end{bmatrix} \quad (4)$$

The input original codeword is multiplied by the generator matrix, and the codeword is changed from the original 4 bits to 7 bits.

$$v = u*G = \begin{pmatrix} u_0 & u_1 & u_2 & u_3 \end{pmatrix} * \begin{Bmatrix} g_0 \\ g_1 \\ g_2 \\ g_3 \end{Bmatrix} = u_0 g_0 + u_1 g_1 + u_2 g_2 + u_3 g_3 \quad (5)$$

This (7,4) Hamming code has a total of 16 correct codewords. In fact, a 7-bit binary number can represent 128 kinds of code words, and for the other 112 kinds of code words, all are wrong. For this array design, the codeword after 4-bit data is encoded is 7 bits, the codeword after 8-bit data is encoded is 14 bits, and there are two LED lights that are not working, so it is agreed that the 8-bit data is divided into two groups , the second row, fourth column and fourth row, fourth column of a 4×4 array are set to "0" by default, and this convention is adopted by default in the following coding process.

The function of error detection and correction at the receiving end is accomplished by judging whether the syndrome s is "0".

$$s = r*H^T \quad (6)$$

When s is not "0", it means that the received codeword is wrong. It is not among the 16 codewords in the above table. The error correction function is determined according to the error pattern. The (7, 4) Hamming code can only correct one error bit.

$$H^T = \begin{pmatrix} 1 & 0 & 1 \\ 0 & 1 & 0 \\ 0 & 0 & 1 \\ 1 & 1 & 0 \\ 0 & 1 & 1 \\ 1 & 1 & 1 \\ 1 & 0 & 1 \end{pmatrix} \quad (7)$$

In order to ensure the synchronization of the transmitter and the receiver and the positioning of the initial LED array and each LED light, this design needs to design a marker symbol agreed with the receiver to synchronize at the beginning of transmission. This design uses an FPGA development board. At startup, the default output pins are all high-level characteristics, and the LED array is fully lit as a convention mark to start data transmission.

B. Receiver preprocessing

The image processing at the receiving end first locates the position of the LED light in the received image, and then uses the method of image binarization to distinguish the on-off state of the LED light. Image binarization is to first grayscale the image captured by the optical camera at the receiving end, scan each pixel of the image to obtain the pixel value of the point, and classify the pixel value of the pixel whose pixel value is lower than a certain threshold. The pixel value is set to 0, and the point is black at this time. Set the pixel position of the pixel value higher than a certain threshold to 255, at this time the point is white. This threshold depends on the characteristics of the image and is expected to be 127 in this design. Using this method, the image after binarization can be obtained at the fastest speed, reducing the amount of calculation and improving the receiving speed of the receiver.

After the receiving end is binarized, the area determined by each LED lamp on the pixel plane is processed separately. When the area of the pixel point is 255 (the image is intuitively white) within the area of a divided LED lamp, the area occupies When the LED light area is a certain area (threshold value), it is judged that the LED is on. Because the OOK modulation is used, the LED light is on, and the data is "1" at this time, that is, the light signal sent by the LED is received as one bit. data "1". Conversely, when the area where the pixel position in the range is 0 (the image is intuitively black) occupies a certain area (threshold) of the area, the LED is judged to be off. According to the principle of OOK modulation, the LED light is off, and the data is It is "0", that is, one bit of data "0" is received. Among them, the threshold for judging its area should be determined by the characteristics of the channel and the image sensor at the receiving end. When there is more interference from other lights in the channel, this threshold should be appropriate, and when there is less other light interference, this threshold should be appropriately reduced. At the same time, this threshold should be adjustable so that the LED array and the OCC system can be used normally in different environments.

C. LED array emitter implementation and result analysis

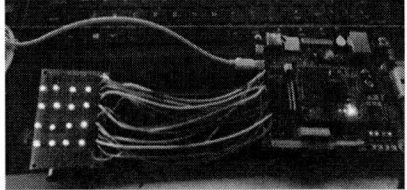

Figure 2. Host computer, FPGA development board, LED array.

The 16 output pins on the FPGA development board are used for driving and power supply. The reason for adopting the common cathode design is to take advantage of the fact that the default output pins of the FPGA development board are all high-level at startup, and the LED array output is all "1" as the sign as the convention sign Start transferring data.

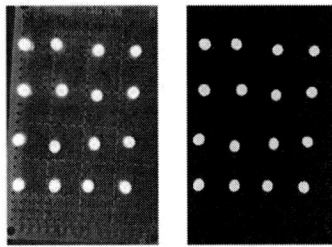

Figure 3. Captured images and preprocessing results when the LED array is fully lit.

- When conducting the information transmission test, firstly in a completely dark environment, at 0.5 meters, 1.5 meters and 3 meters respectively, receive the hexadecimal "87" emitted by the LED array (that is, input the 8-bit data "1 0 0 0" 0 1 1 1"), after preprocessing, the result is shown in the figure.

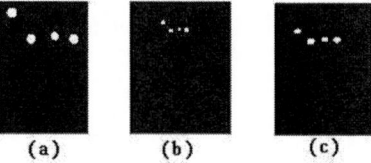

Figure 4. a,b,c are the test results at 0.5m, 1.5m, and 3m respectively.

It can be seen that when the transmission distance is 0.5 meters, the captured image is relatively clear, and the image after preprocessing can clearly see the light-emitting position of the LED light. As the distance increases, the image visible to the naked eye becomes blurred, and the border of the LED light becomes less and less obvious, but after the preprocessed image, the border of the LED does not intersect, but the processed shape of the LED light changes from a circle. The shape tends to be rectangular and the area increases, indicating that when the distance is too far. After the captured image is processed, the borders of LED lights may overlap, resulting in decoding errors.

- To verify the effect of Manchester encoding and Hamming code, further tests were performed. Comparing the preprocessing results of sending hexadecimal "873e" (data "1 0 0 0 0 1 1 1 0 0 1 1 1 1 1 0") data without encoding and Manchester encoding and Hamming code, the test effect is good, the received data is correct.

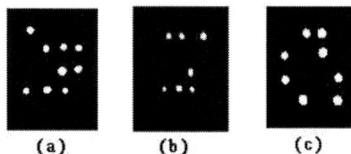

Figure 5. a,b,c are the test results of no coding, Manchester coding and Hamming code respectively.

IV. PREPARE YOUR PAPER BEFORE STYLING

This paper designs an LED array control module based on optical camera communication, drives the LED array circuit, lights up the LED array to send information, and uses the smartphone camera to observe the environment and the influence of different receiving distances on the received signal, so as to judge the sending end. performance. Next, the design needs to be continuously optimized to solve the problem that only one synchronization correction can be achieved, and how to achieve a larger amount of data transmission for further research.

REFERENCES

[1] Jovicic A ,Li J , Richardson T . Visible light communication: opportunities, challenges and the path to market[J]. IEEE Communications Magazine, 2013, 51(12):26-32.

[2] Chow C W, Liu Y, Yeh C H, et al. Display Light Panel and Rolling Shutter Image Sensor based Optical Camera Communication (OCC) Using Frame-Averaging Background Removal and Neural Network[J]. Journal of Lightwave Technology, 2021.

[3] Chow C W, Yeh C H, Liu Y F, et al. Improved modulation speed of LED visible light communication system integrated to main electricity network[J]. Electronics letters, 2011, 47(15): 867-868.

[4] Dansereau D G, Schuster G, Ford J, et al. A wide-field-of-view monocentric light field camera[C]//Proceedings of the IEEE Conference on Computer Vision and Pattern Recognition. 2017: 5048-5057.

[5] Yamazato T, Takai I, Okada H, et al. Image-sensor-based visible light communication for automotive applications[J]. IEEE Communications Magazine, 2014, 52(7): 88-97.

[6] Huang W, Tian P, Xu Z. Design and implementation of a real-time CIM-MIMO optical camera communication system[J]. Optics express, 2016, 24(21): 24567-24579.

[7] Katz M ,Ahmed I . Opportunities and Challenges for Visible Light Communications in 6G[C]// 2020 2nd 6G Wireless Summit (6G SUMMIT). 2020.

[8] Wang J, Huang W, Xu Z. Demonstration of a covert camera-screen communication system[C]//2017 13th International Wireless Communications and Mobile Computing Conference (IWCMC). IEEE, 2017: 910-915.

[9] Zhang M, Shi M, Wang F, et al. 4.05-Gb/s RGB LED-based VLC system utilizing PS-Manchester coded Nyquist PAM-8 modulation and hybrid time-frequency domain equalization[C]//Optical Fiber Communication Conference. Optical Society of America, 2017: W2A. 42.

[10] Wu H and Fan Q,et al. 2020 International Conference on Wireless Communications and Smart Grid (ICWCSG) ,Study on LED Visible Light Communication Channel Model Based on Poisson Stochastic Network Theory, 2020[C]. 5-9.

[11] He P, Huang Z, Huang M, Li W, Liu M and Ji Y ,et al. 2017 16th International Conference on Optical Communications and Networks (ICOCN) ,Indoor intelligent visible light system based on Triple-Domain-Cooperation scheme, 2017[C]. 1-3.

[12] Liu X, et al. 2018 Asia Communications and Photonics Conference (ACP An InGaN micro-LED based photodetector array for high-speed parallel visible light communication, 2018[C]. 1-3.

[13] Li H, et al. in IEEE Photonics Journal, vol. 12, no. 6, A Fast and High-Accuracy Real-Time Visible Light Positioning System Based on Single LED Lamp With a Beacon,2020[C]. 1-12.

[14] Lou G and Shi H, et al. in China Communications, vol. 17, no. 2, Face image recognition based on convolutional neural network,2020[C] .117-12.

[15] Sun E, et al, 2021 IEEE 5th Advanced Information Technology, Electronic and Automation Control Conference (IAEAC), Small-scale image recognition based on Cascaded Convolutional Neural Network, 2021[C]. 2737-2741.

[16] Lee M, Lee J, Kim J, Kim B and Kim J, et al. 2019 International SoC Design Conference (ISOCC)The Sparsity and Activation Analysis of Compressed CNN Networks in a HW CNN Accelerator Model, 2019[C]. 255-256.

[17] Nguyen T, Thieu M D, Jang Y M. 2D-OFDM for optical camera communication: Principle and implementation[J]. IEEE access, 2019, 7: 29405-29424.

[18] Ahmed M F, Hasan M K, Shahjalal M, et al. Design and Implementation of an OCC-based real-time heart rate and pulse-oxygen saturation monitoring system[J]. IEEE Access, 2020, 8: 198740-198747.

[19] Lain J K, Yang Z D, Xu T W. Experimental DCO-OFDM optical camera communication systems with a commercial smartphone camera[J]. IEEE Photonics Journal, 2019, 11(6): 1-13.

[20] Nguyen D T, Park S, Chae Y, et al. VLC/OCC hybrid optical wireless systems for versatile indoor applications[J]. IEEE Access, 2019, 7: 22371-22376.

Self-Supervised Denoising of single OCT image with Self2Self-OCT Network

Chenkun Ge[1], Xiaojun Yu[1], Mingshuai Li[1], and Jianhua Mo[2]

[1]*School of Automation, Northwestern Polytechnical University, Xi'an, Shaanxi, China, 710072*
[2]*School of Electronics and Information Engineering, Soochow University, Suzhou, Jiangsu, China, 215006*
Corresponding Author: Xiaojun Yu   Email: XJYU@nwpu.edu.cn

Abstract—In recent years, supervised deep learning of image denoising has attracted extensive research interests. Those methods usually required numerous pairs of noisy image and its corresponding clean image in training processing. However, in most real situations, it is hard to collect high-quality clean images such as optical coherence tomography (OCT) images. Therefore, it is of great significance to study a effective denoising network without clean images for supervising, which is only trained with noisy image. In this article, for a single OCT image, we propose a self-supervised deep learning model called Self2Self-OCT network by improved the Self2Self network and added a loss function that can effectively remove the background noise of OCT images, which makes the whole training do not need correlative clean images. Specifically, we use gated convolution to replace the partial convolution layer of the encoder's block in Self2Self. The input image and its Bernoulli sampling instance are put into our network respectively, and the background noise attenuation loss is added to loss function during training. The result is estimated based on the average value of multiple prediction outputs. The experiments with different OCT images indicate that proposed model not only has obvious advantages compared with the existing single deep learning methods and non-learning methods, but also surpasses the supervised learning of a small number of sample training.

Keywords-self-supervised;speckle noise;denoising;optical coherence tomography image

I. INTRODUCTION

Optical coherence tomography (OCT) is an optical interference imaging method based on low coherence and interferometry, which can provide high-resolution images of biological tissue microstructure [1]. Due to its non-invasive and high-resolution characteristics, OCT has been widely used in the diagnosis of various diseases, especially in ophthalmology [2]. Speckle noise greatly reduces the quality of the image. Due to its blur effect [3], it reduces the accuracy of OCT disease diagnosis. For improving the quality of OCT image and the accuracy of OCT based diagnosis, various research teams have proposed more and more effective OCT denoising methods in recent ten years.

In recent years, deep learning methods have emerged with excellent performances in OCT image denoising. Deep learning methods are popular because they can retain more details in the denoising process. Specifically, Jain et al., for the first time, proposed the use of a convolutional neural network (CNN) to denoise natural images [4] and achieved better results than the conventional methods. Zhang et al.

proposed a deep network called DnCNN to reduce noise in OCT image [5]. Its denoising performance is boosted because DnCNN adopts the residual learning scheme to improve the network learning ability, and uses the batch normalization layer to solve the gradient dispersion effect. However, at present, most denoising deep learning networks are supervised. Those methods require collecting numerous noise and clean image pairs for training. For the generation of real OCT images, clean images are hard to get. The most effective way is to use multiple averaging strategy(i,e., repeat imaging the same position for many times) to produce relatively clean images. In most cases, relatively clean images will also have problem such as serious blurs by long acquisition time and some motions. Therefore, it is of great significance to study a deep learning denoising model that does not need clean images. In other words, the denoising neural network only learns from the noisy image. By now, little work has been done in OCT image denosing without clean image. Ulyanov et al. [6] proposed a deep learning model named the deep image prior(DIP) for single image recovery. But performance of this method is not competitive to BM3D. Quan et al. [7] proposed a self-supervised learning method named Self2Self with only uses the noisy image for training.

Inspired by the concept that result of Self2Self is obtained by averaging the output images by the trained model with dropout, we attempt to build a network named Self2Self-OCT, which introduces gated convolutional layer [8] to replace partial convolutional layer [9] in encoder's blocks of the original network, and a loss function which can effectively remove the background noise is added to the original loss function. The following is a summary of our technical contributions.

●Gated convolution is utilized to boost the denoising performance of the Self2Slef-OCT network.

●Background noise attenuation loss is added to the loss function, which can effectively remove the background noise of OCT image.

●Compared with existing denoising method based on single image or small sample of images, it has reliable performance improvement.

The rest of this paper is organized as follows. Section 2 introduces the proposed Self2Self-OCT network including training scheme and denoising scheme. Section 3 presents

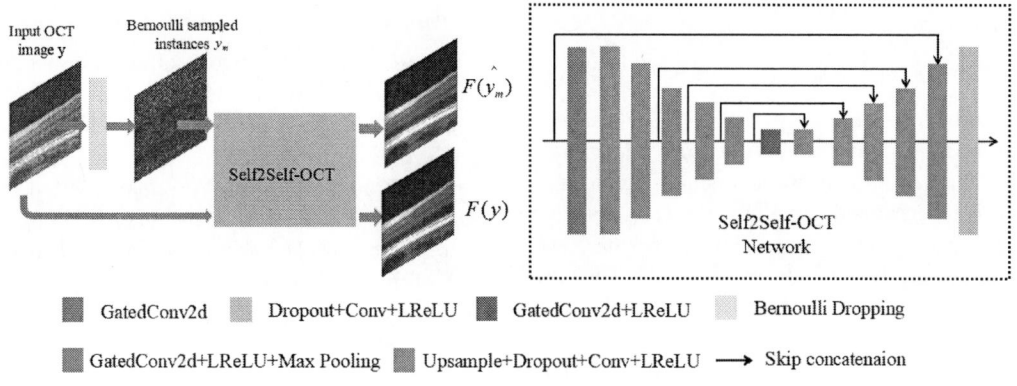

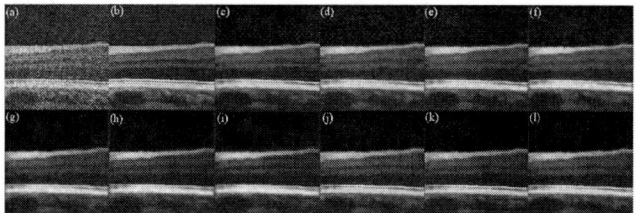

Figure 1. Structure of Self2Self-OCT network.

the denoising results on clinical OCT data set. Section 4 summarizes this paper and puts forward some future work.

II. METHOD

In this paper, we propose a single image denoising method, which can be trained without clean images. The following will introduce the network structure of Self2Self-OCT, and the schemes for self-supervised learning and denoising.

A. Structure of Self2Self-OCT

The framework of Self2Self-OCT is illustrated in Fig.1. The whole network structure is an encoder-decoder. The size of an OCT noisy image is $H \times W \times 1$, it obtains an $H \times W \times 48$ shape through the encoder with gated convolution [8], and then processed by the following six encoder's blocks. The first five encoder's block all contains gated convolution layer, a leaky rectified linear unit(LReLU), and max pooling layer with 2×2 kernel size and stride of 2. The output channel of every encoder's block is set to 48. And the size of final output of the encoder is $H/32 \times W/32 \times 48$. All the gated convolution layers are used 3×3 kernel size, and each LReLU is set as 0.1.

The decoder part has five decoder's blocks. The decoder block contains upsample layer, convolution layer with dropout and LReLU. Every upsample layer used 2 scaling factor for decoding. The output channels of first four decoder blocks are 96. And final decoder block utilizes three convolution layers with dropout and LReLU to obtain output image with size of $H \times W \times 1$, and every output channel in final decoder block are set to 64, 32, 1, respectively.

The structure of our network is basically the same as that of Self2Self network. The difference is that we use gated convolution to replace partial convolution in encoder's block, which further promotes the effectiveness and efficiency of model training.

Figure 2. A single noise OCT image cropped as 256×256 and its clean image and its denoising results. (a) noisy image,(b)clean image,(c)Result of the first round of output,(d)Result of the second round of output,(e)Result of the third round of output,...(l)Result of the 10th round of output.

B. Training Scheme

Simply training a single noise image y has no practical significance, because the learned features can not achieve effective denoising effect. Multiple imaging pairs are needed to produce from y. Therefore, we can use a series of Bernoulli sampling pairs to train the network. A set of image pairs $\{(\hat{y}_m, \bar{y}_m)\}_{m=1}^{M}$ is defined as:

$$\hat{y}_m = b_m \cdot y; \bar{y}_m = (1 - b_m) \cdot y \quad (1)$$

where $\cdot$ denotes the element-wise multiplication and b indicates the mask image obtained after Bernoulli sampling, and its shape is the same as noisy image y.

With each set of produced image pairs, the self-prediction loss could be defined as:

$$Loss_{self-prediction} = \min \sum_{m=1}^{M} \|f(\hat{y}_m) - \bar{y}_m\|_{b_m}^2 \quad (2)$$

where $f(\cdot)$ denotes the training network.

In order to better eliminate the background noise, we also introduce the background noise attenuation loss, it defined as:

$$Loss_{bna} = \min \frac{\sum_{m=1}^{M} |f(y) - f(\hat{y}_m)|}{M} \quad (3)$$

Given the self-prediction loss and background noise attenuation loss, the whole loss function is minimized by following formula:

$$Loss = Loss_{self-prediction} + \alpha \cdot Loss_{bna} \quad (4)$$

From self-prediction loss, only the loss of partially masked pixels by b_m is calculated. And from background noise attenuation loss, it minimizes the output of the original noise image and its Bernoulli sampling instances, in order to weaken the learning ability of the network on the background noise of image. In a word, self-prediction loss can learn the information related to ground truth image x from beginning to end, because training the pairs of complementary Bernoulli sampled images $\{(\hat{y}_m, \bar{y}_m)\}$ is very close to training the pairs of a Bernoulli sampled image $\hat{y}_m$ and the clean image x when many complementary Bernoulli sampled pairs are used for training; background noise attenuation loss can suppress background noise learning. Generally speaking, in the early stage of training, $f(\hat{y}_m)$ first generates the structural details of the image through network learning, and then slowly learns the relevant noise. $f(y)$ and $f(\hat{y}_m)$ are minimized to weaken the learning ability of the network in the background area in the initial learning. Because the background area of OCT image itself is no signal.

C. Denoising Scheme

A trained neural network with dropout could generate a series of neural networks whose weights follow an independent Bernoulli distribution. In our example, the purpose of droput is to reduce the variance of denoised images. We aim to generate multiple neural networks from trained neural networks in order to have multiple estimators that may have some independence. The multiple denoised images $\hat{x}_1, \cdots, \hat{x}_N$ could be outputed by inputting a Bernoulli sampled image of y to every neural networks. The final denoised result x^* is produced by averaging multiple denoised images.

$$x^* = \frac{1}{N} \sum_{n=1}^{N} \hat{x}_n = \frac{1}{N} \sum_{n=1}^{N} f(b_{M+n} \cdot y) \quad (5)$$

In the denoising process, the output denoising results can be generated in the same period with the training process.

III. EXPERIMENTS

A. Datasets

The public OCT image data set is used for denoising experiments. These images are collected through the Bioptigen SDOCT system (Durham, NC, USA) with an axial resolution of 4.5 μm per pixel in tissue [10].

The retinal OCT volume data set is introduced from [11] [12], which includes 18 pairs of noisy OCT images and its corresponding clean images. The clear images are obtained by registering and averaging several B-scans acquired at same position. The 10 pairs of noise/clean image pairs with good quality are selected to experiment (Each is of size 500×950) and and normalized all the images. The clean image is only used to calculate the denoising metrics.

B. Parameters Setting

In Self2Self-OCT network, all the dropout probability and probability of bernoulli sampling are set to 0.5. In the loss function of training, α is set to 0.3. The learning rate of the whole training is 0.0001, and the total number of training steps is 2000. 200 times of dropout is tested to generate the final result. In other words, the whole training will output 10 denoised results. The networks were implemented in Python by PyTorch framework, and all experiments were conducted on a workstation (Intel Xeon W-2145 CPU @3.70GHz) and were accelerated by an NVIDIA GeForce RTX 3060Ti GPU with 8G memory.

C. Results of single image of mini size

Firstly, we use the cropped noise image(size is 256×256) to conduct the experiment. The experimental results are shown in the Fig.2. It can be seen that in the initial output denoised image, the background part of the image has no noise, but there are still some blurring phenomena in the whole structural details. With further training, the structural details of the image become clearer and clearer, and finally get a series of subsequent denoising images with very good quality.

In addition, we also recorded the denoising metrics of the denoised image and clean image in each round of training, as shown in Fig.3. Fig.3 demonstrates that the change degree of PSNR is basically consistent with that of SSIM. Due to the application of dropout, the indicators calculated in each round may be good or bad, but there will always be the best time to achieve the indicators. Moreover, it is worth noting that the two indicators we use are related to the clean image, but we can observe that there are some noise in the background even in the clean image. Ideally, the background of a clean image is noise free. Therefore, the calculated two indicators are only for reference.

D. Results comparison with other method

For verifying the effectiveness of the proposed Self2Self-OCT, experiments were tested on a public OCT retinal image data sets from [11]. And our method is compared with some state-of-the-art denoising methods: BM3D [13], NWSR [14], DnCNN [5], DIP [6] and TSI [3].Ten OCT retinal images from dataset were processed for comparisons in this study. We denoise ten different OCT images, record the denoising metrics of each round, and select the image with the best performance as the final denoised result. Other algorithms are also tested to get the best results. In particular, for the DnCNN with supervised learning, we only use three images, which are crop into 256×256 size by a stride of 50 for training, and the whole training set has only about 210 images. Therefore, for DnCNN, less training sets are used for training.

Here, a visual comparisons of testing OCT image is demonstrated. For the testing image, it can be seen visual

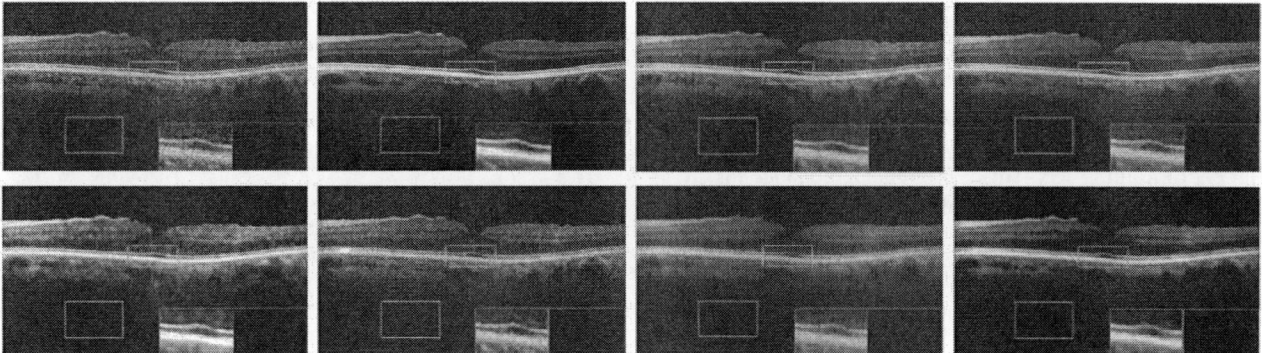

Figure 3. An OCT image selected from data set and its different denoising results. (a)Original, (b)averaged, (c)BM3D, (d)NWSR, (e)DnCNN, (f)DIP, (g)TSI, (h)Self2Self-OCT.

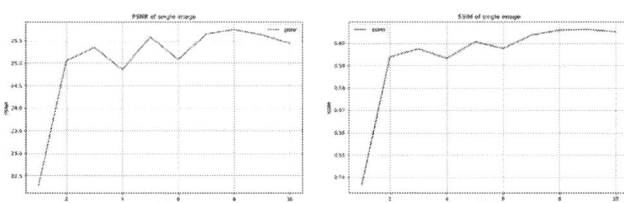

Figure 4. The PSNR and SSIM is calculated in training process in first testing image.

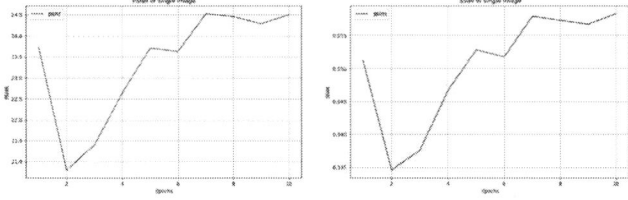

Figure 5. The metrics of denoising result and clean image in each round.

comparison of denoising results obtained from the BM3D, NWSR, DnCNN, DIP, and TSI method in Fig.4. Meanwhile, we also crop the image patch with background noise for observing. As can be seen, BM3D, DnCNN and DIP exhibit serious noisy appearance, NWSR and TSI significantly remove noise, however, there is still residual noise in some areas of the background. In addition, no matter from the complete image or the cropped image patch, the background noise after denoising is well suppressed, and the structural details of the whole image are preserved. Fig. 5 shows the calculation metrics obtained by different rounds in the training process. It can be found from the Fig.5 that the change trend of PSNR and SSIM is consistent in the whole training process, indicating that they have a very high correlation. Moreover, the metrics do not increase all the time and may decline at some time. So we need to manually select the most effective testing result.

Ten OCT images in the data set are used as noise images.

Table I
QUANTITATIVE RESULTS OF THE DATA SET WITH DIFFERENT METHODS.

	PSNR(dB)	SSIM
BM3D	21.0777	0.8841
NWSR	20.6912	0.8824
DnCNN	23.1747	0.9911
DIP	20.0808	0.8729
TSI	20.6395	0.8698
Self2Self-OCT	**24.8002**	**0.9961**

We average the metrics obtained by each method, and get Table 1. Table 1 presents the performances of different methods with the images in data set. Hence, it can be observed from Table 1 that Self2Self-OCT achieves the highest PSNR and SSIM values in all those denoising methods, which shows that our methods has excellent denoising ability compared with other methods and can better remove the background noise. This indicates that our method is superior to the existing well-known single OCT image denoising algorithm, and exceeds the supervised learning method of a small number of samples. It can be widely used in real-time clinical denoising applications to solve the corresponding clinical diagnosis.

IV. CONCLUSION

In summary, a self-supervised deep learning method named Self2Self-OCT for a single OCT image denoising is proposed, which do not require clean image during training. We input the original OCT noise image and its Bernoulli sampling images into the Self2Self-OCT network respectively. Reducing the prediction variance is the focus of denoising problem. Therefore, dropout is used in training and testing. We get the final result by averaging to reduce the prediction variance. In the network, we use gated convolution in the encoder's block and add background noise attenuation loss to the loss function in training. These applications improve the overall denoising level. A large number of experiments show that the performance of the Self2Self-OCT network based on OCT image denoising is far better than other single image denoising methods, and even surpasses the deep learning method based on a small number of training sets. This is of great significance for OCT single image denoising technology.

ACKNOWLEDGMENT

This research is supported in part by the National Natural Science Foundation of China (61705184), the Key Research and Development Program of Shaanxi (2021SF-342), the Fundamental Research Funds for the Central Universities (G2018KY0308), China Postdoctoral Science Foundation (2018M641013), Postdoctoral Science Foundation of Shaanxi Province (2018BSHYDZZ05).

REFERENCES

[1] D. Huang, E. A. Swanson, C. P. Lin, J. S. Schuman, W. G. Stinson, W. Chang, M. R. Hee, T. Flotte, K. Gregory, C. A. Puliafito *et al.*, "Optical coherence tomography," *science*, vol. 254, no. 5035, pp. 1178–1181, 1991.

[2] W. Drexler and J. G. Fujimoto, "State-of-the-art retinal optical coherence tomography," *Progress in retinal and eye research*, vol. 27, no. 1, pp. 45–88, 2008.

[3] X. Wang, X. Yu, X. Liu, S. Chen, S. Chen, N. Wang, and L. Liu, "A two-step iteration mechanism for speckle reduction in optical coherence tomography," *Biomedical Signal Processing and Control*, vol. 43, pp. 86–95, 2018.

[4] V. Jain and S. Seung, "Natural image denoising with convolutional networks," *Advances in neural information processing systems*, vol. 21, 2008.

[5] K. Zhang, W. Zuo, Y. Chen, D. Meng, and L. Zhang, "Beyond a gaussian denoiser: Residual learning of deep cnn for image denoising," *IEEE transactions on image processing*, vol. 26, no. 7, pp. 3142–3155, 2017.

[6] D. Ulyanov, A. Vedaldi, and V. Lempitsky, "Deep image prior," in *Proceedings of the IEEE conference on computer vision and pattern recognition*, 2018, pp. 9446–9454.

[7] Y. Quan, M. Chen, T. Pang, and H. Ji, "Self2self with dropout: Learning self-supervised denoising from single image," in *Proceedings of the IEEE/CVF conference on computer vision and pattern recognition*, 2020, pp. 1890–1898.

[8] J. Yu, Z. Lin, J. Yang, X. Shen, X. Lu, and T. S. Huang, "Free-form image inpainting with gated convolution," in *Proceedings of the IEEE/CVF International Conference on Computer Vision*, 2019, pp. 4471–4480.

[9] G. Liu, F. A. Reda, K. J. Shih, T.-C. Wang, A. Tao, and B. Catanzaro, "Image inpainting for irregular holes using partial convolutions," in *Proceedings of the European conference on computer vision (ECCV)*, 2018, pp. 85–100.

[10] S. Farsiu, S. J. Chiu, R. V. O'Connell, F. A. Folgar, E. Yuan, J. A. Izatt, C. A. Toth, A.-R. E. D. S. . A. S. D. O. C. T. S. Group *et al.*, "Quantitative classification of eyes with and without intermediate age-related macular degeneration using optical coherence tomography," *Ophthalmology*, vol. 121, no. 1, pp. 162–172, 2014.

[11] L. Fang, S. Li, R. P. McNabb, Q. Nie, A. N. Kuo, C. A. Toth, J. A. Izatt, and S. Farsiu, "Fast acquisition and reconstruction of optical coherence tomography images via sparse representation," *IEEE transactions on medical imaging*, vol. 32, no. 11, pp. 2034–2049, 2013.

[12] L. Fang, S. Li, D. Cunefare, and S. Farsiu, "Segmentation based sparse reconstruction of optical coherence tomography images," *IEEE transactions on medical imaging*, vol. 36, no. 2, pp. 407–421, 2016.

[13] K. Dabov, A. Foi, V. Katkovnik, and K. Egiazarian, "Bm3d image denoising with shape-adaptive principal component analysis," in *SPARS'09-Signal Processing with Adaptive Sparse Structured Representations*, 2009.

[14] A. Abbasi, A. Monadjemi, L. Fang, and H. Rabbani, "Optical coherence tomography retinal image reconstruction via nonlocal weighted sparse representation," *Journal of biomedical optics*, vol. 23, no. 3, p. 036011, 2018.

Proceedings of the 7th Optoelectronics Global Conference (OGC 2022)

Adaptive Dynamic Analysis-based Optical Coherence Tomography Angiography for Blood Vessel Tail Artifacts Suppression

Junxiong Zhou
Department of Biomedical Engineering
Southern University of Science and Technology
Shenzhen, Guangdong 518055, China
11810511@mail.sustech.edu.cn

Yuntao Li
Department of Biomedical Engineering
Southern University of Science and Technology
Shenzhen, Guangdong 518055, China
454134703@qq.com

Jianbo Tang*
Department of Biomedical Engineering
Southern University of Science and Technology
Shenzhen, Guangdong 518055, China
*Corresponding author: tangjb@sustech.edu.cn

Abstract—**Optical coherence tomography angiography (OCTA) for blood vessel 3-D structure imaging suffers from blood vessel tail artifacts when using a long decorrelation time (e.g., repeat B-scan acquisition in regular OCTA) or loss of micro vessel signal when using a short decorrelation time. In this work, we developed an adaptive first order field autocorrelation function (g1) analysis-based technique to suppress the tail artifacts under macro vessels while enhancing the dynamic signal of micro vessels. The proposed method is based on the differences of the decorrelation rate and the phase variations of g1 between the vessel voxels and the artifacts regions. A short or a long decorrelation time was applied to obtain the dynamic index of the tail artifacts region or the blood vessel region, respectively. Compared to the post image processing-based techniques, the proposed approach addresses this problem through the physical basis and shows a good performance in suppressing the tail artifacts while enhancing the detection of the micro vessels.**

Keywords- optical coherence tomography angiography; brain imaging; dynamic imaging

I. INTRODUCTION

Optical coherence tomography angiography (OCTA) has become an important tool for studying the micro-vascular structural and dynamic changes [1,2]. OCTA obtains the blood flow information by comparing the signal changes at the same spatial location with the fact that the signals within blood vessels are changing over time due to the flow of red blood cells (RBCs), while the signal reflected from tissue are more stable. Thus, by detecting the changes of OCT amplitude/intensity, phase, or complex signals [3–8], OCTA is able to image the blood vessel with high spatial and temporal resolution. It has been widely used in ophthalmology [5,7,9], dermatology [10,11] and neurosciences [12–14].

Essentially, OCTA imaging of blood vessels is based on the detection of dynamic contrasts between the moving RBCs and the static tissues. As the main scatter particles in the blood vessels, RBCs have very high scattering anisotropy, leading to strong forward multiple scattering [15]. Photons carrying dynamic signals after hitting the moving RBCs may experience single-backward scattering, multiple-forward and backward scattering, and even transmit through the blood vessel and then be reflected back after hitting the tissues underneath the blood vessels. In the last two scenarios, the light paths of the multiple scattered photons are longer than that of the directly reflected photons (Fig.1A). Hence the dynamic signal from photons hit with the moving RBCs may extend to regions underneath the bottom layer of the vessel, resulting in the long-standing challenge of the blood vessel tail artifacts in OCTA imaging. The blood vessel tail artifacts not only extend the blood vessel lumen in the axial direction but also bury the micro vessels' signal in the tail artifacts background.

Several hardware-based and post image processing-based approaches have been proposed to suppress the blood vessel tail artifacts in OCTA [16–18]. Leahy et al. applied a high NA objective to suppress the blood vessel tail artifacts by reducing multiple scattering signal for the out-of-focus region at the expense of a reduced depth of field [17]. Woo et al. calculated the mean intensity of all pixels in each individual A-lines and subtracted it from original image [16]. Stefan et al. introduced a deep learning method for vessel visualization, enhancement and segmentation, which also suppressed the tail artifacts [18]. We previously proposed a g_1-OCTA method which used a shorter decorrelation to suppress the blood vessel tail artifacts [19]. The g_1-OCTA method had a limitation in balancing the goals of suppressing blood vessel tail artifacts and enhancing signals from slowly moving capillary networks. The blood vessel tail artifacts became stronger when using a long decorrelation time, while the micro vessel signals became weaker when using a short decorrelation time.

In this work, we developed an approach to adaptively determine the decorrelation time of the g_1 function for the calculation of the dynamic index (Ag_1-OCTA) with intrinsic contrast. Results suggest that our method can greatly

978-1-6654-8699-6/22 $31.00 © 2022 IEEE

suppress the tail artifacts under macro vessels (large pial vein and large pial artery) and enhance the dynamic signal of micro vessels (arterials, venules, and capillaries).

II. MATERIALS AND METHODS

A. Experimental Set-up and Scanning Protocol

Our OCT system used a dual-SLD light source with a center wavelength of 1550 nm and a bandwidth of 210 nm. The spatial resolution in the brain of the system was 3.7 um in axial and 3.5 um in lateral using a 10× objective. The data was processed offline with MATLAB 2020b (MathWorks, Natick, MA).

B. Animal Preparation

We acquired *in vivo* data from 5 female C57BL/6 mice prepared with a 3 mm diameter chronic cranial window. The surgeries were performed two weeks before the experiments. During the imaging session, the animals were anesthetized with 1-2% isoflurane and a region of interest (ROI) with 600 um×600 um (400×400 pixels in X and Y) area was acquired with our SD-OCT system. The animal experiment protocol was approved and supervised by the Animal Ethics Committee at the Southern University of Science and Technology.

C. Ag_1-OCTA Method

We first obtained the autocorrelation function $g_1()$ from the OCT field signal for each voxel:

$$g_1(\tau) = \frac{\langle R^*(t)R(t+\tau)\rangle_t}{\langle R^*(t)R(t)\rangle_t} \qquad (1)$$

Where $R(t)$ is the complex OCT signal at time t, $R^*(t)$ is the complex conjugate, τ is the time lag, and $\langle\rangle$ represents ensemble averaging. [19]

Then, we defined the dynamic contrast index I_d as the maximum decorrelation during the defined decorrelation time of the after the first time lag:

$$I_{d(1:n_\tau)} = |g_1(1)| - \min|g_1(\tau)_{(1:n_\tau)}| \qquad (2)$$

Where $||$ denotes magnitude, and n_τ is the decorrelation time point. Usually, with a larger n_τ the $g_1(\tau)_{(1:n_\tau)}$ decays to a lower level thus resulting in a larger I_d. The goal of Ag_1-OCTA is to adaptively select n_τ for I_d calculation, i.e. using a short n_τ for the tail artifact regions and a long n_τ for the blood vessel voxels. So, the key is to distinguish the vessel voxels from the tail artifacts. We noted that the g_1 decays differently for voxels within the vessels and in the tail artifacts regions.

The goal of Ag_1-OCTA is to adaptively select n_τ for I_d calculation, i.e. using a short n_τ for the tail artifact regions and a long n_τ for the blood vessel regions. We noted that g_1 decorrelates differently for voxels within the vessels and in the tail artifacts regions. As shown in Fig. 1B, within the big vessels (black dot), blood flows at high speeds which results in fast decorrelation (black curve) in Fig. 1C1. For voxels in

small vessels (blue and green curve), the decorrelation is slower than that in the big vessels but still faster compared to the tail artifacts region (red curve). However, if the decorrelation time becomes long enough the g_1 from tail artifacts region would decay to a similar level to the small vessels, resulting in a non-negligible tail region beneath the big vessels. That's the case of regular OCTA which employs B-scan repeats to detect the dynamic contrast and the time interval is usually longer than 5 ms. Then, by comparing the imaginary signal of g_1 between the artifacts and micro vessels, we further noted that the micro vessel's signal has a clear phase change compared to the artifacts, as shown in Fig. 1C2. This is due to the fact that the signal from micro vessel is mainly from directional and transitional flow of RBCs, which results in a constant frequency shift or broadening. In contrast, the signal from the tail artifacts is a mix of various multiple scattered dynamic signals from the flowing red blood cells and the static signal reflected by the underneath tissue, which is featured with a random dynamic signal plus a static offset.

With these features, we propose a g_1 analysis method to determine the blood vessel from the tail artifacts region, and then adaptively apply a shorter decorrelation time to suppress the tail artifacts and a longer decorrelation time to enhance the blood flow signal.

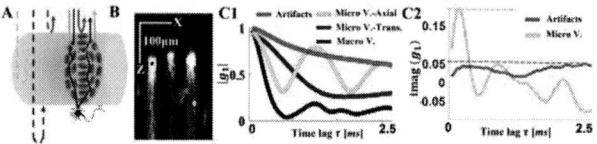

Figure 1. (A) Photon-RBC interaction (solid lines) and the equivalent light path (dashed lines). Revised from[20]. (B) Blood vessel tail artifacts and typical $g_1(\tau)$ decorrelation from different regions (C1). V. stands for vessel, Trans. stands for transverse. (C2) Distinct phase changes were observed between the micro vessels (green) and artifacts (red) in the tail region.

Fig. 2 illustrates the adaptive dynamic approach. Firstly, based on the fact that the tail artifacts primarily appear under big pial blood vessels, we determined the macro vessel regions from the maximum intensity projection (MIP) image calculated with long n_τ, as shown in Fig. 2A. We used the binary map (Fig. 2A1) of blood vessels to obtain a vessel diameter encoded skeletonized vessel map (Fig. 2A2). Then the micro vessels were removed if the diameter is less than 25 μm, which was determined empirically. Finally, the transverse coordinates of the big vessels were obtained as shown in Fig. 2A3, and the tail artifacts suppression and micro flow signal enhancement were performed in the axial projection regions of the big vessels.

The second step is to identify the macro vessel lumen area by using the axial profile analysis of the dynamic index (I_d) in combination with the $g_1(\tau)$ decorrelation. We determined the blood vessel lumen by finding the axial profile peak and the width to peak from the upper boundary of the vessel, then examined the $g_1(\tau)$ decorrelation which shall have similar decay rate.

The third step is to obtain the micro vessel voxels in the tail region from the artifacts background. The signal from the tail artifacts region is a mix of the dynamic signal due to multiple scattering of photons with the moving RBCs and the static signal that reflected from the tissue. Thus, the $g_1(\tau)$ decorrelation in the tail artifacts region would slowly decay and not fully decorrelate to 0, while the signal within the vessel decays faster and would decay to a much lower level. With this feature, we first used a $g_1(\tau)$ decorrelation threshold of 0.5 within the first 1 ms. In addition, we applied the feature that for micro vessels the imaginary part of $g_1(\tau)$ has a constant phase change that featured with a larger amplitude change(Fig. 2D) to determine the micro vessels in the tail regions.(Fig. 2C).

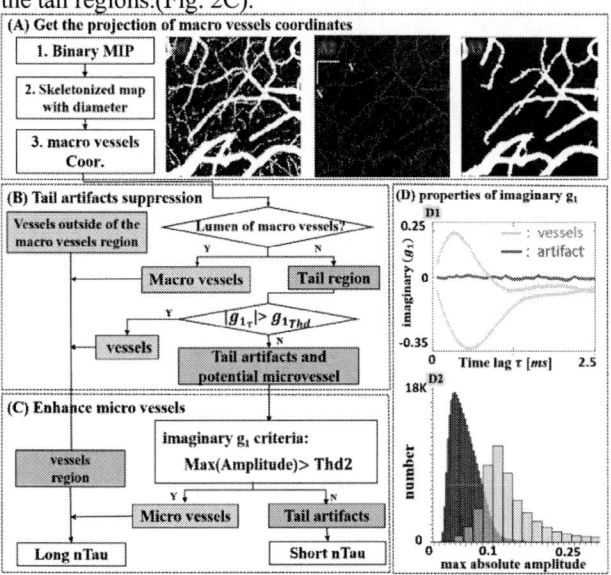

Figure 2. Ag_1-OCTA data processing flowchart. (A) Step 1: identify the large blood vessels' coordinates; (A1) -(A3): Stepwise results to obtain the large vessel binary maximum intensity projection (MIP) map. (B) Step 2: determine macro vessel and the tail regions; Step 3: use the criteria of $|g_1(\tau)|$ decorrelation rate to suppress artifacts in the tail region; (C)Step 4: identify the micro vessels from the tail region by the imaginary part of $g_1(\tau)$ criteria; Step 5: apply long decorrelation time to get enhanced blood vessel signal and short decorrelation time to suppress the tail artifacts. Scale bar: 100um. (D) Quantitative imaginary part of $g_1(\tau)$ between vessels and artifacts; (D1) Representative decorrelation of the imaginary part of $g_1(\tau)$ for vessel voxels (green) and artifact voxel (red); (D2) Histogram distribution shows the maximum change of the imaginary part of $g_1(\tau)$ of the artifacts voxels (red) and the vessel voxels (green) in the tail region. Note: different bin widths were used to draw the histograms.

Finally, to suppress the tail artifacts and enhance blood vessel signal, we applied a short decorrelation time for the dynamic contrast I_d calculation in the artifact region, and a long decorrelation for the identified vessel voxels.

III. RESULTS

Fig. 3 shows the ability to recover the micro vessel signals of the proposed Ag1-OCTA technique from the tail artifacts background. Fig. 3A shows a macro pial vessel and its tail region. We see that strong tail artifacts extend into the deep regions and the underneath micro vessel signals are merely identifiable. The $g_1(\tau)$ decorrelation curves of the

macro vessel (black dot along the red line in Fig. 3A), an artifact voxel (red dot along the red line in Fig. 3A), and a micro vessel (green dot along the red line in Fig. 3A) decay at different rates, as shown in Fig. 3B1. If only apply the decorrelation rate threshold (g_{1Thd}) to distinguish the artifacts and flow voxels, we would obtain a result with many broken-flow-micro vessels in the tail region as indicated by the green arrows in Fig. 3B3. The phase change differences between the artifacts and the blood flows can further help us to recover the micro vessel voxels. As shown in Fig. 3C1, the imaginary signal change of an artifacts voxel (red dot of the en face cross section image in Fig. 3A) and a micro vessel voxel (green dot of the en face cross section image in Fig. 3A) shows that the micro vessel has a dominant frequency change with a much larger phase change. With this feature, we can recover the micro vessel voxels in the tail regions, as shown in Fig. 3C3.

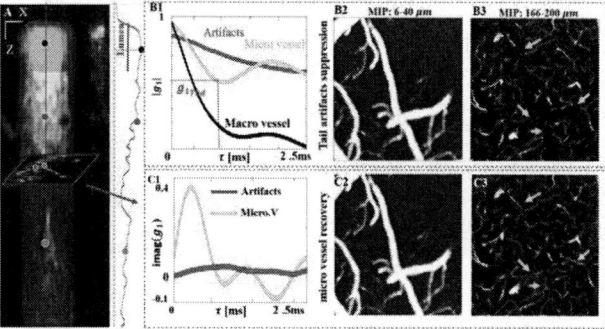

Figure 3. The phase information from $g_1(\tau)$ can be used to recover the micro vessel signal after applied with the decay rate threshold criteria. (A) X-Z view of a macro pial vessel and its tail region. The en face cross sectional image shows a micro vessel in the tail artifacts with the red dot of an artifact voxel and the green dot of a micro vessel voxel. (B1) The magnitude of $g_1(\tau)$ decorrelation of the black, red, and green voxels along the red line in (A). (B2&B3) Results were obtained by applying the decorrelation rate of g_{1Thd} only. (C1) The imaginary signal of $g_1(\tau)$ decorrelation of an artifact voxel and a micro vessel voxel marked by the red and green dots in the en face image in (A). (C2&C3) Results show that with the phase information micro voxels in the tail region can be recovered.

We then compared the results calculated with long-, short-, and the proposed adaptively determined-decorrelation time, as shown in Fig. 4. We see that at the top layer (Fig. 4(A1-C1)) the long and adaptive methods obtained similar results, while the short decorrelation-based method was unable to image many micro vessels and it has much weaker signal intensity at slow flowing macro vessels. In the middle layer (Fig. 4(A2-C2)), the long decorrelation-based calculation has strong tail signal, while both the short decorrelation-based and the proposed adaptive methods can greatly suppress the tail artifacts beneath the macro vessels (red arrows). However, in deeper layers (Fig. 4(A3-C3)) the short decorrelation-based method shows very weak signal in micro vessel regions, whereas the adaptive method shows strong signal which is comparable to that obtained with long decorrelation calculation (blue arrows), suggesting that the proposed method can not only suppress tail artifacts but also preserve blood flow signals of micro vessels.

978-1-6654-8699-6/22 $31.00 © 2022 IEEE

Fig. 4D&E further compare the axial cross-sectional results obtained with these three methods. We see that the proposed adaptive method (Fig. 4(D3)) can greatly suppress the blood vessel tail artifacts under macro vessels, enhance signals in slowly flowing vessels (green arrows & red circles), and extract micro vessel signals buried in the tail artifacts background (yellow arrow, Fig. 4(E1)). The overlapped color-encoded B-scan image (Fig. 4(E1)) shows that a large amount of tail artifacts background is suppressed with the proposed adaptive method while micro vessels are

and then adjusted the signal intensity in the deeper layers accordingly.

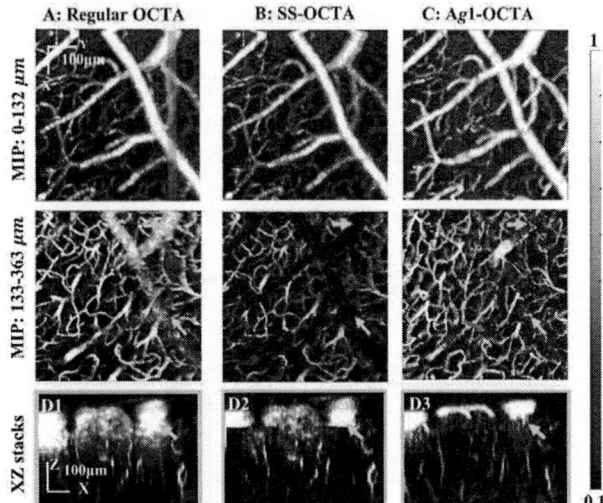

Figure 5. Comparison with the regular OCTA (A), the slab-subtraction based method (B), and the proposed Ag_1-OCTA (C) at two depth layers. (D1-D3): corresponding XZ stacks (MIP over 66 μm in Y) marked by blue shaded region in figure A. The green circles show the ability of suppressing tail artifacts. Scalebar: 100 μm.

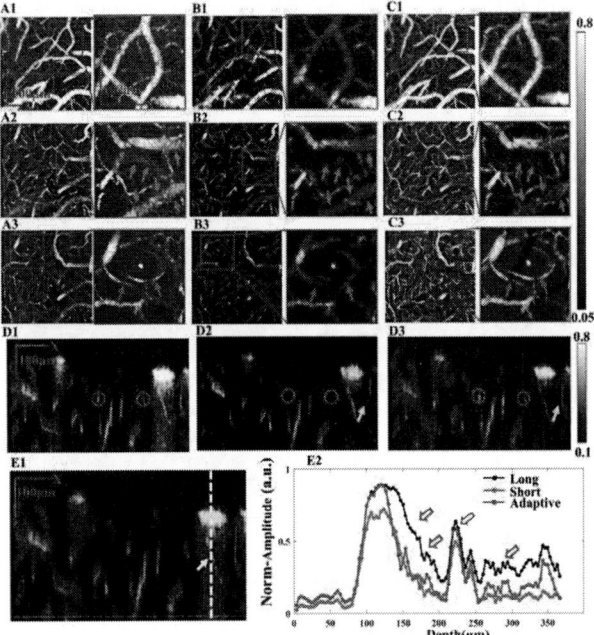

Figure 4. Comparison of the long-, short-, and adaptively determined-decorrelation time obtained g1-OCTA results. En face MIPs obtained with long (A1-A3), short (B1-B3), and the proposed adaptively determined decorrelation time (C1-C3), respectively. (D1-D3) XZ-view stacks (MIP over 25 μ m in Y) marked by red shaded region in A1. (E1) Overlapped OCTA B-scan image of D1(gray background) and D3 (cyan color). (E2) Axial profiles obtained with the three methods at the dashed yellow line marked in E1. Yellow arrow in E1 and E2 indicate preserved micro vessels under macro vessels. Red hollow arrows in E2 indicates suppressed tail artifacts region.

preserved. Fig. 4(E2) further compare the axial profiles obtained with the long, short, and the proposed adaptive decorrelation time methods. We note that the long decorrelation-based calculation has strong tail signal, while both the short decorrelation-based and the proposed adaptive methods can greatly suppress the tail artifacts beneath the macro vessels (red hollow arrows). The micro vessel signal obtained with the proposed method has similar signal intensity to the long decorrelation time-based method but with a suppressed tail artifacts background.

We further compared the proposed method with one of the representative post image processing-based methods, the slab-subtraction approach (SS-OCTA) [21]. For a fair comparison and a better visual assessment, we first normalized the results so that the mean signal intensity of the pial macro vessels is the same of '1' for all three methods,

Compared to the regular OCTA (Fig. 5A), it shows that the SS-OCTA (Fig. 5B) removes the blood vessel tail artifacts and the proposed approach (Fig. 5C) does a good job in suppressing the tail artifacts. In the deeper layers (Middle MIP: 133-363 μm), we noted that the SS-OCTA artefactually produces many disconnected micro vessel segments in the macro vessel tail regions. In contrast, Ag_1-OCTA can not only suppress the tail artifacts but also preserved the micro vessel signal in the tail regions, as indicated by the green arrows. The bottom row of Fig. 5 compares the coronal cross-sectional MIP of the three methods. Clearly, the regular OCTA has the strongest tail artifacts, and the SS-OCTA removes almost all signals beneath the macro vessels. In contrast, the proposed Ag_1-OCTA can preserve and enhance the micro vessel signals while suppressing the tail artifacts.

Worth noting that, we do notice that for some slowly flowing transverse vessel segments, the Ag_1-OCTA is not able to obtain the flow signal compared to the regular OCTA, as indicated by the red arrows in Fig. 5. This is due to the fact that Ag_1-OCTA calculates a much shorter period of decorrelation time (~1 ms) compared to the regular OCTA (>5 ms). In this case, it may not detect the RBC signal due to the discontinuity, single-passage file, and even the RBC stalling effect in the capillaries. In addition, the regular OCTA is usually obtained with 10 times of averaging while the A g_1-OCTA doesn't do averaging. Increasing the decorrelation time for g_1 calculation can greatly improve the micro vessel signal and its continuity, but at the expense of a longer acquisition time.

978-1-6654-8699-6/22 $31.00 © 2022 IEEE

IV. DISCUSSION

To summarize, in this work we show that the proposed Ag_1-OCTA has the ability to suppress the blood vessel tail artifacts and enhance the detectability of the micro vessels in the tail regions. This method first identified the macro vessels and the underneath tail regions, then both the magnitude and phase differences of the $g_1(\tau)$ decorrelation were analyzed in the tail region to differentiate the vessel voxels from the artifact voxels. Finally, a short decorrelation time was applied to suppress the tail artifacts signal, while a longer decorrelation time was used to enhance the dynamic contrast of the blood flow signal.

Noting that we found it's challenging to detect the slowly flowing transverse vessel voxels. This is due to the fact that the OCT signals from micro vessels are usually composed with the slow flowing signal from moving RBCs and static signal from vessel walls, resulting in a quite similar $g_1(\tau)$ decorrelation to the tail artifacts regions. Further, for transverse flowing vessels, the $g_1(\tau)$ doesn't have a dominant phase change of the imaginary $g_1(\tau)$ like the vessels with an axial velocity component do. Thus, the voxels of slowly flowing transverse micro vessels may be treated as tail artifacts region and the signal would be suppressed with Ag_1-OCTA. However, from our observation, most micro vessels close to the macro vessels have an axial velocity component, enabling the differentiation from the artifact region. Further, in deeper layers, the tail artifacts are weaker and usually decay slower compared to flow signal, giving Ag_1-OCTA the ability to different vessel voxel from artifacts region and enhance the detection of micro vessels in deep layers. With a longer decorrelation time, the Ag_1-OCTA would provide a higher signal intensity and improved flow continuity.

ACKNOWLEDGMENT

Authors thank the support from Shenzhen Science and technology Innovation Committee 20210316161406001, the Guangdong Science and Technology Department (2022A1515011984).

REFERENCES

[1] R. K. Wang, S. L. Jacques, Z. Ma, S. Hurst, S. R. Hanson, and A. Gruber, "Three dimensional optical angiography," Opt. Express **15**(7), 4083 (2007).

[2] S. Makita, Y. Hong, M. Yamanari, T. Yatagai, and Y. Yasuno, "Optical coherence angiography," Opt. Express 14(17), 7821 (2006).

[3] A. Mariampillai, B. A. Standish, E. H. Moriyama, M. Khurana, N. R. Munce, M. K. K. Leung, J. Jiang, A. Cable, B. C. Wilson, I. A. Vitkin, and V. X. D. Yang, "Speckle variance detection of microvasculature using swept-source optical coherence tomography," Opt. Lett. 33(13), 1530 (2008).

[4] J. Enfield, E. Jonathan, and M. Leahy, "In vivo imaging of the microcirculation of the volar forearm using correlation mapping optical coherence tomography (cmOCT)," Biomed. Opt. Express 2(5), 1184‐1193 (2011).

[5] Y. Jia, O. Tan, J. Tokayer, B. Potsaid, Y. Wang, J. J. Liu, M. F. Kraus, H. Subhash, J. G. Fujimoto, J. Hornegger, and D. Huang, "Split-spectrum amplitude-decorrelation angiography with optical coherence tomography," Opt. Express 20(4), 4710 (2012).

[6] D. Y. Kim, J. P. Fingler, J. S. Werner, D. M. Schwartz, S. E. Fraser, and R. J. Zawadzki, "In vivo volumetric imaging of human retinal circulation with phase-variance optical coherence tomography.," Biomed. Opt. Express 2(6), 1504‐1513 (2011).

[7] R. K. Wang, L. An, P. Francis, and D. J. Wilson, "Depth-resolved imaging of capillary networks in retina and choroid using ultrahigh sensitive optical microangiography," Opt. Lett. 35(9), 1467 (2010).

[8] V. J. Srinivasan, J. Y. Jiang, M. A. Yaseen, H. Radhakrishnan, W. Wu, S. Barry, A. E. Cable, and D. A. Boas, "Rapid volumetric angiography of cortical microvasculature with optical coherence tomography," Opt. Lett. 35(1), 43 (2010).

[9] T. E. de Carlo, A. Romano, N. K. Waheed, and J. S. Duker, "A review of optical coherence tomography angiography (OCTA)," Int. J. Retin. Vitr. 1(1), 5 (2015).

[10] B. Jeong, B. Lee, M. S. Jang, H. Nam, S. J. Yoon, T. Wang, J. Doh, B.-G. Yang, M. H. Jang, and K. H. Kim, "Combined two-photon microscopy and optical coherence tomography using individually optimized sources," Opt. Express 19(14), 13089 (2011).

[11] L. Huang, Y. Fu, R. Chen, S. Yang, H. Qiu, X. Wu, S. Zhao, Y. Gu, and P. Li, "SNR-Adaptive OCT Angiography Enabled by Statistical Characterization of Intensity and Decorrelation with Multi-Variate Time Series Model," IEEE Trans. Med. Imaging 38(11), 2695‐2704 (2019).

[12] H. S. Sandhu, M. Elmogy, A. Taher Sharafeldeen, M. Elsharkawy, N. El-Adawy, A. Eltanboly, A. Shalaby, R. Keynton, and A. El-Baz, "Automated Diagnosis of Diabetic Retinopathy Using Clinical Biomarkers, Optical Coherence Tomography, and Optical Coherence Tomography Angiography," Am. J. Ophthalmol. 216, 201‐206 (2020).

[13] C.-L. Chen and R. K. Wang, "Optical coherence tomography based angiography [Invited]," Biomed. Opt. Express 8(2), 1056 (2017).

[14] J. F. Zhang, S. Wiseman, M. C. Valdés-Hernández, F. N. Doubal, B. Dhillon, Y. C. Wu, and J. M. Wardlaw, "The Application of Optical Coherence Tomography Angiography in Cerebral Small Vessel Disease, Ischemic Stroke, and Dementia: A Systematic Review," Front. Neurol. 11(September), 1‐12 (2020).

[15] N. G. Ferris, T. M. Cannon, M. Villiger, B. E. Bouma, and N. Uribe-Patarroyo, "Forward multiple scattering dominates speckle decorrelation in whole-blood flowmetry using optical coherence tomography," Biomed. Opt. Express 11(4), 1947 (2020).

[16] W. J. Choi, B. Paulson, S. Yu, R. K. Wang, and J. K. Kim, "Mean-subtraction method for de-shadowing of tail artifacts in cerebral OCTA images: A proof of concept," Materials (Basel). 13(9), (2020).

[17] C. Leahy, H. Radhakrishnan, M. Bernucci, and V. J. Srinivasan, "Imaging and graphing of cortical vasculature using dynamically focused optical coherence microscopy angiography," J. Biomed. Opt. 21(2), 020502 (2016).

[18] S. Sabina and J. Lee, "Deep learning toolbox for automated enhancement, segmentation, and graphing of cortical optical coherence tomography microangiograms," Biomed. Opt. Express 11(12), 7325‐7342 (2020).

[19] J. Tang, S. E. Erdener, S. Sunil, and D. A. Boas, "Normalized field autocorrelation function-based optical coherence tomography three-dimensional angiography," J. Biomed. Opt. 24(3), 036005 (2019).

[20] J. Zhu, C. Merkle, M. Bernucci, S. Chong, and V. Srinivasan, "Can OCT Angiography Be Made a Quantitative Blood Measurement Tool?," Appl. Sci. 7(7), 687 (2017).

[21] U. Baran, W. J. Choi, Y. Li, and R. K. Wang, "Tail artifact removal in OCT angiography images of rodent cortex," J. Biophotonics 10(11), 1421‐1429 (2017). Ş. E. Erdener, J. Tang, A. Sajjadi, K. Kılıç, S. Kura, C. B. Schaffer, and D. A. Boas.

Proceedings of the 7th Optoelectronics Global Conference (OGC 2022)

Ultra-stable and Low-complexity Retiming Technique for Bandwidth-limited 112-Gbps PAM-4 systems

Lin Sun, Luxiao Zhang, Yi Cai, Gangxiang Shen,
and Gordon Ning Liu
Suzhou Key Laboratory of Advanced Optical
Communication Network Technology
Soochow University
Suzhou, China
gordonnliu@suda.edu.cn

Bin Chen
School of Computer Science and Information
Engineering
Hefei University of Technology
Hefei, China
bin.chen@hfut.edu.cn

Abstract—**We proposed an improved retiming algorithm for optical PAM-4 system by introducing a moving average filter into the conventional Gardner loop. It exhibits an enhanced stability especially when system bandwidth is limited.**

Keywords-Optical signal processing; Retiming; timing recovery

I. INTRODUCTION

In recent years, a lot of digital signal processing (DSP) algorithms have been proposed to recover distorted signals in optical communication systems, even including deep learning with a huge complexity and processing latency. However, the induced power consumption and extra latency of these complicated DSPs are not suitable for the high-density optical communication scenarios such as intra-data center communications. Currently, optical intensity modulation and direct detection (IMDD) is still the most favored solution for short-reach communications. To take the low-complexity and low-power consumption advantages of IMDD, the DSP as an assistance of communication should also exhibit low complexity and high potential of real-time deployment. For optical IMDD communications, several simplified DSP flowcharts have been proposed, including retiming [1], equalizing [2] and decoding [3]. As an important part of DSP at the receiver, retiming should perform an optimal sampling at the largest eye-opening with the best signal-to-noise ratio (SNR). Reported retiming methods with the possible real-time deployment for PAM-4 include adding framing header [1], non-integer oversampling performed on frequency domain [4], and Gardner retiming loop [5]. The Gardner retiming loop achieves minimum timing error based on 2-samples/symbol sequences, so it has low complexity and is well compatible with the other 2-samples/symbol equalizers if needed. However, the output of numerically-controlled oscillator (NCO) in a traditional Gardner loop, namely the sampling phase, is not stable for spectrum-filtered PAM-4 signal [6]. When signals are narrowly filtered due to channel bandwidth limitation and chromatic dispersion, the rising/falling edges will be correspondingly distorted, which requires a more precise retiming for a considerable BER performance.

In this work, we focus on the traditional Gardner retiming loop and try to further improve its stability in an optical IMDD system with the presence of channel filtering. The output of NCO controls the sampling phase, which requires to be very stable for ensuring a precise retiming performance. Here, we introduce a moving average filter (MAF) in the traditional Gardner loop to stabilize the interaction of NCO output. Simulations indicate that the optimal tap length of MAF is highly related to channel bandwidth for achieving the lowest sampling jitter. The tap length of MAF with considerable jitter improvement is up to 10, which shows the persistent low complexity for symbol retiming. Due to the low complexity implementation of the traditional Gardner loop on FPGA, we believe that the proposed method can be deployed in the real-time reception of high-speed PAM signals.

II. PRINCIPLE

Realized by a loop to update the sampling phase, minimization of timing error can be achieved without using a prior-known pilot for synchronization. In particular, an interpolation filter is used to obtain the sampled value at the sampling phase of u_k, then the retiming error is calculated using (1).

$$Error(k) = [y(k-1) - \frac{y(k) - y(k-2)}{2}] \cdot [y(k) - y(k-2)]. \quad (1)$$

The loop filter is an integral filter for generating a control word w_k for the numerical controlled oscillator (NCO). Updated w_k is related to the timing error through $w(k+1) = w(k) + C_1[Error(k) - Error(k-1)]$, where C_1 is loop filter coefficient. This formula indicates that when the differential error $Error(k) - Error(k-1)$ is a non-zero value, coefficient w_k varies correspondingly and it will affect the accuracy of the timing phase. To solve this problem, a stabilizing method of timing phase is proposed by multiplying the loop filter output with a scaling factor to reduce the variance of NCO output. However, an additional digital switch is required for deciding when to apply the scaling factor at the loop filter output. In this work, we employ a moving average filter (MAF) after NCO to stabilize the sampling phase.

978-1-6654-8699-6/22 $31.00 © 2022 IEEE

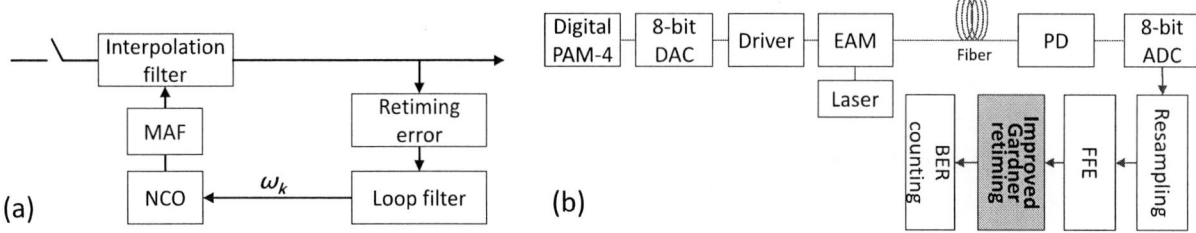

Figure 1. (a) Diagram of the improved Gardner retiming loop. (b) Evaluation setup of the proposed retiming scheme on a 112-Gbps PAM-4 IMDD system.

The basic reason for the use of MAF is that the sampling phase is a stationary random sequence. The tap length of MAF for this can be shown to be short enough for maintaining low complexity and latency. Moreover, the proposed retiming scheme does not reduce convergence speed because the loop filter coefficients C_1 is not changed. Based on the principle of the traditional Gardner loop, we know that the sampling point is interacting with the retiming error which is highly dependent on the slope of rising/falling edges. It indicates that when channel bandwidth is limited, the narrow filtering distorts the rising/falling edges then leads to an enlarged variance of sampling points. One can reduce the loop filter coefficient to stabilize the NCO output, but compromise on a slower iteration of optimal sampling phase. In order to stabilize the traditional Gardner loop to get an improved jitter performance, we introduce an MAF to control the NCO, as shown in Fig. 1(a). Because the MAF performs the averaging of NCO's output, it can produce a balanced sampling point among M symbols where M is the tap length of MAF. Due to the different edge slopes at different bandwidth constraint, the optimal tap length of MAF is also related to the end-to-end bandwidth of an optical IMDD system.

III. RESULTS AND DISCUSSIONS

The evaluation setup for the proposed retiming scheme is depicted in Fig. 1(b). A 112-Gbps PAM-4 IMDD system is realized on the VPItransmissionMakers platform, where the key parameters are set according to the existing equipment in experiment [7]. Digital PAM-4 signal is generated by Gray coding of PRBS-17. An 8-bit resolution digital-to-analogue converter (DAC) is working at 56 Gbaud rate. Following a driver, an electro-absorption modulator (EAM) is used to modulate the laser at 1550-nm wavelength. After 2-km SMF transmission, a photodetector with 0.8-A/W sensitivity is used to receive optical signals. The overall channel bandwidth limitation is emulated by a 4th order Bessel low-pass filter with manually-defined cutting-off frequency, to evaluate the performance of the proposed retiming scheme under different level of narrow filtering. Although DSP is performed offline, every component of the flowchart including FFE, Gardner

retiming can be implemented in FPGA. 4-samples/symbol resampling is performed before FFE. Then the improved Gardner retiming processes the equalized 4-samples/symbol sequences. At last, BER counting and jitter analysis are performed.

Fixing the channel bandwidth at 30GHz, the retiming error interactions of received samples of the conventional Gardner loop and the proposed retiming method with MAF length of 10 are plotted in Fig. 2. Deviation of sampling phase (NCO output) in Fig. 2(a) comes from the imperfect rising/falling edge after the narrow filtering. Figure 2(b) indicates that after using the proposed retiming method, the sampling phase becomes more stable, that indicates a reduced sampling jitter. However, the histograms in Fig. 2(c) and (d) are ambiguous. The reason is the channel bandwidth is set at 30GHz, while there is a great number of the sampling points sharing a similar constellation quality. The circumstances with more stringent constraints of bandwidth will be discussed later.

Then we estimate the jitter under different bandwidth limitations for comparing the stability of the proposed retiming method and the traditional Gardner retiming loop. Jitter is defined as the variance of sampling phase after 20k samples iteration, $Jitter = 20 \log_{10}(\delta\mu_t)$, where μ_t is the sampling phase. Normally, a modulation format with more levels has a larger jitter than those with less levels, due to the more variety of transitions among levels. As discussed in the principle, the channel narrow filtering distorts the raising/falling edges which requires a more precise retiming. Thus, jitter vs. MAF tap length curves are given in Fig. 3 under different channel bandwidths. Insets in Fig. 3 are the received eye-diagrams under different channel bandwidths. Insets show that the eye-closure becomes severe when the channel bandwidth is 20GHz, that requires a more precise sampling. We also observe that the interacted jitter reduces with the decreasing of channel bandwidth. That can be explained by the bandwidth limitation narrowed eyes are much more sensitive to the timing jitter. With the assistance of MAF having a tap length over 10, timing jitters can be significantly reduced.

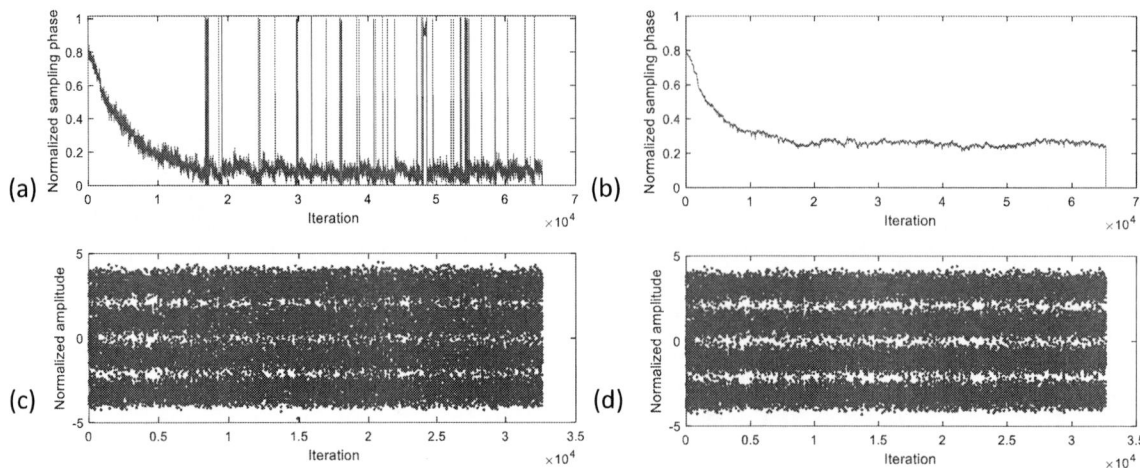

Figure 2. (a) Normalized sampling phase iteration with samples using the traditional Gardner retiming and (b) the proposed stable retiming method. (c) Histogram of samples using the traditional Gardner retiming and (d) the proposed stable retiming method with MAF length at 10.

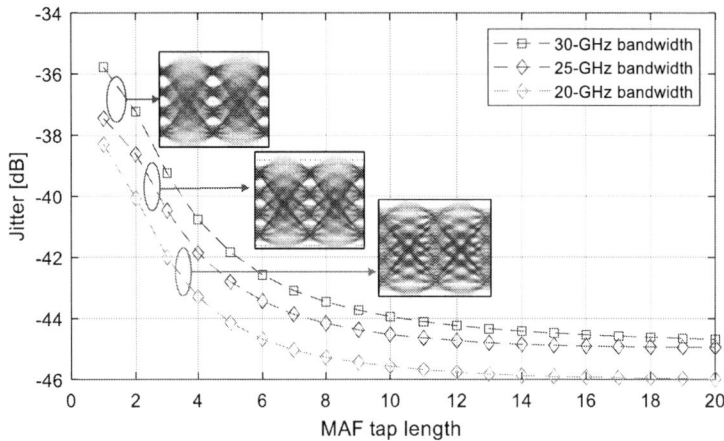

Figure 3. Sampling jitter vs. MAF tap length under different bandwidth constraints.

BER curves are plotted against the received optical power in Fig. 4 by comparing the traditional Gardner retiming and the proposed retiming method. Under 30-GHz bandwidth limitation, there's an ignorable difference between traditional Gardner retiming and the proposed one. While for the case of 25-GHz bandwidth, the proposed stable retiming can offer 1.2-dB sensitivity gain at the BER threshold of 2×10^{-2}. The results indicate that the proposed method is beneficial to the bandwidth-limited systems.

IV. CONCLUSIONS

We proposed an improved retiming algorithm for optical PAM-4 signals with the presence of channel bandwidth limitation. By introducing a short MAF into the traditional Gardner loop, significant reduction of timing jitter has been achieved. For future research, we plan to deploy the proposed algorithm in FPGA to investigate its benefits in real-time DSP implementation.

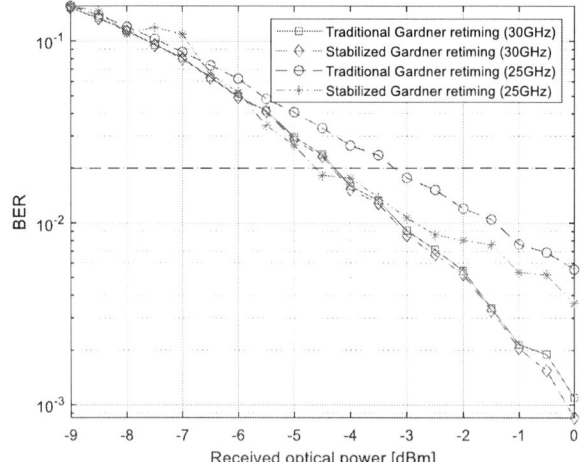

Figure 4. BER vs. received optical power under different bandwidth limitation conditions.

978-1-6654-8699-6/22 $31.00 © 2022 IEEE

ACKNOWLEDGMENT

This work is financially supported by National Key R&D Program of China under grant 2018YFB1801701, and National Natural Science Foundation (NSFC) of China under Grants 62105273.

REFERENCES

[1] J. Zhang, X. Xiao, J. Yu, J. S. Wey, X. Huang, and Z. Ma, "Real-time FPGA demonstration of PAM-4 burst-mode all-digital clock and data recovery for single wavelength 50G PON application," Optical Fiber Communication Conference, Optical Society of America. M1B-7 (2018).

[2] R. Deng, J. He, J. Yu, X. Xiao, K. Lyu, and X. Xin, "Experimental demonstration of 20-Gb/s dual-band Nyquist PAM-4 transmission over a short-reach IMDD system using super-Nyquist sampling technique," Opt. Lett., 43(11), 2640-2643. (2018).

[3] L. Sun, B. Xu, X. Fan, J. Du, Z. He, and C. Lu, "Real-time channel conditional distribution tracking for intelligent decoding of optical IMDD signals," Opt. Lett., 46(17), 4426-4429 (2021).

[4] A. Josten, B. Bäuerle, M. Eppenberger, E. Dornbierer, D. Hillerkuss, and J. Leuthold, "168 Gb/s line rate real-time PAM receiver enabled by timing recovery with 8/7 oversampling in a single FPGA," Optical Fiber Communication Conference, Optical Society of America. W2A-24 (2017).

[5] J. Shi, J. Zhang, Y. Zhou, Y. Wang, N. Chi, and J. Yu, "Transmission performance comparison for 100-Gb/s PAM-4, CAP-16, and DFT-S OFDM with direct detection," Journal of Light. Technol., 35(23), 5127-5133. (2017).

[6] S. Zhang, L. Zhong, Y. Zou, M. Zhu, X. Dai, J. Zhang, and D. Liu, "Improved multiplier-free Mueller-Müller Baud-rate timing error detector for optical IM/DD system," Opt. Exp., 29(26), 44129-44137. (2021).

[7] J. Zhang, X. Wu, L. Sun, J. Liu, A. P. T. Lau, C. Guo, and C. Lu, "C-band 120-Gb/s PAM-4 transmission over 50-km SSMF with improved weighted decision-feedback equalizer," Opt. Exp., 29(25), 41622-41633. (2021).

Proceedings of the 7th Optoelectronics Global Conference (OGC 2022)

Application-aware Configuration of All-optical Interconnects in Hyper-FleX-LION
(*Invited Paper*)

Hao Yang[†‡] and Zuqing Zhu[†]

[†]School of Information Science and Technology, University of Science and Technology of China, Hefei, China
[‡]Department of Information Engineering, Southwest University of Science and Technology, Mianyang, China
[†]Email: {zqzhu}@ieee.org

Abstract—Due to the advantages of optical circuit switching (OCS), all-optical interconnects (AOIs) for data center networks (DCNs) have attracted intensive interests recently. Hyper-FleX-LION is a highly-flexible AOI architecture that operates with the OCS based on wavelength-division multiplexing (WDM). In this paper, we present our recent research activities on Hyper-FleX-LION. First, to prove the superiority of Hyper-FleX-LION, we enumerate various communication patterns of distributed machine learning (DML) in DCNs, and analyze the acceleration effect achieved by Hyper-FleX-LION over existing interconnect architectures, such as the hybrid optical/electrical interconnect based on optical cross-connect (HOE-w/OXC). Results show that Hyper-FleX-LION can better accelerate the tasks of DML. Then, we classify network applications in DCNs as bandwidth-intensive and data-intensive ones, and analyze the operational costs and task completion time that Hyper-FleX-LION brings to them and how to optimize the topology design and traffic routing of Hyper-FleX-LION adaptively. The Analysis results confirm the importance of designing an application-aware configuration scheme for Hyper-FleX-LION.

Index Terms—Data center networks, All-optical interconnects, Application-aware configuration, Distributed machine learning.

I. INTRODUCTION

Recently, the fast emergence of network applications has promoted the research and development (R&D) on data center networks (DCNs) [1, 2]. Hence, the architectures of network interconnects in DCNs are undergoing revolutionary changes to match the capacity/latency/cost of interconnects among computing and storage platforms and the quality-of-service (QoS) demands of network applications. The traditional interconnects that are solely based on electrical packet switching (EPS) have a number of drawbacks, such as ever-increasing power consumption and limited port capacity, which make them difficult to cope with the time-varying traffic patterns and huge traffic volumes of most network applications today [3]. Optical circuit switching (OCS) is a promising replacement for EPS, since it can achieve higher energy efficiency and larger port capacity [4–8]. Therefore, the all-optical interconnects (AOIs) that leverage OCS have been designed and implemented to assist or even replace EPS-based interconnects in DCNs for supporting various network applications better [9–13].

Yoo *et al.* [15, 16] designed and fabricated FleX-LION, which is an integratable OCS device, and with it, large-scale reconfigurable AOIs (namely, Hyper-FleX-LION) can be

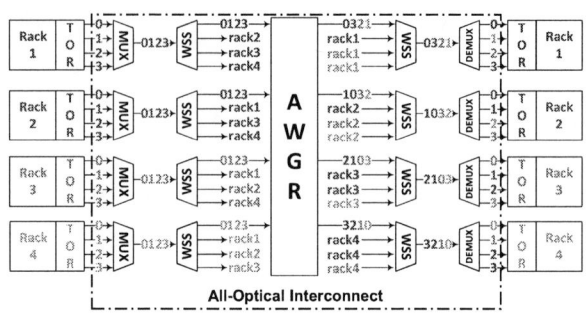

Fig. 1. AOI in Hyper-FleX-LION interconnecting 4 racks, MUX/DEMUX: wavelength multiplexer/demultiplexer, WSS: wavelength selective switch, AWGR: arrayed waveguide grating router (adapted from [14]).

architected to carry different traffic patterns efficiently [13]. Here, the topology of an AOI that interconnects N racks can be referred to as N-Hyper-FleX-LION, and Fig.1 shows that the architecture and operation principle of a 4-Hyper-FleX-LION. There is an arrayed waveguide grating router (AWGR) sitting in the middle of the 4-Hyper-FleX-LION, and the transmitting and receiving structures of each rack are located on its left and right, respectively. As for each rack, its top-of-rack (ToR) switch has 4 ports and each port equips a transceiver (TRX).

As illustrated in Fig. 1, the wavelengths used by the TRXs are labeled by numbers, while the color of each number denotes the source rack of a TRX. In the transmitting structure, all the outputs of a ToR switch are wavelength-multiplexed before entering a wavelength selective switch (WSS). One of the WSS' outputs is connected to the AWGR, and the others go directly to the inputs of the WSS' in the receiving structures of other racks. In the receiving structure of each rack, the optical signals are first groomed by its WSS and then distributed to the TRXs of the ToR switch by a wavelength de-multiplexer. Hence, by utilizing the wavelength switching capability of the AWGR and adjusting the switching states of the WSS', we can get various topologies to interconnect the racks (*e.g.*, the configuration in Fig. 1 leads to a full-mesh topology among the 4 racks). In other words, if we regard each wavelength channel from the TRXs on a ToR switch as the origin of a lightpath, the lightpath can be set up adaptively to use any of the racks in the Hyper-FleX-LION as its destination (*e.g.*,

978-1-6654-8699-6/22 $31.00 © 2022 IEEE

with software-defined networking (SDN) [17, 18]).

Note that, compared with EPS, OCS is actually less adaptive because of the large switching granularity and long reconfiguration latency [19]. Even though these issues can be partially resolved by introducing virtualization techniques [20–25] to slice virtual networks and grooming the traffic of similar network applications with them, the benefits of AOIs cannot be fully explored if their topologies are not managed intelligently or the traffic through them are not scheduled adaptively [10]. Therefore, we need to study the problem of how to effectively operate the AOIs in Hyper-FleX-LION to efficiently serve complicated network applications in DCNs.

In this paper, we describe our recent research activities on the aforementioned problem. Section II discusses our investigation that confirms the acceleration effect on distributed machine learning (DML) achieved by Hyper-FleX-LION over an existing hybrid optical/electrical interconnect based on optical cross-connect (HOE-w/OXC). In Section III, we classify network applications in DCNs as bandwidth- and data-intensive ones, and analyze the operational costs and task completion time that Hyper-FleX-LION brings to them and how to optimize the topology design and traffic routing of Hyper-FleX-LION. Finally, Section IV summarizes the paper.

II. ACCELERATION OF DISTRIBUTED MACHINE LEARNING

As one of the current mainstream network applications, DML not only occupies high communication bandwidth, but also has a variety of traffic patterns with different characteristics, which brings great challenges to network interconnects in DCNs [26]. Previously, researchers added OXC to EPS-based interconnects to realize HOE-w/OXC, and they have proved that this architecture can effectively accelerate DML tasks by reasonably reconfiguring the OXC's connectivity to assist the traffic routing in the EPS-based interconnect [27]. However, limited by the one-to-one connectivity of OXC, HOE-w/OXC still cannot accelerate DML tasks to the maximum extent.

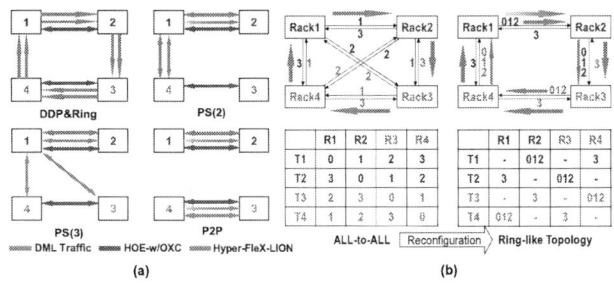

Fig. 2. Acceleration of DML tasks with various traffic patterns using a 4-Hyper-FleX-LION, (a) Traffic patterns of DML architectures, (b) Reconfiguration of Hyper-FleX-LION. DDP: Distributed Data Parallel, Ring: Ring-AllReduce, PS(N): Parameter Server with 1 master node and N worker nodes, P2P: Peer-to-Peer (adapted from [14]).

Fig. 2(a) shows the traffic patterns of different DML tasks, where the colorful arrows denote the traffic, and the purple arrows represent the acceleration bandwidth that can be provided by HOE-w/OXC to the DML tasks [14]. Here,

we define the *acceleration bandwidth* as the bandwidth that can be provisioned to DML in addition to that of the basic EPS-based interconnect. We can find that the acceleration bandwidth provided by the OXC cannot adapt to the traffic patterns of all the DML tasks. For example, OXC can only provision acceleration bandwidth between *Racks* 1 and 2 and between *Racks* 3 and 4 for the DML in *DDP&Ring*. Hence, the other traffic flows that cannot be accelerated will become bottlenecks and affect the overall acceleration of the DML. The same problem also occurs for the DML in *PS*, and the more worker nodes that the DML in *PS* allocates, the more bottlenecks will be generated. On the other hand, the blue arrows represent the acceleration bandwidth that Hyper-FleX-LION can provide. The configuration for the DML in *Ring&DDP* is shown in Fig. 2(b). We configure the wavelength switching in the WSS' according to the table at the bottom-left of the figure, and thus the topology changes from a full-mesh to a ring, which matches to the traffic pattern of the DML in *Ring&DDP* exactly. Moreover, in addition to the wavelength that makes up the basic ring topology, there are two more wavelength channels per rack, which can be leveraged to provide more acceleration bandwidth. Therefore, Hyper-FleX-LION can achieve a better acceleration effect than HOE-w/OXC for the DML in *Ring&DDP*.

Similarly, Hyper-FleX-LION can also provide more acceleration bandwidth to the DML in *PS(2)* than HOE-w/OXC. Specifically, *Rack* 1 uses two of the ports on its ToR switch to carry the basic communications between the server and the two workers, and allocates the remaining two ports to provide acceleration bandwidth. However, the situation becomes different for the DML in *PS(3)*, when an additional worker is placed on *Rack* 3. As Hyper-FleX-LION already uses three ports on *Rack* 1 for the basic communications, it can only use the remaining port to provide acceleration bandwidth, which makes its acceleration effect the same as that of HOE-w/OXC. At last, as for the DML in *P2P*, because its traffic only occurs between 2 racks, the acceleration effects of Hyper-FleX-LION and HOE-w/OXC are the same again, as shown in Fig. 2(a).

III. APPLICATION-AWARE TOPOLOGY CONFIGURATION

Our previous work in [14] has verified that Hyper-FleX-LION possesses the flexibility for dealing with various network applications in DCNs. However, how to better plan its network topology to maximize the advantages for specific network applications is still an unexplored problem. Hence, we need to roughly classify network applications as bandwidth-intensive and data-intensive ones, and study the topology configuration schemes for them, respectively.

Bandwidth-intensive applications mostly exist in tenant-oriented commercial DCs, and their main QoS demands are on bandwidth capacity. Specifically, such an application usually has a traffic matrix to describe the least bandwidth that it needs to occupy among the racks. Therefore, operators usually want to serve these applications with the minimum operational cost (*e.g.*, the least number of active ports). Fig. 3(a) explains how an improper topology design can make traffic routing of

978-1-6654-8699-6/22 $31.00 © 2022 IEEE

a bandwidth-intensive application infeasible. Here, there are three demands among the ToR switches, as 1→2, 1→3, and 2→3 for the bandwidth of 15, 5, and 5 units, respectively, and the capacity of each port assumed to be 10 units. Then, if the topology is designed as the left one in Fig. 3(a) (*i.e.*, the established connections in the Hyper-FleX-LION are marked as black arrows), the demand of 1→2 cannot be satisfied. On the other hand, the right configuration in Fig. 3(a) can serve all the demands properly. Although the two configurations in Fig. 3(a) activate same number of ports (*i.e.*, 3), their supports to the bandwidth-intensive application are different.

Fig. 3(b) explains why proper traffic routing helps to explore the bandwidth in a specific topology design. This time, the demands are still for 1→2, 1→3, and 2→3, but their bandwidth requirements become 5, 11, and 5 units, respectively. Then, to support the bandwidth requirements in the left topology in Fig. 3(b), we need to activate three ports on *ToR* 1, while the right topology only requires two active ports. Hence, the traffic routing in the right topology helps to save one active port.

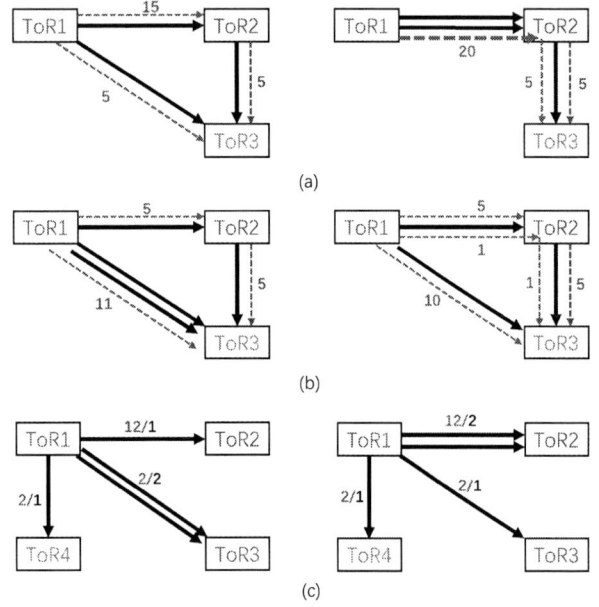

Fig. 3. Examples on correlation of topology design and traffic routing for (a) and (b) bandwidth-intensive applications, and (c) data-intensive application.

On the other hand, data-intensive applications need to transfer certain amounts of data among racks, and such an application is completed only when all of its data has been transferred. Therefore, operators need to plan the topology reasonably to complete all the data transfers as soon as possible. Fig. 3(c) shows a straightforward example to explain why topology management is also important for these applications. Here, we consider an application that needs to transfer data from *Rack* 1 to the other three racks, and the amounts of data that needs to be transferred are 1→2: 12 units, 1→3: 2 units, and 1→4: 2 units, respectively. Here, each black arrow denotes an optical connection with a bandwidth capacity of one unit, and the number on it shows the duration of the data

transfer that uses it. Therefore, according to the left topology configuration in Fig. 3(c), the data transfer of 1→2 will become the bottleneck of the application, *i.e.*, it takes $\frac{12}{1} = 12$ time-units. But if we use the right topology configuration in Fig. 3(c), we can reduce the completion time of the data transfer to $\frac{12}{2} = 6$ time-units without activating more ports. To this end, we can see that a well-designed topology configuration mechanism together with adaptive traffic routing can make Hyper-FleX-LION accelerate both bandwidth- and data-intensive applications better with less operational cost.

IV. Summary

This paper summarized our recent research activities on AOIs in Hyper-FleX-LION. We first proved that Hyper-FleX-LION can provide a better acceleration effect than HOE-w/OXC when serving DML. Then, we introduced the acceleration effects brought by Hyper-FleX-LION on bandwidth- and data-intensive applications, and explained why we should consider the application-aware topology configuration of Hyper-FleX-LION for enhancing the QoS of applications.

Acknowledgments

This work was supported by NSFC project 61871357 and Fundamental Fund for Central Universities (WK3500000006).

References

[1] "Cisco Annual Internet Report (2018-2023)," *Online White Report.* [Online]. Available: https://www.cisco.com/c/en/us/solutions/collateral/executive-perspectives/annual-internet-report/white-paper-c11-741490.html.

[2] P. Lu *et al.*, "Highly-efficient data migration and backup for Big Data applications in elastic optical inter-datacenter networks," *IEEE Netw.*, vol. 29, pp. 36–42, Sept./Oct. 2015.

[3] W. Lu *et al.*, "AI-assisted knowledge-defined network orchestration for energy-efficient data center networks," *IEEE Commun. Mag.*, vol. 58, pp. 86–92, Jan. 2020.

[4] Z. Zhu, W. Lu, L. Zhang, and N. Ansari, "Dynamic service provisioning in elastic optical networks with hybrid single-/multi-path routing," *J. Lightw. Technol.*, vol. 31, pp. 15–22, Jan. 2013.

[5] M. Zhang *et al.*, "Bandwidth defragmentation in dynamic elastic optical networks with minimum traffic disruptions," in *Proc. of ICC 2013*, pp. 3894–3898, Jun. 2013.

[6] W. Shi, Z. Zhu, M. Zhang, and N. Ansari, "On the effect of bandwidth fragmentation on blocking probability in elastic optical networks," *IEEE Trans. Commun.*, vol. 61, pp. 2970–2978, Jul. 2013.

[7] L. Gong *et al.*, "Efficient resource allocation for all-optical multicasting over spectrum-sliced elastic optical networks," *J. Opt. Commun. Netw.*, vol. 5, pp. 836–847, Aug. 2013.

[8] Y. Yin *et al.*, "Spectral and spatial 2D fragmentation-aware routing and spectrum assignment algorithms in elastic optical networks," *J. Opt. Commun. Netw.*, vol. 5, pp. A100–A106, Oct. 2013.

[9] N. Farrington *et al.*, "Helios: a hybrid electrical/optical switch architecture for modular data centers," *ACM SIGCOMM Comput. Commun. Rev.*, vol. 40, pp. 339–350, Oct. 2010.

[10] K. Chen *et al.*, "OSA: An optical switching architecture for data center networks with unprecedented flexibility," *IEEE/ACM Trans. Netw.*, vol. 22, pp. 498–511, Apr. 2013.

[11] H. Ballani *et al.*, "Sirius: A flat datacenter network with nanosecond optical switching," in *Proc. of ACM SIGCOMM 2020*, pp. 782–797, Jul. 2020.

[12] J. Benjamin *et al.*, "PULSE: Optical circuit switched data center architecture operating at nanosecond timescales," *J. Lightw. Technol.*, vol. 38, pp. 4906–4921, May 2020.

[13] G. Liu *et al.*, "Architecture and performance studies of 3D-Hyper-FleX-LION for reconfigurable All-to-All HPC networks," in *Proc. of SC 2020*, pp. 1–16, Nov. 2020.

[14] H. Yang, Z. Zhu, R. Proietti, and B. Yoo, "Which can accelerate distributed machine learning faster: Hybrid optical/electrical or optical reconfigurable DCN?" in *Proc. of OFC 2022*, pp. 1–3, Mar. 2022.

[15] X. Xiao *et al.*, "Multi-FSR silicon photonic Flex-LIONS module for bandwidth-reconfigurable all-to-all optical interconnects," *J. Lightw. Technol.*, vol. 38, pp. 3200–3208, Mar. 2020.

[16] ——, "Silicon photonic Flex-LIONS for bandwidth-reconfigurable optical interconnects," *IEEE J. Sel. Top. Quantum Electron.*, vol. 26, pp. 1–10, Nov. 2020.

[17] Z. Zhu *et al.*, "Demonstration of cooperative resource allocation in an OpenFlow-controlled multidomain and multinational SD-EON testbed," *J. Lightw. Technol.*, vol. 33, pp. 1508–1514, Apr. 2015.

[18] S. Li *et al.*, "Protocol oblivious forwarding (POF): Software-defined networking with enhanced programmability," *IEEE Netw.*, vol. 31, pp. 58–66, Mar./Apr. 2017.

[19] N. Bitar, S. Gringeri, and T. Xia, "Technologies and protocols for data center and cloud networking," *IEEE Commun. Mag.*, vol. 51, pp. 24–31, Sept. 2013.

[20] L. Gong and Z. Zhu, "Virtual optical network embedding (VONE) over elastic optical networks," *J. Lightw. Technol.*, vol. 32, pp. 450–460, Feb. 2014.

[21] W. Fang *et al.*, "Joint spectrum and IT resource allocation for efficient vNF service chaining in inter-datacenter elastic optical networks," *IEEE Commun. Lett.*, vol. 20, pp. 1539–1542, Aug. 2016.

[22] L. Gong, Y. Wen, Z. Zhu, and T. Lee, "Toward profit-seeking virtual network embedding algorithm via global resource capacity," in *Proc. of INFOCOM 2014*, pp. 1–9, Apr. 2014.

[23] Q. Sun, P. Lu, W. Lu, and Z. Zhu, "Forecast-assisted NFV service chain deployment based on affiliation-aware vNF placement," in *Proc. of GLOBECOM 2016*, pp. 1–6, Dec. 2016.

[24] L. Gong, H. Jiang, Y. Wang, and Z. Zhu, "Novel location-constrained virtual network embedding (LC-VNE) algorithms towards integrated node and link mapping," *IEEE/ACM Trans. Netw.*, vol. 24, pp. 3648–3661, Dec. 2016.

[25] J. Liu *et al.*, "On dynamic service function chain deployment and readjustment," *IEEE Trans. Netw. Serv. Manag.*, vol. 14, pp. 543–553, Sept. 2017.

[26] M. Abadi *et al.*, "Tensorflow: A system for large-scale machine learning," in *Proc. of OSDI 2016*, pp. 265–283, Nov. 2016.

[27] C. Wang *et al.*, "Acceleration and efficiency warranty for distributed machine learning jobs over data center network with optical circuit switching," in *Proc. of OFC 2021*, pp. 1–3, Jun. 2021.

978-1-6654-8699-6/22 $31.00 © 2022 IEEE

Proceedings of the 7th Optoelectronics Global Conference (OGC 2022)

Flexible Dispersion Engineering in Thin GaP-OI Frequency Comb Resonator Design

Zhaoting Geng[1], Houling Ji[1], Zhuoyu Yu[1], Weiren Cheng[1], Yi Li[1†], and Qiancheng Zhao[1*]

[1]School of Microelectronics, Southern University of Science and Technology, Shenzhen, Guangdong, China, 518000

e-mail: †liy37@sustech.edu.cn, *zhaoqc@sustech.edu.cn

Abstract—**Thickness-constrained waveguides enable ultralow loss photonics while suffering severe chromatic issues nowadays. We present a concentric ring configuration to reduce the dispersion requirements for the core waveguide thickness. This approach employs a second ring to squeeze light into additional anti-bonding modes to manifest the dispersion profile with lateral coupling parameters such as the 2nd ring width as well as the gap size. We then applied this method to a 200 nm-thick SiO₂-passivated concentric coupled GaP-OI resonator design for Kerr soliton frequency comb generation which gives rise to an anomalous dispersion span of 150 nm. The dispersion profile could be further optimized by changing the gap of the coupled rings, demonstrating the design flexibility. Our concentric microrings approach can find anomalous dispersions on thickness-constrained materials, opening the possibilities for novel integrated nonlinear photonic applications.**

Keywords integrated photonics; Kerr soliton frequency comb; mode coupling; anomalous dispersion;

I. INTRODUCTION

Microresonator-based Kerr soliton frequency comb is an enabling factor for numerous applications such as quantum photonics [1], LIDAR [2], frequency synthesis [3], optical atomic clocks [4], [5], and precision spectroscopy [6]. Benefiting from tight mode confinement and low propagation loss, the threshold power of waveguide resonators has been reduced to μW level [7] for frequency comb generation. Further improvement on four-wave-mixing efficiency triggered the exploration of different material platforms. During the last decade, a variety of nonlinear material platforms have been investigated with varying degrees of success, including silicon [8], silicon nitride [9], [10], aluminum nitride [11], [12], lithium niobate [13], and tantalum pentoxide [14]. Recently, III-V materials, such as AlGaAs [7], [15] and gallium phosphide (GaP) [16], emerge as promising solutions due to their high Kerr coefficients, paving the way towards lower threshold frequency comb operation.

GaP has a non-centrosymmetric crystal structure that is nearly lattice-matched to silicon. It has a high refractive index (~3.05 in the C-band), a conspicuous $\chi^{(3)}$ nonlinearity, a non-zero piezoelectric effect, and a large $\chi^{(2)}$ nonlinearity. It also has a good thermal conductivity for temperature tuning, a broad transparency range from 550 nm to 11 μm, and negligible two-photon absorption above 1.1 μm wavelength [16]–[19]. These optical properties render GaP an excellent candidate for nonlinear photonics [20], [21]. Early GaP devices were fabricated on top of a sacrificial layer which was later removed in the region under the device by wet etching. However, this method is limited to freestanding structures that rely on the refractive index contrast between GaP and air for waveguiding. Recently, the rise of wafer-scale gallium phosphide on insulator (GaP-OI) platform [16], [22] releases the potential for novel integrated nonlinear photonic devices, making it possible to design waveguides, grating couplers and ring resonators.

Anomalous dispersion for resonant modes is essential to realizing frequency comb generation in ring resonators. This is often achieved by optimizing waveguide cross-section or using air as cladding. While this approach enables dispersion engineering, its effectiveness could be weakened by the waveguide thickness restriction. For GaP-OI structure, the GaP layer is limited by the defect-free epitaxial growth thickness, and may be further thinned for surface smoothening. Such a thickness limitation hinders waveguides from having anomalous dispersion. The problem could be mitigated by using avoided mode crossing, which leverages the hybridization of adjacent waveguide modes and brings the antisymmetric mode into an anomalous dispersion region. This method can be implemented in coupled concentric ring resonators [23]. Since the resonant mode dispersion is tuned by optimizing the lateral dimension (in-plane), it relaxes the waveguide thickness requirement, and could achieve similar amount of anomalous dispersion in a thinner waveguide.

Figure 1. The concept figure of a concentric coupled GaP-OI resonator.

Here, we present a SiO₂-passivated concentric coupled resonator design in GaP-OI platform based on our previous work [24] for Kerr soliton frequency comb formation at telecom wavelengths with a quite thin device layer thickness of 200 nm. Such a thickness only yields normal dispersion in strip waveguides in the infrared. Our design, however, can engineer the resonator dispersion with a high degree of freedom to achieve a strong anomalous dispersion with over 150 nm spanning, and afterward coherent soliton frequency

978-1-6654-8699-6/22 $31.00 © 2022 IEEE

comb generation. The dispersion-engineered resonator supports a free spectrum range (FSR) of 100 GHz. The theoretical pump threshold power is 13.5 mW. Under the optimized and reasonable design parameters, the Kerr soliton frequency comb can be generated with the peak of anomalous dispersion profile precisely at 1550 nm. The dispersion curve peak, width, and flatness could be fine-tuned with just the waveguide lateral dimensions, demonstrating its design flexibility. This approach could also be applied to other material platforms to generate arbitrary dispersion profile at desired wavelengths.

II. WAVEGUIDE DISPERSION ENGINEERING

For Kerr soliton frequency comb generation, anomalous dispersion is required to facilitate four-wave-mixing process. For waveguides, both material dispersion and waveguide dispersion contribute to the total dispersion. Waveguide geometry strongly determines the value and sign of waveguide dispersion. In a strip waveguide, the usual way to engineer the dispersion is by changing its width and height [16]. However, it has a limited effect on engineering dispersion signs from normal to anomalous because material dispersion turns out to be strongly normal which dominates the total sign of dispersion. To be specific, we investigated three types of cladding design at a core height of 200 nm, which all have strong normal dispersion at telecom wavelengths (Figure 2). Therefore, a strip waveguide finds it incapable to achieve anomalous dispersion at this thickness. Novel waveguide cross-section design is required to overcome the challenge.

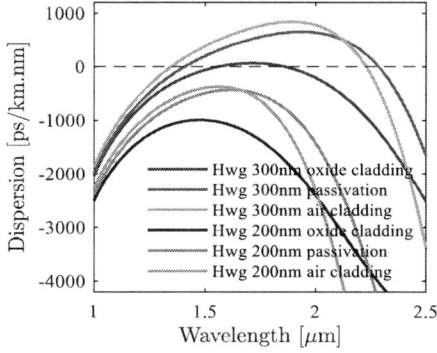

Figure 2. The dispersion profiles of the waveguides. The reference waveguide cross-section is 200 nm × 750 nm and the thicker-core waveguide cross-section is 300 nm × 750 nm. All the waveguides have 5 nm thick Al_2O_3 passivation layers above the GaP surface.

Inspired by Ref.[10], the devices patterned and etched in the epitaxial GaP layer are composed of two concentric coupled rings. The GaP layer has been direct-bonded to a thermal oxide wafer with the help of a layer of 5 nm-thick ALD Al_2O_3 [19]. Another 5 nm ALD Al_2O_3 layer is deposited after etching to passivate the GaP surface. An extra 100 nm SiO_2 cladding layer is deposited for further passivation [16] and sidewall smoothening [25], as shown in Figure 3. The key point of our proposed design is the coupled structure. The fundamental TE modes ($TE_{00,in}$ and $TE_{00,out}$ for the inner and outer waveguide, respectively) in the inner and outer waveguides couple with each other, which results in mode hybridization. To achieve this mode hybridization, the phase-matching condition should be considered. When the round-trip optical path length (OPL) of the inner and outer rings are equal, the phase-matching condition is met. The OPL can be calculated by OPL = $2\pi R n_{eff}$ [10], [23]. Figure 4 shows the OPL of the inner and outer rings cross corresponding to different values of W_{gap}. The position of the OPL crossing point can be adjusted by simply changing values of W_{gap}, which shows our design's flexibility. Around the OPL crossing point, $TE_{00,in}$ and $TE_{00,out}$ transform into a pair of antisymmetric mode and symmetric mode. The symmetric mode has a slightly higher effective refractive indices, while that of the antisymmetric mode is lower. The small deviations in the effective refractive indices alter the dispersion profile of the two modes, making the antisymmetric mode dispersion anomalous (Figure 5).

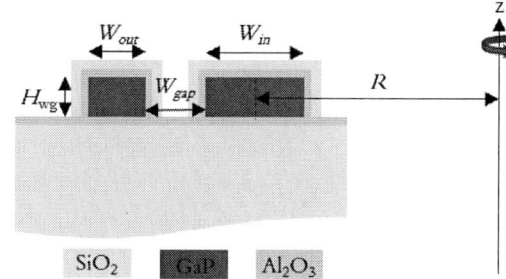

Figure 3. The cross-section of the surface passivated, concentric coupled, and direct-bonded GaP-OI waveguide. The waveguide height and width refer to the GaP core dimension. The gap refers to the distance between two GaP cores.

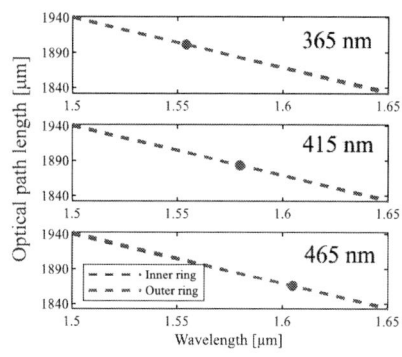

Figure 4. The optical path length of the inner (blue) and outer (red) rings cross points corresponding to different values of W_{gap}.

To meet the phase-matching condition in a ring structure, the effective refractive indices of the inner waveguide should be slightly larger than that of the outer waveguide. It means that the width of the inner ring should be larger than that of the outer ring, as shown in Figure 3. The inner and outer waveguide core cross-sections are simulated to be 200 nm×750 nm and 200 nm×720 nm respectively, along with the gap designed to be 415 nm to realize the peak of anomalous dispersion precisely at 1550 nm (Figure 5). The position of the peak of the anomalous dispersion provided by the antisymmetric mode is tuned by the OPL crossing point in

978-1-6654-8699-6/22 $31.00 © 2022 IEEE

Figure 4. As shown in Figure 6, when the inner ring is fixed, widening W_{gap} leads to a red shift of the anomalous dispersion peak along with increased peak values as well as anomalous dispersion bandwidth. The inner-ring radius is designed to be 140 μm to support an FSR of 100 GHz at 1550 nm, and the bending effect has been taken into consideration in the dispersion calculation. The dispersion profiles of the coupled modes and TE_{00} mode in the single waveguide are shown in Figure 5. One can see that the antisymmetric mode provides the anomalous dispersion approximately spanning from 1.50 μm to 1.65 μm, while the single ring naturally has normal dispersion. The insets show that the mode profiles at 1550 nm. It is apparent that the electric field orientation in two coupled rings is in opposite direction for the antisymmetric mode, whereas the electric field is in the same direction for the symmetric mode.

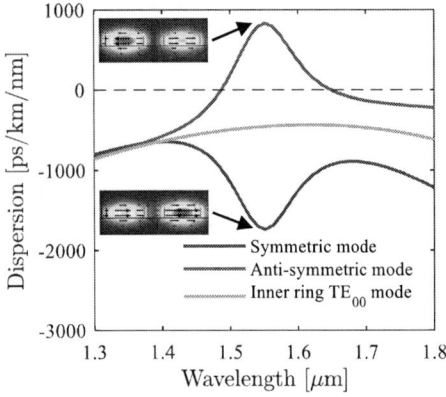

Figure 5. The dispersion profile of the concentric coupled waveguide for symmetric mode (red) and antisymmetric mode (blue); Dispersion in the single waveguide of 200 nm height for TE_{00} mode (orange). Insets show the mode profiles at each dispersion peak.

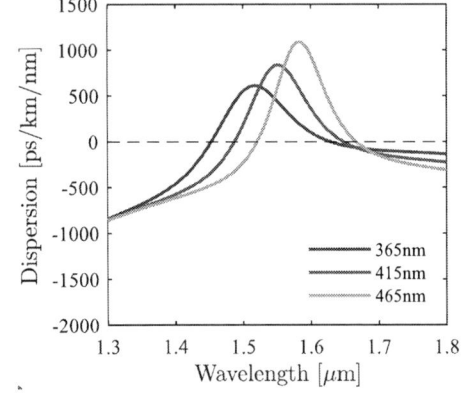

Figure 6. The dispersion profile for antisymmetric modes of different values of W_{gap}.

III. KERR FREQUENCY COMB GENERATION

We investigated the feasibility of generating soliton frequency combs in our optimized resonator using the antisymmetric mode. The antisymmetric resonant frequency can be expanded in Taylor series as $\omega_\mu = \omega_0 + \mu D_1 + 1/2 \cdot$

$\mu^2 D_2 + 1/6 \cdot \mu^3 D_3 + \cdots$, where ω_0 is pump frequency, μ is the mode number related to a ω_0, $D_1/2\pi$ refers to the FSR around the pump frequency ω_0, D_2 is related to the group velocity dispersion β_2 by $D_2 = -c/n \cdot D_1^2 \beta_2$. The integrated dispersion relative to the pump mode is given by $D_{int}(\mu) = \omega_\mu - (\omega_0 + D_1\mu)$. The waveguide integrated dispersion D_{int} is drawn in Figure 7. The second order dispersion $D_2/2\pi$ is extracted to be 2.62 MHz through a parabolic fitting, confirming the resonator is operating at the anomalous dispersion region. Such a small D_2 indicating a relatively flat anomalous dispersion curve, rendering the comb teeth around the pump frequency quite smooth (Figure 7 inset).

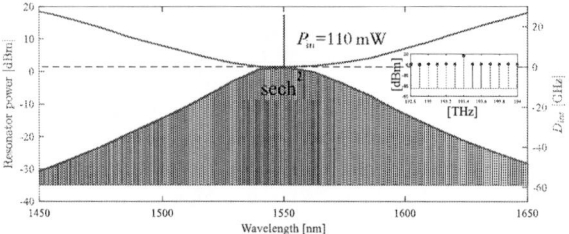

Figure 7. The left y-axis (blue) shows the simulated frequency comb spectrum inside the resonator. The spectrum follows the $sech^2$ shape near the pump wavelength. The right y-axis (orange) represents the integrated dispersion of the resonator. The inset shows the zoomed-in comb teeth around the pump frequency with a spacing of 0.1 THz.

The Kerr soliton frequency comb can be solved by Lugiato-Lefever Equation [26] with proper selections of pump power, laser detuning rate, intrinsic Q factor, and coupled Q factor. The waveguide nonlinear parameter γ is estimated to be 65.9 $m^{-1}W^{-1}$ by simulating the mode area $A_{eff} = 0.6765$ μm^2 and assuming the nonlinear refractive index $n_2 = 11 \times 10^{-18}$ m^{2}W^{-1} [16]. The spectral profile exhibits a $sech^2$ envelope, which indicates that temporally there is a single soliton circulates through the cavity, sitting on a continuous wave background.

For Kerr soliton frequency comb generation, in addition to the need for anomalous dispersion, low propagation loss is another critical factor that should be considered. The 200 nm height of the waveguide allows a considerable amount of light energy to travel in the SiO_2 cladding, which should lower down the propagation loss caused by the loss in the core. The high Q factor supports enough optical power in the cavity and determines P_{th} to a large extent. In our analysis, we assume the waveguide propagation loss to be 1.5 dB/cm, yielding an intrinsic Q factor of 4×10^5. This assumption is valid since the state of the art of GaP-OI waveguide propagation loss ranges from 1.2 dB/cm to 52 dB/cm in literature [16], [18], [19]. The frequency comb pump threshold power is calculated to be 13.5 mW based on the equation $P_{th} = 1.54(\pi n \omega A_{eff})/(2\eta n_2 D_1 Q_t^2)$ [16].

IV. CONCLUSION

In summary, we present a dispersion-engineering approach using concentric ring structures for integrated nonlinear photonic applications. We demonstrate this approach on a 200 nm-thick SiO_2-passivated concentric coupled GaP-OI resonator design for Kerr soliton frequency comb generation with a tooth spacing of 100 GHz and 13.5 mW theoretic threshold power. This approach offers a flexible dispersion-engineering design which breaks through the thickness requirement of the anomalous dispersion and achieves a similar amount of that in a thinner waveguide. This is especially useful for thickness-constrained materials or high-loss materials, where a thinner waveguide layer is preferred.

ACKNOWLEDGMENT

This work is supported by Guangdong Basic and Applied Basic Research Foundation under the Grant Number 2021B1515120074.

REFERENCES

[1] M. Karpov, M. H. P. Pfeiffer, J. Liu, A. Lukashchuk, and T. J. Kippenberg, "Photonic chip-based soliton frequency combs covering the biological imaging window," *Nat. Commun.*, vol. 9, no. 1, p. 1146, Dec. 2018, doi: 10.1038/s41467-018-03471-x.

[2] J. Riemensberger *et al.*, "Massively parallel coherent laser ranging using a soliton microcomb," *Nature*, vol. 581, no. 7807, pp. 164–170, May 2020, doi: 10.1038/s41586-020-2239-3.

[3] D. T. Spencer *et al.*, "An optical-frequency synthesizer using integrated photonics," *Nature*, vol. 557, no. 7703, pp. 81–85, May 2018, doi: 10.1038/s41586-018-0065-7.

[4] Z. L. Newman *et al.*, "Architecture for the photonic integration of an optical atomic clock," *Optica*, vol. 6, no. 5, p. 680, May 2019, doi: 10.1364/OPTICA.6.000680.

[5] A. D. Ludlow, M. M. Boyd, J. Ye, E. Peik, and P. O. Schmidt, "Optical atomic clocks," *Rev. Mod. Phys.*, vol. 87, no. 2, pp. 637–701, Jun. 2015, doi: 10.1103/RevModPhys.87.637.

[6] M.-G. Suh, Q.-F. Yang, K. Y. Yang, X. Yi, and K. J. Vahala, "Microresonator soliton dual-comb spectroscopy," *Science*, vol. 354, no. 6312, pp. 600–603, Nov. 2016, doi: 10.1126/science.aah6516.

[7] L. Chang *et al.*, "Ultra-efficient frequency comb generation in AlGaAs-on-insulator microresonators," *Nat. Commun.*, vol. 11, no. 1, p. 1331, Dec. 2020, doi: 10.1038/s41467-020-15005-5.

[8] L. Zhang, Q. Lin, Y. Yue, Y. Yan, R. G. Beausoleil, and A. E. Willner, "Silicon waveguide with four zero-dispersion wavelengths and its application in on-chip octave-spanning supercontinuum generation," *Opt. Express*, vol. 20, no. 2, p. 1685, Jan. 2012, doi: 10.1364/OE.20.001685.

[9] G. Moille, Q. Li, S. Kim, D. Westly, and K. Srinivasan, "Phased-locked two-color single soliton microcombs in dispersion-engineered Si_3N_4 resonators," *Opt. Lett.*, vol. 43, no. 12, p. 2772, Jun. 2018, doi: 10.1364/OL.43.002772.

[10] S. Kim *et al.*, "Dispersion engineering and frequency comb generation in thin silicon nitride concentric microresonators," *Nat. Commun.*, vol. 8, no. 1, p. 372, Dec. 2017, doi: 10.1038/s41467-017-00491-x.

[11] Z. Gong *et al.*, "High-fidelity cavity soliton generation in crystalline AlN micro-ring resonators," *Opt. Lett.*, vol. 43, no. 18, p. 4366, Sep. 2018, doi: 10.1364/OL.43.004366.

[12] A. W. Bruch *et al.*, "Pockels soliton microcomb," *Nat. Photonics*, vol. 15, no. 1, pp. 21–27, Jan. 2021, doi: 10.1038/s41566-020-00704-8.

[13] X. Wang *et al.*, "2 µm optical frequency comb generation via optical parametric oscillation from a lithium niobate optical superlattice box resonator," *Photonics Res.*, vol. 10, no. 2, p. 509, Feb. 2022, doi: 10.1364/PRJ.432076.

[14] H. Jung, S.-P. Yu, D. R. Carlson, T. E. Drake, T. C. Briles, and S. B. Papp, "Tantala Kerr nonlinear integrated photonics," *Optica*, vol. 8, no. 6, p. 811, Jun. 2021, doi: 10.1364/OPTICA.411968.

[15] M. Pu, L. Ottaviano, E. Semenova, and K. Yvind, "Efficient frequency comb generation in AlGaAs-on-insulator," *Optica*, vol. 3, no. 8, p. 823, Aug. 2016, doi: 10.1364/OPTICA.3.000823.

[16] D. J. Wilson *et al.*, "Integrated gallium phosphide nonlinear photonics," *Nat. Photonics*, vol. 14, no. 1, pp. 57–62, Jan. 2020, doi: 10.1038/s41566-019-0537-9.

[17] A. Anthur *et al.*, "Demonstration of second harmonic generation in gallium phosphide nano-waveguides," 2020, doi: 10.1364/iprsn.2020.itu2a.3.

[18] M. Billet *et al.*, "Gallium phosphide on insulator photonics enabled by micro-transfer printing," in *OSA Advanced Photonics Congress (AP) 2020 (IPR, NP, NOMA, Networks, PVLED, PSC, SPPCom, SOF)*, Washington, DC, 2020, p. ITu2A.6. doi: 10.1364/IPRSN.2020.ITu2A.6.

[19] K. Schneider, P. Welter, Y. Baumgartner, H. Hahn, L. Czornomaz, and P. Seidler, "Gallium Phosphide-on-Silicon Dioxide Photonic Devices," *J. Light. Technol.*, vol. 36, no. 14, pp. 2994–3002, Jul. 2018, doi: 10.1109/JLT.2018.2829221.

[20] J. Cambiasso, G. Grinblat, Y. Li, A. Rakovich, E. Cortés, and S. A. Maier, "Bridging the Gap between Dielectric Nanophotonics and the Visible Regime with Effectively Lossless Gallium Phosphide Antennas," *Nano Lett.*, vol. 17, no. 2, pp. 1219–1225, Feb. 2017, doi: 10.1021/acs.nanolett.6b05026.

[21] A. Martin, S. Combrié, A. de Rossi, G. Beaudoin, I. Sagnes, and F. Raineri, "Nonlinear gallium phosphide nanoscale photonics [Invited]," *Photonics Res.*, vol. 6, no. 5, p. B43, May 2018, doi: 10.1364/PRJ.6.000B43.

[22] A. D. Logan *et al.*, "400%/W second harmonic conversion efficiency in 14 µm-diameter gallium phosphide-on-oxide resonators," *Opt. Express*, vol. 26, no. 26, p. 33687, Dec. 2018, doi: 10.1364/OE.26.033687.

[23] M. Soltani, A. Matsko, and L. Maleki, "Enabling arbitrary wavelength frequency combs on chip: Enabling arbitrary wavelength frequency combs on chip," *Laser Photonics Rev.*, vol. 10, no. 1, pp. 158–162, Jan. 2016, doi: 10.1002/lpor.201500226.

[24] Q. Zhao, Z. Geng, H. Ji, and Y. Li, "GaP-OI Resonator Design for Octave-spanning Kerr Soliton Frequency Comb Generation," in Conference on Lasers and Electro-Optics (2022) (Optica Publishing Group, 2022), p. JW3B.16.

[25] A. Khanna *et al.*, "Impact of ALD grown passivation layers on silicon nitride based integrated optic devices for very-near-infrared wavelengths," *Opt. Express*, vol. 22, no. 5, p. 5684, Mar. 2014, doi: 10.1364/OE.22.005684.

[26] G. Moille, Q. Li, L. Xiyuan, and K. Srinivasan, "pyLLE: A Fast and User Friendly Lugiato-Lefever Equation Solver," *J. Res. Natl. Inst. Stand. Technol.*, vol. 124, p. 124012, May 2019, doi: 10.6028/jres.124.012.

Proceedings of the 7th Optoelectronics Global Conference (OGC 2022)

A Solid-state FMCW Lidar System Based on Lens-assisted Beam Steering

Xianyi Cao, Kan Wu*, Chao Li, Tianyi Li, Jiaxuan Long, Jianping Chen
The State Key Laboratory of Advanced Optical Communication Systems and Networks.
Shanghai Jiao Tong University
Shanghai, China
*kanwu@sjtu.edu.cn

Abstract—**We propose and demonstrate an all solid-state light detection and ranging (Lidar) system based on lens assisted beam steering (LABS) technology. A frequency modulated continuous wave (FMCW) coaxial Lidar system at 1550 nm with 16 scanning directions, 0.35° beam steering step and 1.05° total steering angle is demonstrated. The function of beam steering and 3D imaging are demonstrated. In FMCW ranging experiments, the Lidar system is measured to have 210 m ranging capability. The ranging accuracy can be further improved by applying phase noise compensation. This work proves the potential application of a new integrated beam steering technology in Lidar and demonstrates the way for a fully integrated Lidar system.**

Keywords-component; Lidar; beam steering; 3D imaging; FMCW detection.

I. INTRODUCTION

In the past decades, the Lidar technology has attracted enormous interest in numerus fields as the capacities of high resolution and immunity to electromagnetic interference, such as sensing, 3D mapping, autonomous driving, autonomous drones, wind detection, etc. It is known that solid-state non-mechanical Lidar systems have the potential to achieve unparalleled performance compared with the traditional bulky mechanical Lidar systems. In recent years, various all solid-state beam steering devices with compact size and high stability have been demonstrated, including microelectromechanical (MEMS) systems [1, 2], liquid crystals[3, 4] and optical phased array (OPA)[5-8]. However, the above-mentioned technologies also suffer from the fundamental limits while solving the problem of non-mechanical beam steering. For example, the OPA technology has relatively high control complexity. Meanwhile, the inherent side lobes and grating lobes of the OPA reduce the emitted power. Liquid crystals technology usually has a limited response speed and MEMS technology has limited service life. Recently, the LABS technology is widely perceived as an outstanding beam steering strategy owing to its salient advantages that includes low divergent beam, flexible 2D beamforming, low power consumption, high background suppression and ultra-low control complexity[9]. Therefore, LABS is considered as a prominent candidate for Lidar technology.

Recently, various works of LABS technology have been reported with different configurations, such as metalens[10, 11], photonic crystal waveguide (PCW) grating[12-14], MEMS optical switch[15, 16], etc. In our previous work[17], a Time-of-Flight (ToF) Lidar system based on this technology has been demonstrated, with a detection range of 20 m and ranging error of 3 cm. The FMCW detection method has various obvious advantages compared with ToF detection method, such as high resolution, low power budget and the capability of measuring the velocity of moving target without any additional complexity. Therefore, the FMCW detection method has a border application prospect. However, it is necessary for light source in FMCW detection to maintain a high linearity of frequency sweep to guarantee the accuracy of measurement[18, 19]. Meanwhile, due to the fact that the spectral purity of beat signal affects the spatial resolution, the intrinsic spectral broadening of laser source spectrum modifies the sweep linearity even if the sweeping mechanics is perfect. Studies have attempted to reduce the phase noise of the laser to extend ranging distance. Nonetheless, this approach inevitably leads to an increase in cost. So far, a LABS based FMCW Lidar has shown a maximum ranging distance of 75 m[9].

In this work, we have demonstrated an all solid-state Lidar system with the longest ranging distance up to 210 m. A directly modulated distributed feedback (DFB) laser is applied to this Lidar system for the purposes of high bandwidth and low budget. With elaborative adjusted predistortion driving voltage of DFB laser, the high linearity frequency sweep with residual nonlinearity $1-r^2$ of 5.19×10^{-8} has achieved[20]. A FMCW coaxial Lidar system at 1550 nm with 16 scanning directions, 0.35° beam steering step and 1.05° total steering angle is verified. This Lidar system can be easily extended to larger number of scanning points and larger scanning angle by properly arranging the emitter array and lens parameters. Meanwhile, by utilizing phase noise compensated approach in our ranging system, the spatial resolution is powerfully guaranteed even if measurement beyond the coherence length of laser.

II. PRINCIPLE OF BEAM STEERING

This section provides a brief review of LABS principle. As illustrated in Fig. 1(a), the focal plane of a lens is overlapped with the emission plane of an emitter array. The light emitted from a certain emitter illuminate the lens and will be collimated and steered by the lens.

978-1-6654-8699-6/22 $31.00 © 2022 IEEE

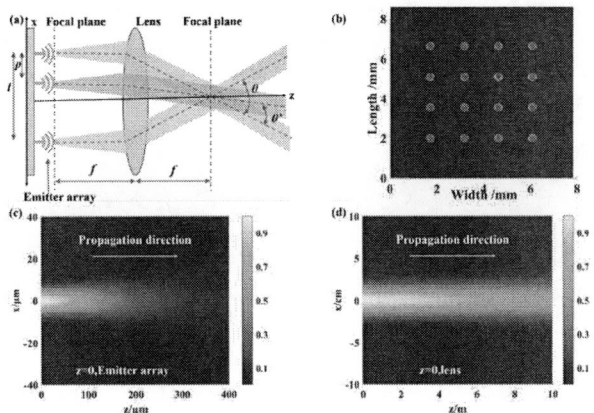

Figure 1(a). Principle of lens assisted beam steering. (b) Simulation of beam pattern in far field. Numerical simulation of beam (c) before collimation and (d) after collimation.

To realize beam steering function, the incident light is routed to a certain emitter/antenna through the optical switch network and emitted to the free space. The beam is collimated by the lens and then its propagation direction is steered. By switching the light to different emitter/antenna, beam steering is performed, i.e. the location of emitter/antenna in the focal plane of the lens corresponds to the angle to which the beam is directed. It should be emphasized that only one emitter emits light at any time. A more detailed analysis on LABS theory may refer to[21]. Therefore, the maximum beam-steering angle θ and the beam steering step θ' can be expressed as follows [22]:

$$\theta = 2\tan^{-1}(\frac{l}{2f}) \qquad (1)$$

$$\theta' = 2\tan^{-1}(\frac{p}{2f}) = \frac{\theta}{N-1} \qquad (2)$$

where l is the size of the emitter array, p is the pitch between two adjacent emitters, f is the focal length of the lens, and N is the number of emitters/antennas. As described above, each emitter corresponds to a beam direction, thus N emitters can realize N discrete angle scanning. According to the principle of LABS, the existence of blind zone in the field of view (FOV) is inherent. The blind zone can also be suppressed well as shown in [23].

Simulation of the beam propagation is performed in Zemax to confirm the aforementioned analysis. A lens with a focal length of 40 mm and diameter of 20 mm at 1550 nm are utilized in simulation. As the Gaussian waist of emitted beam is much smaller than the focal length of the lens, thus the far-field condition is satisfied, the light point source model is adopted to simplify the problem. The pitch of a 4 × 4 two dimensionally point light source array is 250 μm. Each single-spot source emits light with a divergence angle of 7°. The simulated beam pattern for a plane located 0.5 m behind the lens' rear focal plane is shown in Fig. 1(b), indicating the

beams are routed to different directions. No obvious aberration is observed as all the point sources are located in the paraxial region of the lens. If the scale of emitters array is further increased, a multi-lens system is requisite to eliminate the aberration.

To further confirm the beam property in this Lidar system, light beam propagation is calculated theoretically based on Gaussian beam theory. The calculated Gaussian beam pattern before and after traversing lens are shown in Fig. 1(c) and Fig. 1(d) respectively. It can be clearly witnessed that the beams divergence angle is significantly decreased after the collimation of lens, which is conducive to long-distance ranging.

III. BEAM STEERING DEVICE AND CHARACTERIZATION

As illustrated in Fig. 2 (a), an integrated beam-steering device is proposed. The beam steering device, which is composed of an integrated 1 × 16 switch array chip, a 4 × 4 fiber array and a lens, is presumed to realize a long ranging distance. The switch chip, which is fabricated by standard planar lightwave circuit (PLC) technology is connected to the fiber array. The lens is meticulously placed beside the fiber array, to deflect the beam emanating from the emission area, and thereby boosting the beam steering and reception performance.

The switch chip is fabricated on a material stack of 400 μm silicon substrate with a standard PLC progress. The propagation loss of the silica waveguide is less than 0.05 dB/cm, and the total insertion loss of this photonic chip is measured to be ~3 dB, which is low enough to ensure a low link loss. Figure 2(b) shows the photograph of the chip. It consists binary tree structure of four cascaded 1×2 Mach-Zehnder interferometer (MZI) switches. The lead on chip is made of gold with low resistivity, and the heater is made of titanium with high resistivity. The thermal heater has a length of 2000 μm and a resistance of ~80 Ω. When direct current (DC) voltage is applied to the thermal heater on one arm of the MZI switch, the refractive index of waveguide is changed, thereby realizing channel selection. Figure 2 (c) and (d) shows histograms of the extinction ratios of all switches and the insertion losses of each channel respectively. Due to the fabrication error, each MZI switch has slightly different extinction ratio and insertion loss. All the channel has a small insertion loss of ~3 dB, which ensures the realization of enough output power for long-distance ranging. The high background suppression of ~20 dB guarantees high beam quality, which means that the leakage light from other non-working emitters is negligible.

All the 16 outputs of the switch chip are connected to a 4 × 4 fiber array to convert this 1D array into a 2D array, which can be replaced by 3D waveguide. The interval of fiber array is 0.25 mm. A lens with a focal length of 40 mm and a diameter of 20 mm is fixed beside the fiber array, as shown in Fig. 2(a). According to Eq. (1) and Eq. (2), the corresponding beam steering step is 0.35° and the total steering angle is 1.05°×1.05°. A larger scanning angle can be

978-1-6654-8699-6/22 $31.00 © 2022 IEEE

easily realized by choosing a lens with a smaller focal length and a larger size emitting array.

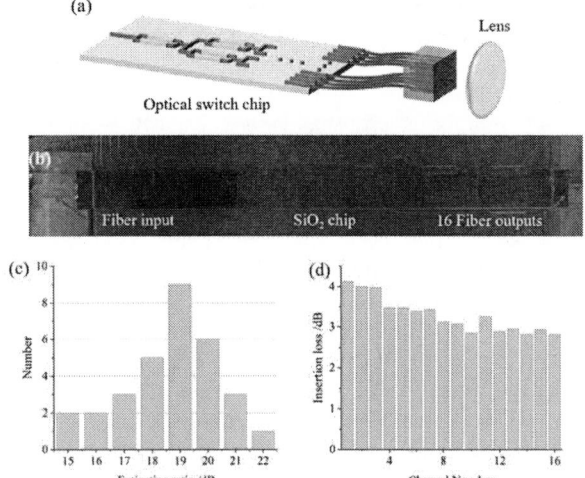

Figure 2(a). Photos of the SiO₂ switch chip. (b) Measured extinction ratio for all 15 MZM units' outputs. (c) Measured insertion loss of the 16 optical channels. (d) Measured insertion loss of 16 channels.

IV. EXPERIMENTS

A. Setup and Result

The experimental setup of the FMCW Lidar is shown in Fig. 3(a). The Lidar system is based on FMCW detection approach. A DFB laser is driven by a pre-distortion triangle-like waveform generated by an arbitrary function generator (AFG). The output optical frequency of this DFB laser varies with the driving voltage. To overcome the nonlinear relationship between driving voltage and laser output optical frequency, a high efficiency iteration method is proposed in our previous work [20] to generate pre-distortion voltage to maintain high linearity frequency sweep. The maximal frequency sweep range is ~26 GHz with a residual nonlinearity of 5.19×10^{-8}, which prove the nonlinearity is effectively compensated. Such a high optical frequency sweep linearity is comparable to external modulation and helpful for the realization of long-distance ranging.

A 1550 nm continuous light with average power of 0 dBm and a period of 1 ms (triangle frequency sweep) is generated by DFB laser. The laser output is guided into the beam steering device through a polarization controller (PC) and a circulator successively. By manipulated the thermal phase shifters of MZI switches, the light is routed to a certain fiber emitter. At any time, only one fiber emitter irradiates light. Beam steering is achieved by guiding light to different position of the emitter. Accordingly, the continuous light is emitted to free space by the beam steering device. The emitted light illuminates the target object (a retroreflective sheet) and is then reflected back. It is noteworthy that the emitted light is partially (4%) reflected by the fiber/air interface. This part of the reflected light can be directly served as reference light. The coaxial design allows the transmitting and receiving light to share the same light path. The reference light and the scattered light pass through the circulator successively and beat in an avalanche photodetector (APD). The distance information can be inferred from the beat signal:

$$R = \frac{c f_b}{2\gamma} \qquad (3)$$

where R is the ranging distance, c is the speed of light in vacuum, f_b is the beat frequency, and γ is the frequency sweep rate.

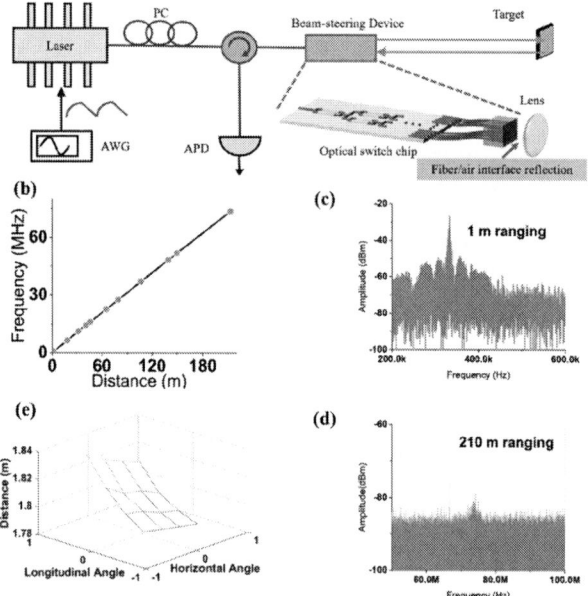

Figure 3(a). Experimental setup of the LABS based FMCW Lidar. (b) Beat frequency versus distance. (c-d) Spectra of beat signal at 1 m and 210 m respectively. (e) Measured point cloud of a globe.

Fig. 3(b) shows the relationship between ranging distance and beat frequency. The linear relationship is obvious, which is consistent with Eq. (3). Figure 3(c) shows a typical spectral of the beat signal when the target is placed at distance of 1 m. The beat signal has a sharp peak with a 3dB linewidth of 1 kHz when the ranging distance is within the laser coherent length. The beat signal is broadened and the signal-to-noise (SNR) drops sharply when beyond the coherent length. The maximum ranging distance is ~210 m, and the spectral of the beat signal is illustrated in Fig. 3(d). To the best of our knowledge, it is the longest ranging distance in LABS based Lidar system. Further extending the ranging distance can be realized with higher optical power. Consequently, it is categorically claimed that the proposed LABS Lidar system has unique advantage compared with counterpart in terms of the long-distance ranging.

Furthermore, the beam steering function of the Lidar system is verified. Benefited from the high frequency sweep bandwidth, the resolution of this Lidar system is on the order

978-1-6654-8699-6/22 $31.00 © 2022 IEEE

of millimeters. As illustrated in Fig. 3(e), a point cloud 3D image of a spherical object placed 1.8 m away is obtained with notable features of sphere. Meanwhile, more points can be obtained by increasing the number of channels. The 2D beam steering capability of the proposed Lidar system is confirmed.

B. Phase Noise Compensation

The existence of inevitable phase fluctuation in laser will seriously deteriorate the SNR of beat signal, as well as broaden the beat signal, resulting in a decrease in accuracy for long distance ranging. Studies have attempted to reduce the phase noise of the laser to extend ranging distance. For example, some researchers have effectively compressed the laser linewidth by self-injection locking[24]. Nonetheless, this approach inevitably leads to an increase in cost. To realize a low-cost and large-ranging Lidar system, a phase noise compensation method is utilized.

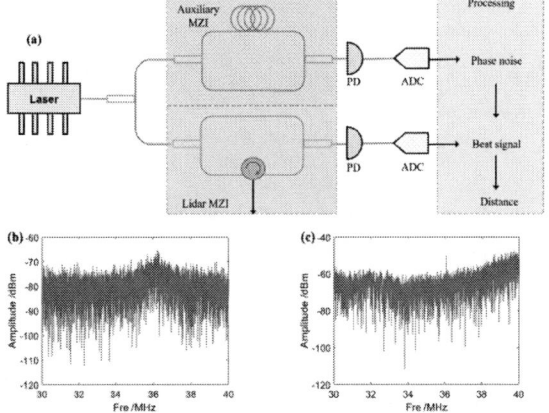

Figure 4. (a) Phase noise compensation process. (b-c) Spectra of beat signal before compensation and after compensation respectively.

The proposed phase noise compensation process is illustrated in Fig. 4 (a). A portion of light from laser is fed into an additional auxiliary fiber-based MZI with a fix time delay τ_a for real-time phase noise extraction. The instantaneous frequency of a linearly chirp laser can be expressed as $f(t) = f_0 + \gamma t + f_{nl}$, where f_0 is the carrier frequency, and nonlinear term f_{nl} is the frequency fluctuation result from the phase noise. Due to the existence of this nonlinear term, for the auxiliary fiber-based MZI and the Lidar MZI with an unknown round-trip τ, the beat signal can be respectively expressed as[25, 26]:

$$i_a(t) = A_a \cos[2\pi\gamma\tau_a t + \Delta\varphi_n(t, \tau_a)] \qquad (4)$$

$$i(t) = A \cos[2\pi\gamma\tau t + \Delta\varphi_n(t, \tau)] \qquad (5)$$

where A_a and A are the beat signal amplitudes of auxiliary MZI and Lidar MZI respectively, and $\Delta\varphi_n$ is the beat signal phase noise. If the values of τ is integral multiples of τ_a, i.e. $\tau = k\tau_a$, the Lidar MZI phase noise $\Delta\varphi_n(t, \tau) = \varphi_n(t) - \varphi_n(t-\tau)$ can

be computed from the auxiliary MZI phase noise $\Delta\varphi_n(t, \tau_a) = \varphi_n(t) - \varphi_n(t-\tau_a)$, and the laser phase noise effects can be fully removed. The phase noise $\Delta\varphi_n(t, k\tau_a)$ can be expressed as follow:

$$\Delta\varphi_n(t, k\tau_a) = \begin{cases} \sum_{p=0}^{k-1} \Delta\varphi_n(t - p\tau_a), k \geq 1; \\ 0, k = 0. \end{cases} \qquad (6)$$

If the round-trip time τ is not the integral multiples of τ_a, i.e. $\tau = k\tau_a + \delta\tau$ where $\delta\tau \leq \tau_a/2$, the phase noise in beat signal $i(t)$ can be partially compensated by multiplying the term of $\exp[-j\Delta\varphi_n(t, k\tau_a)]$ to $i(t)$ in signal processing, and the following equation can be easily derived:

$$i_k(t) = \frac{A}{2}\exp[j2\pi\gamma\tau t + j\Delta\varphi_n(t - k\tau_a, \delta\tau)] + cc. \qquad (7)$$

Compared with Eq. (5), it can be observed that the phase noise is reduced from $\Delta\varphi_n(t, \tau)$ to $\Delta\varphi_n(t-k\tau_a, \delta\tau)$, indicating that the effective distance for phase noise is reduced from $c\tau/2$ to $c\delta\tau/2$, thus the beat signal can be narrowed significantly.

Even though the shape of beat signal spectrum is deteriorated in incoherent regime, the power spectral density (PSD) of beat signal is still dependent on the ranging distance. The prior knowledge about the ranging distance information is obtained from the PSD of beat signal degenerated to Lorentzian distribution, so that the k value in Eq. (6) can be estimated more quickly. Then the phase noise compensation processing is applied. The beat signal power spectrum of the 105 m ranging is shown in Fig. 4(b-c). It is noted that the coherent length of DFB laser used in this experiment is about 130m by measuring the laser linewidth, which means that the ranging resolution will extremely degrade for an object at a larger than 65 m distance. It is obvious that the beat signal is significantly sharpened after the phase noise compensation process, even if the coherent length is exceeding, which prove this method effectively improves the accuracy of this Lidar system in long ranging. In brief, at the expense of an additional fiber-based MZI and back-end data processing, the measurement accuracy of the proposed Lidar system is greatly improved.

V. CONCLUSION

In conclude, a solid-state LABS Lidar system at 1550 nm based on FMCW detection method has been demonstrated on silica platform. Compared with conventional non-mechanical beam steering scheme, this Lidar system has the advantages of low control complexity of $O(\log_2 N)$, low power consumption and excellent background suppression. As a proof of conception, we have achieved 4×4 two-dimension beam steering and detection in a FMCW mode. The emission beam has $1.05° \times 1.05°$ FOV and ~20 dB background suppression. A larger number of scanning points and larger scanning angle can be realized by properly arranging the scale of emitter array and the lens parameters. A maximum ranging distance of 210 m is achieved with 0 dBm emitted power when the target is set as retroreflective sheet. It is believed that a longer ranging distance can be achieved by increasing the emitted power. Furthermore, a phase noise compensation method is applied in the proposed Lidar system to improve the detection accuracy in long-distance ranging. This work overcomes the main challenge of ranging distance in current integrated Lidar systems, and could facilitate the development of practical solid-state beam steering.

ACKNOWLEDGMENT

The authors acknowledge the National Natural Science Foundation of China for funding.

REFERENCES

[1] B. Smith, B. Hellman, A. Gin, A. Espinoza, and Y. Takashima, "Single chip lidar with discrete beam steering by digital micromirror device," Opt Express 25, 14732-14745 (2017).

[2] C. Errando-Herranz, N. Le Thomas, and K. B. Gylfason, "Low-power optical beam steering by microelectromechanical waveguide gratings," Opt Lett 44, 855-858 (2019).

[3] J. A. Frantz, J. D. Myers, R. Y. Bekele, C. M. Spillmann, J. Naciri, J. Kolacz, H. G. Gotjen, V. Q. Nguyen, C. C. McClain, L. Brandon Shaw, and J. S. Sanghera, "Chip-based nonmechanical beam steerer in the midwave infrared," Journal of the Optical Society of America B 35(2018).

[4] M. G. de Blas, J. P. Garcia, S. V. Andreu, X. Q. Arregui, M. Cano-Garcia, and M. A. Geday, "High resolution 2D beam steerer made from cascaded 1D liquid crystal phase gratings," Sci Rep 12, 5145 (2022).

[5] D. N. Hutchison, J. Sun, J. K. Doylend, R. Kumar, J. Heck, W. Kim, C. T. Phare, A. Feshali, and H. Rong, "High-resolution aliasing-free optical beam steering," Optica 3(2016).

[6] C. V. Poulton, M. J. Byrd, M. Raval, Z. Su, N. Li, E. Timurdogan, D. Coolbaugh, D. Vermeulen, and M. R. Watts, "Large-scale silicon nitride nanophotonic phased arrays at infrared and visible wavelengths," Opt Lett 42, 21-24 (2017).

[7] W. Xu, L. Zhou, L. Lu, and J. Chen, "Aliasing-free optical phased array beam-steering with a plateau envelope," Opt Express 27, 3354-3368 (2019).

[8] J. Sun, E. Timurdogan, A. Yaacobi, E. S. Hosseini, and M. R. Watts, "Large-scale nanophotonic phased array," Nature 493, 195-199 (2013).

[9] C. Rogers, A. Y. Piggott, D. J. Thomson, R. F. Wiser, I. E. Opris, S. A. Fortune, A. J. Compston, A. Gondarenko, F. Meng, X. Chen, G. T. Reed, and R. Nicolaescu, "A universal 3D imaging sensor on a silicon photonics platform," Nature 590, 256-261 (2021).

[10] Y. C. Chang, M. Chul Shin, C. T. Phare, S. A. Miller, E. Shim, and M. Lipson, "2D beam steerer based on metalens on silicon photonics," Opt Express 29, 854-864 (2021).

[11] W.-B. Lee, C.-S. Im, C. Zhou, B. Bhandari, D.-Y. Choi, and S.-S. Lee, "Metasurface doublet-integrated bidirectional grating antenna enabling enhanced wavelength-tuned beam steering," Photonics Research 10(2021).

[12] H. Ito, Y. Kusunoki, J. Maeda, D. Akiyama, N. Kodama, H. Abe, R. Tetsuya, and T. Baba, "Wide beam steering by slow-light waveguide gratings and a prism lens," Optica 7(2020).

[13] H. Abe, M. Takeuchi, G. Takeuchi, H. Ito, T. Yokokawa, K. Kondo, Y. Furukado, and T. Baba, "Two-dimensional beam-steering device using a doubly periodic Si photonic-crystal waveguide," Opt Express 26, 9389-9397 (2018).

[14] T. Baba, T. Tamanuki, H. Ito, M. Kamata, R. Tetsuya, S. Suyama, H. Abe, and R. Kurahashi, "Silicon Photonics FMCW LiDAR Chip With a Slow-Light Grating Beam Scanner," IEEE Journal of Selected Topics in Quantum Electronics 28, 1-8 (2022).

[15] X. Zhang, K. Kwon, J. Henriksson, J. Luo, and M. C. Wu, "A large-scale microelectromechanical-systems-based silicon photonics LiDAR," Nature 603, 253-258 (2022).

[16] E. H. Cook, S. J. Spector, M. G. Moebius, F. A. Baruffi, M. G. Bancu, L. D. Benney, S. J. Byrnes, J. P. Chesin, S. J. Geiger, D. A. Goldman, A. E. Hare, B. F. Lane, W. D. Sawyer, and C. R. Bessette, "Polysilicon Grating Switches for LiDAR," Journal of Microelectromechanical Systems 29, 1008-1013 (2020).

[17] X. Cao, G. Qiu, K. Wu, C. Li, and J. Chen, "Lidar system based on lens assisted integrated beam steering," Opt Lett 45, 5816-5819 (2020).

[18] X. Zhang, J. Pouls, and M. C. Wu, "Laser frequency sweep linearization by iterative learning pre-distortion for FMCW LiDAR," Opt Express 27, 9965-9974 (2019).

[19] G. Zhang, Z. Ding, K. Wang, C. Jiang, J. Lou, Q. Lu, and W. Guo, "Demonstration of high output power DBR laser integrated with SOA for the FMCW LiDAR system," Opt Express 30, 2599-2609 (2022).

[20] X. Cao, K. Wu, C. Li, G. Zhang, and J. Chen, "Highly efficient iteration algorithm for a linear frequency-sweep distributed feedback laser in frequency-modulated continuous wave lidar applications," Journal of the Optical Society of America B 38(2021).

[21] S. J. Spector, "Review of lens-assisted beam steering methods," Journal of Optical Microsystems 2(2022).

[22] D. Inoue, T. Ichikawa, A. Kawasaki, and T. Yamashita, "Demonstration of a new optical scanner using silicon photonics integrated circuit," Opt Express 27, 2499-2508 (2019).

[23] C. Li, X. Cao, K. Wu, G. Qiu, M. Cai, G. Zhang, X. Li, and J. Chen, "Blind zone-suppressed hybrid beam steering for solid-state Lidar," Photonics Research 9(2021).

[24] L. Tang, H. Jia, S. Shao, S. Yang, H. Chen, and M. Chen, "Hybrid integrated low-noise linear chirp frequency-modulated continuous-wave laser source based on self-injection to an external cavity," Photonics Research 9(2021).

[25] F. Ito, X. Fan, and Y. Koshikiya, "Long-Range Coherent OFDR With Light Source Phase Noise Compensation," Journal of Lightwave Technology 30, 1015-1024 (2012).

[26] X. Zhang, "Laser Chirp Linearization and Phase Noise Compensation for Frequency-modulated Continuouswave LiDAR." University of California, Berkeley (2021).

Proceedings of the 7th Optoelectronics Global Conference (OGC 2022)

Silicon Photonic Integrated Reservoir Computing Processor with Ultra-high Tunability for High-speed IM/DD Equalization

Aolong Sun[1], An Yan[1], Penghao Luo[1], Junwen Zhang[1, 2]* and Nan Chi[1,2]

[1]Key Laboratory for Information Science of Electromagnetic Waves (MoE), Department of Communication Science and Engineering, Fudan University, No. 220, Handan Road. Shanghai 200433, China
[2]Peng Cheng Laboratory, Shenzhen, 518055, China
junwenzhang@fudan.edu.cn

Abstract—**Intensity modulation and direct detection (IM/DD) technology still dominates the optical fiber communication region for the sake of cost and energy efficiency. Reservoir Computing (RC), a special machine learning algorithm suitable for sequence models, has recently been applied to reduce the inter-symbol interference (ISI) caused by dispersion and Kerr nonlinearity in IM/DD systems. In this paper, we designed and numerically simulated a Photonic Integrated Reservoir Computing Processor (PIRCP) with two recurrent nodes using a standard silicon-on-insulator platform. The PIRCP exhibits ultra-high tunability of phase, intensity, delay time and detuning frequency of the optical carrier, which greatly facilitates parameter sweeping for the obtaining of the optimal processing performance. To validate the efficiency of our design, we implemented the PIRCP along with a Feed Forward Equalizer (FFE) in the receiver-end, and finally achieved sub HD-FEC performance for 112 Gbps/λ transmission over 60 km standard single-mode fiber (SSMF) with an ROP of -15 dBm, showing an improvement of 5 dBm compared with non-RC scheme.**

Keywords-Photonic Reservoir Computing; silicon photonics; IM/DD equalization

I. INTRODUCTION

In the interconnection of data centers, intensity modulation and direct detection (IM/DD) technology is still the mainstream technology due to the consideration of cost and power efficiency. 400G Ethernet is achieved based on four lanes of 4-level pulse amplitude modulation (PAM4) with 100 Gbps/λ [1]. In metro datacenter interconnection (metro-DCI), interconnection distance can be up to 80 km [2], which inevitably introduces fiber chromatic dispersion (CD) and nonlinear interference greatly.

Nowadays reservoir computing (RC) has become an effective candidate for dispersion compensation and nonlinear equalization in optical communication due to its ability to efficiently tackle the complex problem of inter-symbol interference (ISI) in sequence model [3]. In addition, RC greatly reduces the training complexity, which is conducive to the realization of various functionalities based on photonic integrated circuits [4-6]. Previous work adopts two-filter-loop based RC to slice the optical spectrum into two different parts [7], which prominently removes CD and Kerr nonlinearity. However, they only simulated such design with ideal parameter model of devices, without designing

any specific photonic device to validate the feasibility of photonic integration.

In this work, we propose a highly tunable Photonic Integrated Reservoir Computing Processor (PIRCP) with two recurrent nodes on a silicon-on-insulator (SOI) platform. The phase, intensity, and frequency detuning of signal light can be flexibly tuned to obtain the optimal processing performance, which is validated through device-level along with system-level simulations targeting one task: the mitigation of transmission effects (i.e. CD and Kerr effect) for a 56 Gbaud PAM4 signal after transmission over up to 60 km standard single-mode fiber (SSMF) in C-band. Simulation results demonstrate the superior performance and robustness of our PIRCP for reducing BER in high bit-rate IM/DD systems.

II. SYSTEM CONFIGURATION AND PRINCIPLE

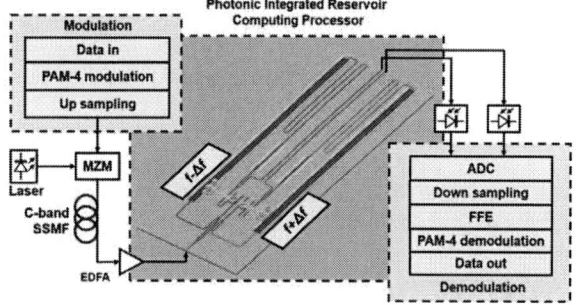

Figure 1. The system set-up. The input of PIRCP is PAM4 signal transmitted through a C-band single-mode fiber, while the outputs of PIRCP are transported to photodetectors (PDs) and Analog to Digital Converters (ADCs). Finally, a Feed Forward Equalizer (FFE) is employed to process the data.

The system set-up is shown in Fig. 1. The proposed PIRCP is implemented as an equalizer in the receiver end, consisting of two integrated recurrent nodes with identical configuration and opposite frequency detuning. The configuration of a single recurrent node is clearly illustrated in Fig. 2, which consists of four core functional components, namely a thermally tunable microring resonator (MRR) filter, a variable optical attenuator (VOA) based on the singe-layer graphene (SLG), a waveguide delay line and a waveguide phase shifter based on PIN junction. The entire device is designed on a standard silicon-on-insulator (SOI) platform

978-1-6654-8699-6/22 $31.00 © 2022 IEEE   227

with a 220-nm-thick silicon layer and a 3-µm-thick buried oxide (BOX) layer. The waveguide profiles in three different parts of the circuit are pointed out by black dotted arrows shown in Fig. 2, indicating that the silicon layer of the whole circuit is fully etched except for the phase shifter, which is partially etched by 130 nm to form a ridge waveguide. As for the VOA part, a 30-nm-thick Al_2O_3 layer is deposited on the silicon waveguide to act as an electric buffer area for SLG. All the waveguides are 500 nm wide and cladded by SiO_2. The waveguide mode used in this design and simulation is fundamental TE mode.

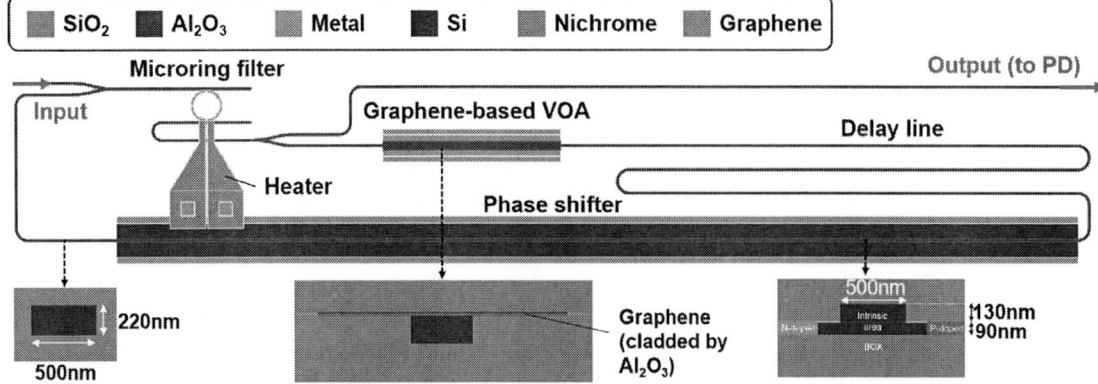

Figure 2. One of the recurrent nodes, composed of a MRR filter, a VOA, a waveguide delay line, a phase shifter, a Y-branch combiner and a 3-dB Y-branch splitter. The insets are layer profiles with structure parameters in three different parts of PIRCP, namely the transmission waveguide, VOA and phase shifter (from left to right).

As shown in Fig.2, when signal light with a frequency of f enters the recurrent node, it first goes through the MRR filter with a resonant frequency of $f+\Delta f$ ($f-\Delta f$ for the other node). The frequency detuning Δf can be tuned in real time by changing the voltage applied to the heater. Next, a 3-dB power splitter is implemented to equally split the light, half of which is directly transmitted to the output port and received by a PD, while the other half continues propagating in the node and its power and phase are adjusted by the VOA and phase shifter, respectively. The phase shifter provides phase compensation during the recurrent loop, while the function of VOA can be interpreted as to adjust the weight of the recurrent component back to the node. Besides, the delay line is employed to increase the delay of the loop in order to achieve a total delay of $T_{symbol}/2$ (i.e. 9 ps), which can obtain the lowest BER as proven in [5]. The light finally returns to the node and combines with the input light through a Y-branch combiner, marking the beginning of next loop. Therefore, the previous symbols can interact with current symbol, realizing reservoir computing.

III. DEVICE DESIGN AND CHARACTERIZATION

The parameters and materials of each device must be accurately designed to meet required need. First, the graphene-based VOA is set to be 40 µm long for a 3-dB variation range. Fig. 3(a) shows the cross-section of VOA. The SLG is 10 nm above the Si waveguide, and cladded by a 30-nm-thick Al_2O_3 buffer. Changing the voltage applied to the graphene will shift its Fermi level, and thus alter its optical absorption [8]. The main reason for choosing SLG is that it only produces a small additional phase shift while

changing optical absorption, reducing the burden of phase shifter. Fig. 3(b) shows that the total attenuation can be tuned from 0.18 dB to 3 dB by changing the Fermi level from 0.6 eV to 0.2 eV.

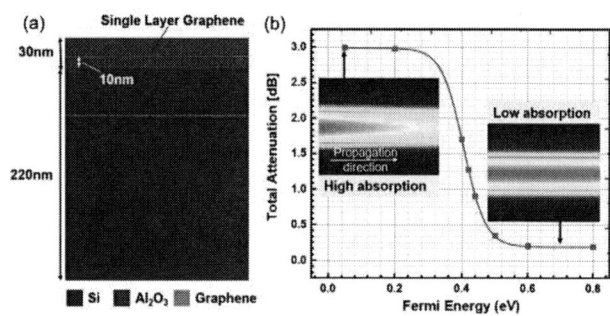

Figure 3. (a) Cross-section of VOA waveguide (with SLG marked with a blue line). (b) Total attenuation of VOA as a function of the Fermi energy level of graphene, with two insets illustrating the optical power distribution along the propagation direction under different optical absorption. The length of VOA is 40 µm.

As for the phase shifter, it is constructed with a PIN junction on a ridge waveguide, of which the doping profile is shown in the inset of Fig. 2. Doping concentration is set to be 2×10^{18} cm^{-3}. The 90-nm thick slab can benefit the transfer of free carriers. Fig. 4 exhibits that a positive voltage applied to PIN reduces the refractive index due to plasma dispersion effect [9]. The length of phase shifter is set to be 215 µm to realize a tunable range from 0 to π rad under the voltage of 2 V.

978-1-6654-8699-6/22 $31.00 © 2022 IEEE

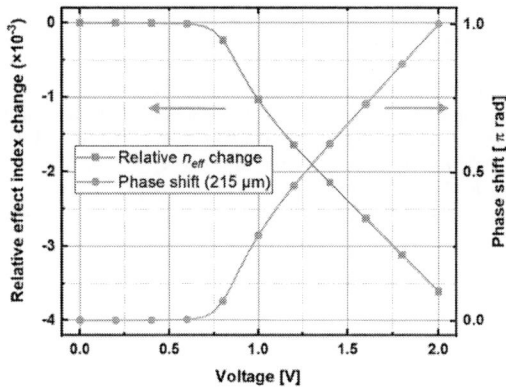

Figure 4. Simulated refractory index change (orange) and total phase shift (blue) of PIN phase shifter as a function of voltage. The length of phase shifter is 215 μm.

The MRR filter is based on a racetrack resonator with a radius of 3.2 μm and a straight coupling length of 400 nm as shown in Fig. 5(a). The coupling coefficient between ring and bus waveguide is about 0.04. The nichrome heater is 150 nm above the waveguide and controlled by voltage through Au electrodes. Results show that the 3-dB bandwidth of MRR filter is 60 GHz, approximately equal to the baudrate used in this design (i.e. 56 Gbaud). Fig. 5(b) shows that the resonant wavelength (λ_r, μm) increases linearly with the square of voltage (U, volt), which can be expressed by the following fitting equation (1):

$$\lambda_r = 8.09U^2 + 0.074U + 1548.45 \qquad (1)$$

The MRR filter can realize a frequency detuning range over 250 GHz even at a voltage less than 0.5 V. The spectra can be shifted in the whole C-band by adding DC bias voltages to the heater. Since the shape of heater is matched with MRR, the heat is almost completely limited in MRR (Fig. 4(b) lower inset and Fig. 4(c)), thus only producing trivial thermal crosstalk to other devices.

The combiner and splitter are based on the same Y-branch structure as shown in Fig. 2. The gap and angle of Y-branch are optimized through particle swarm optimization (PSO) to obtain the lowest insertion loss, and finally set to be 2.39 μm and 11.77° respectively. Simulation results show that the excess losses of combiner and splitter are 0.13 and 0.05 dB, respectively.

Lastly, the length of waveguide delay line (L_d) is set to be 348 μm after other parts of the processor has been designed and simulated, according to the equation (2):

$$L_d = \frac{c}{n_g}(T_{symbol}/2 - T_{rest}) \qquad (2)$$

where c is velocity of light in vacuum, n_g is the group index (i.e. 4.2), and T_{rest} is the total delay of other devices (i.e. 4.13 ps). Therefore, a total loop delay of $T_{symbol}/2$ is successfully realized.

IV. RESULTS

To obtain the best processing performance, we sweep the parameters of VOAs, MRR filters and phase shifters of both nodes. The optimal optical parameters and their corresponding electrical parameters are listed in Table I, which are well within the tunable range of proposed devices, proving the feasibility of PIRCP.

TABLE I. DEVICE PARAMETERS FOR OPTIMAL PERFORMANCE.

Device	Optical		Electrical	
	Quantity	Value	Quantity	Value
VOA₁	Total Attenuation (dB)	0.52	Fermi Level (eV)	0.47
VOA₂		0.18		0.63
Phase Shifter₁	Phase shift (π rad)	0.075	Applied Voltage (V)	0.813
Phase Shifter₂		0.575		1.364
MRR₁	Frequency detuning (GHz)	+7		0.399
MRR₂		-7		0.416

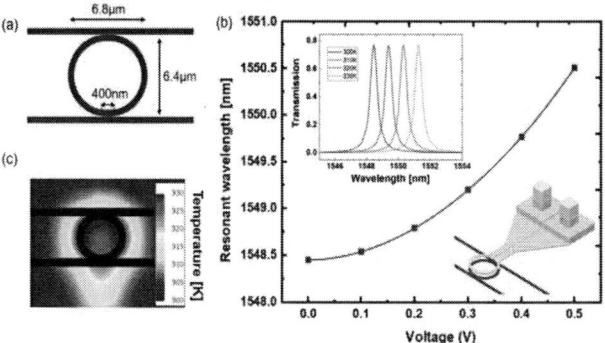

Figure 5. (a) 2D view and parameters of proposed thermally tunable MRR filter. (b) Simulated resonant wavelength of MRR as a function of applied voltage. The upper inset shows that the transmission spectrum of MRR is red-shifted linearly with the increase of temperature. The lower inset is the 3D view of MRR and nichrome heater. (c) Temperature distribution of MRR at a voltage of 0.5V. The device structure is superimposed in black.

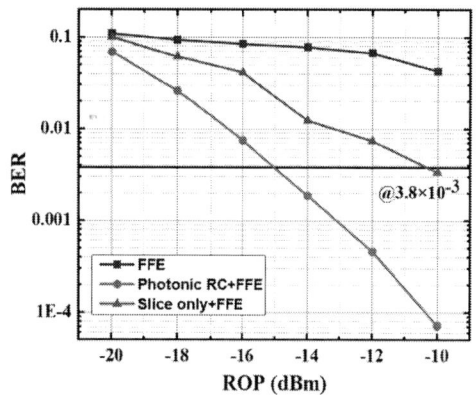

Figure 6. Simulated Bit Error Rate (BER) results for various equalization schemes as a function of Received Optical Power (ROP). BER of photonic RC scheme (red) shows a remarkable decline compared with non-RC scheme. The simulated transmission distance is 60 km.

978-1-6654-8699-6/22 $31.00 © 2022 IEEE

Additionally, three dispersion equalizers, namely a single FFE, a non-recurrent two-filter FFE, and PIRCP-assisted FFE, are simulated and evaluated through 60 km transmission of 56 Gbaud PAM4 signal. Fig. 6 reveals that our PIRCP remarkably reduces BER compared with non-RC schemes, achieving a notable BER performance under HD-FEC limit with an ROP of only -15 dBm, which is a significant improvement of 5 dBm compared with the non-recurrent two-filter FFE scheme.

V. CONCLUSIONS

In summary, a Photonic Integrated Reservoir Computing Processor (PIRCP) for dispersion equalization with ultra-high tunability of phase, power and frequency, is designed and validated through simulations of 60 km C-band transmission of 56 Gbaud PAM4 signal. Our PIRCP dramatically reduces BER compared with non-RC schemes, representing great robustness and efficiency for alleviating chromatic dispersion and Kerr effect in high bit-rate IM/DD systems.

ACKNOWLEDGMENT

This work is partially supported by National Natural Science Foundation of China (61925104, 62031011, 62171137); Natural Science Foundation of Shanghai (21ZR1408700); The Major Key Project of PCL (PCL2021A14).

REFERENCES

[1] M. Filer, S. Searcy, Y. Fu, R. Nagarajan and S. Tibuleac, "Demonstration and performance analysis of 4 Tb/s DWDM metro-DCI system with 100G PAM4 QSFP28 modules," 2017 Optical Fiber Communications Conference and Exhibition (OFC), 2017, pp. 1-3.

[2] J. Cheng, C. Xie, Y. Chen, X. Chen, M. Tang and S. Fu, "Comparison of Coherent and IMDD Transceivers for Intra Datacenter Optical Interconnects," 2019 Optical Fiber Communications Conference and Exhibition (OFC), 2019, pp. 1-3.

[3] L. Appeltant, M.C. Soriano, G. Van der Sande, J. Danckaert, S. Massar, J. Dambre, B. Schrauwen, C.R. Mirasso and I. Fischer. "Information processing using a single dynamical node as complex system". Nature Communication, 2: 468, 2011.

[4] C. Mesaritakis, K. Sozos, D. Dermanis and A. Bogris, "Spatial Photonic Reservoir Computing based on Non-Linear Phase-to-Amplitude Conversion in Micro-Ring Resonators," 2021 Optical Fiber Communications Conference and Exhibition (OFC), 2021, pp. 1-3.

[5] M. Borghi, S. Biasi and L. Pavesi, "Reservoir computing based on a silicon microring and time multiplexing for binary and analog operations", Scientific Reports, 11: 15642, 2021.

[6] S. Sackesyn, C. Ma, J. Dambre, P. Bienstman, "Experimental Demonstration of Nonlinear fibre Distortion Compensation with Integrated Photonic Reservoir Computing", 2021 European Conference on Optical Communication (ECOC), 2021, pp. 1-4.

[7] K. Sozos, A. Bogris, P. Bienstman and C. Mesaritakis, "Photonic Reservoir Computing based on Optical Filters in a Loop as a High Performance and Low-Power Consumption Equalizer for 100 Gbaud Direct Detection Systems," 2021 European Conference on Optical Communication (ECOC), 2021, pp. 1-4.

[8] M. Romagnoli, V. Sorianello, M. Midrio, F. Koppens, C. Huyghebaert, D. Neumaier, P. Galli, W. Templ, A. D'Errico and A. Ferrari, "Graphene-based integrated photonics for next-generation datacom and telecom," Nature Reviews Materials, vol. 3, no. 10, pp. 392–414, 2018.

[9] R. Soref and B. Bennett, "Electrooptical effects in silicon," IEEE Journal of Quantum Electronics, vol. 23, pp. 123–129, 1987.

Proceedings of the 7th Optoelectronics Global Conference (OGC 2022)

Double-tip Scandium Aluminum Nitride Edge Couplers at 1550 nm Wavelength

Hengyu Wang[1], Xingyan Zhao[1], Shaonan Zheng[1], Zhengji Xu[2], Yuan Dong[1]*, Ting Hu[1]

[1]School of Microelectronics, Shanghai University, Shanghai, P. R. China
[2]School of Microelectronics Science and Technology, Sun Yat-sen University, Zhuhai, P. R. China
*e-mail: dongyuan@shu.edu.cn

Abstract—**Double-tip scandium aluminum nitride ($Al_{1-x}Sc_xN$) edge couplers with three different Sc concentrations ($x = 0$, 0.09, 0.23) working at 1550 nm wavelength are designed. The geometric parameters of the devices are optimized by the particle swarm optimization (PSO) algorithm. It is shown that the double-tip edge couplers have higher coupling efficiencies for both transverse-electric (TE) and transverse-magnetic (TM) modes compared with the single-tip ones with the same tip width and device length.**

Keywords-scandium aluminum nitride; edge coupler; integrated photonics; double-tip

I. INTRODUCTION

Aluminum nitride (AlN) is a wide bandgap semiconductor which has been widely used in microelectronics, opto-electronics and micro-electro-mechanical systems (MEMS) [1-3]. As AlN is transparent from ultraviolet to mid-infrared wavelengths, it has been considered as one promising material for optical waveguides in integrated photonics. In addition, the relatively strong piezoelectric coefficient makes AlN a suitable material for acousto-optic modulation, which could be utilized to realize on-chip optical isolators or circulators [4-6].

Incorporating scandium (Sc) element into AlN has been proved to greatly enhance the piezoelectric performances [7-10]. This makes $Al_{1-x}Sc_xN$ a promising candidate for integrated photonic circuits with acousto-optic devices. $Al_{1-x}Sc_xN$ optical waveguides have been demonstrated theoretically and experimentally [11,12]. Recently, Huang et al. have reported acousto-optic modulators in Si waveguides based on $Al_{0.6}Sc_{0.4}N$ film [13]. Although progresses have been made, there are still missing building blocks in the $Al_{1-x}Sc_xN$ photonic circuits. Edge coupler is commonly used in integrated photonic systems to couple light between fiber tips and optical waveguides. To the best of our knowledge, the detailed study of $Al_{1-x}Sc_xN$ edge coupler has not been reported yet.

In this work, double-tip $Al_{1-x}Sc_xN$ edge couplers operating at 1550 nm wavelength are designed by numerical simulation. The Sc concentration x varies from 0 to 0.23. The simulation results show that the double-tip couplers present higher coupling efficiency than the single-tip ones at $\lambda = 1550$ nm for both transverse-electric (TE) and transverse-magnetic (TM) modes.

II. DESIGN OF DOUBLE-TIP EDGE COUPLERS

The three-dimensional (3D) schematic of an $Al_{1-x}Sc_xN$ double-tip edge coupler is shown in Fig. 1(a). Light couples from the tip of a lensed fiber into the coupler along X-direction. Fig. 1(b) shows the Y-Z cross-sectional schematic of the edge coupler at the facet. Si substrate is not shown in this figure. Silicon dioxide (SiO_2) is used as the lower and the upper cladding layers, with thicknesses of 3 μm and 2 μm respectively. The thickness of $Al_{1-x}Sc_xN$ is fixed at 400 nm. The refractive indices of $Al_{1-x}Sc_xN$ are obtained using the same approach in [11]. The sidewall angle of $Al_{1-x}Sc_xN$ waveguides is fixed at 80° in our simulation according to the experimental results of $Al_{1-x}Sc_xN$ dry etching process in [14,15]. The tip width at the top of the waveguide is denoted as W_{tip}, and the gap between the tips is represented by W_{g1}. Similarly, other geometric parameters in this work are defined based on the top of the structure (top CD rather than bottom CD). Shown in Fig. 1(c) is the top-view of the double-tip edge coupler. The width of the two inverse tapers increases linearly from W_{tip} to W_e after a length of L_1. Similarly, the gap between the two tapers changes from W_{g1} to W_{g2}. The waveguide width W_{wg} is 0.95 μm, 0.85 μm, 0.75μm for Sc concentrations $x = 0$, 0.09, 0.23, respectively in order to fulfill the single-mode conditions for both TE and TM polarizations. The overall coupling efficiency η of a double-tip edge coupler is defined by

$$\eta = \eta_1 \times \eta_2, \qquad (1)$$

where η_1 is the modal overlap efficiency between the coupler facet and the light source, and η_2 is the power transmission of the edge coupler to waveguide.

The modal overlap efficiency η_1 of the device is firstly optimized by using the finite-difference eigenmode (FDE) method. The mode field diameter (MFD) of the light source is fixed at 2.5 μm. It should be noted that the minimum critical dimension (CD) of the device structure in this simulation is set to be no less than 500 nm to make the fabrication process compatible to the platforms with less sophisticated photolithography tools, e.g. g-line or i-line steppers. The particle swarm optimization (PSO) method is used to find the optimal W_{tip} and W_{g1} values to realize the highest η_1 for TE mode. The optimized W_{tip} and W_{g1} values are summarized in Table I, with the optimized TE mode η_1 of

978-1-6654-8699-6/22 $31.00 © 2022 IEEE

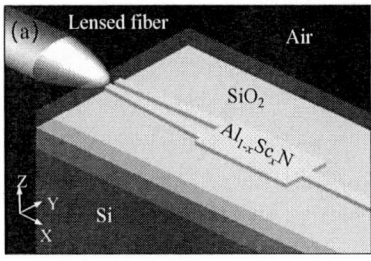 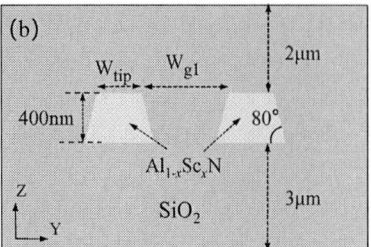 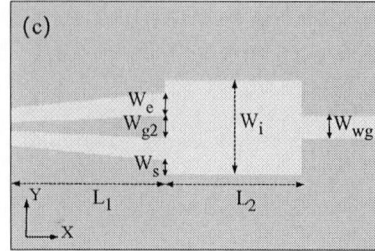

Figure 1. Schematic diagram of the $Al_{1-x}Sc_xN$ double-tip edge coupler: (a) 3D view, (b) cross-sectional view, and (c) top view of the double-tip edge coupler.

74.05%, 69.33% and 63.59% for the devices with $x = 0$, 0.09 and 0.23, respectively. In addition, the TM mode η_1 is calculated based on the same W_{tip} and W_{g1} shown in Table I, with the values of 81.95%, 73.48% and 72.50% for Sc concentrations of 0, 0.09 and 0.23, respectively.

After determining the two parameters of tip width and tip gap, we employ another PSO process in finite domain time

difference (FDTD) simulator to optimize the overall coupling efficiency η for the whole structure. Five parameters W_e, W_{g2}, W_s, L_1 and L_2 are optimized to achieve highest TE mode η at 1550 nm wavelength. All of the optimized geometric parameters of the $Al_{1-x}Sc_xN$ ($x = 0$, 0.09, 0.23) double-tip edge couplers are summarized in Table I.

TABLE I. THE OPTIMIZED GEOMETRIC PARAMETERS OF THE $AL_{1-x}SC_xN$ ($X = 0$, 0.09, 0.23) DOUBLE-TIP EDGE COUPLERS.

Sc concentration x	W_{tip} (nm)	W_{g1} (nm)	W_{g2} (nm)	W_e (nm)	W_s (nm)	L_1 (μm)	L_2 (μm)
0	500	590	996	802	500	17.2	11
0.09	500	540	1117	1150	534	15.4	14.7
0.23	500	540	978	1039	500	19.5	12.2

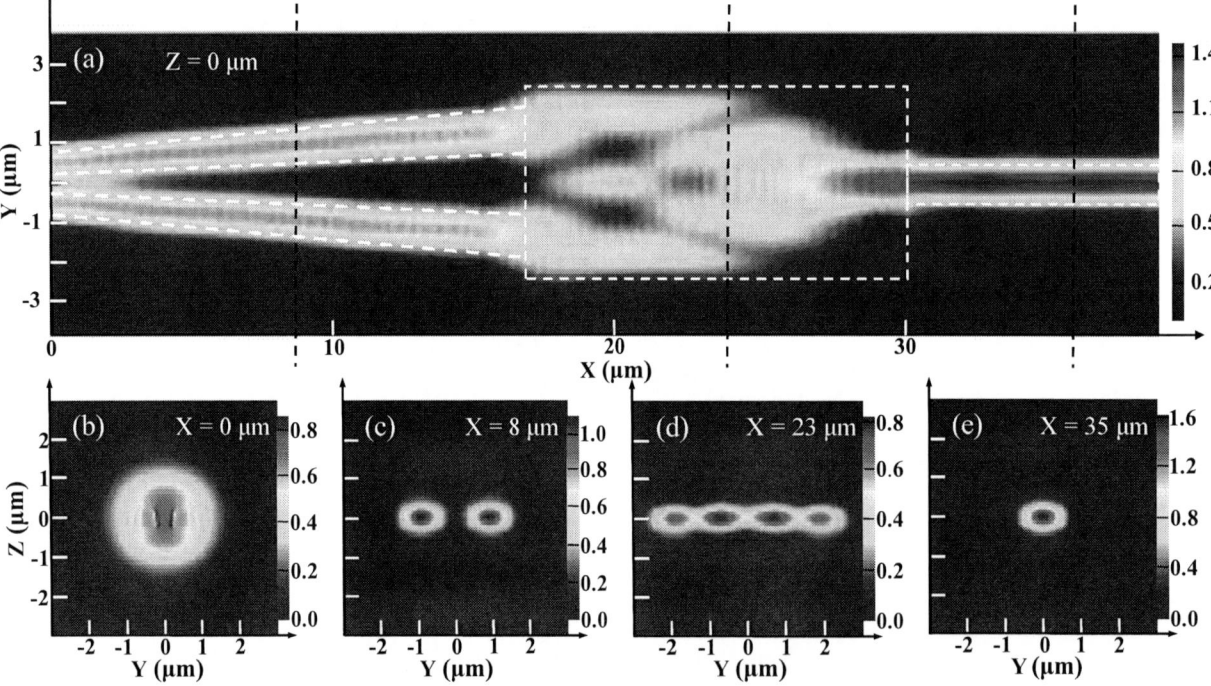

Figure 2. (a) Electric field distribution of an $Al_{0.91}Sc_{0.09}N$ double-tip edge coupler for TE mode; (b)-(e) The electric field distributions of the cross sections at different positions (X = 0, 8, 23, 35 μm). The unit of colorbar is V/m.

978-1-6654-8699-6/22 $31.00 © 2022 IEEE

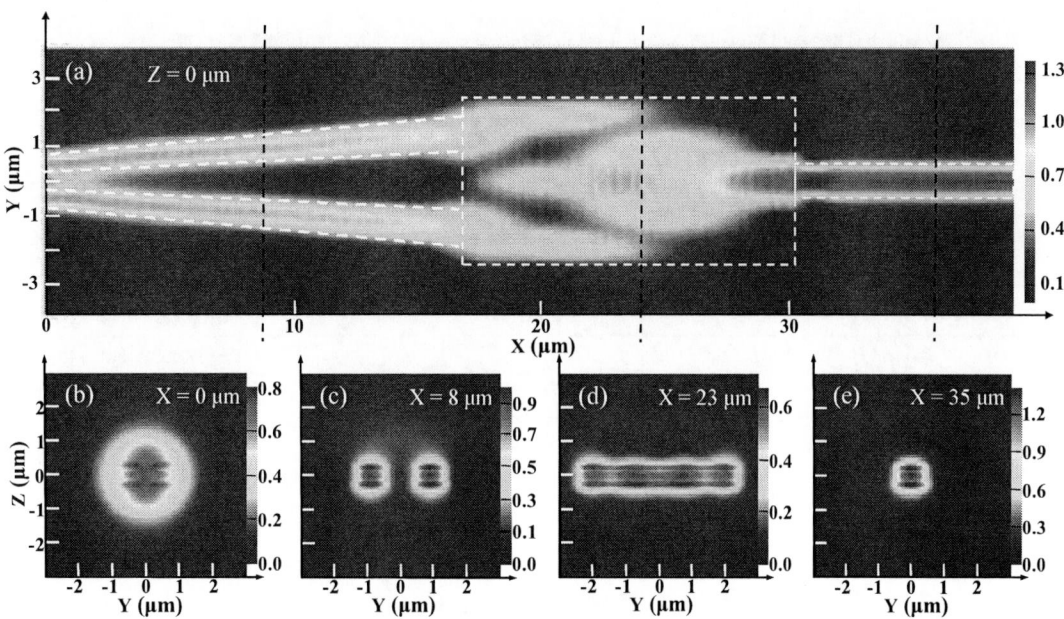

Figure 3. (a) Electric field distribution of an $Al_{0.91}Sc_{0.09}N$ double-tip edge coupler for TM mode; (b)-(e) The electric field distributions of the cross sections at different positions (X = 0, 8, 23, 35μm). The unit of colorbar is V/m.

III. RESULTS AND DISCUSSION

Modal profiles of an $Al_{0.91}Sc_{0.09}N$ double-tip edge coupler are simulated to provide insights on how light is coupled into the optical waveguides. The wavelength is fixed at 1550 nm. The TE modal profiles of the device are plotted in Fig. 2. The electric field distribution in the X-Y plane along the light propagation direction at Z = 0 μm is shown in Fig. 2(a). Figs. 2(b) to 2(e) show the electric field distributions in the Y-Z planes at X = 0, 8, 23, and 35 μm, respectively. It can be observed that the TE mode size is relatively large at the facet (X = 0 μm) to match the MFD of the fiber. Next the light is separated into two arms [Fig. 2(c)], and then entered into the multimode interference region [Fig. 2(d)]. Finally, the light is coupled into the TE mode

waveguide [Fig. 2(e)]. It should be noted that although the double-tip edge coupler is optimized for TE mode coupling, it is able to couple TM mode light as well, as shown in Fig. 3.

The coupling efficiency vs. wavelength characteristics of the optimized $Al_{1-x}Sc_xN$ (x = 0, 0.09, 0.23) double-tip edge couplers for both TE and TM modes are plotted in Fig. 4. The wavelength range is 1500 to 1600 nm. The geometric parameters of each device are summarized in Table I. It can be observed that the η spectra are relatively flat within the wavelength range, indicating a large bandwidth of the edge coupler for both TE and TM modes.

Next, the comparison of coupling efficiencies is made between the double-tip edge coupler and the single-tip one with the same tip width (W_{tip}) and total device length ($L_1 + L_2$). The wavelength is fixed at 1550 nm. The results are

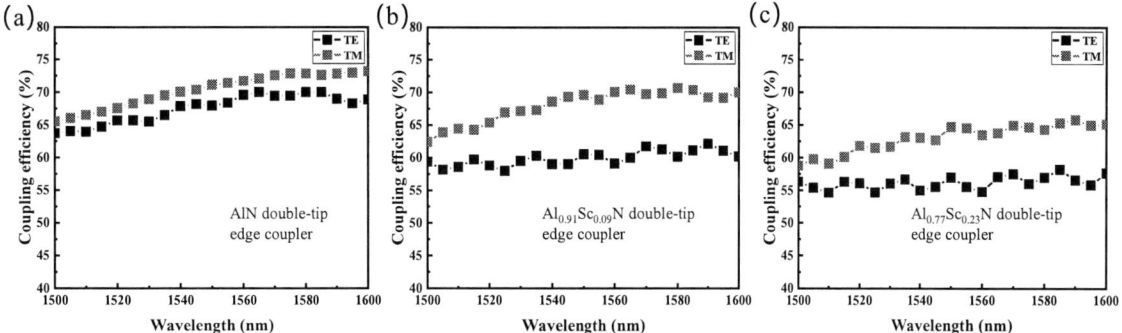

Figure 4. Simulated coupling efficiency for both TE and TM modes at 1500 nm to 1600 nm wavelength range for the (a) AlN, (b) $Al_{0.91}Sc_{0.09}N$ and (c) $Al_{0.77}Sc_{0.23}N$ double-tip edge couplers.

summarized in Table II. It is shown that the double-tip edge couplers have higher η values for both TE and TM modes. For example, for Sc concentration x = 0.23, the efficiency of

the double-tip coupler is 8.18% higher for TE mode and 9.88% higher for TM mode as compared to the single inverse-tapered edge coupler.

TABLE II. THE SIMULATED RESULTS OF TWO TYPES OF $AL_{1-x}SC_xN$ ($x = 0, 0.09, 0.23$) EDGE COUPLERS.

Sc concentration x	Device type	Total length (μm)	Tip width (nm)	Coupling efficiency at λ =1550 nm
0	Double-tip	28.2	500	TE: 67.94% TM: 71.13%
0	Single-tip	28.2	500	TE: 61.92% TM: 66.86%
0.09	Double-tip	30.1	500	TE: 60.57% TM: 69.61%
0.09	Single-tip	30.1	500	TE: 55.95% TM: 61.50%
0.23	Double-tip	31.7	500	TE: 57.00% TM: 64.70%
0.23	Single-tip	31.7	500	TE: 48.82% TM: 54.82%

IV. CONCLUSION

In summary, double-tip $Al_{1-x}Sc_xN$ edge couplers operating at 1550 nm wavelength are designed using FDE and FDTD simulation. Three Sc concentrations $x = 0, 0.09, 0.23$ are used in this work. PSO methods are utilized to optimize the geometric parameters of the devices. The as-designed edge couplers work for both TE and TM modes. In addition, the double-tip edge couplers exhibit higher coupling efficiencies than the conventional single-tip ones with the same tip width and total device length.

REFERENCES

[1] S. Ghosh, and G. Piazza, "Elasto-optic modulator integrated in high frequency piezoelectric MEMS resonator," Proc. IEEE 29th International Conference on Micro Electro Mechanical Systems (MEMS 2016), Jan. 2016, pp. 9-12, doi:10.1109/MEMSYS.2016.7421544.

[2] S. Zhu, Q. Zhong, T. Hu, Y. Li, Z. Xu, Y. Dong, and N. Singh, "Aluminum nitride ultralow loss waveguides and push-pull electro-optic modulators for near infrared and visible integrated photonics," Optical Fiber Communications Conference and Exhibition (OFC 2019), Mar. 2019, pp. 1-3, doi:10.1364/OFC.2019.W2A.11.

[3] X. Liu, A. Bruch, Z. Gong, J. Lu, J. B. Surya, L. Zhang, J. Wang, J. Yan, and H. X. Tang, "Ultra-high-Q UV microring resonators based on a single-crystalline AlN platform," Optica, vol. 5, Jul. 2018, pp. 1279-1282, doi: 10.1364/OPTICA.5.001279.

[4] F. A. Kittlaus, W. M. Jones, P. T. Rakich, N. T. Otterstrom, R. E. Muller, and M. Rais-Zadeh, "Electrically driven acousto-optics and broadband non-reciprocity in silicon photonics," Nat. Photonics, vol. 15, Jan. 2021, pp. 43–52, doi:10.1038/s41566-020-00711-9.

[5] D. B. Sohn, O. E. Örsel, and G. Bahl, "Electrically driven optical isolation through phonon-mediated photonic Autler–Townes splitting," Nat. Photonics, vol. 15, Oct. 2021, pp. 822-827, doi://10.1038/s41566-021-00884-x.

[6] H. Tian, J. Liu, A. Siddharth, R. N. Wang, T. Blésin, J. He, T. J. Kippenberg, and S. A. Bhave, "Magnetic-free silicon nitride integrated optical isolator," Nat. Photonics, vol. 15, Nov. 2021, pp. 828-836, doi:10.1038/s41566-021-00882-z.

[7] M. Akiyama, K. Kano, and A. Teshigahara, "Influence of growth temperature and scandium concentration on piezoelectric response of scandium aluminum nitride alloy thin films," Appl. Phys. Lett., vol. 95, Jul. 2009, p. 162107, doi: 10.1063/1.3251072.

[8] W. Wang, P. M. Mayrhofer, X. He, M. Gillinger, Z. Ye, X. Wang, A. Bittner, U. Schmid, and J. K. Luo, "High performance AlScN thin film-based surface acoustic wave devices with large electromechanical coupling coefficient," Appl. Phys. Lett., vol. 105, Sep. 2014, pp. 133502, doi:10.1063/1.4896853.

[9] A. Lozzi, E. Ting-Ta Yen, P. Muralt, and L. G. Villanueva, "$Al_{0.83}Sc_{0.17}N$ contour-mode resonators with electromechanical coupling in excess of 4.5%," IEEE Trans. Ultrason. Ferroelectr. Freq. Control, vol. 66, Jan. 2019, pp. 146-153, doi: 10.1109/TUFFC.2018.2882073.

[10] J. Zhou, Y. Liu, Q. W. Xu, Y. Xie, Y. Cai, J. Y. Liu, W. J. Liu, A. Tovstopyat, and C. L. Sun, "ScAlN/AlN film-based Lamé mode resonator with high effective electromechanical coupling coefficient," J. Microelectromech. Syst., vol. 30, Oct. 2021, pp. 677-679, doi:10.1109/JMEMS.2021.3102145.

[11] X. Zhang, S. Zheng, Q. Zhong, L. Jia, Z. Xu, Y. Dong, T. Hu, and Y. Gu, "Aluminum scandium nitride waveguide in the near-infrared," Proc. SPIE 13th International Photonics and OptoElectronics Meetings (POEM 2021), vol. 12154, Jan. 2022, pp. 78-85, doi:10.1117/12.2626712.

[12] S. Zhu, Q. Zhong, N. Li, T. Hu, Y. Dong, Z. Xu, Y. Zhou, Y. H. Fu, and N. Singh, "Integrated ScAlN photonic circuits on silicon substrate," Conference on Lasers and Electro-Optics (CLEO 2020), May. 2020, pp. 1-2, doi:10.1364/CLEO_SI.2020.STu3P.5.

[13] C. Huang, H. Shi, L. Yu, K. Wang, M. Cheng, Q. Huang, W. Jiao, and J. Sun, "Acousto-optic modulation in silicon waveguides based on piezoelectric aluminum scandium nitride film," Adv. Opt. Mater, vol. 10, Jan. 2022, p. 2102334, doi:10.1002/adom.202102334.

[14] X. Liu, C. Sun, B. Xiong, L. Niu, Z. Hao, Y. Han, and Y. Luo, "Smooth etching of epitaxially grown AlN film by $Cl_2/BCl_3/Ar$-based inductively coupled plasma," Vacuum, vol. 116, Jun. 2015, pp. 158-162, doi:10.1016/j.vacuum.2015.03.030.

[15] R. James, Y. Pilloux, and H. Hegde, "Reactive ion beam etching of piezoelectric ScAlN for bulk acoustic wave device applications," J. Phys.: Conf. Ser., vol. 1407, Nov. 2019, p. 012083, doi:10.1088/1742-6596/1407/1/012083.

Proceedings of the 7th Optoelectronics Global Conference (OGC 2022)

Optomechanical Cavity for Electrical Voltage Sensing

Qiong Yao[1], Xia Ji[1], Fuyin Wang[1], Chunyan Cao[1], and Shuidong Xiong[1]

College of Meteorology and Oceanography, National University of Defense Technology, Changsha 410073, P. R. China
Email: jbhsjmj@163.com; blovexiaji@163.com; fuyin_wang@126.com (F. W.), ccy_nudt@163.com (C. C.), and
nudtxsd@163.com (S. X.)

Abstract—**Devices for measuring physical, chemical, and biological phenomena must be able to detect electrostatic charge with high accuracy. A strong interaction between an optical cavity and a mechanical resonator is accomplished by a nanophotonic optomechanical cavity, which limits the light field at the nanoscale. It offers promise for applications in precision sensing thanks to its strong optomechanical coupling and high optical quality factor cavity. Using a zipper cavity and a suspended photonic crystal nanobeam (PCN) that functions as a moveable mechanical resonator, an integrated optomechanical electrometer for electrical voltage measurement is presented here.**

Keywords-Cavity optomechanics; nanophotonics; photonic crystal nanobeam; eletrometer; sensitivity

I. INTRODUCTION

Ultrasmall particle detection, bimolecular charge detection, and electrochemical charge detection seem to be a handful of the precise applications for electrometers [1]. Miniaturization of electrometers at on-chip platforms attract research interest due to advanced characteristics of low power consumption, compact package size and high detection resolution. Currently, two typical chip-scale electrometers are based on single-electron tunneling transistor (SET) [2] and micro and nano electromechanical system (M/NEMS) technologies [3]. A percentage of the elementary (sub-single electron) charge may be easily detected by SET, which is regarded as a very sensitive, miniature electrometer. Low-temperature operation is necessary to witness single-electron tunneling. One of the greatest technological barriers preventing the development and widespread use of these SET devices as useful charge sensors is the cryogenic temperature functioning, which necessitates large and expensive equipment. M/NEMS electrometers can convert the perturbation of input charge to the change in output periodical signal of the core resonator oscillators, e.g., frequency. Due to their inherent mechanical damping loss, M/NEMS electrometers now only have the capability to measure charges with a maximum resolution of 0.17 eHz$^{-1/2}$ [3]. Inspired by this obstacle that existed in the M/NEMS electrometers, we proposed and experimentally demonstrated an optomechanical electrometer based on a photonic crystal cavity. Benefitting from strong optomechanical interaction between the high-quality optical cavity and mechanical resonator, the effective mechanical damping loss can be modified (greatly reduced in the self-sustained oscillation state) by the cavity optomechanical dynamical backaction.

Therefore, optomechanical electrometry in this work provides a technical advance and can potentially contribute to the development of a high-resolution nanovoltmeter.

II. PCN DESIGN AND MEASUREMENT SETUP

Figure 1 shows the proposed optomechanical electrometer based on the coupled PCN cavities. It is a simple sensing structure established on the SOI chip for CMOS compatibility, as indicated in Fig. 1 (b). One of the two PCN cavities in Fig. 1 (a) is mechanically suspended from four clamping beams to function as a mechanical resonator. The other one is named as the fixed PCN to physically contacted and coupled with a fiber taper for light launching and collection, as shown in Fig. 1 (c). The sensing element is the movable PCN with its coupling strength modulated by the perturbated electrostatic effect and the cavity optomechanics induced optical force. Movement of movable PCN can modulate the variation of cavity resonance modes when it is forced by the external electric voltages, and the strong optomechanical interaction contributes to and the mechanical resonant frequency shift is used to monitor the variation of electrical voltage. Benefiting from the strong optomechanical interaction, this chip-scale electrometer can potentially detect electrical signals with a resolution reaching up to the sub-electron level. The optical cavity consists of two coupled PCN cavities, and the geometric determination and optical validation are according to the proposed method in the reference [4]. The proposed optomechanical electrometer is developed on an SOI wafer that is 220 nm thick and has buried oxide (BOX) layers that are 2 μm thick. The manufacturing process is compatible with CMOS.

III. RESULTS AND ANALYSIS

This in-plane motion is optically recorded by an electrical signal analyzer, which has the laser wavelength detuned at the shoulder of the optical resonance of TE$_{1,e}$ mode. The moveable PCN's in-plane fundamental and the other two harmonic peaks are visible in the RF power spectrum (also known as the power spectrum density (PSD) spectrum), as illustrated in Fig. 2a. The background thermal noise level of this eigenmode's theoretical calculation is shown by the green line. A Lorentzian fitting of the eigenmode peak in red demonstrates that the mechanical quality reaches Q$_m$≈1280

978-1-6654-8699-6/22 $31.00 © 2022 IEEE   235

under a vacuum pressure of 5×10^{-3} mBar. The right axis of Fig. 2a describes the converted displacement PSD deducted from the power PSD, which indicates a displacement noise floor of ≈ 0.8 fm/$\sqrt{\text{Hz}}$. By sweeping mechanical spectra through a laser-cavity detuning, it can be observed from Fig. 2b that a spring stiffening occurs in the blue-detuned regime and a spring softening occurs in the red-detuned regime, respectively. Under pump powers of 30 μW and 120 μW, Figure 2c extracts the mechanical frequency shift across the laser-cavity detuning range. The mechanical resonance peaks under 30 W are altered across the laser detuning range of 65 pm, and the highest mechanical frequency shift of (Ω'_m-Ω_m)/$2\pi \approx 3.5$ kHz is accomplished at the detuning of 0.8 GHz. The optomechanical coupling strength may be calculated as $g_{OM} \approx 13.2$ GHz/nm by fitting the mechanical frequency shift-detuning. The mechanical frequency shift is improved up to (Ω'_m-Ω_m)/$2\pi \approx 26$ kHz when the pump power is increased to 120 W. Due to the higher slope of the RF frequency shift caused by increasing intracavity energy compared to laser-cavity detuning, the detection sensitivity can be increased.

Given a fixed power and wavelength of the tunable laser, the optomechanical electrometer can work by the readout of RF shift to detect the applied electrical voltages. When the bias voltage exists, an electrostatic force exerts on the nanoscale capacitance to modulate the mechanical stiffness.

The bias electrode is subjected to a rising voltage in the experiments, from zero to 120 mV with a step voltage of 5 mV, which causes the RF shift to change as a function of applied bias voltages, as seen in Figs. 2d and 2e. A zoom-in image in Fig. 2e is recorded from the RF peak areas in Fig. 2d to show a modest RF shift of roughly 100 Hz at the highest voltage of 120 mV due to the mechanical resonance's high mechanical linewidth (greater than 1 kHz). To determine the mechanical resonance's peak point, a Lorentzian fitting is needed. An equivalently linearized electrical sensitivity of $\beta = 0.007$ Hz·mV^{-2} (or a scale factor of $S_1 = \beta^{-1} = 142.8$ mV^2·Hz^{-1}) is obtained in Fig. 2f. Additionally, the RF frequency is monitored in order to determine the Allan deviation and assess the optomechanical electrometer's sensing capabilities. As shown in Fig. 3, when the optomechanical electrometer works below the threshold power, an Allan deviation of $\sigma_y = 2.1\times10^{-6}$ at the zero slope (bias stability) is measured for mechanical resonant frequency $\Omega_m/2\pi = 1.286$ MHz at 2 s integration. As a result, the minimum detectable electric voltage for this optomechanical electrometer is $R = (S_1 \times \Omega_m \times \sigma_y)\sqrt{t_{inter}} = 545.38$ mV^{2}Hz$^{-1/2}$. Converting to the form of a square of electric voltage by $S \times \sigma_y(\tau) \times \Omega_m/2\pi$, the electrical bias stability is 385.57 mV^2 with a white noise level obtained at the -1/2 slope of 478.62 mV^2/Hz$^{1/2}$.

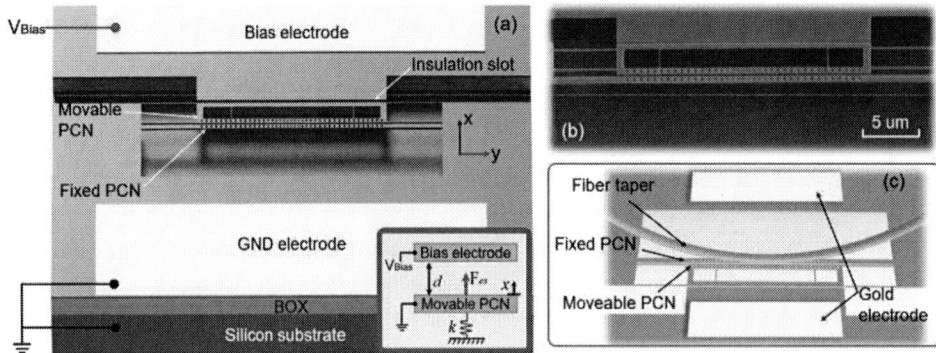

Figure 1. Configuration of optomechanical electrometer. (a) Schematic of optomechanical electrometer with a schematic of a parallel-plate capacitor shown in the inset; (b) SEM image of chip-scale optomechanical electrometer; (c) Experimental system of fiber taper coupled with the PCN cavities.

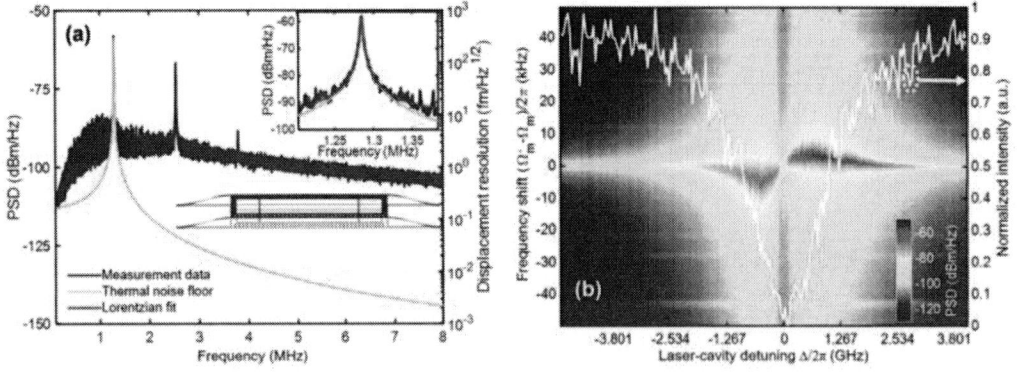

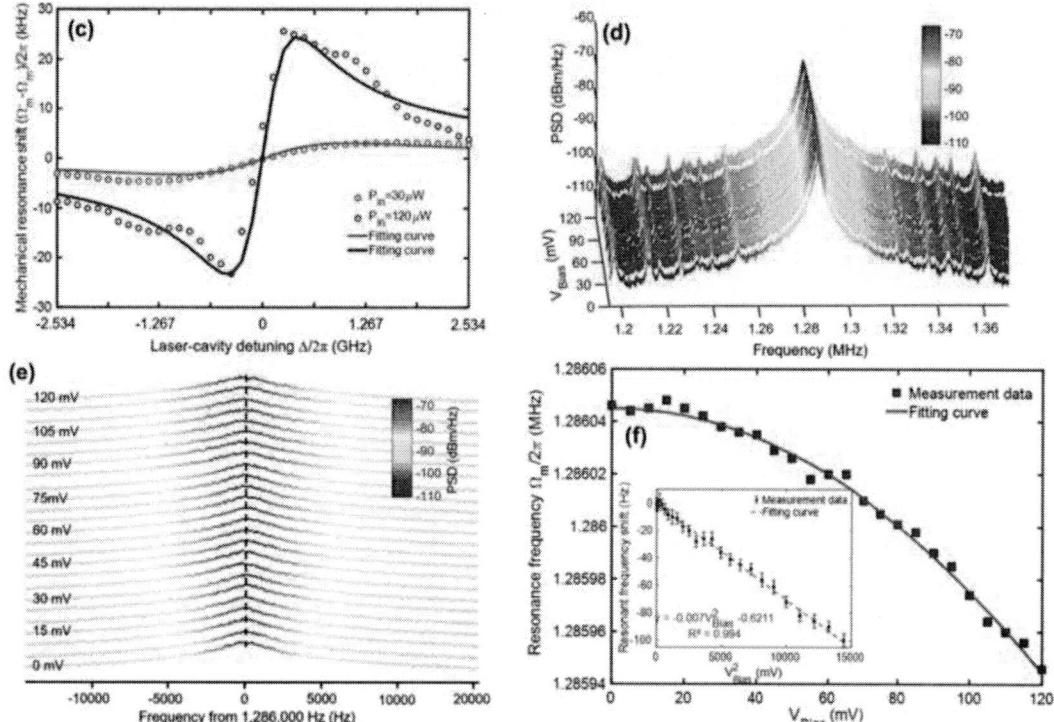

Figure 2. Electrical voltage actuated motion transduction of the optomechanical electrometer. (a) Measured RF spectrum of moveable PCN. Insets are FEM mode of the movable PCN and the zoom-view on the spectrum of fundamental vibrating mode. (b) Intensity image of measured RF spectrum as a function laser-cavity detuning to verify the optical spring effect of the optomechanical electrometer. (c) RF shift versus laser-cavity detuning for different pump powers (30 μW for red data and 120 μW for blue data). (d) RF spectra vary for measurement of applied voltages from 0 mV to 120 mV. (e) Zoom-in view on the RF peaks of spectra under applied voltages. (f) RF shift versus applied voltages.

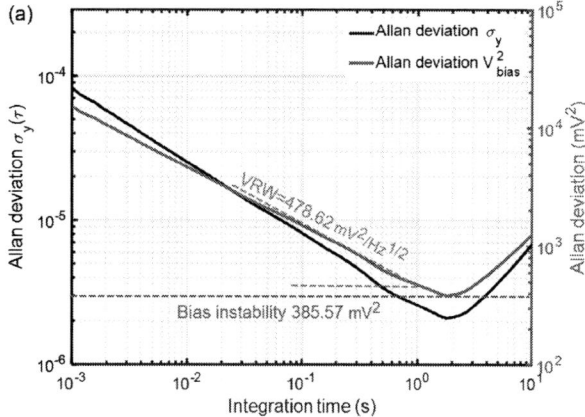

Figure 3. Allan deviation σ_y as a function of integration time τ for the optomechanical electrometer operating below threshold power.

IV. CONCLUSION

For the first time, an optomechanical electrometer is suggested in this paper to detect applied electrical voltages. The optical readout of mechanical resonant frequency shift permits a practical measurement of the change of applied electrical voltages thanks to the strong optomechanical interaction between the optical cavity and mechanical resonator. This optomechanical electrometer provided a technical advance to develop the next generation of chip-scale high precise electrical measurement instruments.

ACKNOWLEDGMENT

This research was funded by National Natural Science Foundation of China (NSFC) with grant number 61905283.

REFERENCES

[1] M. Yuan, Z. Yang, D. Savage, M. Lagally, M. Eriksson, and A. Rimberg. Charge sensing in a Si/SiGe quantum dot with a radio

frequency superconducting single-electron transistor. Appl. Phys. Lett., 101, 142103 (2012).

[2] Chen, H. Zhang, J. Sun, et al. Ultrasensitive resonant electrometry utilizing micromechanical oscillators. Phys. Rev. Appl., 14, 014001 (2020).

[3] M. Eichenfield, R. Camacho, J. Chan, K. J. Vahala, and O. Painter. A picogram-and nanometre-scale photonic-crystal optomechanical cavity. Nature, 459, 550–555 (2009).

[4] M. Eichenfield, R. Camacho, J. Chan, K. J. Vahala, and O. Painter. A picogram-and nanometre-scale photonic-crystal optomechanical cavity. Nature, 459, 550–555 (2009).

Proceedings of the 7th Optoelectronics Global Conference (OGC 2022)

A Solver for Devices of Subwavelength Lamellar Gratings

Zhuang Wang, Chuan Shen, Liu Wang, Bin Wang, Daofeng He, Sui Wei[*]

Key Laboratory of Intelligent Computing & Signal Processing, Ministry of Education

Anhui University

Hefei, China

e-mail: swei@ahu.edu.cn

Abstract—A solver for subwavelength lamellar gratings is presented synchronized with the development of a light modulator for holographic video display Grating Liquid Crystal on Silicon (GLCoS) and is regarded as an important part of the whole R&D work. In this way it not only gives computational support in the whole design process including physical concept descriptions, verifications, predictions, fabrications and other experimental activities but also provides a support platform for further product development. Based on the generic Fourier modal method, we focus on the electrodynamics specific to conductors (Au, Al) and charge carriers in semiconductors (Indium Tin Oxide, ITO) in SPPs (Surface Plasmon Polaritons) and permittivity characteristics of the materials to make the solver oriented towards subwavelength (metal, semiconductor) lamellar gratings. Further, we analyze and calculate the optical characteristics of subwavelength lamellar gratings composed of ITO and Au. The results of these calculations not only agree with the results of actual GLCoS device tests to the same order of accuracy, but also demonstrate the validity and accuracy of the solver.

Keywords-subwavelength; lamellar gratings; liquid crystal on silicon structure; Fourier modal method; Drude model; reflection efficiency

I. INTRODUCTION

Recognized as the ultimate tool for 3D visualization, holographic video display requires spatial light modulators (SLMs) that are more suitable, for human visual perception, than flat panel display, virtual reality technology and 3D film technology can provide, which is a very challenging technology. Such SLMs should have a very high space bandwidth product and the capability of dynamical modulation. It should have a pixel pitch of at least less than 1μm and the feature size of the device should necessarily be in subwavelength order.

Current advanced surface plasmon polaritons (SPP) [1,2] and metasurfaces [3,4] can provide subwavelength modulation techniques, but, so far, most of them especially all-dielectric metasurfaces, lack dynamic control ability, e.g. [5] noted.

We have produced a light modulator (Grating Liquid Crystal on Silicon, GLCoS) for holographic video display, the structure of which is illustrated in Fig. 1. In the GLCoS, the upper electrode consists of an indium tin oxide (ITO) and a subwavelength gold grating, and the lower electrode, an aluminum plate, acts both as a light-reflecting backplane and

as a waveguide with the upper electrode slab and a thin liquid crystal box to induce a zero-order phase modulation of the reflected light. While manipulating the optical wavefront, the GLCoS also acts as an electronically controlled device, applying a voltage to change the refractive index of the liquid crystal, thus achieving active $0-2\pi$ phase modulation. We have carried out the entire conceptual design process including model analysis, numerical simulations and the fabrication and testing of the principle experimental device [6-10]. The results are qualitatively consistent with the conceptual design.

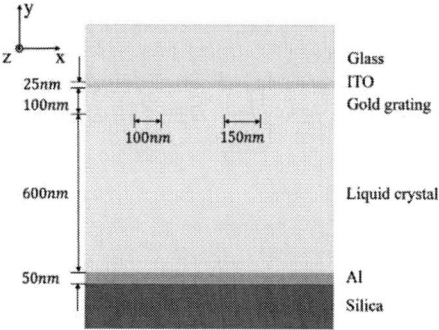

Figure 1.   GLCoS structure.

The choice of such an SPP structure is the result of systematic consideration of our system design metrics, as well as calculations and technical feasibility. The GLCoS hierarchy is similar to the current Liquid Crystal on Silicon (LCoS) which is compatible with the Complementary Metal Oxide Semiconductor (CMOS) process. Although the physical principles are radically different, the processes and equipment for LCoS in assembling liquid crystal cells and fabricating the driver circuits are available to employ More importantly, we must place an important consideration on the design tool software at the outset of development, which has to give guidance on the design process physical concept description, calculation, verification, prediction, fabrication and other experimental activities to provide support for further product development. Therefore, the most likely technical solutions for the theoretical calculations should be considered, taking into account achievable specifications such as high spatial bandwidth, modulated phase or intensity of depth frame rates for holographic video displays. Otherwise, we risk going down the same road as in the past: articles are published, prototypes have qualitative results but

978-1-6654-8699-6/22 $31.00 © 2022 IEEE

engineering cannot proceed smoothly due to changes in the actual parameters, or even the previous work is abandoned.

We know that even subwavelength devices involving quantum effects or rather including nonlocal effects can still be described in many cases by the phenomenological Maxwell's equations plus appropriate constitutive equation model [11]. Several general computational methods for Maxwell's equations exist, such as the finite difference in the time domain (FDTD) method, the finite difference in the frequency domain (FDFD) method, the finite element method (FEM) and finite integration techniques (FIT). Such calculations are accurate when the infinitely small grid limit is assumed. However, the calculations are time-consuming when the device feature scale is much smaller than the entire calculation domain. Furthermore, when the device structure is geometrically and materially complex, such simulations are cumbersome or even computationally infeasible and lack a clear physical interpretation. Of course, the exact theoretical calculation of the universal Maxwell equations is currently a challenge, as E. Popov says in the editorial preface to his book [12]: "A typical question that almost all of us (the authors' team and other colleagues) has been asked not only once has in general the meaning (although usually being shorter): "What is the best method for modeling of light diffraction by periodic structures?".....The best method has not yet been invented. Maybe it never will be."

The exception is for lamellar gratings, which can be modeled by modal methods. A near-100% prediction of the optical transmission of 1D subwavelength grating arrays has been predicted by Porto et al. [13] using a modal method. The Fourier modal method is a widely used model method [14]. Our solver presented here for devices is oriented to subwavelength laminated gratings based on the classical electrodynamics and semi classical models of charge carriers in conductors (gold, aluminum) and semiconductors (ITO) and permittivity of the materials although learning from classical Fourier modal method. To illustrate the effectiveness, accuracy and flexibility of this software tool, the calculation results of optical performance for different ITO films are compared with the measurement results for our actual fabricated GLCoS.

II. METHODS AND PRINCIPLES

Lamellar gratings, with a specific geometry invariant in the direction perpendicular to the periodicity, can be expanded the electromagnetic field by the eigenmode method, which has been used extensively for diffraction analysis and design of structures. There are two main branches of the mode method, the exact mode method [14] and the Fourier Mode Method (FMM) [15]. The Fourier modal method has the advantage of involving less mathematical theory and relatively few programming steps.

The eigenmode method uses interfaces parallel to the grating layer to divide the space of devices, and expands the electric or magnetic fields of each layer with orthonormal basis functions of unknown coefficients, then matching boundary conditions, further solving for the unknown coefficients using the relevant matrix inverse, and finally calculating reflection, transmission and efficiency. From an electromagnetic point of view, this is equivalent to solving a boundary value problem. Its programming sticking point is the calculation of the field in the tooth groove. A brief description is given below.

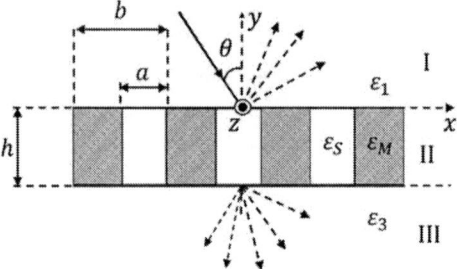

Figure 2. Lamellar grating under non-conical diffraction..

The definition and notation of the lamellar gratings used in the calculations are shown in Fig. 2. A right-angle coordinate system of x and y axes are perpendicular and parallel to the grating grooves respectively. The origin of the coordinate system is at the midpoint of the upper surface of one of the grating grooves. The grating period, groove depth and groove width are h, b and a. The Fourier Modal Method, also known as Rigorous Coupled Wave Analysis (RCWA), assumes that the grating structure has translational invariance in the z-direction parallel to the grooves, and lets the electromagnetic field be expressed in the form of

$$E(x,y,z,t) = E(x,y)exp(i\gamma z - i\omega t). \tag{1}$$

$$H(x,y,z,t) = H(x,y)\exp(i\gamma z - i\omega t). \tag{2}$$

where γ is the propagation constant and ω is the angular frequency of the incident light.

Since surface plasmon polarization excitations can only be generated at Transverse Magnetic (TM) polarized incidence, consider TM polarized visible light incidence, the nonzero field components are H_z, E_x and E_y. Thus Maxwell equations in the Gaussian unit system for the two included curl operator can be written as

$$\partial_y E_x - \partial_x E_y = ik_0 H_z, \tag{3}$$

$$\partial_x H_z = -ik_0 \varepsilon E_y, \tag{4}$$

$$\partial_y H_z = ik_0 \varepsilon E_x. \tag{5}$$

Eliminate E_x and E_y yields,

$$k_0 \varepsilon H_z(x,y) + \varepsilon \partial_x H_z(x,y) = -\partial_y^2 H_z(x,y), \tag{6}$$

where k_0 is the magnitude of the vacuum wave vector, and ε is permittivity, noting in particular that it is in region II with respect tox the periodic function, and H_z is the z component of the magnetic field, which is a function of x and y.

Since the coefficients of (6) do not depend on y, applying the separation of variables yields that H_z has a solution of the following form:

$$H_z(x,y) = H_z(x)\exp(i\beta y), \qquad (7)$$

where β is the propagation constant. The variable y can be eliminated by substituting (7) into (6):

$$k_0\varepsilon H_z(x) + \varepsilon\, d\left\{\left[dH_z(x)/dx\right]/\varepsilon\right\}/dx = \beta^2 H_z(x). \quad (8)$$

In both regions I and III, the permittivity is a constant, and we can easily obtain $H_z(x,y)$ by (6) and (8). Note that the permittivity is a piecewise constant function of spatial variables, and in the grating region (Region II) it is a periodic function of x only. (8) is difficult to solve. Each β^2 of the eigenvalue spectra of (5) has a corresponding β and $-\beta$, which correspond respectively to $\exp(i\beta y)$ and $\exp(-i\beta y)$ which are the two fundamental modes propagating upwards and downwards along the y axis. Assuming that the number of β^2 satisfying (5) is q, the total field of the magnetic field can be expressed as

$$H(x,y) = \sum_{q=1}^{\infty}\left[u_q(y) + d_q(y)\right]H_{z,q}(x), \qquad (9)$$

where $u_q(y) = u_q\exp(i\beta y)$, $d_q(y) = d_q\exp(-i\beta y)$, u_q and d_q are arbitrary constants.

Writing ε and $H_z(x)$ as a Fourier series, the inverse rule [16] gives

$$\left(\mathbf{T}(1/\varepsilon)\right)^{-1}\left[k_0^2\mathbf{I} - \boldsymbol{\alpha}\left(\mathbf{T}(\varepsilon)\right)^{-1}\boldsymbol{\alpha}\right]\mathbf{H} = \beta^2\mathbf{H}. \quad (10)$$

where $\mathbf{T}(\varepsilon)$ denotes the Toeplitz matrix consisting of the Fourier coefficients of ε, and $\mathbf{I}$ is the unit matrix, and $\boldsymbol{\alpha}$ is the diagonal matrix consisting of $\alpha_n = k_0\sin\theta + 2n\pi/d$, $\mathbf{H}$ is a column vector consisting of the Fourier coefficient of $H_z(x)$. It can be found that (10) is a defining equation for the eigenvalues and eigenvectors, from which the eigenvalues β^2 and the eigenvector $\mathbf{H}$.

Through the continuity of the magnetic field in the y-direction and the pseudo-periodic nature of the x-direction,

$u_q(y)$ and $d_q(y)$ were calculated by the S-matrix algorithm [17]. At this point, the magnetic field of region II is completely derived, and its electric field can then be calculated according to (4) and (5), again with the Fourier series as for the magnetic field. In this case, we can calculate the reflected and transmitted fields as well as the total field even if region II contains multiple layers or periodic grating structures of arbitrary shape, as long as region II can be divided into enough layers when using the S-matrix algorithm.

In the calculation of metallic (semiconductor) gratings, permittivity of the material is wavelength dependent and the calculation must take into association it associated the classical electrodynamics and semi classical models [18, 19]. According to classical electrodynamic theory, the intrinsic structure of a metal is related to

$$\mathbf{D} = \varepsilon_0\mathbf{E} + \mathbf{P}, \qquad (11)$$

where $\mathbf{D}$ is the Electric displacement vector, $\mathbf{E}$ is the electric field, and $\mathbf{P}$ describes the macroscopic polarization generated by the $\mathbf{E}$ field driving all displaced electrons within the metal. Under the assumptions of linearity, localization and time dependence depending only on the time difference, $\mathbf{P}$ and $\mathbf{E}$ are linearly related by the electric susceptibility χ:

$$\mathbf{P} = \varepsilon_0\chi\mathbf{E}. \qquad (12)$$

Incorporating the microscopic response into relative permittivity $\varepsilon(\omega)$:

$$\mathbf{D} = \varepsilon_0\mathbf{E} + \mathbf{P} = \varepsilon_0(1+\chi)\mathbf{E} = \varepsilon_0\varepsilon(\omega)\mathbf{E}. \qquad (13)$$

The general characteristics of metals can be characterized by $\varepsilon(\omega)$.

But the optical properties of gold in the visible domain are mainly determined by free electrons and interband jumping. One analytic form of permittivity of gold is the so-called modified Drude's free electron model:

$$\varepsilon(\omega) = \varepsilon_\infty - \left[\omega_p^2/\left(\omega^2 + i\gamma\omega\right)\right], \qquad (14)$$

This is usually $1 \leq \varepsilon_\infty \leq 10$. (14) is the Drude model When $\varepsilon_\infty = 1$, which considers only free electron effects. In the Drude model, $\omega_p = \left(ne^2\right)/\left(\varepsilon_0 m\right)$ is the frequency of plasma collective oscillations (for silver, copper, gold and 1-electron alkali metals, $\hbar\omega_p = 9\text{eV}$ is independent of operating frequency), m is the effective mass of each electron, the collision frequency $\gamma = 1/\tau$, and τ is called the relaxation time of the free electron gas (controlled mainly by electron-phonon scattering at room temperature). This equation can reasonably be applied in the red and longer wavelength parts, but not in the yellow to blue parts, where the D-band transition is important.

Also, when the permittivity of the material at the grating interface is nonconstant, then (7) - (9) need to be modified. In the next section we will discuss the analysis and calculation of ITO and gold laminated grating structures.

III. EXAMPLE OF COMPUTATIONAL AND ANALYSIS-OPTICAL CHARACTERISTICS OF ITO AND GOLD LAMINATED GRATING STRUCTURES

The structure of the analyzed device as shown in Fig. 1 is divided from top to bottom into four layers: ITO layer, gold grating layer, E7 liquid crystal layer and Al layer. The dispersion data for the E7 liquid crystals used are from [20]. The topmost layer of glass and the bottom layer of silica have thicknesses on the order of millimeters, much greater than the thicknesses of the other layers, and their role is to act as an attachment for the gold grating and a carrier for the Al film, so the glass and silica are set as the incident medium for the transmission medium respectively. The electrodynamics model and dielectric function of the gold grating have been described in the previous section. In contrast to precious metals, the average energy of visible light photons is not sufficient to induce intrinsic excitation of carriers within the ITO film and therefore the ITO film has a high transmittance in the visible range. However, it is also a dispersive material, and the optical properties of the films in the visible to near-infrared spectrum follow in principle the Drude model (14) based on classical dynamics [21,22]. This suggests that ITO films exhibit selectivity towards electromagnetic waves. Therefore, it is necessary to analyze the effect of the ITO layer on the diffraction efficiency of GLCoS at different incident light wavelengths. Also since ITO is a nonchemometric compound, its optical properties depend heavily on the growth/deposition process and annealing conditions, it is necessary to give special attention to the solver.

Next, we will calculate the reflection and transmission efficiencies of each order of GLCoS using the Fourier modal method for four different ITOs, where the No.0 ITO is from our laboratory, the data for the No.1 ITO is from [23], and the No.2 and No.3 ITOs are from [24].

A. Convergence Test

Before using the Fourier modal method for the calculations, we have to test the convergence of the number of orders of the Fourier series. Because it is impossible to write the Fourier series in the form of an infinite series in a computer, instead we can only approximate the Fourier series by the sum of finite terms. Therefore, the question of how many orders of Fourier series can be retained to keep the error within acceptable limits must be considered first.

For different numbers of orders, we calculated the zero-order reflection efficiency for incident light at 532 nm incident vertically on the GLCoS, and the results are shown in Fig. 3, which shows that the Fourier series need to be expanded by at least 70 orders to achieve good convergence.

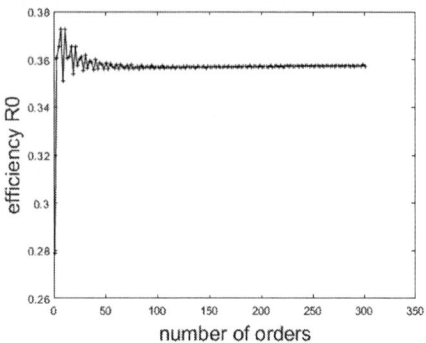

Figure 3. Convergence of zero-order reflection efficiency with number of orders at wavelength 532 nm for No.0 ITO.

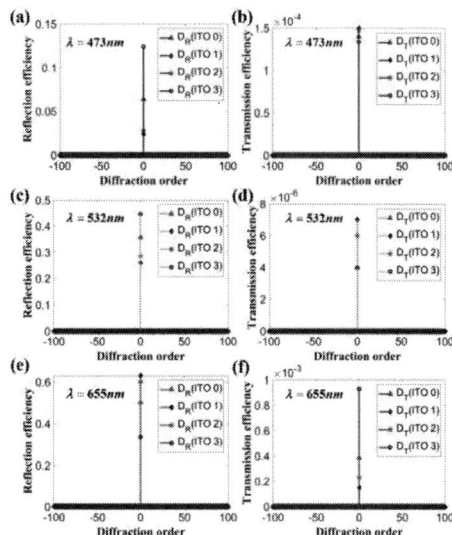

Figure 4. The diffraction efficiencies of each diffraction order corresponding to the four ITOs at three wavelengths.

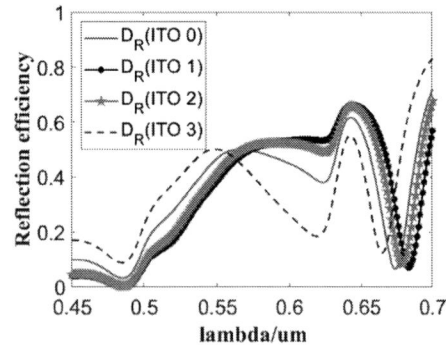

Figure 5. Zero-order reflection efficiency as a function of wavelengths for four ITOs.

Thus, we calculated the diffraction efficiency of each diffraction order at 201 Fourier orders when visible light with wavelengths of 473 nm, 532 nm and 655 nm incident ITOs vertically, respectively. The results are shown in Fig. 4, where the three subplots on the left and right represent the reflection and transmission efficiencies, respectively, and the three subplots from the top to the bottom represent the diffraction efficiencies at the incident wavelengths of 473 nm, 532 nm, and 655 nm, respectively. It can be seen from Fig. 4 that different ITOs affect both reflection and transmission efficiencies, but since our device is in the subwavelength magnitude, only zero-order reflection or transmission plays a role (values less than 10 have been neglected), which is consistent with the theoretical analysis, and the reflection efficiency is much greater than the transmission efficiency, which is a great advantage for designing reflective SLMs.

We also show on Fig. 4 the diffraction efficiency for each diffraction order corresponding to the four ITOs. The reflection efficiency at 473 nm is relatively low compared to the other two wavelengths, while at 532nm and 655 nm, the reflection efficiency of different ITOs at these two wavelengths is variable. Therefore, it is necessary to find suitable wavelengths for different ITOs. In addition, considering the convergence and computational time compatibility, we use 101 orders to expand the Fourier series for subsequent calculations.

B. Calculating the Diffraction Efficiency of Each Order

After the above analysis, we can focus only on the zero-order reflection efficiency. In order to analyze the reflection efficiency of GLCoS at each wavelength, we set the incident light wavelength from 450nm to 700nm respectively, in steps of 1nm, with vertical incidence and other settings kept constant. The results can be obtained as shown in Fig. 5.

The results corresponding to the ITOs No.0 to No.3 are indicated by the red solid line, the blue dashed line, the black star line and the magenta pentagram line respectively. We can see that although each of the four ITOs differs, the zero-order reflectance curve has a very small value in the blue band tested and is relatively low around 473 nm, while the red part has at least a very small value and 655 nm is close to a very small value, all four ITOs have a relatively stable and not low reflectance efficiency in the wider green band. In the ITO material we use, the red light near 655 nm has a narrow high-reflection efficiency zone, so when the frequency of the laser used is not very stable, the phenomenon is easy to disappear. 473 nm is a low reflection efficiency zone, so it is not easy to observe the phenomenon. The reflection efficiency at 532 nm is not very high, but it still has a stable reflection efficiency within a certain range, so it is easy to observe the phenomenon. In contrast, when making new devices, we should choose an ITO material as close as possible to the No.1 ITO (black curve) to be more conducive to the realization of our function.

As can be seen from the above experimental procedure, the solver we have built for subwavelength laminated grating devices is a good guide to the actual experimental results, reflecting to a large extent the actual test results and providing a more flexible test of the influencing factors. In summary, the solver is a good guide to the design of subwavelength lamellar grating devices.

IV. CONCLUSION

Learning from the generic Fourier modal, we develop a solver for subwavelength lamellar gratings. Using the structure of the GLCoS developed by us, we have tested the solver. we analyze the effect of different ITO materials on the optical properties of the GLCoS, and the results obtained are in good agreement with the theoretical analysis and tests of the actual GLCoS. The significance and usefulness of the solver are proven with these examples. In writing the solver, special attention has been paid to the properties of the local and nonlocal natures of permittivities, so the method can be easily expanded to methods involving nonlocal issues (e.g. hydrodynamic models) for nano-devices and abnormal SPP. The ability of SPP to handle both electronic and photonic signals has important applications in all-optical computing, data storage, signal switching, modulation and selection, and the solver devised here can also provide lessons for the analysis and computation of other devices with similar structures.

REFERENCES

[1] M. Ozaki, J. Kato, S. Kawata, "Surface-Plasmon Holography with White-Light Illumination," Science, vol. 332, Apr. 2011, pp. 218-220, doi: 10.1126/science.1201045.

[2] S. Kawata and M. Ozaki, "Surface-plasmon holography," Iscience, vol. 23, Dec. 2020, p. 101879, doi: 10.1016/j.isci.2020.101879.

[3] L. Huang, S. Zhang and T. Zentgraf, "Metasurface holography: from fundamentals to applications," Nanophotonics, vol. 7, Jun. 2018, pp. 1169-1190, doi: 10.1515/nanoph-2017-0118.

[4] Z. Deng, Z. Wang, F. Li, M. Hu, X. Li, "Multi-freedom metasurface empowered vectorial holography," Nanophotonics, vol. 11, Jun. 2022, pp. 1725-1739, doi: 10.1515/nanoph-2021-0662.

[5] W. Yang, G. Qu, F. Lai, Y. Liu, Z. Ji, Y. Xu, Q. Song, J. Han, and S. Xiao, "Dynamic bifunctional metasurfaces for holography and color display," Advanced Materials, vol. 33, Jul. 2021, 33, p. 2101258, doi: 10.1002/adma.202101258

[6] W. Zhu; C. Shen, M. Zhang, S. Wei, X. Wang, and Y. Wang, "Architectural design of deep metallic sub-wavelength grating for practical holography display," Proc. SPIE. AOPC 2017: Optoelectronics and Micro/Nano-Optics, Oct. 2017, SPIE-INT SOC OPTICAL ENGINEERING, vol. 10460, pp. 337-343. doi:10.1117/12.2285193

[7] M. Zhang, C. S., W, Zhu and W. Sui, "Fabry-perot resonance on aluminum sub-wavelength gratings for holographic display device," Chinese Journal of Electron Devices, vol. 41, Aug. 2018, pp. 833-838, doi:10.3969/j.issn.1005-9490.2018.04.004.

[8] H. Hu., G. Wang, S. Wei, C. Shen, R. Rong, and M. Bu, "liquid crystal phase modulator based on deep sub-grating structure," Proc. SPIE. Three-Dimensional Image Acquisition and Display Technology and Applications, SPIE-INT SOC OPTICAL ENGINEERING, Dec. 2018, vol. 10845, pp. 84-89. doi:10.1117/12.2505111

[9] H. Yu, G. Wang, S. Wei, C. Shen, and J. Chen, "Research on active SLM device structure of deep sub-wavelength FP resonance," Proc. SPIE. AOPC 2019: Display Technology and Optical Storage, SPIE-INT SOC OPTICAL ENGINEERING. Dec. 2019, vol. 11335, pp. 98-103. doi:10.1117/12.2547993.

[10] B. Tao, G. Wang, S. Wei, C. Shen, and Z. Xu, "Colors display with spatial light modulator based on sub-wavelength metal grating," Proc. SPIE. AOPC 2020: Optoelectronics and Nanophotonics; and Quantum Information Technology, SPIE-INT SOC OPTICAL ENGINEERING, Nov. 2020, vol. 11564, pp. 108-113. doi:10.1117/12.2580475.

[11] M. Gonçalves, H. Minassian and A. Melikyan, "Plasmonic resonators: fundamental properties and applications," Journal of Physics D:

Applied Physics, vol. 53, Oct. 2020, p.443002, doi:10.1088/1361-6463/ab96e9.

[12] E. Popov, Gratings: Theory and Numeric Applications. 2nd ed., vol. 1. Provence: Universitaires, 2014, pp.1-2.

[13] J. Porto, F. García-Vidal, and J. Pendry. "Transmission resonances on metallic gratings with very narrow slits." Physical review letters, vol 83, Oct. 1999, p. 2845, doi:10.1103/PhysRevLett.83.2845.

[14] I. Botten, M. Craig, R. McPhedran, J. Adams and J. Andrewartha, "The Dielectric Lamellar Diffraction Grating," Optica Acta: International Journal of Optics, vol 28, Nov. 2010, pp. 413-428, doi:10.1080/713820571.

[15] L. Li, "Fourier modal method," in Gratings: Theory and Numeric Applications, E. Popov, Ed. Provence: Universitaires, 2014, pp. 537-578.

[16] L. Li, "Use of Fourier series in the analysis of discontinuous periodic structures," Journal of the Optical Society of America A, vol. 13, Sep. 1996, pp. 1870-1876, doi: 10.1364/JOSAA.13.001870.

[17] L. Li, "Formulation and comparison of two recursive matrix algorithms for modeling layered diffraction gratings," Journal of the Optical Society of America A, vol. 13, May. 1996, pp. 1024-1035, doi:10.1364/JOSAA.13.001024.

[18] S. Maier, Plasmonics: Fundamentals and Applications, 1st ed., vol. 1. New York: springer,2007, pp. 5-19.

[19] S. Raza, S. Bozhevolnyi, M. Wubs and N. Mortensen, "Nonlocal optical response in metallic nanostructures," Journal of Physics: Condensed Matter, vol 27, Apr. 2015, p. 183204, doi:10.1088/0953-8984/27/18/183204.

[20] V. Tkachenko, G. Abbate, A. Marino, F. Vita, M. Giocondo and A. Mazzulla et al., "Nematic liquid crystal optical dispersion in the visible-near infrared range," Molecular Crystals and Liquid Crystals, vol. 454, Sep. 2006, pp. 263-665, doi:10.1080/15421400600655816.

[21] J. Ederth, P. Heszler, A. Hultåker, G. Niklasson and C. Granqvist, "Indium tin oxide films made from nanoparticles: models for the optical and electrical properties," Thin solid films, vol. 445, Dec. 2003, pp. 199-206, doi: 10.1016/S0040-6090(03)01164-7.

[22] Y. Lu, X. Feng, Q. Wang, X. Zhang, M. Fang, and W. Sha et al., "Integrated Terahertz Generator-Manipulators Using Epsilon-near-Zero-Hybrid Nonlinear Metasurfaces," Nano Letters, vol. 21, Sep. 2021, pp. 7699-7707, doi: 10.1021/acs.nanolett.1c02372.

[23] R. Moerland and J. Hoogenboom, "Subnanometer-accuracy optical distance ruler based on fluorescence quenching by transparent conductors," Optica, vol. 3, Feb. 2016, pp. 112-117, doi: 10.1364/OPTICA.3.000112.

[24] T. König, P. Ledin, J. Kerszulis, M. Mahmoud, M. El-Sayed, J. Reynolds and V. Tsukruk, "Electrically tunable plasmonic behavior of nanocube–polymer nanomaterials induced by a redox-active electrochromic polymer," ACS nano, vol 8, May. 2014, pp. 6182-6192, doi:10.1021/nn501601e.

Proceedings of the 7th Optoelectronics Global Conference (OGC 2022)

Optically Levitated Conveyor Belt Based on Specular-Reflection Photonic Nanojet

Feng Xu
College of Engineering and Applied Sciences, Nanjing
University
Nanjing, China
e-mail: xufeng1025@126.com

Song Zhou
Faculty of Mechanical and Material
Engineering, Huaiyin Institute of Technology
Huai'an, China
e-mail: zs41080218@126.com

Jiahui Zhang
College of Engineering and Applied Sciences, Nanjing
University
Nanjing, China
e-mail: 854790335@qq.com

Guanghui Wang
College of Engineering and Applied Sciences, Nanjing
University
Nanjing, China
e-mail: wangguanghui@nju.edu.com

Fei Xu
College of Engineering and Applied Sciences, Nanjing
University
Nanjing, China
e-mail: feixu@nju.edu.cn

Abstract—**We have proposed an optically levitated conveyor belt based on specular-reflection photonic nanojet (s-PNJ). The trapped particles can suspend in the water and transport with the movement of hot spots caused by the deflection of exciting light. In addition, the rate of successful particle transporting can be up to 96.3% after the distance of 50 μm when the power intensity is as low as 0.2 mW/μm².**

Keywords-optical tweezers; microsphere; particle transport; photonic nanojet

I. Introduction

Optical tweezers (OTs) have a wide range of applications in micro-robotics, biomedical field and nano-fabrication [1-3] because of the noninvasive and versatile performance for micro- or nano-scale objects. Gradient force is the dominant factor for optical trapping, so traditional OTs are always dependent on the focusing spots of Gaussian beam. However, diffraction can not be ignored in this situation. In the recent decade, OTs based on waveguide and plasmatic nanostructure have been proposed and deeply researched because of its low damage on manipulating samples. In addition, the location can also be determined more accurately. Simultaneously, other functions such as particle transporting and sorting have also been investigated in the past years. However, most of the particle manipulating rely on the evanescent or coupled wave on or near the surface of nanostructures which cannot guarantee the particles suspended in fluid to be trapped. We have ever successfully tried the approach of far-field trapping and transporting while a cover plate is necessary because of the impact of

scattering force. In addition, nanoscale particle cannot be trapped stably [4].

Photonic nanojet (PNJ) used for particle trapping have attracted a lot of interests of researchers due to the simple structure and unique focusing effect [5, 6]. In order to avoid the influence of phase and form semi-symmetrical and counter-propagating optical fields, a metal or dielectric flat mirror can be used to reflect the PNJ [7]. Meanwhile, the standing wave can reduce the disadvantages of single-direction scattering force and realize trapping particles and keeping them suspending in the fluid in the true sense.

Here, we have utilized microsphere to generate PNJ and then reflected by a gold plate so that the gradient field can be formed in the gap. With the deflection of exciting light, hot spots will move directionally and particles will follow the hot spots just like a conveyor belt. Furthermore, we have investigated the rate of successful particle transporting with the consideration of Brownian motion.

II. Structure and Design

The schematic diagram of the specular-reflection model based on microspheres and gold plates is shown in Fig.1(a). Several 5-μm diameters microparticles are arranged tightly over a plate and the fluid channel is below the microspheres. The refractive index of microspheres is set as 1.6. At the bottom of fluid channel, a gold plate is used to realize specular-reflection PNJ. The distance between microspheres and gold plate is 500 nm. Parallel exciting light subject to the microspheres will be focused in the fluid channel and nanoparticles will be trapped in the hot spots. Furthermore, with the deflection of exciting directions, nanoparticles will follow the hot spots directionally. Fig. 1(b) shows the

978-1-6654-8699-6/22 $31.00 © 2022 IEEE

electrical field distributions in the x-y plane when the rotation angles of exciting light is 0°, 20° and 40°. The focused hot spots are always located at -1.73 μm (coordinates of the spherical center in x-y plane is (-4.5, -7) μm) whatever the deflection angles are. However, the focused point in y-direction will always put forward with the transformation of deflection angles.

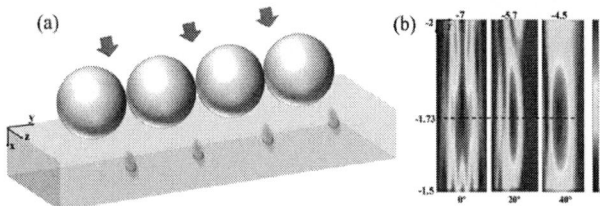

Figure 1. (a) Schemeatic diagram of optically levitated conveyor belt based on specular-reflection PNJ. (b) The normalized electric field distributions with different deflection angles of 0°, 20° and 40°.

III. SIMULATIONS AND DISCUSSIONS

The force and potential wells on a 200-nm diameter nanoparticle with refractive index of 1.5 in x- and y-direction are shown in Fig. 2. Firstly, we have simulated the x-direction force when nanoparticles are placed in the gap between microspheres and gold plates (from -1.95 to -1.55 μm) with the method of Maxwell Stress Tensor (MST) in FDTD. Then, the potential well can be calculated by integrating the force along x- or y- direction. It can be seen from Fig. 2(a) and (b) that the nanoparticles will be exerted positive and negative forces when they are located at the place smaller and larger than about -1.73 μm. Therefore, the bottom of potential well in x-direction is -1.73 μm which means the particle can be stably trapped in the direction perpendicular to the motion direction.

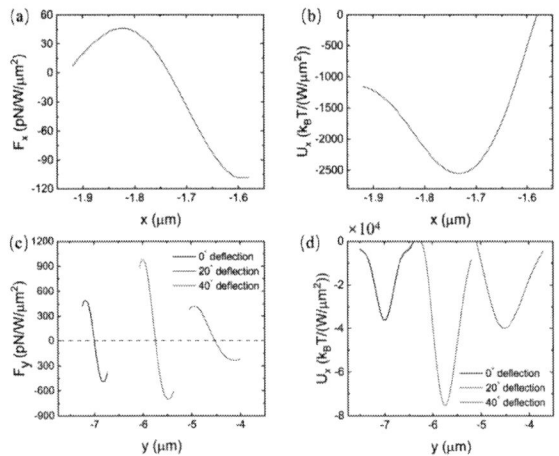

Figure 2. (a) F_x and (b) the corresponding potential well in x direction. (b) F_y and (d) the corresponding potential well in y direction under the deflection angles of 0° (blue lines), 20° (pink lines) and 40° (green lines), respectively.

In order to realize the particle transporting in y-direction, the deflection angle of exciting light is changed with the step

of 2°. Therefore, we have also calculated F_y with different deflection angles of irritating light and the exerted force and potential wells of 0°, 20° and 40° have been selected to shown in Fig. 2(c) and (d). All the three curves of F_y have undergone 0-point which is corresponding to the bottom of potential wells. The dept of potential well with 20° deflection angle is deeper than that of 0° and 40° because the electric field is more concentrated and the absolute value of electric field intensity is higher. In addition, with the increase of deflection angles, the bottom of potential wells will also increase gradually which represents the directional movement of nanoparticles.

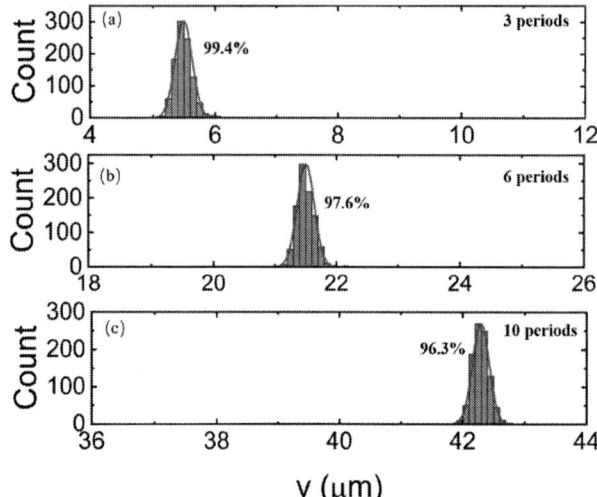

Figure 3. Analysis of the transporting within 10 periods (each period corresponds to the microsphere diameter of 5 μm). the nanoparticle distributions after (a) 3 periods, (b) 6 periods, (c) 10 periods. The red curves in the figures represent the fitting Gaussian equation according to the particle distributions. The numbers in the figures show the rate of successful transport.

In order to verify the effect of nanoparticles transporting in our microsphere systems. We have taken Brownian motion into account. Therefore, the PS particles will suffer optical gradient force, Brownian force and viscous resistance simultaneously. We can calculate the y-direction shift according to Einstein's Brownian motion theory [8]. Here, when the power intensity is fixed as low as 0.2 mW/μm², we have simulated the particle distributions after 3, 6 and 10 periods corresponding to the distance of 15, 30 and 50 μm. The rate of successful particle transporting can be up to 99.4%, 97.6% and 96.3%. Therefore, most of the particles can be conveyed to the target location. The red lines in the figures represent the fitting Gaussian curves according to particle distributions. It can be found that the fitting curves have never become deformed as shown in Fig. 3 which denote that the transporting effect can maintain an excellent level.

IV. CONCLUSIONS

We have realized nanoparticle transporting based on specular-reflection photonic nanojet numerically. in our design, particles can suspend in the fluid both along and perpendicular to the transport directions. At last, we have also calculated the particle distributions when the power intensity is fixed as 0.2 mW/μm^2. Furthermore, the rate of successful transporting can reach to 96.3% even after 50-μm transport and still maintain a good level. We believe our design will be beneficial to the biomolecule transport in biomedical field.

ACKNOWLEDGMENT

This work was sponsored by National Natural Science Foundation of China (nos. 61875083, 61535005), and Social Development Project of Jiangsu Province (BE2019761).

REFERENCES

[1] P. Darwin, and J. GlÜckstad. "Gearing up for optical microrobotics: micromanipulation and actuation of synthetic microstructures by optical forces." Laser & Photonics Reviews, Vol. 7(4), pp. 478-494, 2013.

[2] Q Zhong, H Ding, B Gao, et al. "Advances of Microfluidics in Biomedical Engineering." Advanced Materials Technologies, Vol. 4(6). pp. 1800663, 2019

[3] L Lin, J Zhang, X Peng, et al. "Opto-thermophoretic assembly of colloidal matter." Science advances, Vol. 3(9), pp. e1700458, 2017

[4] F Xu, Y Liu, C Zhang, et al. "Optically levitated conveyor belt based on polarization-dependent metasurface lens arrays." Optics Letters, Vol. 47(9), pp. 2194-2197, 2022.

[5] A. A. R. Neves, "Photonic nanojets in optical tweezers." J. Quant. Spectrosc. Radiat. Transfer, Vol.162, pp. 122–132, 2015.

[6] Y. E. Geints and A. A. Zemlyanov, "Metalens optical 3D-trapping and manipulating of nanoparticles." J. Opt. Vol. 20(7), pp. 075102, 2018.

[7] I. V. Minin, Y. E. Geints, A. A. Zemlyanov, et al. "Specular-reflection photonic nanojet: physical basis and optical trapping application." Optics Express, Vol. 28(15), pp. 22690-22704, 2020.

[8] L. Verslegers, P. B. Catrysse, Z. Yu, et al. "Planar Lenses Based on Nanoscale Slit Arrays in a Metallic Film", Nano Lett. Vol. 9, pp. 235 2009.

Proceedings of the 7th Optoelectronics Global Conference (OGC 2022)

Optimal Raman Spectral Classifcation Model Based on Differentiable Architecture Search of Hybrid Structure Network for Disease Diagnosis

Jiaqi Hu[1], Jinna Chen[1,*], Chenlong Xue[1], Yanqun Xiang[2], Guoying Liu[2], Hong Dang[1], Dan Lu[1], Huanhuan Liu[1], Longqing Cong[1], Zhen Gao[1], Haibin Su[3], Perry Ping Shum[1,*]

[1]*Southern University of Science and Technology, Department of Electronic and Electrical Engineering, Shenzhen, China*
[2]*Sun Yat-sen University Cancer Centre, Department of Nasopharyngeal Carcinoma, Guangzhou, China*
[3]*The Hong Kong University of Science and Technology, Department of Chemistry, Clear Water Bay, Hong Kong, China*
e-mail: shenp@sustech.edu.cn;chenjn@sustech.edu.cn

Abstract—Identification and classification are important application areas of surface-enhanced Raman spectroscopy (SERS). Substance is identified via the chemical finger-print function of Raman spectroscopy. Diseases can be diagnosed through bio-fluidic Raman spectrum analysis and classification accordingly. Since bio-fluidic, such as serumurineand tissue fluid contains various substances, Raman spectrum is too complex to be classified manually. The optimization of deep learning classification model is critical in diagnosis accuracy improvement. Here we propose, for the first, applying DARSHN algorithm in automatic diagnosis model design and optimization. DARSHN was applied to serialize the discrete search space. Optimal structural solution was generated through approximate gradient descent subsequently. This research suggested that DARSHN can be used in the optimization of classification models automatically and effectively. Its advantages in the application of serum SERS-based cancer diagnosis compared to residual network spectral classification models were shown in this paper.

Index Terms—Raman spectra, Cancer, Classfiication, Differentiable, Search, Model

I. INTRODUCTION

Raman spectroscopy is considered to be a molecular fingerprint spectrum, which making it essential for the non-destructive analysis of various biological samples. [1]. For example, Kast et al. [2] demonstrated that Raman spectroscopy could distinguish malignant tumors from normal breast tissue, and detect early tumour changes in experimental mice. Jess et al. [3] showed that Raman spectroscopy could be a valuable tool for the early detection of cells exposed to human papillomavirus (HPV). HPV (especially HPV16) infection can lead to the development of uterine tumors. Therefore, identifying HPV infection in clinical samples is of great importance. The signal of Raman spectroscopy is weak and highly susceptible to noise and fluorescence. In addition, the label-free nature of Raman spectroscopy represents a complexity of information that requires machine learning for spectral classification.

The most crucial function of Raman spectroscopy is the identification of substances. Therefore, a suitable method must be selected after obtaining a Raman spectrum to process the spectral data and achieve an accurate characterization. As a dimensionality reduction algorithm, Principal Component Analysis (PCA) [4] [5] is widely used in the Raman spectral data processing field. The extraction of Raman spectral features through PCA can improve the classification accuracy in the analysis process. Support vector machines (SVM) [6], as a binary classification model, are also frequently used to identify Raman spectra. Machine learning has become an integral part of the analytical processing of Raman spectra [7]. Although conventional machine learning algorithms have performed well, they still suffer from a lack of accuracy for some complex spectra, especially those obtained from portable instrument measurements.

In recent years, with the continuous development of deep learning techniques, the processing of Raman spectra using convolutional neural networks has become an excellent method. Hao Yan [8], Wu Yin [9], Kaidi Wang [10], Kirchberger-Tolstik [11] et al. have achieved notable results on Raman spectra prediction using CNN. In addition to convolutional layers, Transformer [12] and ConvNeXt [13] in recent years have been more fit and robust in their classification algorithms. However, the design of classification models for Raman spectra relies for the most part on the experience of the designer and a great deal of trial and error, and the application of new structures such as Transformer and ConvNeXt to Raman spectra has not yet been seen. How best to integrate the results of these models to obtain optimal Raman spectral classification results remains a challenge at this stage.

This paper proposes a Differentiable Architecture Search of Hybrid Structure Network (DARSHN), a combination of CNN, Transformer and ConvNeXt deep learning network architectures using gradient descent methods, and a search for Raman spectral classification. It applys algorithms related to automatic search networks developed in the context of computer vision and combines a variety of one-dimensional model structures to automate the network search [14]. It uses algorithms developed in the context of computer vision for automatic network search and incorporates a variety of one-dimensional model structures for automated network search. For the model, three basic network structures and three different connection sides are predefined: residual convolu-

978-1-6654-8699-6/22 $31.00 © 2022 IEEE   248

tion module, transformer module, residual ConvNeXt module, maxpooling side, avepooling side and convolution side. We created an efficient structural search method to search four cells and six sides, selecting the most relevant results for each cell and side. We selected the open-source mineral Raman spectroscopy dataset as the search dataset for DARSHN. In addition, models obtained from the DARSHN search were validated on serum SERS dataset for the application of classification. The results showed that the model we obtained using the DARSHN search provided the best accuracy for disease diagnosis.

II. PROBLEM FORMULATION

We define a search space of basic structures (Section II.A) where each architecture and each edge computation process are defined as a directed acyclic graph. In the next step, we refer to the DARTS relaxation method and introduce a simple continuous relaxation method for the search space, which makes the learning process differentiable (Section II.A). Finally we use an approximate approximate distribution optimization method, which allows model to reduce the computational effort and increase the effectiveness of the search (Section II.B).

A. Search cells and sides

Several simple structures for deep learning can already be applied to classification. We have chosen three basic forms and three different types of edges. The model automatically learns the cell and side that best fits the model optimally.

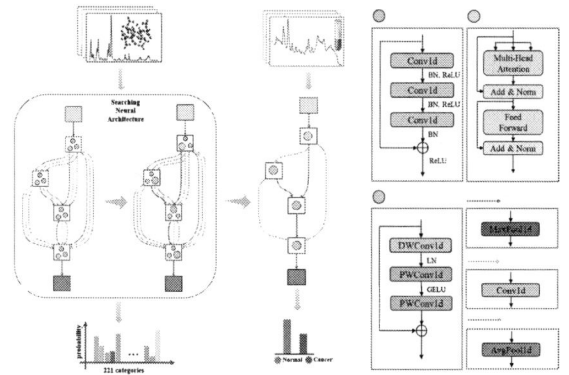

Fig. 1. Flow chart of the overall design of DARSHN. A basic structure chart is constructed. Three cells and four sides are selected for the search. As the search progresses, the weights of different cells and sides change. After search completed, select cells and edges with highest weights and re-evaluate the results. When the evaluation was completed, this optimal model was applied to serum SERS cancer diagnosis. Because Raman spectra has both long span and local features. DARSHN used a combination of Conv1d, Transformer and ConvNeXt1d in this search.

Each cell and side are directed acyclic graphs consisting of an ordered sequence of N nodes. Each node is a potential representation (e.g. a spectrum of features in a convolutional network) and each directed edge (i,j) is associated with some o(i,j) operation that transforms the pair [15]. The output is

obtained by performing a reduction operation (e.g. concatenation) on all intermediate nodes. Each side and cell node are calculated based on forward nodes:

$$C^{(j)} = \sum_{i<j} o^{\mathrm{i,j}}(c^{(i)}) \tag{1}$$

$$S^{(j)} = \sum_{i<j} o^{\mathrm{i,j}}(s^{(i)}) \tag{2}$$

Let O be a set of search operations, where each operation will be applied to a function. C and S optimization process operations are collectively referred to as α. In order to make the search space continuous, we perform Softmax on the operations [14]:

$$\bar{o}^{(i,j)}(x) = \sum_{o \in O} \frac{\exp(\alpha_o(i,j))}{\sum_{o' \in O} \exp(\alpha_{o'}(i,j))} o(x) \tag{3}$$

$$o(i,j) = \arg\max_{o \in O} \alpha_o(\mathrm{i,j}) \tag{4}$$

where a vector parameterization is applied to the mixed weights of (i,j) of the nodes. Thus, as shown in Fig. 1, at the end of the search, each hybrid mixing operation o(i,j) is replaced by the most probable operation. Our optimization model aims to learn the weights of cells and sides in all hybrid operations. (e.g. Pham et al. [16]).

B. Approximate gradient optimization

The goal of DARSHN is optimal validation loss, where training and validation losses are computed separately in gradient descent. As the gradient optimization process in accuracy evaluation if performed internally would be computationally expensive, sides use the approximation scheme proposed by Liu et al [14].

$$\nabla_\alpha L_{val}(w*(\alpha), \alpha) \approx \frac{\nabla_\alpha L_{val}(w - \xi \nabla_w L_{train}(w, \alpha), \alpha)}{2\varepsilon} \tag{5}$$

where w is the current weight maintained by the algorithm and f is the internal learning up. A training step was mainly applied for approximate gradient solving. In addition, we found $\varepsilon = 0.0001/||\nabla_{w'} L_{val}(w, \alpha)||_2$ to be the best performer in our optimization process.

Algorithm 1: DARSHN-Differentiable Architecture Search of Hybrid Structure Network

---

Creat a mixed $\bar{o}^{(i,j)}$ parametrized by $\alpha^{(i,j)}$ for each cells and sides
While not converged **do**:
    1. Updata architecture α_c by descending $\nabla_{\alpha_c} L_{val}(w*(\alpha_c), \alpha_c)$ for cells
    2. Update architecture α_s by descending $\nabla_{\alpha_s} L_{val}(w*(\alpha_s), \alpha_s)$ for sides
    3. Update weights w by descending $\nabla_\alpha L_{train}((\alpha), \alpha)$
Derive the final architecture based on architecture

III. MAIN RESULTS

Our experiments on Raman spectral prediction model construction consisted of two parts, cells search and sides search. In this phase, the training and validation sets are divided into a 1:1 ratio to quickly determine the results of models. Further, we redivided the dataset into a 7:3 ratio, trained them from scratch, and reported their performance in the validation set. We also investigated classification models accuracy by comparing them with three module compositions.

A. Search for cells and sides on RRUFF

We design four cells and five sides in O for the operation. To reduce the search volume, only model structure is optimized, not all the parameters are searched in this paper. The design of model follows He [17], Vaswani [12] and Liu [13], where both Conv and ConvNeXt are reconstructed as one-dimensional spectral operations based on previous study. For each of these cells, the input and output channels and number of features are unchanged, and sides input and output are changed depending on the position of the model.

We set three different cells and three different sides, and ensure that each part could be connected. The input and output features and channels were kept constant for each cell. While each side doubled the input channels and halved the features. In this research, a total of four cells and six sides are set to be searched. In addition, the c and s initialization methods in each cell and side are all chosen to be randomly initialized, followed by Softmax to ensure the sum of the individual weights to one. All operations are performed in step two for the three methods on the sides. Therefore, steps in this process are as follows. 1. define the cells and sides for the entire search. 2. partition the dataset 5:5 for the training and validation sets, and input it into the model for training search to obtain the weight distribution in each cell and side. 3. select the cell and side with the most significant weight. 4. partition the dataset 8:2 for the training and validation sets, and input it into the training evaluation model.

Results of the model search are shown in Fig. 2. We repeated these search for 20 times, retrained from scratch using the training set (30 epochs on RRUFF), and then evaluated models with the validation set. We repeated the experiment four times for each task using different random seeds and used the final accuracy as the evaluation result. Three models with highest accuracies were finally selected. The search results show that the structure for all cells are identical, with slightly different sides. The accuracy of these three models was 84.91%, 83.88% and 83.82%. In addition, to prevent gradient explosion during the search process, each cell and side were processed using batch normalization during the search. Two optimizations were alternated to make the search process smoother.

B. ARCHITECTURE EVALUATION

To evaluate the architecture obtained from the final search, we repeated the model four times to restart training and evaluation. To better assess our resulting architecture, we randomly initialize its weights, train it from scratch, and report

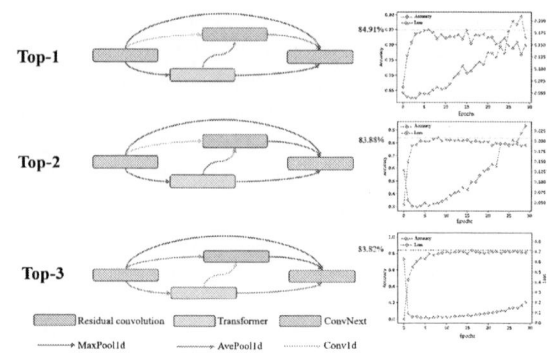

Fig. 2. Search results for DARSHN. We repeated the search 20 times and retrained from scratch using the training set (50 epochs on RRUFF), and then evaluated on the validation set. For each task, we repeated the experiment 4 times using different random seeds and used the final accuracy as the evaluation result.

TABLE I
COMPARISON WITH THREE MANUALLY DESIGNED MODELS ON RRUFF

Architecture	Accuracy (test)	Params (M)	Search Cost (GPU days)	Search Method
ResNet	93.87%	39	-	manual
Transformer	94.98%	41	-	manual
Res-ConvNeXt	96.12%	41	-	manual
DARSHN	98.23%	42	0.2	gradient-based

performance in the validation set. In addition, ResNet model, Transformer model and ConvNeXt model constructed with similar trainable parameters were selected for evaluation under the same conditions.

The results of DRASHN (top-1) for classification prediction are given in Table I. For all models, there was no pre-training or other special operations during the training process. All models have an approximate number of trainable parameters. With only a small amount of training resources (0.2 GPU days), the searched DARSHN was used to classify mineral Raman spectra (221 classes). These results show that DARSHN outperforms ResNet, Transformer and Res-ConvNext for spectral classification, and the variation of the accuracy of the validation set for each training process in Fig. 3 also shows that DATSHN has highest accuracy of 98.23%. At this stage, there is a lack of a unified and fully validated classification model structure in 1D Raman spectral classification models. This paper proposes the structure search method for Raman spectroscopy classification models.

C. DARSHN FOR CANCER SERS DIAGNOSIS

For serum SERS spectra, 25 nasopharyngeal carcinoma cases(NPC) and 25 normal human sera were collected. For each sample, 5 spectral data were collected. Test sets were selected by randomly drawing five normal serum SERS and nasopharyngeal carcinoma serum sers each. Serum SERS spectrum is shown in Fig. 4, with different patterns of peaks between normal individuals and cancer patients.. However, there are also large fluctuations in the spectra, and the algorithm needs to be trained to predict the spectra. Here, all

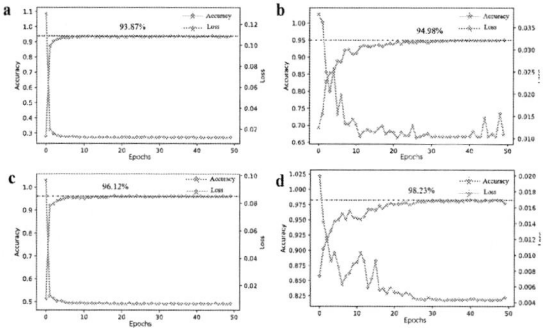

Fig. 3. Variation of accuracy and loss values in the training process for 4 different model validation sets. (a) ResNet; (b) Transformer; (c) Res-ConvNeXt; (d) DARSHN.

model parameters are used directly for training without any adjustment. As in the previous section, we simultaneously used the serum SERS spectrum for the training and evaluation cycle for each of the four models. From the results in Table II, DARSHN was found to achieve optimal diagnostic results. However, the accuracy improvement is less on this side compared to the effects on RRUFF. This is because of that serum spectra are more complex than mineral spectra, so the overall accuracy serum spectra are low. It is found that the accuracy of DARSHN is still significantly improved compared to the traditional residual network. This result suggests that DARSHN has higher accuracy when used for serum SERS diagnosis.

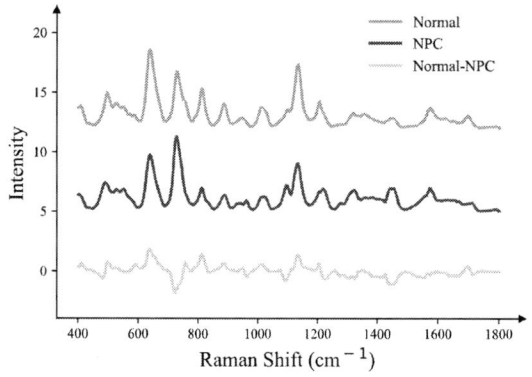

Fig. 4. Normalized averaged SERS spectra comparison of the 25 normal and 25 NPC serum samples, where the shaded areas represent the mean standard deviations. Normal-NPC is the difference between SERS spectra of Normal and NPC.

IV. EXPERIMENTS

A. RRUFF

The RRUFF dataset is an open-source dataset of mineral spectra [18] (211 spectra, 3684 entries in total). Because the framework of the model changes during the search, we used normalization against each spectrum rather than batch-to-batch normalization. To perform the architectural search, we took

TABLE II
COMPARISON WITH THREE MANUALLY DESIGNED MODELS ON SERUM SERS

Architecture	Accuracy (test)	Params (M)	Search Cost (GPU days)	Search Method
ResNet	85.68%	39	-	mannual
Transformer	87.13%	41	-	mannual
Res-ConvNext	87.71%	41	-	mannual
DARSHN	89.32%	42	0.2	gradient-based

out half of the RRUFF data as the validation set. One was trained using DARSHN with 50 epochs and a Batch size of 32 (for both the training and validation sets), these numbers were chosen to ensure that the network did not take up too much memory. We used Adam to optimize the weights with an initial learning rate of 0.0001 (annealed to zero according to the cosine schedule without restarting.) We use zero initialization of the architecture variables to ensure that all operations are weighted the same after Softmax. In the early stages, this ensures that the weights of each candidate operation receive sufficient learning signals. We use Adam as the optimizer for α, with an initial learning rate of 0.0001, momentum $\beta = (0.5, 0.999)$ and weights decaying by 3×10^{-8}. The search takes about 4h on a single GPU.

B. Serum SERS dataset

Serum SERS is measured using Portable Raman Spectrometer. In total, there were 25 cases of normal control, and 25 cases of nasopharyngeal carcinoma. Raman spectra were acquired at 785nm, with an integration time of 8s, two integrations and 5 data for each sample.

C. ResNet, Transformer and ConvNeXt

The structure of ResNet, Transformer and ConvNeXt is shown in Fig. 1. only the core module of the Transformer module is used, and the encoding module is replaced by convolution. It is worth noting that the design process of ConvNeXt was performed according to the study of Liu et al. [13] In simple terms, this means that the two-dimensional image transformation process in the original model is rewritten as a one-dimensional spectral operation. To evaluate the performance of three models on the application of classification, the connections between all three modules were made using residual connections, which were repeated 8, 6 and 8 times, respectively. In the evaluation process, the training and validation sets were with the ratio of 8:2.

Depending on the characteristics of the model, the CNN focuses on local areas of the spectrum and the Transformer is more concerned with the effect of changes in the peak shape of the spectrum during the classification process. In contrast, the Transformer focuses more on the Raman shift and not so much on the peak shape itself. These features also provide the basis for a free combination of search networks.

D. Limitations

During DARSHN search, it is easy to generate a gradient explosion if the initial learning rate is significant. After conducting 20 investigations, we found that the results of each probe were not unique, but the final structure was highly similar. This may also mean that the model structure is still not optimal. Moreover, these structures of deep learning have not been explored in depth at this stage of applied research in spectroscopy (only CNNs and LSTMs have emerged at this stage, not yet the Transformer). With the increasing complexity of the network structure, it isn't easy to search the entire model at once with this method. As an alternative, the structure can be explored in steps and combined. Although, models obtained by DARSHN searching can replace manual design to a certain extent and improve the accuracy. We also believe that as the technology develops, more and better model design methods will be used in constructing spectral models in the future.

V. Conclusion and future problems

We present DARSHN, a simple and efficient architectural search algorithm for multi-structural search. DARSHN performs efficient model selection and reconstruction for Raman spectral classification tasks by searching in continuous space. Compared to manually designed models, DARSHN can effectively improve SERS diagnosis accuracy. Within limits, DARSHN reduces time cost for model design and combines experience to obtain optimized models automatically. There are many exciting directions, in which DARSHN could be further improved. For example, the current algorithm only searches for model classes, while further algorithm could be designed to search for parameters, structural details, etc. Moreover, annealed softmax temperatures can be involved in thermal selection for the improvement of DARSHN. To be noted, the gradient tending to disappear during model training, when the learning rate is set unsuitable. With further development, there will also be better methods and better models.

ACKNOWLEDGMENT

This work was partially supported by Guangdong Basic and Applied Basic Research Foundation (Grant No. 2021B1515120013); the Stable Support Program for Higher Education Institutions from Shenzhen Science, Technology Innovation Commission (SSTIC, Grant No. 20200925162216001); Special Funds for the Major Fields of Colleges and Universities by the Department of Education of Guangdong Province (Grant No. 2021ZDZX1023); and Open Fund of State Key Laboratory of Information Photonics and Optical Communications (Beijing University of Posts and Telecommunications), P. R. China (Grant No. IPOC2020A002).This study was approved by the ethics committee of Southern University of Science and Technology (2022GZR048) and every subject provided informed consent before serum collection.

References

[1] Y. Xu, P. Zhong, A. Jiang, X. Shen, X. Li, Z. Xu, Y.-D. Shen, Y. ming Sun, and H. Lei, "Raman spectroscopy coupled with chemometrics for food authentication: A review," *Trends in Analytical Chemistry*, vol. 131, p. 116017, 2020.

[2] R. E. Kast, S. C. Tucker, K. Killian, M. Trexler, K. V. Honn, and G. W. Auner, "Emerging technology: applications of raman spectroscopy for prostate cancer," *Cancer and Metastasis Reviews*, vol. 33, pp. 673–693, 2013.

[3] P. Jess, D. D. W. Smith, M. Mazilu, K. Dholakia, A. C. Riches, and C. S. Herrington, "Early detection of cervical neoplasia by raman spectroscopy," *International Journal of Cancer*, vol. 121, 2007.

[4] E. Gurian, A. Di Silvestre, E. Mitri, D. Pascut, C. Tiribelli, M. Giuffrè, L. S. Crocè, V. Sergo, and A. Bonifacio, "Repeated double cross-validation applied to the pca-lda classification of sers spectra: a case study with serum samples from hepatocellular carcinoma patients," *Analytical and bioanalytical chemistry*, vol. 413, no. 5, pp. 1303–1312, 2021.

[5] E. Gurian, A. D. Silvestre, E. Mitri, D. Pascut, C. Tiribelli, M. Giuffré, L. S. Crocè, V. Sergo, and A. Bonifacio, "Repeated double cross-validation applied to the pca-lda classification of sers spectra: a case study with serum samples from hepatocellular carcinoma patients," *Analytical and Bioanalytical Chemistry*, vol. 413, pp. 1303 – 1312, 2020.

[6] C. He, X. Wu, J. Zhou, Y. Chen, and J. Ye, "Raman optical identification of renal cell carcinoma via machine learning," *Spectrochimica Acta Part A: Molecular and Biomolecular Spectroscopy*, vol. 252, p. 119520, 2021.

[7] F. Lussier, V. Thibault, B. Charron, G. Q. Wallace, and J.-F. Masson, "Deep learning and artificial intelligence methods for raman and surface-enhanced raman scattering," *TrAC Trends in Analytical Chemistry*, vol. 124, p. 115796, 2020.

[8] H. Yan, M. Yu, J. Xia, L. Zhu, T. Zhang, Z. Zhu, and G. Sun, "Diverse region-based cnn for tongue squamous cell carcinoma classification with raman spectroscopy," *IEEE Access*, vol. 8, pp. 127 313–127 328, 2020.

[9] W. Yin, C. He, and Q. Wu, "A tomato quality identification method based on raman spectroscopy and convolutional neural network," in *Journal of Physics: Conference Series*, vol. 1438, no. 1. IOP Publishing, 2020, p. 012029.

[10] K. Wang, L. Chen, X. Ma, L. Ma, K. C. Chou, Y. Cao, I. U. Khan, G. Gölz, and X. Lu, "Arcobacter identification and species determination using raman spectroscopy combined with neural networks," *Applied and environmental microbiology*, vol. 86, no. 20, pp. e00 924–20, 2020.

[11] T. Kirchberger-Tolstik, P. Pradhan, M. Vieth, P. Grunert, J. Popp, T. W. Bocklitz, and A. Stallmach, "Towards an interpretable classifier for characterization of endoscopic mayo scores in ulcerative colitis using raman spectroscopy," *Analytical Chemistry*, vol. 92, no. 20, pp. 13 776–13 784, 2020.

[12] A. Vaswani, N. Shazeer, N. Parmar, J. Uszkoreit, L. Jones, A. N. Gomez, Ł. Kaiser, and I. Polosukhin, "Attention is all you need," *Advances in neural information processing systems*, vol. 30, 2017.

[13] Z. Liu, H. Mao, C.-Y. Wu, C. Feichtenhofer, T. Darrell, and S. Xie, "A convnet for the 2020s," in *Proceedings of the IEEE/CVF Conference on Computer Vision and Pattern Recognition*, 2022, pp. 11 976–11 986.

[14] H. Liu, K. Simonyan, and Y. Yang, "Darts: Differentiable architecture search," *arXiv preprint arXiv:1806.09055*, 2018.

[15] B. Zoph, V. Vasudevan, J. Shlens, and Q. V. Le, "Learning transferable architectures for scalable image recognition," in *Proceedings of the IEEE conference on computer vision and pattern recognition*, 2018, pp. 8697–8710.

[16] R. Shin, C. Packer, and D. Song, "Differentiable neural network architecture search," 2018.

[17] K. He, X. Zhang, S. Ren, and J. Sun, "Deep residual learning for image recognition," in *Proceedings of the IEEE conference on computer vision and pattern recognition*, 2016, pp. 770–778.

[18] B. Lafuente, R. T. Downs, H. Yang, and N. Stone, "1. the power of databases: The rruff project," in *Highlights in mineralogical crystallography*. De Gruyter (O), 2015, pp. 1–30.

Chiral-Selective Transmission of Edge States in Terahertz Valley Topological Photonic Crystals

Hongyi Li
College of Precision Instrument and Optoelectronics
Engineering
Tianjin University
Tianjin, China
e-mail: lihongyi@tju.edu.cn

Jiajun Ma
College of Precision Instrument and Optoelectronics
Engineering
Tianjin University
Tianjin, China
e-mail: majiajun@tju.edu.cn

Shilei Liu
College of Precision Instrument and Optoelectronics
Engineering
Tianjin University
Tianjin, China
e-mail: slliu@tju.edu.cn

Yi Liu
College of Precision Instrument and Optoelectronics
Engineering
Tianjin University
Tianjin, China
e-mail: lyi@tju.edu.cn

Chunmei Ouyang
College of Precision Instrument and Optoelectronics Engineering
Tianjin University
Tianjin, China
e-mail: cmouyang@tju.edu.cn

Abstract—**Topological photonic systems provide a novel platform for controlling the flow of light. Recently, valley photonic crystals have been favored for their friendly of implement. Here, we demonstrate the chiral-selective terahertz transmission in a valley photonic crystal comprised of dielectric rods. The unidirectional propagation of chiral edge states indicated by the slope of energy band dispersion is numerically observed. The proposed regime has potential in multi-functional photonic broadband communication and other on-chip applications in the terahertz region, particularly for 6G communications.**

Keywords-topological photonics; valley photonic crystals; unidirectional edge states.

I. Introduction

Topology originated from condensed matter [1] has been gradually introduced into classical wave systems such as acoustics [2] and photonics [3]-[7]. The topological insulator system is favored by researchers because of its bulk insulation while edge conducted feature. When two materials with different topological invariants are put in contact, there must exist spatially localized edge states at the interface, with the energies that lying within the energy gap of the surrounding bulk materials, this principle is called bulk-edge correspondence [4]. The generated edge states are characterized by unidirectional transmission, no backscattering and robustness to impurities, defects and disorders. As photonic analogues of topological insulators,

topological photonic insulators provide a more convenient platform for topology research due to their simple implementation conditions. Researchers have proposed a two-dimensional photonic crystal array composed of ferrites, which is the photonic analogue of the quantum Hall effect [8], [9]. However, this method requires an external magnetic field and materials with magneto-optical response to break the time-reversal symmetry, which is difficult to achieve in the traditional optical frequency range. In recent years, the photonic analogues of the quantum valley Hall effect realize topological insulators by constructing valley degrees of freedom and breaking mirror inversion symmetries while maintaining time-reversal symmetry [5], [10]-[23]. This method is easy to implement and combined with modern micro-nano fabrication technology, which has aroused the interest of researchers. It has been implemented on a variety platforms including resonant rings [20], optical waveguides [14], [19], [24]-[27] and photonic crystals [11], [28]-[30], and provides a powerful platform for exploreing topological systems.

The unidirectional propagation properties of topological edge states have always been the focus of research and have been verified in a variety of quantum systems, including the quantum Hall system [8], quantum spin Hall system [31]-[34] and quantum valley Hall system [17], [20], [24], [35]. In the quantum Hall system, due to time-reversal symmetry breaking, the edge state does has the characteristics of unidirectional propagation, while in quantum spin Hall and quantum valley Hall systems, the time-reversal symmetry is maintained, so the unidirectional propagation characteristics can only be shown in the case of the chiral source excitation.

In this paper, we propose a valley photonic crystal with an all-silicon structure based on the valley Hall effect. Valley photonic crystals with different topological states are obtained by the rotation of the original structure, and a line shaped channel is constructed to verify the existence and transmission of their edge states. The propagation direction of the chiral edge states is indicated by the slope of the edge state dispersion. We verify the unidirectional transmission characteristics in the frequency range of 0.52-0.59 THz and the bi-directional transmission characteristics within 0.59-0.61 THz excited by the same chiral source. This work enriches the exploration of topological systems and is easy to integrate with photonic chips, which has potential application prospects in 6G communication.

II. DESIGN AND SIMULATION

As shown in Fig. 1(a), the proposed VPC consists of a honeycomb lattice composed of three silicon rods, the diameter of the rod is d, the edge length of the hexagonal cell is a, the lattice constant is a_0, and the distance from the center of the rod to the center of the cell is r. The three rods can rotate around the center of the cell, and the rotation angle is represented by α. Figure 1(a) shows the case when $\alpha = \pm 30°$. When $\alpha = 0°$ (see the inset in Fig. 1 (b)), the double degenerate Dirac point protected by symmetry at K point is obtained. When α changes, the mirror inversion symmetry is broken and the C_{3v} symmetry decreases to C_3 symmetry, so that the original twofold Dirac degeneracy at K (K') is opened, forming a bandgap.

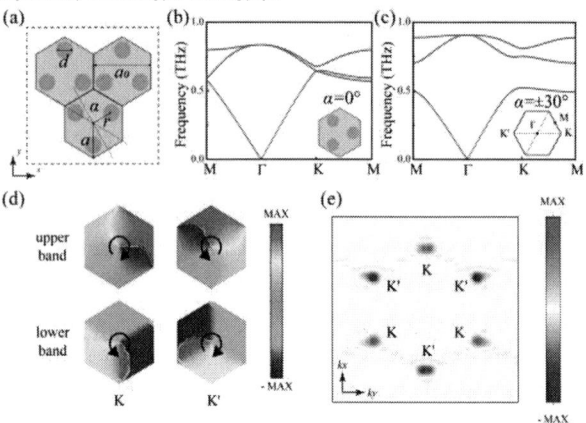

Figure 1. Schematic of the proposed VPC. (a) The unit structure when $\alpha = \pm 30°$. Band structures when $\alpha = 0°$(b) and $\alpha = \pm 30°$(c), respectively, insets: the unit structure when $\alpha = 0°$ (b) and the first Brillouin zone (c). (d) Simulated spatial distributions of the phase vortex of Ez at K and K' points in the first and second bands. (e) Calculated berry curvature of the first band when $\alpha = +30°$.

As shown in Fig. 1(b), when $\alpha = 0°$, there is a double degenerate Dirac cone at 0.644 THz at K point of the first Brillouin zone. When α changes to 30°, we obtain a bandgap of 0.524-0.704 THz, as shown in Fig. 1(c). The phase vortices of the eigenmode Ez under and above the bandgap at

K point are shown in Fig. 1(d), respectively. There are different vortices at $\alpha = +30°$ and $\alpha = -30°$, indicating a topological related phase transition occured. Furthermore, the topological invariant of VPCs is described by valley Chern number, which is defined as $C_V = C_K - C_{K'}$, where C_K is the integral of the berry curvature around K. Figure. 1(e) shows the calculated berry curvature distribution in the first band when $\alpha = +30°$, indicating that the berry curvature at K and K' is opposite, so that $C_V = +1$ is calculated when $C_K = +1/2$ and $C_{K'} = -1/2$, similarly, $C_V = -1$ can be calculated when $\alpha = -30°$.

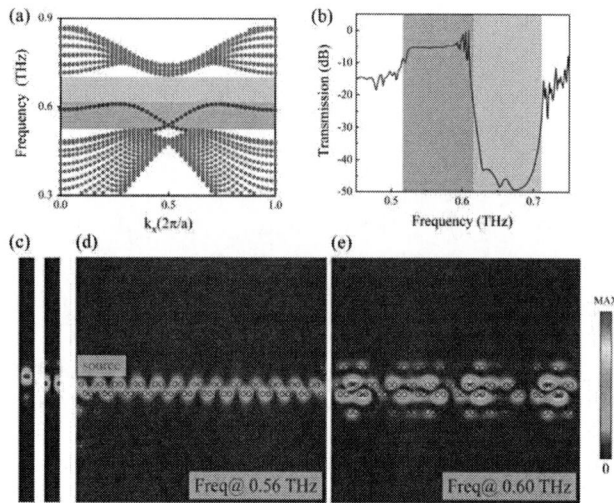

Figure 2. (a) Band dispersion of edge states. (b) Transmission spectrum of edge state. (c) Eigenmodes of the localized edge states when $k_x = 0.5$. (d), (e) Ez distributions of the edge state at 0.56 THz and 0.60 THz, respectively.

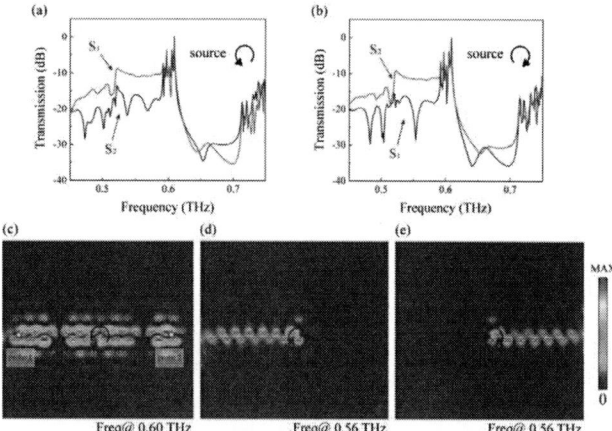

Figure 3. (a), (b) Transmission spectra of edge states excited by a clockwise and counter-clockwise chiral source, respectively. S_1 and S_2 are the transmission spectra obtained at probe 1 and probe 2 labeled in (c).(c) Ez distribution of edge state at 0.60 THz under counter-clockwise chiral source excitation. (d), (e) Ez distributions of edge states at 0.56 THz under counter-clockwise and clockwise chiral source excitation, respectively.

The energy band dispersion of the edge state has been calculated, as shown in Fig. 2(a), and we can see that the edge state appears in the frequency range of 0.52-0.61 THz (red dotted lines in Fig. 2(a)), which is bandgap region in

978-1-6654-8699-6/22 $31.00 © 2022 IEEE

bulk state, but there is still no permissible state when it is greater than 0.61 THz (orange box in Fig. 2a). In the case of $k_x=0.5$, the Ez field diagram of the edge state mode is shown in Fig. 2(c), and the localized edge states between the two kinds of VPCs can be seen. We construct a line shaped transmission channel and verify the transmission spectra and field distributions numerically. The edge state transmission peak is obtained at 0.52-0.61 THz, as shown in Fig. 2(b), and there is still a bandgap at 0.61-0.71 THz. The field distributions at 0.56 THz and 0.60 THz are shown in Figs. 2(d) and 2(e), respectively.

The propagation direction of the chiral edge state is indicated by the slope of the edge state dispersion. The edge state, as shown in Fig. 2(a), has the same dispersion slope (green box) at 0.52-0.59 THz, so it has the characteristic of unidirectional propagation. At 0.59-0.61THz, it has a different dispersion slope (blue box) and therefore, the edge state can propagate in both directions. We verify this numerically and excite it with a chiral source. The transmission spectra excited by clockwise and counter-clockwise chiral sources are shown in Figs. 3(a) and 3(b), respectively.

III. CONCLUSION

We numerically demonstrate a VPC with an all-silicon structure at terahertz frequencies, which breaks the mirror inversion symmetry by rotating the silicon rods and has different topological properties at K and K 'points. A line shaped transmission channel is constructed to verify the transmission of the excited edge state. The propagation direction of the chiral edge state is indicated by the slope of the edge state dispersion. we numerically verify the unidirectional and bidirectional transmission characteristics of the edge states under the excitation of a chiral source. Efficient and diverse terahertz functional devices will play a great role in 6G communications. This work proposes a terahertz waveguide, which is easy to integrate with other silicon-based photonic devices and has potential application prospects.

ACKNOWLEDGMENT

This work was supported by the National Natural Science Foundation of China (Grant Nos. 62175180, 61875150, 61805129, 62005193, and 11874245)..

REFERENCES

[1] M. Z. Hasan and C. L. Kane, "Colloquium: Topological insulators," *Reviews of Modern Physics,* vol. 82, no. 4, pp. 3045-3067, Nov. 2010, doi: 10.1103/RevModPhys.82.3045.

[2] Y. Liu, X. Chen, and Y. Xu, "Topological Phononics: From Fundamental Models to Real Materials," *Advanced Functional Materials,* vol. 30, no. 8, Feb. 2019, Art no. 1904784, doi: 10.1002/adfm.201904784.

[3] G. J. Tang, X. T. He, F. L. Shi, J. W. Liu, X. D. Chen, and J. W. Dong, "Topological Photonic Crystals: Physics, Designs, and Applications," *Laser & Photonics Reviews,* vol. 16, no. 4, Feb. 2022, Art no. 2100300, doi: 10.1002/lpor.202100300.

[4] L. Lu, J. D. Joannopoulos, and M. Soljacic, "Topological photonics," *Nature Photonics,* vol. 8, no. 11, pp. 821-829, Nov. 2014, doi: 10.1038/Nphoton.2014.248.

[5] J.-W. Liu et al., "Valley photonic crystals," *Advances in Physics: X,* vol. 6, no. 1, 2021, Art no. 1905546, doi: 10.1080/23746149.2021.1905546.

[6] M. Segev and M. A. Bandres, "Topological photonics: Where do we go from here?," *Nanophotonics,* vol. 10, no. 1, pp. 425-434, Jan. 2020, doi: 10.1515/nanoph-2020-0441.

[7] A. B. Khanikaev and G. Shvets, "Two-dimensional topological photonics," *Nature Photonics,* vol. 11, no. 12, pp. 763-773, Dec. 2017, doi: 10.1038/s41566-017-0048-5.

[8] Z. Wang, Y. D. Chong, J. D. Joannopoulos, and M. Soljacic, "Observation of unidirectional backscattering-immune topological electromagnetic states," *Nature,* vol. 461, no. 7265, pp. 772-775, Oct. 2009, doi: 10.1038/nature08293.

[9] F. D. M. Haldane and S. Raghu, "Possible realization of directional optical waveguides in photonic crystals with broken time-reversal symmetry," *Physical Review Letters,* vol. 100, no. 1, Jan. 2008, Art no. 013904, doi: 10.1103/PhysRevLett.100.013904.

[10] X. X. Wu et al., "Direct observation of valley-polarized topological edge states in designer surface plasmon crystals," *Nature Communications,* vol. 8, no. 1, Nov. 2017, Art no. 1304, doi: 10.1038/s41467-017-01515-2.

[11] Y. H. Kang, X. Ni, X. J. Cheng, A. B. Khanikaev, and A. Z. Genack, "Pseudo-spin-valley coupled edge states in a photonic topological insulator," *Nature Communications,* vol. 9, no. 1, Aug. 2018, Art no. 3029, doi: 10.1038/s41467-018-05408-w.

[12] M. P. Makwana and R. V. Craster, "Designing multidirectional energy splitters and topological valley supernetworks," *Physical Review B,* vol. 98, no. 23, Dec. 2018, Art no. 235125, doi: 10.1103/PhysRevB.98.235125.

[13] K. Qian, D. J. Apigo, C. Prodan, Y. Barlas, and E. Prodan, "Topology of the valley-Chern effect," *Physical Review B,* vol. 98, no. 15, Oct. 2018, Art no. 155138, doi: 10.1103/PhysRevB.98.155138.

[14] M. I. Shalaev, W. Walasik, A. Tsukernik, Y. Xu, and N. M. Litchinitser, "Robust topologically protected transport in photonic crystals at telecommunication wavelengths," *Nature Nanotechnology,* vol. 14, no. 1, pp. 31-34, Jan. 2018, doi: 10.1038/s41565-018-0297-6.

[15] X. D. Chen, X. T. He, and J. W. Dong, "All-Dielectric Layered Photonic Topological Insulators," *Laser & Photonics Reviews,* vol. 13, no. 8, Aug. 2019, Art no. 1900091, doi: 10.1002/lpor.201900091.

[16] Z. G. Geng, Y. G. Peng, P. Q. Li, Y. X. Shen, D. G. Zhao, and X. F. Zhu, "Mirror-symmetry induced topological valley transport along programmable boundaries in a hexagonal sonic crystal," *Journal of Physics: Condensed Matter,* vol. 31, no. 24, Jun. 2019, Art no. 245403, doi: 10.1088/1361-648X/ab0fcc.

[17] X. T. He et al., "A silicon-on-insulator slab for topological valley transport," *Nature Communications,* vol. 10, no. 1, Feb. 2019, Art no. 872, doi: 10.1038/s41467-019-08881-z.

[18] H. Saito, D. Yoshimoto, Y. Moritake, T. Matsukata, N. Yamamoto, and T. Sannomiya, "Valley-Polarized Plasmonic Edge Mode Visualized in the Near-Infrared Spectral Range," *Nano Letters,* vol. 21, no. 15, pp. 6556-6562, Aug. 2021, doi: 10.1021/acs.nanolett.1c01841.

[19] H. Xiong et al., "Topological Valley Transport of Terahertz Phonon-Polaritons in a LiNbO3 Chip," *ACS Photonics,* vol. 8, no. 9, pp. 2737-2745, Sep. 2021, doi: 10.1021/acsphotonics.1c00860.

[20] H. F. Yang et al., "Optically Reconfigurable Spin-Valley Hall Effect of Light in Coupled Nonlinear Ring Resonator Lattice," *Physical Review Letters,* vol. 127, no. 4, Jul. 2021, Art no. 043904, doi: 10.1103/PhysRevLett.127.043904.

[21] J. W. You, Q. Ma, Z. H. Lan, Q. Xiao, N. C. Panoiu, and T. J. Cui, "Reprogrammable plasmonic topological insulators with ultrafast control," *Nature Communications,* vol. 12, no. 1, Sep. 2021, Art no. 5468, doi: 10.1038/s41467-021-25835-6.

[22] X. J. Zhang, L. Liu, M. H. Lu, and Y. F. Chen, "Valley-Selective Topological Corner States in Sonic Crystals," *Physical Review Letters,* vol. 126, no. 15, Apr. 2021, Art no. 156401, doi: 10.1103/PhysRevLett.126.156401.

[23] Z. S. Zhang *et al.*, "Azimuthally and radially polarized orbital angular momentum modes in valley topological photonic crystal fiber," *Nanophotonics,* vol. 10, no. 16, pp. 4067-4074, Nov. 2021, doi: 10.1515/nanoph-2021-0395.

[24] M. Jalali Mehrabad *et al.*, "Chiral topological photonics with an embedded quantum emitter," *Optica,* vol. 7, no. 12, pp. 1690-1696, 2020, doi: 10.1364/optica.393035.

[25] Y. H. Yang *et al.*, "Terahertz topological photonics for on-chip communication," *Nature Photonics,* vol. 14, no. 7, pp. 446-451, Jul. 2020, doi: 10.1038/s41566-020-0618-9.

[26] S. Arora, T. Bauer, R. Barczyk, E. Verhagen, and L. Kuipers, "Direct quantification of topological protection in symmetry-protected photonic edge states at telecom wavelengths," *Light: Science & Applications,* vol. 10, no. 1, Jan. 2021, Art no. 9, doi: 10.1038/s41377-020-00458-6.

[27] Q. L. Chen *et al.*, "Photonic Topological Valley-Locked Waveguides," *ACS Photonics,* vol. 8, no. 5, pp. 1400-1406, May. 2021, doi: 10.1021/acsphotonics.1c00029.

[28] Z. Gao *et al.*, "Valley surface-wave photonic crystal and its bulk/edge transport," *Physical Review B,* vol. 96, no. 20, Nov. 2017, Art no. 201402, doi: 10.1103/PhysRevB.96.201402.

[29] H. Yoshimi, T. Yamaguchi, R. Katsumi, Y. Ota, Y. Arakawa, and S. Iwamoto, "Experimental demonstration of topological slow light waveguides in valley photonic crystals," *Optics Express,* vol. 29, no. 9, pp. 13441-13450, Apr. 2021, doi: 10.1364/Oe.422962.

[30] G. Arregui, J. Gomis-Bresco, C. M. Sotomayor-Torres, and P. D. Garcia, "Quantifying the Robustness of Topological Slow Light," *Physical Review Letters,* vol. 126, no. 2, Jan. 2021, Art no. 027403, doi: 10.1103/PhysRevLett.126.027403.

[31] L. H. Wu and X. Hu, "Scheme for Achieving a Topological Photonic Crystal by Using Dielectric Material," *Physical Review Letters,* vol. 114, no. 22, Jun. 2015, Art no. 223901, doi: 10.1103/PhysRevLett.114.223901.

[32] B. Y. Xie *et al.*, "Higher-order quantum spin Hall effect in a photonic crystal," *Nature Communications,* vol. 11, no. 1, Jul. 2020, Art no. 3768, doi: 10.1038/s41467-020-17593-8.

[33] H. B. Huang, S. Y. Huo, and J. J. Chen, "Reconfigurable Topological Phases in Two-Dimensional Dielectric Photonic Crystals," *Crystals,* vol. 9, no. 4, Apr. 2019, Art no. 221, doi: 10.3390/cryst9040221.

[34] M. Honari-Latifpour and L. Yousefi, "Topological plasmonic edge states in a planar array of metallic nanoparticles," *Nanophotonics,* vol. 8, no. 5, pp. 799-806, May 2019, doi: 10.1515/nanoph-2018-0230.

[35] Y. H. Han, H. M. Fei, H. Lin, Y. M. Zhang, M. D. Zhang, and Y. B. Yang, "Design of broadband all-dielectric valley photonic crystals at telecommunication wavelength," *Optics Communications,* vol. 488, Jun. 2021, Art no. 126847, doi: 10.1016/j.optcom.2021.126847.

Proceedings of the 7th Optoelectronics Global Conference (OGC 2022)

Tuning Performance and Mechanism of Gate-tuned Graphene Grating for Dynamically Controlling Terahertz Wavefront

Qianqian Wang, Xiaotong Li, Jie Liang, Runze Li
Shanghai University
Key Laboratory of Specialty Fiber Optics and Optical Access Networks
Shanghai 200444, China
e-mail: WangQianqian@shu.edu.cn

Abstract—**Recently, dynamically controlling of the terahertz wavefront has attract tremendous attention due to both scientific curiosity and potential applications in many fields. However, the available tunable components in the terahertz band are low efficiency and difficult to integrated with other terahertz components, due to the low light-matter interaction in this frequency range. Here, we design a tunable grating fabricated of graphene combined with metasurfaces. Coupled-mode-theory (CMT) analyses reveal the underlying physics, that is through the control of the applied voltage, graphene as tunable absorption loss, can realize dynamically controlling on the amplitude and phase of THz wavefront.**

Keywords-terahertz wavefront; tunable grating; graphene; metasurface; coupled-mode theory

I. INTRODUCTION

Terahertz technology is the foundation of new generation of IT, communications (6G/7G) and radar technologies. Especially, dynamically controlling of terahertz wavefront has attracted wide attention recently. Unfortunately, most of the devices that currently to tune THz wavefronts are bulky and low efficient due to the low light-matter interaction in this frequency range. Very recently, the rapid development of metasurfaces pave a novel way to dynamically control terahertz wavefront. Here, metasurfaces are ultra-thin metamaterials contains with subwavelength planar microstructures (e.g., meta-atoms) exhibiting tailored EM responses [1-3], showing many fascinating wave-manipulation effects, such as anomalous reflection or refraction, dynamically wavefront control [4-9].

In this paper, combining graphene with metasurface, we design a tunable grating in THz regime. The simulation results show that under the control of gating voltage, the designed grating can realize the controlling on amplitude and phase of the THz wavefront, with the phase resolution be 1 bit (1 bit means 180°).

II. DESIGN, SIMULATIONS AND CMT ANALYSES

A. Design

A schematic diagram of the designed grating is shown in Fig. 1 (a), where V_A is the added gating voltage. Figure 1 (b) shows the metasurface unit structure of A in Fig.1(a), which consists of metal plate, graphene, ION GEL and quartz. We employed an ION GEL gating technique to change the carrier density in graphene metasurface.

(a)

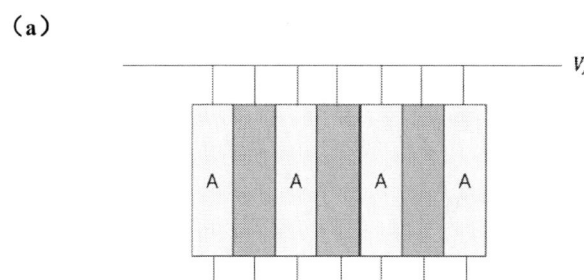

(b)

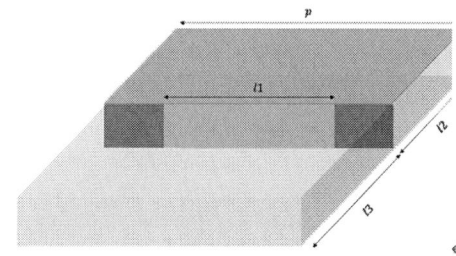

Figure 1. (a) Schematic diagram of the designed grating. (b) unit structure.

B. Simulations

We performed numerical simulations using the commercial Finite difference time domain method (FDTD) solver CST Microwave Studio. The geometric parameters of unit structure are: $p = 80\,\mu m$, $h = 10\,\mu m$, $l_1 = 50\,\mu m$, $l_2 = 60\,\mu m$, $l_3 = 50\,\mu m$. The simulation boundary conditions are unit cell boundary conditions, the frequency range is set as 0.1 THz ~ 0.4 THz, and the incident wave is set as a plane wave. The simulation result is shown in Fig. 2.

From the simulation results of amplitude in Fig. 2, we can see that: with the increase of voltage, the resonance peak becomes deep, but the resonance peak changes to a critical

978-1-6654-8699-6/22 $31.00 © 2022 IEEE

voltage of 0.2eV, and the amplitude at the resonance point starts to reverse upward. Its physical image analysis is shown in section C.

From the simulation results of phase in Fig. 2, we can also see that: in particular, at a frequency of 0.33 THz, the phase difference at 0.1eV and 0.4eV can reach 180 degrees.

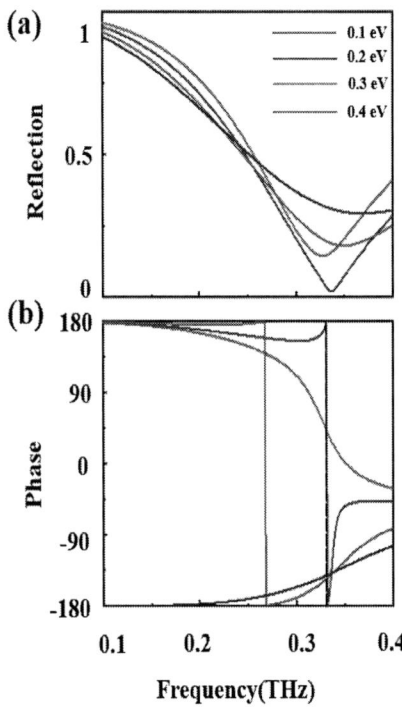

Figure 2. Voltage-modulated reflection amplitude and phase.

The top row of the graph in Fig. 2 corresponds to the modulation of the reflection amplitude, and the bottom row corresponds to the modulation of the reflection phase. Fig. 2(a) can be seen that the reflection decreases when the voltage increases from 0.1eV to 0.2eV; however, it increases again when the voltage increases from 0.2 eV to 0.4 eV. So we can modulate the amplitude of the terahertz wavefront by changing the V_A. The frequency and voltage corresponding to the purple line with the smallest reflection in the figure is the resonant frequency at which the critical damping occurs. The phase undergoes a discontinuous change at this point. From the phase distribution in Fig. 2(b), we can find that at the resonant frequency of about 0.33 THz, the phase of the red line representing the applied gate voltage of 0.1 eV is about 45°, while the phase of the blue line representing the applied gate voltage of 0.4 eV is about 135°. The phase difference between the two reaches 180 degrees, which means that we can achieve 1-bit modulation of the phase by changing the applied gate voltage.

C. CMT Analyses

To understand the physics underlying our simulated results, we employ a CMT model to analyze the EM properties of such a tunable structure. The main mechanism of critical phenomena induced by the applied gate voltage in the graphene metasurface system can be formed by the critical damping phenomenon specific to the single-mode single-port resonance with the coupled-mode-theory resonance frequency located at f_0. In our design, the metal backplate blocks the transmission of electromagnetic waves, eliminating the transmission channel of the system and making the system possess only the reflection channel, which is also known as the single port. The loss of the system can be divided into two parts: absorption loss and radiation loss. The so-called absorption loss, which we label as Γ_i, and the radiation loss, which we label as Γ_r. The absorption loss is caused by the dissipation of the material endowment and the energy is consumed in the form of thermal dissipation. The radiation loss, on the other hand, can be understood as the efficiency of the coupling between the resonant and scattering modes. According to the coupled mode theory, the reflection coefficient r of the system can be expressed as

$$r = -1 + \frac{2\Gamma_r}{-i(f - f_0) + \Gamma_r + \Gamma_i} \qquad (1)$$

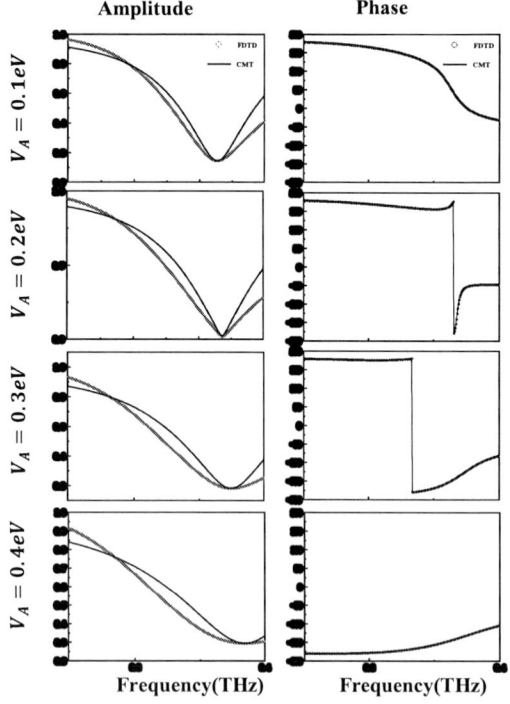

Figure 3. The fitting spectral lines of amplitudes and phase when different voltages are applied.

TABLE I. CMT PARAMETERS

V_A(eV)	f_0 (THz)	Γ_r	Γ_i
0.1	0.327	0.06	0.0447
0.2	0.336	0.06	0.0576
0.3	0.349	0.085	0.0588
0.4	0.369	0.12	0.0653

We performed the FDTD simulations of meta-atoms and fitted the simulation results with the theoretical results obtained by CMT. Fig. 3 depicts the fitting spectral lines of intensity and phase when different voltages are applied. Table I shows the fitting results of some parameters. It can be seen that the CMT model can fit the simulation results very well. With the change of the applied gate voltage, the response of the unit structure to electromagnetic waves also changes significantly.

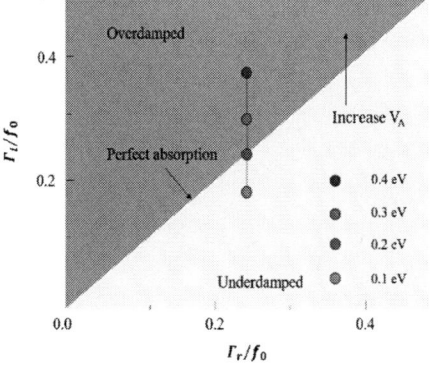

Figure 4. Phase diagram

In the following we understand the above results by means of a phase diagram, as in Fig. 4. We know that increasing the carrier concentration in graphene increases almost only Γ_i, while Γ_r remains almost constant. When the applied voltage is small, at 0.1 eV, the carrier concentration in graphene is low, corresponding to the case of $\Gamma_r > \Gamma_i$ in the coupled mode theory, and the upper surface of the metasurface is relatively highly permeable, and the terahertz wave has penetrated into the resonant cavity of the structure before being absorbed, and the coupling of the near field causes the formation of a reverse current in the resonant cavity, resulting in magnetic resonance. However, as the voltage increases, at 0.3 eV and 0.4 eV, the carrier concentration in graphene becomes very high, corresponding to the case of $\Gamma_r < \Gamma_i$ in the coupled-mode theory, and the strong absorption prevents the terahertz wave from penetrating deeply into the resonant cavity and forming a current loop in the resonant cavity, when the system will exhibit an electrical resonance-like behavior with very limited phase changes on both sides of the resonant frequency. When the voltage is equal to 0.2 eV in the critical case, $\Gamma_r = \Gamma_i$.

When $\Gamma_r > \Gamma_i$, we call the system underdamped; when $\Gamma_r = \Gamma_i$, the system is critically damped, when the system

absorption is maximum, also known as perfect absorption; when $\Gamma_r < \Gamma_i$, the system is overdamped. In fact, graphene is almost mainly affecting the absorption loss coefficient of the system, while it has almost no effect on the radiation loss coefficient of the system. The higher the applied voltage, the higher the carrier concentration in graphene and the higher the absorption loss coefficient of the system.

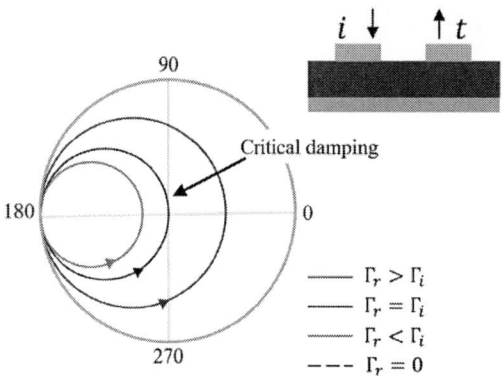

Figure 5. Reflection Smith curve for single-mode single-port model.

Fig. 5 illustrates the reflected Smith curve for the single-mode, single-port model [10]. The variation of r with f can be precisely represented in the Smith diagram, that is, the trace of r in the complex plane as the frequency f is increased from 0 to ∞. The distance from the center of the circle to the point on the curve is the magnitude of r, and the angle between the line and the positive semi-axis of the real axis is the phase of r. All curve approximations start and end at $r = -1$. However, they cross the real axis (when in resonance, i.e., $f = f_0$) at locations determined by r, where $r = (\Gamma_r - \Gamma_i)/(\Gamma_r + \Gamma_i)$; the sign of $(\Gamma_r - \Gamma_i)$ dictates whether the curves enclose the origin. Thus the relationship between Γ_r and Γ_i determines the behavior of the phase and amplitude when the frequency f crosses the resonant frequency. When $\Gamma_r < \Gamma_i$, the Smith curve does not enclose the origin; therefore, the resulting phase variation is less than $180°$ across the resonance (Fig. 5, red curve). However, a behavioral transition occurs at $\Gamma_r > \Gamma_i$, and the Smith curve encloses the origin, causing the reflection phase to undergo a full $360°$ variation (Fig. 5, blue curve). A transition from an underdamped to an overdamped oscillator occurs at the critical damping ($\Gamma_r = \Gamma_i$) as Γ_i is increased. At this point, since the Smith curve crosses the origin, and the reflection corresponding to the resonant frequency is located at the origin. So at this point, the phase of its reflection is undefined. This makes it understandable why the phase changes so dramatically near the critical damping region. If we look at the Smith curve, it is really nothing more than the behavior of the resonance point across the origin. It is only the singular nature of the origin that makes the phase change abruptly. Such a transition fully explains the critical transition that we observe in our metasurfaces.

978-1-6654-8699-6/22 $31.00 © 2022 IEEE

III. CONCLUSION

In summary, we have proposed a tunable graphene grating to achieve dynamical terahertz wavefront control and experimentally demonstrate them in the THz regime (0.1 THz ~ 0.4 THz). We employ the CMT analysis to reveal that the gated graphene as tunable loss can drive the whole structure to vary the magnitude and phase of terahertz wavefront, particularly, the phase modulation is capable of reaching 180°.

REFERENCES

[1] N. Yu, and F. Capasso, "Flat optics with designer metasurfaces," *Nat. Mater.* 13(2), 139-150 (2014).

[2] S. Sun, Q. He, J. Hao, S. Xiao, and L. Zhou., "Electromagnetic metasurfaces: physics and applications," *Adv. Opt. Photonics* 11(2), 380-479 (2019).

[3] S. Xiao, J. Wang, F. Liu, S. Zhang, X. Yin, J. Li, "Spin-dependent optics with metasurfaces," *Nanophotonics* 6(1), 215-234 (2017).

[4] N. Yu, P. Genevet, M. A. Kats, F. Aieta, J. P. Tetienne, F. Capasso, and Z. Gaburro, "Light propagation with phase discontinuities: generalized laws of reflection and refraction," *Science* 334(6054), 333-337 (2011).

[5] S. Sun, Q. He, S. Xiao, Q. Xu, X. Li, and L. Zhou, "Gradient-index meta-surfaces as a bridge linking propagating waves and surface waves," *Nat. Mater.* 11(5), 426-431 (2012).

[6] Z. Li, J. Hao, L. Huang, H. Li, H. Xu, Y. Sun, and N. Dai, "Manipulating the wavefront of light by plasmonic metasurfaces operating in hider order modes," *Opt. Express* 24(8), 8788-8796 (2016).

[7] B. Yang, T. Liu, H. Guo, S. Xiao, and L. Zhou, "High-performance meta-devices based on multilayer meta-atoms: interplay between the number of layers and phase coverage," *Sci. Bull.* 64(12), 823-835 (2019).

[8] Y. Luo, C. H. Chu, S. Vyas, H. Y. Kuo, Y. H. Chia, M. K. Chen, X. Shi, T. Tanaka, H. Misawa, Y. Y. Huang, and D. P. Tsai, "Varifocal metalens for optical sectionig fluorescence microscopy," *Nano Lett.* 21(12), 5133-5142 (2021).

[9] X. Cai, R. Tang, H. Zhou, Q. Li, S. Ma, D. Wang, T. Liu, X. Ling, W. Tan, Q. He, S. Xiao, and L. Zhou, "Dynamically controlling terahertz wavefronts with cascaded metasurfaces," *Adv. Photon.* 3(3), 036003 (2021).

[10] Z. Miao, Q. Wu, X. Li, Q. He, K. Ding, Z. An, Y. Zhang, and L. Zhou, "Widely Tunable Terahertz Phase Modulation with Gate-Controlled Graphene Metasurfaces," *Physical Review X.* 5(4), 041027 (2015).

Switchable Multifunctional Metasurfaces Based on Vanadium Dioxide in the Terahertz Region

Shilei Liu
College of Precision Instrument and Optoelectronics
Engineering, Tianjin University
Tianjin, China
e-mail: slliu@tju.edu.cn

JiaJun Ma
College of Precision Instrument and Optoelectronics
Engineering, Tianjin University
Tianjin, China
e-mail: majiajun@tju.edu.cn

Hongyi Li
College of Precision Instrument and Optoelectronics
Engineering, Tianjin University
Tianjin, China
e-mail: lihongyi@tju.edu.cn

Yi Liu
College of Precision Instrument and Optoelectronics
Engineering, Tianjin University
Tianjin, China
e-mail: lyi@tju.edu.cn

Chunmei Ouyang
College of Precision Instrument and Optoelectronics
Engineering, Tianjin University
Tianjin, China
e-mail: cmouyang@tju.edu.cn

Abstract—**Terahertz metasurfaces attract more and more attention in the recent years due to their unique properties and promising application prospects. Here we propose an active multi-functional metasurface at terahertz frequencies, whose functionalities can switch between the highly efficient absorption and broadband polarization conversion by changing the ambient temperature, based on the phase-transition effect of vanadium dioxide (VO_2). The designed metasurface is polarization independent, and the simple design is easy to fabricate in experiment, which is expected to enrich the terahertz modulators and functional devices.**

Keywords-Terahertz; Active modulation; Vanadium Dioxide; Polarization conversion; Absorption

I. INTRODUCTION

The frequency range of terahertz (THz) refers to the 0.1 - 10 THz located between the microwave and infrared ranges. THz science and technology shows greatly potential application prospects in 6G communications[1], biomedical science[2, 3], security check[4] and THz imaging for its specific characteristics such as high transmittivity, high coherence, low energy and unique fingerprint spectrum, etc. However, lack of compact THz sources, highly sensitive detectors and effective modulators constrains its larger scale practical applications. The emergence of artificial subwavelength metasurfaces promotes the rapid development of functional devices and modulators in the THz range.

The concept of metamaterials can be traced back to the left-handed materials proposed by Veselago in 1986. Metamaterials show potential application prospects in

invisibility cloak[5], holographic imaging, meta-lens and so on. The applications of metamaterials were first realized in the microwave range[6], near infrared range and even optical range[7], and their operation ranges were soon extended to the THz region, which opens the way to be much easier to integrate with other optical equipments. As planar metamaterials, metasurfaces have the characteristics of lower fabrication costs and negligible subwavelength thickness, providing more choices in structure design and samples fabrication. Various THz metasurfaces and multifunctional devices[8-10] have been proposed to satisfy specific practical application requirements and conditions. Until now, the reported realized functionalities include plasma-induced transparency (PIT)[11, 12], polarization conversion[13], beam deflection and absorption[14], etc. THz metasurfaces can be divided into reflection-type, transmission-type, active manipulation and passive transmission. Integrating with the information science theory, various kinds of coding metasurfaces[15], digital metasurfaces[16], programmable metasurfaces[17-19] and information metasurfaces[20] were proposed more recently, which is expected to ease the burden of scientific researchers in structure design and simulation calculations.

The functionalities of conventional passive metasurfaces are unchangeable once their structures are confirmed and/or fabricated. Therefore, active metasurfaces are proposed, which become a research hotspot for their flexibility and convenience. The active manipulation ways include electrically control[21], thermally control and optical control[22]. Integrating with the phase transition materials (VO_2[23], Graphene[24] and GST, etc) is a widely investigated method to realize active manipulation. Among them, VO₂ is a popular option due to its relatively lower phase transition temperature (68°C). For THz wave, VO$_2$ at room temperature is almost transparent with a transmission coefficient of 0.9995. As the ambient temperature where the sample is placed in rises beyond the phase transition temperature, the transmission coefficient of VO$_2$ would attenuate to 0.2135, functioning as a metal with ultra-high conductivity (2.6×10^6 S/m) through the phase transition effect. The characteristic of phase transition provides possibility to realize active manipulation on THz metasurfaces with much smaller cost in detection and sample fabrication.

Here we propose an active THz multifunctional device based on VO$_2$, and this active metasurface consists of the top constructional layer, the dielectric layer and the bottom reflection layer, which can be equivalent to a metal-insulator-metal (MIM) cavity. The unit cell of the top constructional layer is composed of a 200-nm-thick VO$_2$ square ring and three 200-nm-thick gold oblique cut-wires. The function of the active metasurface can be switched between broadband polarization conversion and effective absorption by changing the ambient temperature. The absorptivity of the active metasurface reaches beyond 98% and the bandwidth of polarization conversion ratio (PCR) reaches 0.37 THz (0.91 - 1.28 THz).

II. MATERIALS AND DESIGN

A schematic illustration of the proposed active metasurface is shown in Fig. 1(a). The metal elements in the top and bottom layers are made of gold (4.56×10^7 S/m) with plasmon frequency of 1.37×10^{16} rad/s, and the thicknesses of these two gold layers are both $t_1 = 200$ nm. The thickness of the dielectric layer sandwiched between the metallic layers is $t_2 = 25$ μm, and its material is polyimide (PI) with a relative permittivity $\varepsilon = 3.23$ and loss tangent $\delta = 0.044$. The commercial numerical full-wave simulation software CST Microwave Studio is employed to characterize, design and optimize the active metasurface. Utilizing the variation of the conductivity of VO$_2$ set in the simulation to simulate the phase transition process in the experiment. The unit cell boundary conditions along the x and y directions and open boundary condition along the z direction are set in the simulation. We employ the Floquet port to set the distance between the metasurface and the incident port to greater more than two wavelengths to simulate the far-field detection. We simulate the phase transition effect in experiments by changing the conductivity of VO$_2$ set in simulations. As shown in Fig. 1, the detailed geometrical parameters are choosen as follows: the period of the unit-cell is $p = 100$ μm, outer diameter and width of the square ring are $l = 90$ μm and 5 μm, respectively. Side lengths of the cut-wires are $a = 40$ μm, $b = 10$ μm, $c = 80$ μm and $d = 15$ μm, and the distance between the cut-wires are both $w = 15.5$ μm approximately. To simulate the active metasurfaces fabricated in experiments, a dielectric substrate layer made of sapphire with a relative permittivity of 9.67 is added above the top constructional layer. The fabrication of the proposed active metasurface is easy to realize due to its simple structure design and high tolerance of geometric parameters errors.

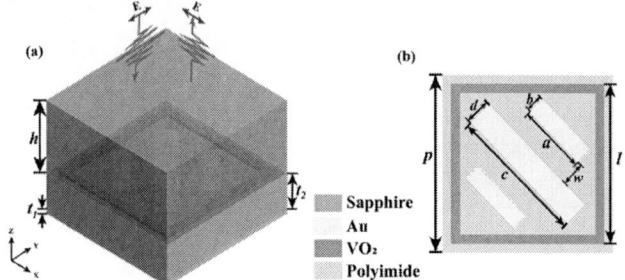

Figure 1. (a) Illustration of the proposed active metasurface. (b) Vertical view of the constructional layer and the detail geometric parameters.

Absorption and polarization conversion are two important applications of metasurfaces. Integrating these two functionalities in one single metasurface is a research hotspot, and can enhance the flexibility of metasurfaces. Here in this paper, the absorption efficiency is derived from:

$$A_x = 1 - |R_{xx}|^2 - |R_{yx}|^2 - |T_{xx}|^2 - |T_{yx}|^2, \quad (1)$$

$$A_y = 1 - |R_{yy}|^2 - |R_{xy}|^2 - |T_{yy}|^2 - |T_{xy}|^2, \quad (2)$$

Since the thickness of the bottom reflective layer is greater than the skin depth of the incident electromagnetic wave, therefore the formula of the absorptivity could be simplified to:

$$A_x = 1 - |R_{xx}|^2 - |R_{yx}|^2, \quad (3)$$

$$A_y = 1 - |R_{yy}|^2 - |R_{xy}|^2, \quad (4)$$

The polarization conversion ratio (PCR) is demonstrated as follow:

$$PCR_x = \frac{|R_x|^2}{|I_y|^2} \text{ and } PCR_y = \frac{|R_y|^2}{|I_x|^2}. \quad (5)$$

where the R_{yx}/T_{yx} describes the reflection or transmission amplitude of the received y-polarized THz wave under the x-polarized normal incidence, and R_x/I_x represents the reflection or incidence amplitude of the x-polarized THz wave.

As stated in the Introduction section, when VO$_2$ stays at metallic state, its transmission coefficient is 0.2135, while possessing a ultra-high conductivity of 2.6×10^5 S/m. Here the VO$_2$ could be equivalent to a metallic material when it undergoes a phase transition to the metallic state. The equivalent structure is shown in Fig. 2(a). The absorption efficiency of the active metasurface under high ambient temperature is given in Fig. 2(b), showing that it can act as a highly efficient absorber with an absorptivity of more than 98% at 0.43 THz.

978-1-6654-8699-6/22 $31.00 ©2022 IEEE

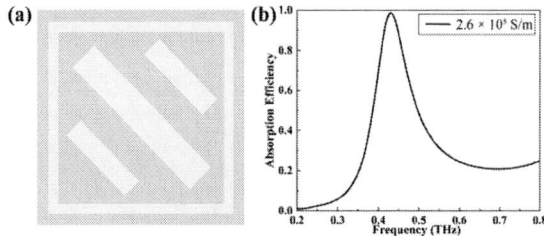

Figure 2. (a) The diagram of the equivalent structure of the active metasurface when VO$_2$ stays at fully metallic state with a conductivity of 2.6×10^5 S/m. (b) The absorption efficiency of the designed active metasurface.

When VO$_2$ stays at dielectric state where its transmission coefficient soars to 0.9995 with a relative permittivity of $\varepsilon = 2.93$ and a conductivity of 10 S/m, here the VO$_2$ could be ignored at THz frequencies. Thus the proposed active metasurface has an equivalent structure as shown in Fig. 3(a), functioning as a broadband polarization converter (see Fig. 3b). The peak of the PCR reaches 76% at 1.0 THz and the bandwidth of the PCR with a more than 60% efficiency reaches 0.37 THz (0.91 – 1.28 THz). It is also worthy noting that the active metasurface is isotropic for the normal incident linear polarized THz waves, that is, the metasurface shows the same effects for x/y-polarized THz waves.

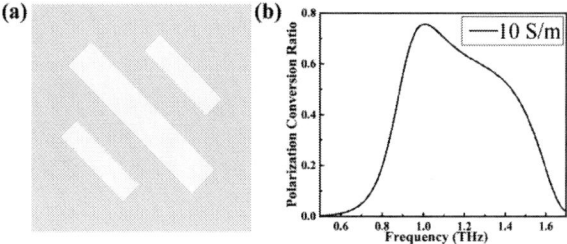

Figure 3. (a) The diagram of the equivalent structure of the active metasurface when VO$_2$ stays at dielectric state with a conductivity of 10 S/m. (b) The polarization conversion ratio of the active metasurface.

III. CONCLUSION

Here we proposed an active multifunctional metasurface at THz frequencies, which can switch between the highly efficient absorption and broadband polarization conversion by changing the ambient temperature. The peak of absorptivity reaches above 98% at 0.43 THz, the peak of PCR reaches 76% at 1.0 THz, and the bandwidth with an efficiency above 60% can reach 0.37 THz. The structure of the active metasurface is polarization-independent, and the design of the structure adopts MIM cavity-like structure, which is easy to fabricate in experiments. The scheme provides an efficient way to designing active THz modulators and functional devices.

ACKNOWLEDGMENT

This work was supported by the National Natural Science Foundation of China (Grant Nos. 62175180, 61875150, 61805129, and 62005193).

REFERENCES

[1] Z. Chen *et al.*, "A survey on terahertz communications," *Communications, China,* 2019, doi: 10.12676/j.cc.2019.02.001.

[2] J. Zhang *et al.*, "Highly sensitive detection of malignant glioma cells using metamaterial-inspired THz biosensor based on electromagnetically induced transparency," *Biosens Bioelectron,* vol. 185, p. 113241, Aug 1 2021, doi: 10.1016/j.bios.2021.113241.

[3] S. Lin *et al.*, "Using Antibody Modified Terahertz Metamaterial Biosensor to Detect Concentration of Carcinoembryonic Antigen," *IEEE Journal of Selected Topics in Quantum Electronics,* vol. 27, no. 4, pp. 1-7, 2021, doi: 10.1109/jstqe.2020.3038308.

[4] E. Grossman *et al.*, "Passive terahertz camera for standoff security screening," *Applied optics,* vol. 49, no. 19, pp. E106-E120, 2010.

[5] B. Orazbayev, N. Mohammadi Estakhri, A. Alù, and M. Beruete, "Experimental Demonstration of Metasurface-Based Ultrathin Carpet Cloaks for Millimeter Waves," *Advanced Optical Materials,* vol. 5, no. 1, 2017, doi: 10.1002/adom.201600606.

[6] S. Yu, L. Li, G. Shi, C. Zhu, and Y. Shi, "Generating multiple orbital angular momentum vortex beams using a metasurface in radio frequency domain," *Applied Physics Letters,* vol. 108, no. 24, 2016, doi: 10.1063/1.4953786.

[7] Y. Cheng, F. Chen, and H. Luo, "Plasmonic chiral metasurface absorber based on bilayer fourfold twisted semicircle nanostructure at optical frequency," *Nanoscale Research Letters,* vol. 16, no. 1, pp. 1-9, 2021.

[8] Y. Yang, W. Wang, P. Moitra, Kravchenko, II, D. P. Briggs, and J. Valentine, "Dielectric meta-reflectarray for broadband linear polarization conversion and optical vortex generation," *Nano Lett,* vol. 14, no. 3, pp. 1394-9, Mar 12 2014, doi: 10.1021/nl4044482.

[9] F. Ding, S. Zhong, and S. I. Bozhevolnyi, "Vanadium Dioxide Integrated Metasurfaces with Switchable Functionalities at Terahertz Frequencies," *Advanced Optical Materials,* vol. 6, no. 9, 2018, doi: 10.1002/adom.201701204.

[10] Y. Li, J. Lin, H. Guo, W. Sun, S. Xiao, and L. Zhou, "A Tunable Metasurface with Switchable Functionalities: From Perfect Transparency to Perfect Absorption," *Advanced Optical Materials,* vol. 8, no. 6, 2020, doi: 10.1002/adom.201901548.

[11] M. Liu *et al.*, "Tailoring the plasmon-induced transparency resonances in terahertz metamaterials," *Opt Express,* vol. 25, no. 17, pp. 19844-19855, Aug 21 2017, doi: 10.1364/OE.25.019844.

[12] S. Xiao, T. Wang, T. Liu, X. Yan, Z. Li, and C. Xu, "Active modulation of electromagnetically induced transparency analogue in terahertz hybrid metal-graphene metamaterials," *Carbon,* vol. 126, pp. 271-278, 2018, doi: 10.1016/j.carbon.2017.10.035.

[13] J. Fan and Y. Cheng, "Broadband high-efficiency cross-polarization conversion and multi-functional wavefront manipulation based on chiral structure metasurface for terahertz wave," *Journal of Physics D: Applied Physics,* vol. 53, no. 2, 2020, doi: 10.1088/1361-6463/ab4d76.

[14] Z. Song and J. Zhang, "Achieving broadband absorption and polarization conversion with a vanadium dioxide metasurface in the same terahertz frequencies," *Opt Express,* vol. 28, no. 8, pp. 12487-12497, Apr 13 2020, doi: 10.1364/OE.391066.

[15] S. Liu *et al.*, "Anisotropic coding metamaterials and their powerful manipulation of differently polarized terahertz waves," *Light Sci Appl,* vol. 5, no. 5, p. e16076, May 2016, doi: 10.1038/lsa.2016.76.

[16] W.-M. Pan and J.-S. Li, "Anisotropic digital metasurfaces relying on silicon Mie resonance," *Optics Communications,* vol. 493, 2021, doi: 10.1016/j.optcom.2021.127033.

[17] T. J. Cui, M. Q. Qi, X. Wan, J. Zhao, and Q. Cheng, "Coding metamaterials, digital metamaterials and programmable metamaterials," *Light: Science & Applications,* vol. 3, no. 10, pp. e218-e218, 2014, doi: 10.1038/lsa.2014.99.

[18] S. Liu and T. J. Cui, "Concepts, Working Principles, and Applications of Coding and Programmable Metamaterials," *Advanced Optical Materials,* vol. 5, no. 22, 2017, doi: 10.1002/adom.201700624.

[19] B. Ren, Y. Feng, S. Tang, L. Wang, H. Jiang, and Y. Jiang, "Dynamic control of THz polarization modulation and multi-channel beam generation using a programmable metasurface," *Opt Express,* vol. 29, no. 11, pp. 17258-17268, May 24 2021, doi: 10.1364/OE.426645.

[20] T. J. Cui, S. Liu, and L. Zhang, "Information metamaterials and metasurfaces," *Journal of Materials Chemistry C,* vol. 5, no. 15, pp. 3644-3668, 2017, doi: 10.1039/c7tc00548b.

[21] J. Zhang, X. Wei, I. D. Rukhlenko, H.-T. Chen, and W. Zhu, "Electrically Tunable Metasurface with Independent Frequency and Amplitude Modulations," *ACS Photonics,* vol. 7, no. 1, pp. 265-271, 2019, doi: 10.1021/acsphotonics.9b01532.

[22] L. Cong, Y. K. Srivastava, H. Zhang, X. Zhang, J. Han, and R. Singh, "All-optical active THz metasurfaces for ultrafast polarization switching and dynamic beam splitting," *Light Sci Appl,* vol. 7, p. 28, 2018, doi: 10.1038/s41377-018-0024-y.

[23] J. Zhao *et al.*, "Temperature-controlled terahertz polarization conversion bandwidth," *Opt Express,* vol. 29, no. 14, pp. 21738-21748, Jul 5 2021, doi: 10.1364/OE.431622.

[24] W. Yao *et al.*, "Electrically tunable graphene metamaterial with strong broadband absorption," *Nanotechnology,* vol. 32, no. 7, p. 075703, Feb 12 2021, doi: 10.1088/1361-6528/abc44f.

978-1-6654-8699-6/22 $31.00 © 2022 IEEE

Proceedings of the 7th Optoelectronics Global Conference (OGC 2022)

Numerical Simulation of C+L Broadband Single-mode Fiber

G. H. Zhang, W. Sun, F. Lei,
R & D department
Jiangsu Hengtong Optic-electric Co., Ltd.
Suzhou, China
e-mail: zhanggh@htgd.com.cn

W. Chen,
Key Laboratory of Specialty Fiber Optics and Optical
Access Networks
Shanghai, China
e-mail: chenweiSD@shu.edu.cn

L. Wang, Y. L. Wang, Y. T. Li
Key Laboratory of Novel Specialty Fiber Optics and
Optical Fiber Preform
Suzhou, China
e-mail: wanglin@jsalpha.com

H. F. Guan, J. C. Yuan, Z. Jiang, Q. Y. Liu
R & D department
Jiangsu Alpha Optic-Electric Technology Co.,Ltd.
Suzhou, China
e-mail: guanhf@jsalpha.com

Abstract—A novelty C+L single-mode fiber (CL fiber for short) which can expand L-band communication is proposed to meet the urgent demand of optical fiber transmission bandwidth for large capacity communication system. The influence of L-band attenuation on the optical signal-to-noise ratio (OSNR) of the system is analyzed and calculated; through the simulation of dense wavelength division multiplexing (DWDM) system, the requirements for fiber attenuation warping degree (FAWD) of CL fiber in 100Gbit/s rate DWDM system are obtained; and the CL fiber and its main attenuation parameters are defined. The results show that, the OSNR of L-band for conventional single-mode fiber will be 2dB-2.5dB worse than traditional C-band on 100km span system, so it is very important to limit the FAWD of L-band; if the L-band transmission can meet the link requirements in relevant standards, it is recommended that the FAWD $\triangle\alpha_{1625}$ of CL fiber is suitable to be controlled below 0.015dB/km, and the fiber attenuation warping degree difference (FAWDD) $\triangle\alpha$ is suitable to be controlled below 0.005dB/km.

Keywords-CL fiber; C+L band; L-band; Attenuation flatness; Fiber attenuation warping degree; C+L wavelength division multiplexing; Fiber bandwidth; DWDM

I. INTRODUCTION

With the gradual maturity, popularization and application of 5G communication technology, Internet users' requirements for network access bandwidth are further improved. According to Cisco estimates, the global fixed access rate and mobile access rate will maintain a compound annual growth rate of 20% and 27% respectively from 2018 to 2023 [1]. The demand of data flow and information consumption for network transmission capacity is increasing rapidly, which makes the existing dense wavelength division multiplexing (DWDM) system show capacity bottleneck.

To continue to improve the transmission capacity of DWDM system, we can only increase the number of optical fiber wavelength division multiplexing channels, improve the transmission rate or spectral efficiency of a single channel, and expand the number of fiber cores in a single fiber [2]. Due to the limitations of optical fiber material properties and optical equipment performance, it is

increasingly difficult to simply improve the transmission rate or spectral efficiency of a single channel, and it has approached the Shannon limit; however, the multi-core optical fiber with expanded core number is not mature in terms of manufacturing level and coupling devices. Therefore, increasing the number of optical fiber wavelength division multiplexing channels has become an important and feasible method to greatly expand the transmission capacity of DWDM system. Restricted by the dispersion performance of the optical fiber and the communication light source, it is extremely limited to increase the number of channels by reducing the channel spacing [3], so we have to consider widening the effective band width of the optical fiber to achieve this purpose.

This paper mainly takes single-mode fiber as the main research object, focuses on the problems faced by fiber transmission in L-band, puts forward a novelty CL fiber to expand L-band transmission, determines the feasibility of its application and the requirements for attenuation flatness through system simulation.

II. L-BAND TRANSMISSION CHARACTERISTICS OF SMF

A. Attenuation Difference in the C+L Band

By analyzing the spectral loss (wavelength-attenuation) curve of SMF, it can be seen that its attenuation trend in C-band (1530nm-1565nm) and L-band (1565nm-1625nm) shows an approximate U-shaped state. The attenuation slope of L-band rises rapidly with the increase of wavelength, and the increase of attenuation is greater than the decrease of C-band. This is determined by the infrared absorption and design parameters of optical fiber glass materials, which actually leads to the degradation of L-band transmission performance.

In order to broaden the L-band bandwidth of the existing single-mode optical fiber and realize ultra-wideband (UWB) communication, the optical fiber needs to have the same attenuation level in the L-band as in the C-band while maintaining the low loss of the C-band, and a novelty C+L-band attenuation flat optical fiber (CL fiber) is explored. In order to facilitate large-scale use and replacement, the more

978-1-6654-8699-6/22 $31.00 © 2022 IEEE

appropriate choice is that the fiber still belongs to class G.652 SMF. The spectral loss curves of typical SMF and CL fiber are shown in Fig. 1. It can be seen that it has a relatively gentle attenuation increase in the L-band.

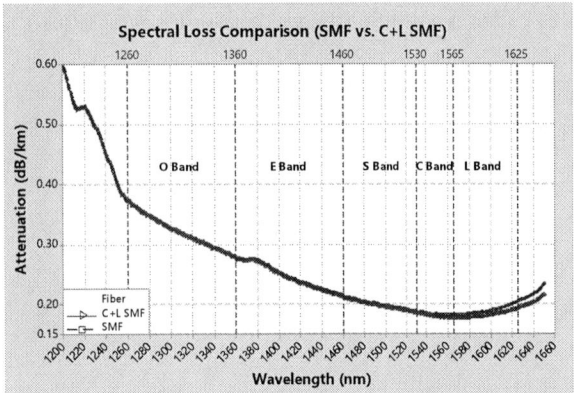

Figure 1. Comparison of spectral loss curves of SMF and CL fiber.

B. L-Band Transmission Performance

According to Shannon's theorem, in the optical transmission channel with random noise, the channel capacity is restricted by OSNR. The higher the OSNR, the greater the transmission capacity of the channel.

$$C = B\log_2(1 + S/N) \tag{1}$$

Where, C is the capacity of the channel, B is the channel bandwidth, and S/N is the signal-to-noise ratio. The relevant ITU standard — ITU-T G.692 [5] gives the calculation equation of OSNR:

$$OSNR = P_{out} - L - NF - 10\log N - 10\log(hv\Delta v_0) \tag{2}$$

Where, P_{out} is the channel output optical power, dBm; L is the span loss between amplifiers, dB; N is the number of spans in the link; NF is the light amplification noise figure; h is Planck constant, (mJ·s); v is the optical transmission frequency; v_0 is the reference bandwidth. For general DWDM system with equal optical amplification section, when the optical band width at 1550nm is 0.1nm, OSNR calculation equation [5] can be simplified as:

$$OSNR = 58 + P_{ch} - L - NF - \log N \tag{3}$$

Where, P_{ch} is the channel optical power (fiber input power), which is related to the optical end injection power and fiber design parameters. Span loss L is positively correlated with fiber attenuation coefficient α.

Assuming that the channel optical power P_{ch}, optical amplification noise factor NF and optical amplification span number N remain unchanged, the effect of increasing L-band attenuation on OSNR is:

$$OSNR_{L-C} = L_L - L_C = (\alpha_C - \alpha_L) \cdot S = \Delta\alpha_{C-L} \cdot S \tag{4}$$

Where, $OSNR_{L-C}$ represents the signal-to-noise ratio difference between L-band and C-band, L_L and L_C are the link loss of L-band and C-band respectively, α_L and α_C is the optical fiber attenuation coefficient of L-band and C-band respectively, S is the span length, $\Delta\alpha_{C-L}$ is the difference of attenuation coefficient between C-band and L-band.

For conventional SMF, compared with the traditional C-band, the attenuation slope of L-band is steeper, which makes the attenuation at 1625nm greater than that at 1520nm. Assuming that the attenuation difference is 0.01dB/km and the system span is 100km, the $OSNR_{L-C}$ can be calculated from (5):

$$OSNR_{L-C} = \Delta\alpha_{C-L} \cdot S = -0.01 \cdot 100 = -1.0 \text{(dB)} \tag{5}$$

From (5), it can be seen that, due to the difference in attenuation, the OSNR of single L-band may be about 1dB worse than that of traditional C-band; considering the additional insertion loss introduced by the C/L combiner and splitter in the C+L system, the OSNR in the C+L system will eventually deteriorate by 2dB-2.5dB.

III. SIMULATION ANALYSIS OF THE CL FIBER

A. Definition of the CL Fiber

In the wavelength range of C+L, the curve of fiber attenuation coefficient changing with wavelength presents a U-shaped state, and there is a minimum attenuation value and its corresponding minimum attenuation wavelength. To describe the difference of attenuation trend between C-band and L-band, we can evaluate it from the attenuation difference of U-shaped curve on both sides of the minimum attenuation wavelength. Therefore, the optical fiber attenuation warping parameter is added in this paper: within the specified wavelength range, the difference between the attenuation coefficient at a specific wavelength and the minimum attenuation coefficient within the wavelength range when the optical fiber attenuation coefficient presents a U-shaped state with the wavelength, is called fiber attenuation coefficient warping degree (FAWD for short); the absolute value of the arithmetic difference of the FAWD at two different specific wavelengths is called the fiber attenuation warping degree difference (FAWDD for short). Obviously, the smaller FAWD in the C+L band, the better the attenuation flatness; it means that the OSNR of L-band transmission will be close to the level of C-band, so it can be used for the transmission of existing DWDM systems with more than 80 waves like C-band. Therefore, CL fiber can be defined as a single-mode fiber with optimized attenuation performance in C-band and L-band, which can be used to support DWDM systems above 80 waves.

In order to obtain the minimum attenuation value required for calculating the FAWD, it is necessary to test the spectral loss of each optical fiber, and the test scanning time is relatively long, which seriously affects the efficiency of batch obtaining CL fibers. In order to simplify the calculation, it is suggested to define the wavelength at the minimum attenuation of SMF from the statistical point of view and fix it as the minimum attenuation wavelength of the same kind of fiber. For example, the batch statistics of 80 measured CL fiber samples in this paper show that the lowest attenuation wavelength is not in the traditional 1550nm transmission window, but mainly distributed in 1570nm-1580nm, and the average value of the minimum attenuation wavelength is 1576nm, as shown in Fig. 2.

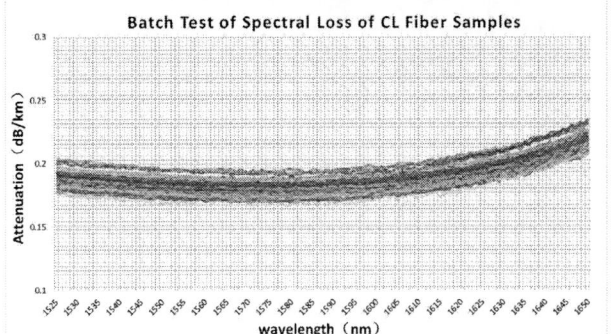

Figure 2. Spectrum loss curve of batch statistical CL fibers.

For this kind of CL fiber, 1576nm can be taken as its minimum attenuation wavelength, and FAWD and FAWDD can be calculated by testing the corresponding window value. Of course, for CL fibers manufactured by different manufacturers, their materials and structures are slightly different, and the minimum attenuation wavelength may not be 1576nm, but this does not affect the judgment of the minimum attenuation wavelength, and this statistical method is still feasible.

B. System Simulation Analysis

1) Simulations based on the existing EDFA

In the simulation test, two types of EDFA are used for C-band and L-band respectively, and the overall wavelength range is 1525nm-1610nm. Both EDFA can cover all 104 optical channels of C-band (namely, noise figure optimization and flatness optimization); both EDFA can only cover 94 optical channels of short wavelength of the L-band, and one of them suppresses the gain bulge at 1576nm, but also depresses the gain at 1570nm (bulge suppression). The EDFA parameters are shown in Fig. 3.

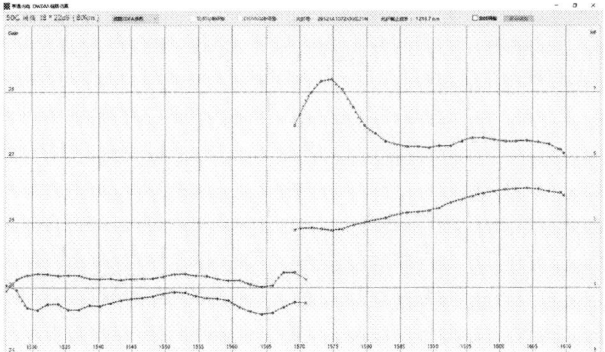

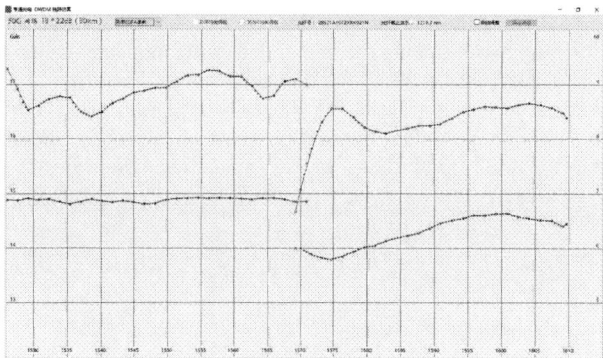

Figure 3. Power parameters of the existing EDFA.

The minimum attenuation wavelength is 1576nm and FAWD at 1610nm ($\triangle \alpha_{1610}$) is 0.0087dB/km of the fiber selected for optical simulation.

Simulation shows that only the C-band noise figure optimized EDFA can meet the requirements of the industry standard YD/T 2485-2013 (2017) [6] after realizing the 18× 22dB optical amplification section. The 22×22dB span system built with this EDFA can barely meet the requirements of the industry standard YD/T 3070-2016 (2017) [7] (there is no 22 span in the link standard, refer to the 20 span type), but cannot build 26×22dB, 12×27dB, 10×27dB and other span systems that meet the requirements, as shown in Fig. 4.

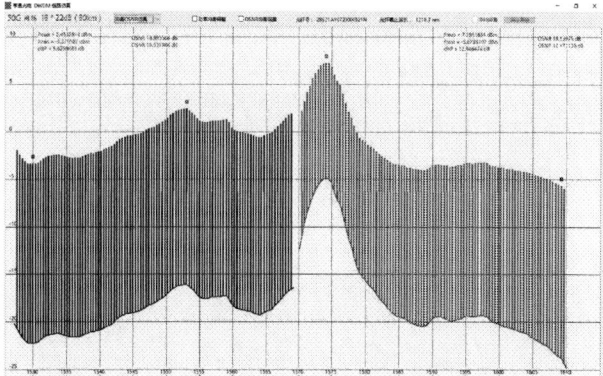

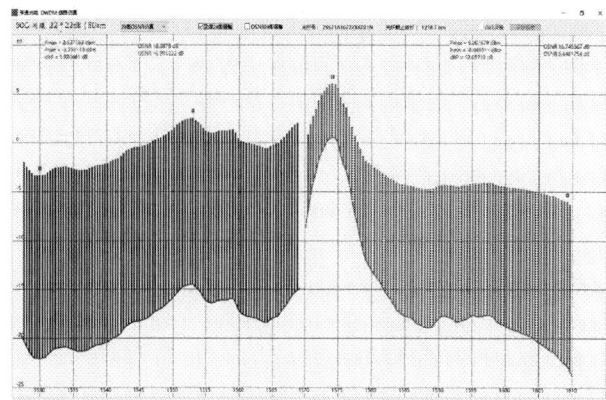

Figure 4. Performance in 18×22dB and 22×22dB span systems (C-band noise optimized EDFA).

978-1-6654-8699-6/22 $31.00 © 2022 IEEE

C-band gain flattening optimized EDFA cannot meet any link type requirements recommended in the industry standard YD/T 2485-2013 (2017) [5] due to its high noise figure. For L-band bulge suppression EDFA, only the shortest link type of 10×22dB can meet the requirements of industry standard YD/T 2485-2013 (2017) [7] for C-band, as shown in Fig. 5

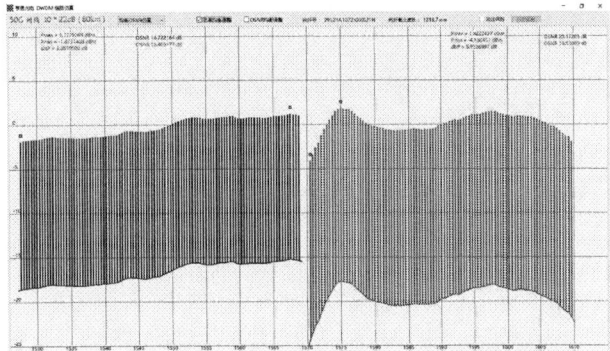

Figure 5.   Performance in 10×22dB span system (L-band bulge suppression EDFA).

2) Simulation based on the optimized EDFA

Assuming that after continuous technological development, the technical level of L-band EDFA can also reach today's C-band level, we will transfer the gain curve of C-band EDFA to L-band for simulation. It can be found that on the 14×22dB span, the system can meet the requirements of 18.5dB received OSNR and 6dB maximum channel power difference specified in relevant industry standards, as shown in Fig. 6. At the same time, it can be seen that the FAWD of the fiber at the long wavelength has a great impact on the flatness of the optical channel.

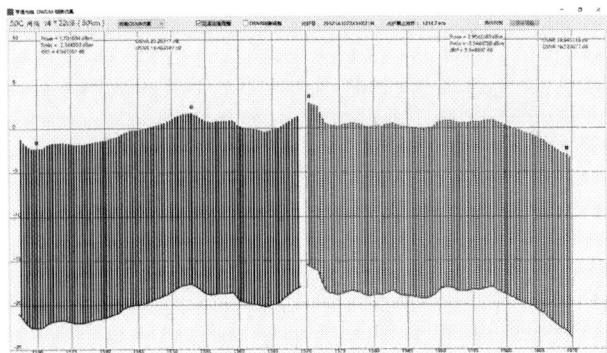

Figure 6.   Performance in 14×22dB span system (L-band optimized EDFA, $\triangle\alpha_{1610}$ = 0.0087dB/km).

When the $\triangle\alpha_{1610}$ is adjusted from 0.0087dB/km to 0.010dB/km, the simulation result shows that for the 14×22dB system, the received OSNR and the maximum channel received optical power difference are somewhat deficient in line with the requirements of the industry standard, as shown in Fig. 7.

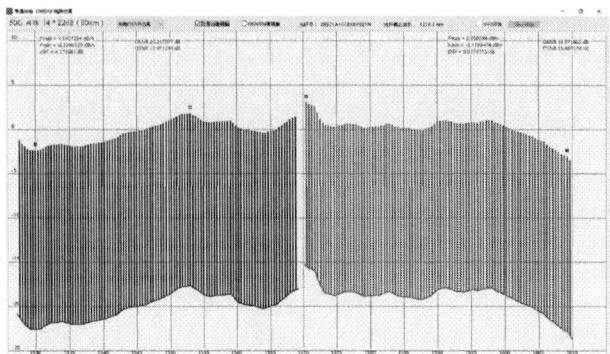

Figure 7.   Performance in 14×22dB span system (L-band optimized EDFA, $\triangle\alpha_{1610}$ = 0.010dB/km).

Continue to improve the $\triangle\alpha_{1610}$ to 0.012dB/km. It can be found that the performance of the wavelength division system is degraded more seriously in terms of both the received OSNR and the channel flatness. At this time, it is necessary to reduce the number of spans to 12 to meet the system requirements, as shown in Fig. 8.

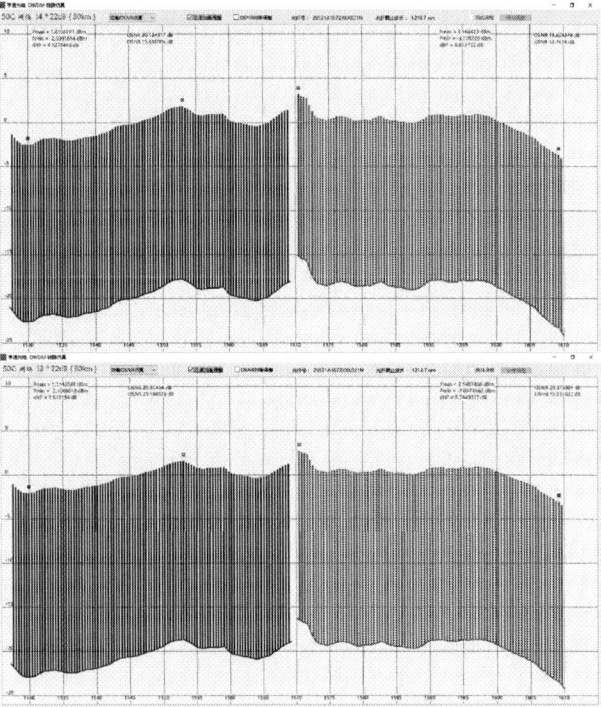

Figure 8.   Performance in 14×22dB and 12×22dB span systems (L-band optimized EDFA, $\triangle\alpha_{1610}$ = 0.012dB/km).

3) Simulation results

Because the fiber can only meet the requirements of receiving OSNR of 18.5dB and maximum channel power difference of 6dB on the shortest link specified in the industry standard, when the $\triangle\alpha_{1610}$ is 0.012dB/km. Therefore, it is better to limit the $\triangle\alpha_{1610}$ less than 0.012dB/km, when selecting the optical fiber used to open the wavelength division system in the C+L band. If it is extended to 1625nm wavelength, the $\triangle \alpha_{1625}$ is suitable to be limited to less than 0.015dB/km.

C. Attenuation Recommendations for CL Fiber

For the optical fiber itself, it is appropriate to specify the maximum FAWD of C-band and L-band to fully meet the C+L band extended transmission. As the novelty CL fiber will support the maximum 240 wave multiplexing in the future, the FAWD will not be limited to the wavelength range covered by the above system simulation. It is recommended to add two FAWD values in the C+L band, which are $\triangle \alpha_{1520}$ and $\triangle \alpha_{1625}$; at the same time, the absolute value $\triangle \alpha$ is specified. Referring to the above simulation results and the wavelength attenuation characteristics of SMF, it is suggested that the attenuation parameters of the novelty CL fiber is shown in Table I.

TABLE I. ATTENUATION RECOMMENDATIONS FOR CL FIBER

Parameters	Units	Recommended Values
Attenuation @1550nm	dB/km	≤0.20
Attenuation @1625nm	dB/km	≤0.25
$\triangle \alpha_{1520}$	dB/km	≤0.010
$\triangle \alpha_{1625}$	dB/km	≤0.015
$\triangle \alpha$ (Absolute Value)	dB/km	≤0.005

In order to obtain CL fibers that meet the above requirements, it is necessary to analyze the relevant parameters that may affect the FAWD of SMF. It is expected that stable CL fibers can be obtained quickly and in batches through the selection of certain conditions or the adjustment and control of some parameters, which will be carried out in the future work.

IV. CONCLUSION

C+L single mode fiber in L band is defined, simulated and analyzed. The OSNR in the L-band system alone will be significantly worse than the traditional C-band system, which makes the attenuation warping and attenuation warping difference of the CL fiber should be controlled at a reasonable level to meet the requirements of the signal to noise ratio and power difference of the 100Gbit/s wave division multiplexing system.

ACKNOWLEDGMENT

This work was supported by National Key Research and Development Program (No. 2021-YFB2900700), Suzhou Major Innovation Team Program (ZXT2019001).

Thanks to F. Lei, senior engineer, who provided the independently developed simulation software.

REFERENCES

[1] Cisco, Cisco annual Internet report (2018-2023), 2018.

[2] Yu S H and He W, "Latest survey on optical fiber communication (in Chinese)," Sci Sin Inform., vol. 50, 2020, pp. 1361–1376, doi:10.1360/SSI-2020-0093.

[3] LIU Bo and Li Linan, "Development status of large-capacity optical transmission systems," Science & Technology Review, vol. 34(16), 2016, pp. 20–33.

[4] OIF-Micro-ITLA-01.1[EB/OL](2019-01), Sep. 2019. "https://www.oi forum.com/wp-content/uploads/2019/01/OIF-Micro-ITLA-01.1.pdf..

[5] ITU-T Study Group, Series G. Transmission system and media，digit al system and networks, optical system design and engineering consid erations: ITU-T G.692, 2005.

[6] Ministry of Industry and Information Technology of the People's Republic of China, Technical Requirements for N×100Gbits Optical Wave Division Multiplexing (DWDM) System: YD/T 2485-2013 (2017), 2013.

[7] Ministry of Industry and Information Technology of the People's Republic of China, Technical Requirements for N×100Gbits Ultra-Long Distance Optical Wave Division Multiplexing (DWDM) System: YD/T 3070-2016 (2017), 2016.

Proceedings of the 7th Optoelectronics Global Conference (OGC 2022)

Coherence Retrieval and Multi-contrast Microscopy Imaging by Transport of Intensity Stack

Runnan Zhang
Smart Computational Imaging Laboratory (SCILab),
School of Electronic and Optical Engineering
Nanjing University of Science and Technology,
Nanjing, China
e-mail: runnanzhang@njust.edu.cn

Zewei Cai
Smart Computational Imaging Laboratory (SCILab),
School of Electronic and Optical Engineering
Nanjing University of Science and Technology,
Nanjing, China
e-mail: zeweicai@njust.edu.cn

Chao Zuo
Smart Computational Imaging Laboratory (SCILab), School of Electronic and Optical Engineering
Nanjing University of Science and Technology,
Nanjing, China
e-mail: zuochao@njust.edu.cn

Abstract—We report a new coherence retrieval method based on transport of intensity stack. We capture intensity images at different focal planes, corresponding to phase-space shearing projection, into four-dimensional (4D) phase space to create both high spatial and spatial frequency resolution phase space. We represent the optical field of an arbitrary coherent mode with the help of a typical function in phase space, i.e., the Wigner distribution function. Previous studies mainly focus on coherence measurement by means of interferometry. In addition, despite the sacrifice of both spatial and spatial frequency resolution, wavefront sensor is also a standard tool for coherence measurement. To overcome the existing problems, our method only requires z-scanning to achieve full-resolution coherent retrieval of the optical field without interferometric detection. Furthermore, without requiring hardware modifications, we introduced our coherence retrieval method into microscopic imaging. We apply our method to multi-contrast microscopic imaging with only intensity stack, which have the potential to allow clinicians to view a sample with three separate contrast methods at once, enhancing the information available for diagnosis and disease discrimination.

Keywords-coherence retrieval; transport of intensity; multi-contrast microscopy imaging

I. INTRODUCTION

Coherence is one of the most fundamental properties of an optical field, and it implies how closely an optical field oscillates in unison at the same position at different times (temporal coherence) or different positions at the same time (spatial coherence). Since the advent of the first ruby laser[1] in the 1960, the laser with high spatial and temporal coherence has emerged as an essential illumination engine in disciplines such as traditional interferometry and holography[2]–[4]. In numerous applications, including adaptive optics[5]–[8], X-ray diffraction imaging[9]–[12],

transmission electron microscope[13]–[15], neutron radiography[16], [17], where it is often inevitable or desirable that fields are partially coherent. In contrast, in the fields of computational microscopy imaging, far-field passive detection, and optical communication, which have developed rapidly in recent years, research has demonstrated that the use of partially coherent sources can effectively enhance information throughput and signal-to-noise ratio. Consequently, the representation, measurement, and application of the coherence are gaining in significance, especially for partially coherent optical fields.

The partially (spatially) coherent optical field is represented by the cross-spectral density function (CSDF) in classical coherence theory and the Wigner distribution function (WDF) in phase space optics. Numerous methods have been studied for coherence retrieval[18]–[21]. Here, we propose a novel coherence retrieval method based on transport of intensity, termed transport of intensity based coherence retrieval (TICR). In accordance with the Hamiltonian description of wave evolution, the joint evolution mechanism of space-frequency information during intensity transport is revealed. Due to the inherent redundancy of phase space, it is possible to rebuild 4D phase space by capturing 3D transport of intensity. The elegance of this approach lies in the fact that it can produce a completer and high-resolution 4D coherence retrieval with a single intensity stack without interference and scanning apparatus. Based on the straightforward implementation of the TICR technique, we expand its application to microscopy and realize the capability of multi-contrast microscopy for biological samples. To our knowledge, this is the first time that coherence retrieval has been applied to microscopic imaging. We perform coherence retrieval of dandelion fluff under bright field with partially coherent illumination and realize the multi-contrast imaging of multi angles of view,

978-1-6654-8699-6/22 $31.00 © 2022 IEEE

synthetic aperture and differential phase contrast (DPC) through light field manipulation. It proves that TICR is promising for computational microscopic imaging, and is anticipated to obtain more valuable applications in life science and biomedicine.

II. METHODS

Coherent light is fully described by its 2D amplitude and phase at every position, while partially coherent light at a given point has a 2D distribution of propagation directions, corresponding to a local ensemble of different phases, therefore a 4D representation is required. We concentrate on phase space optics, of which WDF is a typical representation function. It can be considered as a more rigorous light ray model, defined as generalized radiance[37], with the difference that the generalized radiance may be negative originate from phase space interference[38]. The description of a partially coherent optical field by a coherent function serves as the foundation for WDF definition. By a Fourier transform in the differential space variable $\mathbf{x}'$, the CSDF $G(\mathbf{x}_1, \mathbf{x}_2)$ is turned into WDF in the space-frequency domain of phase space.

$$W(\mathbf{x}, \mathbf{u}) = \int G\left(\mathbf{x} + \frac{\mathbf{x}'}{2}, \mathbf{x} - \frac{\mathbf{x}'}{2}\right) \exp(-i2\pi\mathbf{u}\mathbf{x}') d\mathbf{x}' \quad (1)$$

where $\mathbf{u}$ is the spatial frequency vector corresponding to $\mathbf{x}$. The classical coherence function is computationally complicated due to its bilinear nature, and it is often difficult to provide an intuitive explanation of the physical meaning behind the equation. Since the affine canonical covariance in phase space, WDF simplifies the description of the imaging process, the system, and provides a more intuitive and profound analysis of the underlying physical mechanism.

According to the spatial marginal property, the detectable intensity is:

$$I(\mathbf{x}) = \int W(\mathbf{x}, \mathbf{u}) d\mathbf{u} \quad (2)$$

The propagation equation of the WDF is:

$$W(\mathbf{x}, \mathbf{u}, z) = W(\mathbf{x} - \lambda z\mathbf{u}, \mathbf{u}, 0) \quad (3)$$

Combining the spatial marginal property [Eq.(2)] with the phase space propagation [Eq.(3)], we can obtain the relationship between the intensity transmission and the phase space transformation in the main text as follows:

$$I(\mathbf{x}, z) = \int W(\mathbf{x}, \mathbf{u}, z) d\mathbf{u} = \int W(\mathbf{x} - \lambda z\mathbf{u}, \mathbf{u}, 0) d\mathbf{u} \quad (4)$$

The overview of the iterative TICR reconstruction algorithm are as follows:

Step 1. Make an initial guess of the WDF. The initial value for the final results is not critical. Here we initialize the WDF W_1 by the intensity I_1 at the first focal plane. The back-projection angle is in direct proportion to $(\lambda z)^{-1}$.

Step 2. The iteration process starts. Propagate the initial WDF W_1 (according to Eq. (3)) from the first plane to the next plane W_2 due to the axial difference Δz between two planes.

Step 3. Calculate the update matrix $\Delta I_{k+1} = I_{k+1} - I'_{k+1}$, which is the residual between two intensities. The I'_{k+1} is calculated by Eq. (2) according to the W'_{k+1} at the $(k+1)^{th}$ focal plane. The I_{k+1} is the captured intensity at the corresponding focal plane.

Step 4. Update the W'_{k+1} by ΔI_{k+1}. Redistribute the residual to the phase space by the weighting function, we can obtain $\quad W_{k+1}(\mathbf{x}, \mathbf{u}_i) = W'_{k+1}(\mathbf{x}, \mathbf{u}_i) + \Delta I_{k+1}(\mathbf{x}) \cdot P(\mathbf{x}, \mathbf{u}_i) \quad$. Introduce *stepsize* to control the proportion of update and retention:

$$W_{k+1}(\mathbf{x}, \mathbf{u}_i) = stepsize \cdot W_{k+1}(\mathbf{x}, \mathbf{u}_i) + (1 - stepsize) \cdot W'_{k+1}(\mathbf{x}, \mathbf{u}_i)$$

Step 5. Propagate the updated WDF to the next plane and repeat steps 3-4 until all the intensity images are used, and the iteration completes one cycle.

Step 6. The whole iteration scheme is repeated over M cycles to achieve a self-consistent solution. The cost function is calculated to judge whether iteration stops.

III. MULTI-CONTRAST MICROSCOPY IMAGING.

Moreover, we extend coherence retrieval to biological microscopic imaging. We link the motor stage of Olympus IX81 microscope with sCMOS camera for axial scanning. We capture a 3D intensity stack of dandelion fluff under bright field (quasi-monochromatic partially coherent illumination), with 40× 0.95NA objective lens. The captured raw data is displayed in Fig. 1a. The full 4D WDF reconstruction results are displayed in order of view angles (Fig. 1c). After acquiring the 4D WDF, we may do further manipulations on the light field, such as adjusting the depth and aperture. Among these, there are various types of aperture adjustment, including view angles regulation, differential phase contrast (DPC) imaging, synthetic aperture. To develop further image modes, these settings can be mixed and matched in any way. Only limited imaging modes are illustrated in Fig. 1.

Fig. 1d displays the digital refocusing results under bright field. The zoom-in region is selected in Fig. 1d6. The left-right 3D DPC results are shown in Fig.1e, and the up-right 3D DPC results are shown in Fig. 1f.

The significance of reconstructing the spatial and spatial frequency information of the light field at the same time is demonstrated by the fact that all of the multi-contrast imaging results are derived from only one intensity stack. Traditional measurements that are restricted to the spatial or frequency domain are unable to perform this purpose. Our approach is anticipated to give life scientists a powerful tool for doing multimodal imaging studies.

978-1-6654-8699-6/22 $31.00 © 2022 IEEE

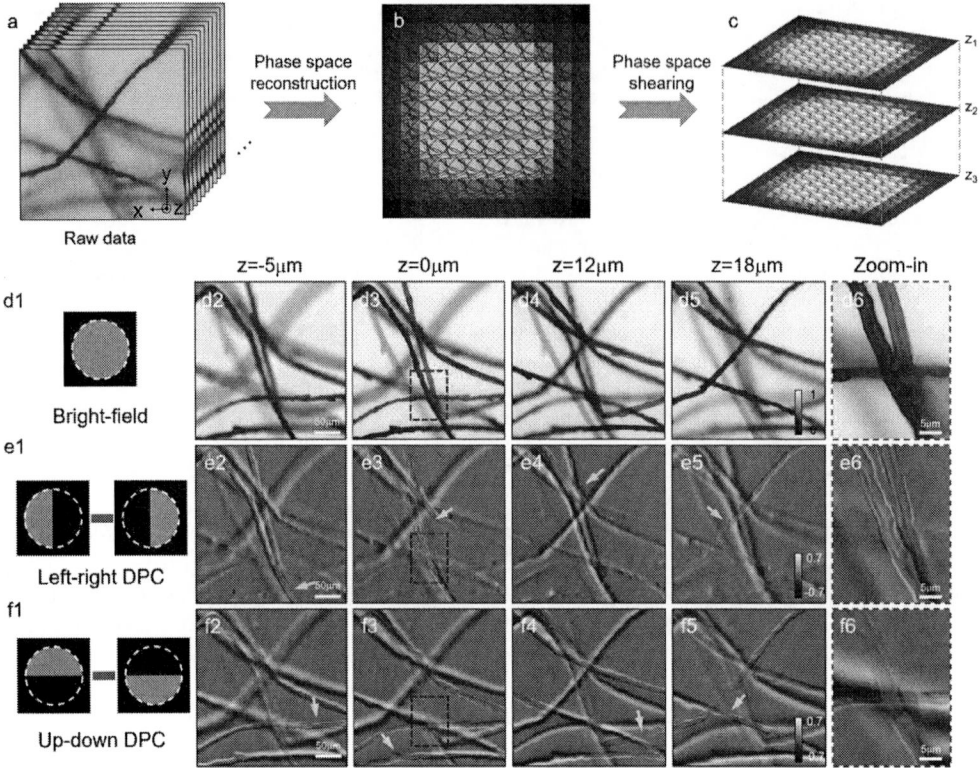

Figure 1. Multi-contrast microscopy imaging. (a) Captured raw data. (b) 4D phase space reconstruction. (c) Phase space shearing to different axial positions. (d) Bright field digital refocusing, focus at z=-5,0,12,18μm respectively. (e) The left-right 3D DPC results. (f) The up-down 3D DPC results.

IV. CONCLUSION

We propose a coherent retrieval method based on 3D transport of intensity, termed TICR. This method can retrieve complete and high-resolution 4D coherence function, and is first applied to microscopic biological imaging. It proves that TICR is promising for computational microscopic imaging, and is anticipated to obtain more valuable applications in life science and biomedicine.

ACKNOWLEDGMENT

This work was supported by the National Natural Science Foundation of China (61905115, 62105151, U21B2033), Leading Technology of Jiangsu Basic Research Plan (BK20192003), Youth Foundation of Jiangsu Province (BK20190445, BK20210338), Fundamental Research Funds for the Central Universities (30920032101), and Open Research Fund of Jiangsu Key Laboratory of Spectral Imaging & Intelligent Sense (JSGP202105).

REFERENCES

[1] T. H. Maiman, "Stimulated Optical Radiation in Ruby," Nature, vol. 187, no. 4736, Art. no. 4736, Aug. 1960, doi: 10.1038/187493a0.

[2] U. Schnars, C. Falldorf, J. Watson, and W. Jüptner, "Digital holography," in Digital Holography and Wavefront Sensing, Springer, 2015, pp. 39–68.

[3] E. Cuche, P. Marquet, and C. Depeursinge, "Simultaneous amplitude-contrast and quantitative phase-contrast microscopy by numerical reconstruction of Fresnel off-axis holograms," Applied optics, vol. 38, no. 34, pp. 6994–7001, 1999.

[4] A. Abramovici et al., "LIGO: The laser interferometer gravitational-wave observatory," science, vol. 256, no. 5055, pp. 325–333, 1992.

[5] F. Roddier, C. Roddier, and N. Roddier, "Curvature Sensing: A New Wavefront Sensing Method," 1988, vol. 0976, pp. 203–209. doi: 10.1117/12.948547.

[6] F. Roddier, "Curvature sensing and compensation: a new concept in adaptive optics," Appl. Opt., AO, vol. 27, no. 7, pp. 1223–1225, Apr. 1988, doi: 10.1364/AO.27.001223.

[7] F. Roddier, "Wavefront sensing and the irradiance transport equation," Applied optics, vol. 29, no. 10, pp. 1402–1403, 1990.

[8] N. A. Roddier, "Algorithms for wavefront reconstruction out of curvature sensing data," 1991, vol. 1542, pp. 120–129. doi: 10.1117/12.48799.

[9] K. Giewekemeyer et al., "Quantitative biological imaging by ptychographic x-ray diffraction microscopy," Proceedings of the National Academy of Sciences, vol. 107, no. 2, pp. 529–534, Jan. 2010, doi: 10.1073/pnas.0905846107.

[10] A. Maiden, G. Morrison, B. Kaulich, A. Gianoncelli, and J. Rodenburg, "Soft X-ray spectromicroscopy using ptychography with randomly phased illumination," Nature communications, vol. 4, no. 1, pp. 1–6, 2013.

[11] P. Thibault, V. Elser, C. Jacobsen, D. Shapiro, and D. Sayre, "Reconstruction of a yeast cell from X-ray diffraction data," Acta Crystallographica Section A: Foundations of Crystallography, vol. 62, no. 4, pp. 248–261, 2006.

[12] J. M. Rodenburg et al., "Hard-x-ray lensless imaging of extended objects," Physical review letters, vol. 98, no. 3, p. 034801, 2007.

[13] J. M. Rodenburg, A. C. Hurst, and A. G. Cullis, "Transmission microscopy without lenses for objects of unlimited size," Ultramicroscopy, vol. 107, no. 2–3, pp. 227–231, Feb. 2007, doi: 10.1016/j.ultramic.2006.07.007.

[14] F. Hue, J. M. Rodenburg, A. M. Maiden, and P. A. Midgley, "Extended ptychography in the transmission electron microscope: Possibilities and limitations," Ultramicroscopy, vol. 111, no. 8, pp. 1117–1123, Jul. 2011, doi: 10.1016/j.ultramic.2011.02.005.

[15] F. Hue, J. M. Rodenburg, A. M. Maiden, F. Sweeney, and P. A. Midgley, "Wave-front phase retrieval in transmission electron microscopy via ptychography," Physical Review B, vol. 82, no. 12, Sep. 2010, doi: 10.1103/PhysRevB.82.121415.

[16] B. Allman et al., "Phase radiography with neutrons," nature, vol. 408, no. 6809, pp. 158–159, 2000.

[17] P. McMahon, B. Allman, D. L. Jacobson, M. Arif, S. Werner, and K. Nugent, "Quantitative phase radiography with polychromatic neutrons," Physical review letters, vol. 91, no. 14, p. 145502, 2003.

[18] L. Waller, G. Situ, and J. W. Fleischer, "Phase-space measurement and coherence synthesis of optical beams," Nat. Photonics, vol. 6, no. 7, pp. 474–479, Jul. 2012, doi: 10.1038/nphoton.2012.144.

[19] L. Tian, Z. Zhang, J. C. Petruccelli, and G. Barbastathis, "Wigner function measurement using a lenslet array," Opt. Express, vol. 21, no. 9, pp. 10511–10525, May 2013, doi: 10.1364/OE.21.010511.

[20] D. L. Marks, R. A. Stack, D. J. Brady, D. C. Munson, and R. B. Brady, "Visible Cone-Beam Tomography With a Lensless Interferometric Camera," Science, vol. 284, no. 5423, pp. 2164–2166, Jun. 1999, doi: 10.1126/science.284.5423.2164.

[21] A. I. González and Y. Mejía, "Nonredundant array of apertures to measure the spatial coherence in two dimensions with only one interferogram," J. Opt. Soc. Am. A, vol. 28, no. 6, pp. 1107–1113, Jun. 2011.

978-1-6654-8699-6/22 $31.00 © 2022 IEEE

Proceedings of the 7th Optoelectronics Global Conference (OGC 2022)

GS Iterative Phase Retrieval Algorithm Based on Fusion of Spatial Phase Gradient Descent and Frequency Domain Amplitude Linear Weighting

Hong Cheng[*], Haonan Zheng, Siwei Sun

School of Electronic and Information Engineering
Anhui University
Hefei, China
e-mail: chenghong@ahu.edu.cn

Abstract—**For the problems of slow convergence and low accuracy of the traditional linear weighted GS iterative phase retrieval algorithm, a GS iterative phase retrieval algorithm based on the fusion of spatial phase gradient descent and frequency domain amplitude linear weighting is proposed. By zero-padding the image, and then applying phase gradient descent in each iteration of the space domain, the algorithm invokes linear weighting in the frequency domain space, thereby avoiding iterative stagnation while ensuring the convergence speed and improving the accuracy of phase retrieval.**

Keywords-phase retrieval; Gerchberg–Saxton algorithm; phase gradient descent; amplitude Linear Weighting; zero-padding

I. INTRODUCTION

The square of the amplitude part of the light wave describes the intensity of the light wave field, which is the intensity of light that we can perceive with our human eyes, so the amplitude part of the light wave is still easy to understand and accept. However, we usually do not have an intuitive sense of the phase part of light because the human eye or existing imaging devices can only detect the intensity information of light, but cannot record the phase information of light. Phase holds information about the surface profile or depth of an object compared to intensity. It is of great importance to retrieve the phase from the intensity information.

Non-interference phase retrieval algorithms can be divided into two categories. One is the phase retrieval algorithm based on the Transport of intensity equation (TIE)[1-2], and the other is an iterative-based phase retrieval algorithm[3]. The phase retrieval algorithm based on TIE uses the collected light intensity information and solves the equation to obtain the phase. This method can obtain rapid and effective phase information but is limited by the propagation distance of the light field. Another class of iterative phase retrieval algorithms [4]. The most classic of these is a phase retrieval technique based on iterative operations proposed by Gerchberg and Saxton in 1972, but GS algorithms often suffer from iterative stagnation or fall into local optima. In order to solve the iterative stagnation problem of GS algorithm, many scholars have developed improvements based on it, such as Errorreduction (ER)

algorithm [5] and Hybrid Input-Output (HIO) algorithm [6]. The former is more widely applicable but its convergence rate is slow and even stagnation occurs in the correct solution space. The latter solves the problem of slow convergence of the Errorreduction algorithm, but it does not solve the case where the algorithm converges to local minima. Many other scholars have proposed the theory of multi-planar iterative phase retrieval[7-10], where phase retrieval is achieved by cyclic iteration of multiple images measured, but this increases the amount of data to be processed and the complexity of the system while decreasing the convergence speed.

In addition to the above algorithms, in recent years, many scholars have developed some improvements based on the GS iterative algorithm one after another. For example, the Adaptive-additive (AA) algorithm proposed by Soifer [11] et al. The algorithm applies the idea of linear weighting to the GS algorithm and adds a linear weighting factor to the frequency domain constraint. Although the iterative stagnation problem of the GS algorithm is effectively avoided, its convergence accuracy and speed are still not optimistic. Guo et al [12] proposed the spatial phase perturbation GS algorithm (SPP GSA) and the combined GS hybrid input-output algorithm (GS/HIOA) in 2015, both of which are good at avoiding iterative stagnation during phase retrieval, but the former converges slowly and the latter converges with low accuracy in the presence of noise. Wang et al [13] proposed a hybrid Gerchberg–Saxton-like algorithm for DOE and CGH calculation in 2017, which achieved fast convergence of the reconstructed images by gradient descent and weighting strategy, but the final convergence accuracy remained low due to noise effects.

In this paper, by zero-padding the images [14-15] and introducing two convergence accelerations in the GS iterations, finally achieves higher accuracy phase retrieval. The core idea is to invoke linear weighting in the frequency domain space while using phase gradient descent in each iteration in the spatial domain, which greatly speeds up the convergence and also avoids iteration stagnation. The algorithm proposed in this paper greatly reduces the influence of noise on the phase retrieval results and improves the accuracy of phase retrieval while ensuring the convergence speed.

978-1-6654-8699-6/22 $31.00 © 2022 IEEE

II. PRINCIPLE OF THE ALGORITHM

The Adaptive-additive algorithm proposed in reference [11], which is based on the GS algorithm with only linear weighting, does not greatly improve the convergence accuracy and speed of the algorithm, although iterative stagnation is avoided to some extent. In order to solve the problems of low convergence accuracy and slow speed of the above algorithms, this paper proposes a GS iterative phase retrieval algorithm with the fusion of spatial domain phase gradient descent and frequency domain amplitude linear weighting. The algorithm is divided into two major modules, the first module is for zero-padding processing of the image, and the second module is for GS hybrid iteration[13]. The schematic diagram of the algorithm proposed in this paper is shown in Fig. 1.

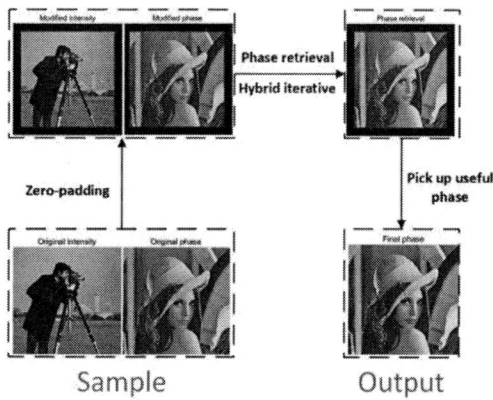

Figure 1. Schematic diagram of the algorithm

First, the original intensity and the original phase in the above figure are zero-padded, respectively, so they are divided into the noise domain and the signal domain, where the noise domain is the zero-padded domain. The purpose of zero padding processing on the image is to remove the background noise in the signal domain to the noise domain and achieve a high signal-to-noise ratio. The two are then substituted into the hybrid iterative module by the GS hybrid iterative module, and finally, the phase containing the noise domain and the signal domain is retrieved as a whole by the GS hybrid iterative algorithm, and the noise domain is removed to get the phase to be retrieved.

The flow chart of the GS hybrid iterative module algorithm is shown in Fig. 2.

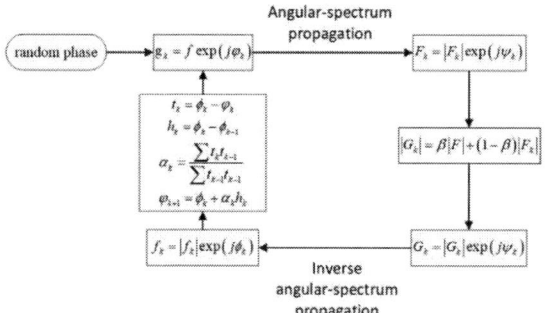

Figure 2. Flow chart of GS hybrid iterative module algorithm

The GS hybrid iteration module is divided into two steps: phase gradient descent in the spatial domain and linear weighting of amplitudes in the frequency domain. The spatial domain part, in addition to introducing phase gradient descent, also performs amplitude substitution. The phase gradient decreases, i.e., the phase ϕ_k obtained in the kth iteration is modulated to φ_{k+1}, which is given by:

$$\varphi_{k+1} = \phi_k + \alpha_k h_k \tag{1}$$

h_k is defined as the gradient direction, which is actually proportional to the difference between the phase obtained in the current iteration and the phase obtained in the previous iteration, as defined below:

$$h_k = \phi_k - \phi_{k-1} \tag{2}$$

α_k is the acceleration factor, and its formula is shown below:

$$\alpha_k = \frac{\sum t_k t_{k-1}}{\sum t_{k-1} t_{k-1}} \tag{3}$$

$$t_k = \phi_k - \varphi_k \tag{4}$$

The phase result φ_{k+1} is then obtained by the above formula and then multiplied with the input surface amplitude f to form the new complex amplitude $g_{k+1} = f \exp(j\varphi_{k+1})$, which is used as the input for the next iteration.

The complex amplitude g_k is obtained from the angular spectral propagation distance d to get F_k, defined as

$$F_k = g_k * H \tag{5}$$

$$H = \exp(jkd)\exp\left[-j\pi\lambda d\left(f_x^2 + f_y^2\right)\right] \tag{6}$$

Where H is the angular spectral transfer function, k is the wave number, λ is the wavelength, f_x, f_y are denoted as the spatial frequencies in the x-axis and y-axis directions, respectively. $f_x = \cos\gamma / \lambda, f_y = \cos\gamma / \lambda$.

A constraint defined in this way alone causes the error to drop sharply only in the first few iterations, and the error then remains very slowly decreasing. To further reduce the error, here we set the frequency domain amplitude linear weighting as the second step of the GS hybrid iteration module to avoid the algorithm from stagnation.

The main idea of the linear weighting of frequency domain amplitudes is to add a linear weighting factor to the frequency domain constraint, and the specific formula of the frequency domain constraint is as follows:

$$|G_k| = \beta|F| + (1-\beta)|F_k| \tag{7}$$

β is a weighting constant between 0 and 1. According to the flow chart of the above algorithm, the error corresponding to the reference value of all weighting constants is obtained using the definition of the error criterion to get an error value, and the smallest value in the error array is found as the final weight. $|F|$ is the collected ideal amplitude, and $|F_k|$ is the amplitude calculated by diffraction in the kth iteration. This part of the improvement achieves the frequency domain "mapping" while retaining part of the information in the previous iteration of the solution, effectively avoiding the iterative stagnation problem of the GS algorithm.

III. ANALYSIS OF EXPERIMENTAL RESULTS

In order to verify the performance of the algorithm proposed in this paper, we conducted further experimental validation. The following Fig. 3(a) shows the original intensity image, and Fig. 3(b) shows the original phase image with the image size of 256pixel*256pixel, propagation distance d=500 mm, and wavelength of 635 nm.

(a) intensity image; (b) phase

Figure 3. Original image

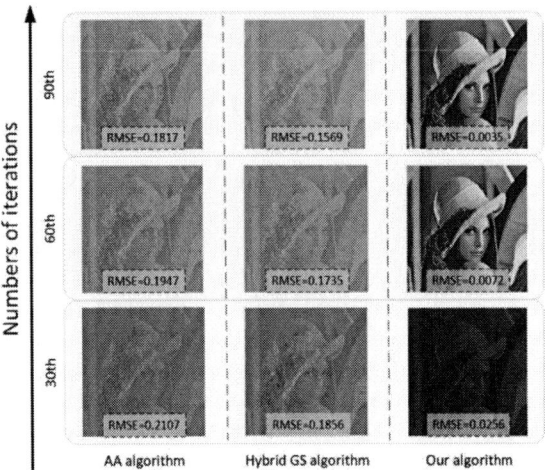

Figure 4. Comparison of experimental results

The experimental results are shown in Fig. 4. We compare the algorithm in this paper with the Adaptive-additive algorithm proposed in reference [11] and the GS hybrid iterative algorithm without zero-padding treatment [13], respectively. And the root mean square error (RMSE) of the retrieved image is calculated accordingly. Let the

original phase be φ and the retrieved phase be φ_k, then the RMSE equation is:

$$RMSE = \sqrt{\frac{\sum_{x,y}(\varphi_k - \varphi)^2}{M \times N}} \qquad (8)$$

From the experimental results, the Adaptive-additive algorithm is still blurred after 90 iterations and the RMSE value is 0.1817. The retrieved phase of the GS hybrid iterative algorithm without the zero-padding process is also blurred at the same number of iterations, and the RMSE value is 0.1569. The algorithm proposed in this paper can achieve high retrieval accuracy after 90 iterations, and its RMSE value is 0.0035.

We have compared the curves of RMSE values for the images retrieved by 100 iterations for the three different methods mentioned above, and the results of the curve comparison of RMSE values are shown in Fig. 5.

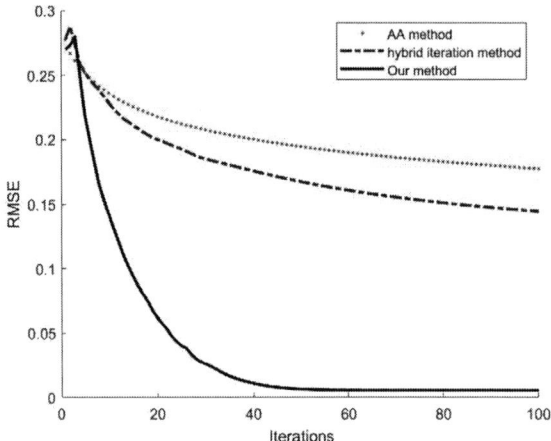

Figure 5. Curve comparison of RMSE values

Fig. 5 gives the RMSE comparison curves of the above three algorithms after retrieval through 100 iterations. In the above figure, the horizontal axis represents the number of iterations and the vertical axis represents the root mean square error. The error curves of the three algorithms are shown by red dots, blue dotted lines, and black solid lines, respectively. Compared with the above two methods, the algorithm proposed in this paper converges faster and with higher accuracy. From the results, the GS hybrid iterative algorithm retrieved better results than the Adaptive-additive algorithm as the number of iterations increased, but its convergence accuracy and speed were still not optimistic. Compared with these two algorithms, the proposed algorithm in this paper converges faster and with higher accuracy. At about 60 iterations, the convergence of the algorithm iterations in this paper has leveled off.

In order to compare the phase results retrieved by the above three methods more obviously. We compared the phase retrieval results of the three methods with the original phase.

(a) Location of the transverse profile of the original phase image

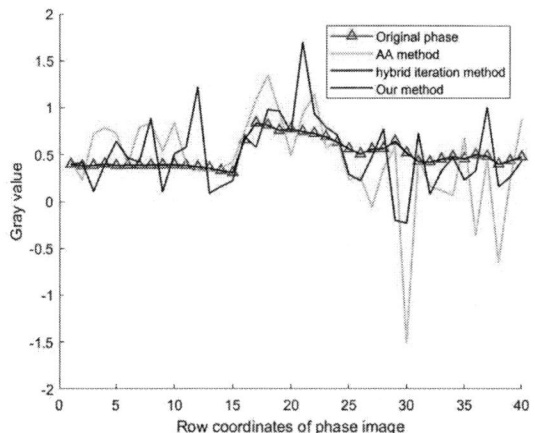

(b)Comparison of the retrieval results of different methods and the profile lines of the original phase image

Figure 6.   Comparison of profile curves

Fig. 6 gives a comparison of the profile lines between the retrieval results of the three methods and the original phase image. Fig. 6(a) shows the location of the original phase transverse profile line. In Fig. 6(b), the red triangle is the profile line of the original phase, the green solid line is the result profile line of the phase retrieval by the Adaptive-additive algorithm, and the blue solid line is the result profile line of the phase retrieval by the hybrid iterative algorithm. The black solid line is the profile line of the phase recovery result of the proposed algorithm. From the above comparison results, the profile line of the proposed method overlaps with the profile line of the original phase image, which proves that the phase retrieval results of the proposed method are better than the other algorithms mentioned above.

IV.   CONCLUSION

In this paper, we propose a GS iterative phase retrieval with the fusion of spatial domain phase gradient descent and frequency domain amplitude linear weighting to address the problems of slow convergence speed and low accuracy of the traditional linearly weighted GS iterative phase retrieval algorithm. The algorithm not only improves the convergence speed but also improves the accuracy of phase retrieval. The

comparison of the experiments in this paper illustrates the superiority of the algorithm proposed in this paper.

ACKNOWLEDGEMENT

The research was supported by the Natural Science Project of Anhui Higher Education Institutions of China (No.KJ2020ZD02,KJ2019ZD04), Natural Science Foundation of Anhui Province, China (No. 2008085MF209).

REFERENCES

[1] Chao Zuo,Jiaji Li,Jiasong Sun,et al.Transport of intensity equation: a tutorial[J].Optics and Lasers in Engineering,2020,16(14), doi.org/10.1016/j.optlaseng.2020.106187.

[2] J.M.D. Martino, G.A. Ayubi, E.A. Dalchiele, Single-shot phase recovery using two laterally separated defocused images[J]. Optics Communications, 2013, 293:1-3, doi.org/10.1016/j.optcom.2012.11.084.

[3] Bon P, Monneret S, Wattellier B. Noniterative boundary-artifact-free wavefront reconstruction from its derivatives[J]. Applied optics, 2012, 51(23): 5698-5704, doi.org/10.1364/AO.51.005698.

[4] Zheng S, Shangguan H, Zeng X, et al. Coherent diffractive imaging via a rotatable cylindrical lens[J]. Optics and Lasers in Engineering, 2020, 124: 105820, doi.org/10.1016/j.optlaseng.2019.105820.

[5] Fienup J R. Reconstruction of an object from the modulus of its Fourier transform[J]. Optics letters, 1978, 3(1): 27-29, doi.org/10.1364/OL.3.000027.

[6] Fienup J R. Phase retrieval algorithms: a comparison[J]. Applied optics, 1982, 21(15): 2758-2769, doi.org/10.1364/AO.21.002758.

[7] Maiden A M, Rodenburg J M. An improved ptychographical phase retrieval algorithm for diffractive imaging[J]. Ultramicroscopy, 2009, 109(10): 1256-1262,doi.org/10.1016/j.ultramic.2009.05.012.

[8] Maiden A M, Humphry M J, Rodenburg J M. Ptychographic transmission microscopy in three dimensions using a multi-slice approach[J]. JOSA A, 2012, 29(8): 1606-1614, doi.org/10.1364/JOSAA.29.001606.

[9] Rodrigo J A, Duadi H, Alieva T, et al. Multi-stage phase retrieval algorithm based upon the gyrator transform[J]. Optics Express, 2010, 18(2): 1510-1520, doi.org/10.1364/OE.18.001510.

[10] Liu Z, Guo C, Tan J, et al. Iterative phase-amplitude retrieval with multiple intensity images at output plane of gyrator transforms[J]. Journal of Optics, 2015, 17(2): 025701, doi:10.1088/2040-8978/17/2/025701.

[11] Soifer, Victor A., V. Kotlar, and Leonid Doskolovich. Iterative Methods For Diffractive Optical Elements Computation. CRC Press, 2014.

[12] Guo, Changliang, Shi Liu, and John T. Sheridan. "Iterative phase retrieval algorithms. I: optimization." Applied optics 54.15 (2015): 4698-4708, doi.org/10.1364/AO.54.004698.

[13] Wang, Haichao, et al. "A hybrid Gerchberg–Saxton-like algorithm for DOE and CGH calculation." Optics and Lasers in Engineering 89 (2017): 109-115, doi.org/10.1016/j.optlaseng.2016.04.005.

[14] Zhai, Tingting, et al. "An approach for holographic projection with higher image quality and fast convergence rate." Optik 159 (2018): 211-221, doi.org/10.1016/j.ijleo.2018.01.055.

[15] Shimobaba, Tomoyoshi, et al. "Reducing computational complexity and memory usage of iterative hologram optimization using scaled diffraction." Applied Sciences 10.3 (2020): 1132, doi:10.3390/app10031132.

Proceedings of the 7th Optoelectronics Global Conference (OGC 2022)

Laser Weld Seam Tracking Sensing Technology Based on Swing Mirror

Tian Changyong
Technical Institute of Physics &
Chemistry, CAS
Beijing, China
12032532@qq.com

Song Xuhao
Institute of Optical Physics and
Engineering Technology, Qilu Zhongke
Beijing, China
songxuh94@163.com

Yin Tie, Zhang Yi
China Petroleum and Natural Gas
Pipeline Research Institute Co., Ltd.
Langfang, China

Abstract—Based on the principle of laser triangulation measurement, a laser welding seam tracking sensor is developed to track U-shaped, V-shaped and flat bottom grooves of pipeline welds in real time. Aiming at the problems of reflection of pipeline welding groove and strong welding arc, which affect the accuracy of U-shaped groove feature recognition, a laser welding seam tracking sensing technology based on swing mirror is developed. The image sensor collects the weld features of the spot-shaped spot area in a cycle, and connects to form a cross-sectional weld feature, which can effectively improve the signal-to-noise ratio of the weld feature identification signal. the actual machine test of pipeline weld tracking is carried out. In the identification test of U-shaped, V-shaped, flat bottom groove, extremely deep and wide weld (40mm depth&30mm width) and extremely deep and narrow weld (40mm depth&12mm width), the sensor has good recognition accuracy and stability.

Keywords-pipe welding; vision sensing; laser seam tracking; swing mirror

I. INTRODUCTION

The pipeline welding work intensity is high, the working environment is harsh, and the precision is strict. Therefore, the pipeline all-position automatic welding technology with the advantages of fast speed, good stability and low cost came into being[1]. The key technology of automatic pipeline welding is the feature recognition of weld seam. The existing welding seam identification methods include arc sensing, ultrasonic sensing, photoelectric sensing, and visual sensing, among which arc sensing and laser vision sensing are the main research directions [2][3].

The arc sensor uses the characteristic that the arc current and voltage change with the distance between the welding torch and the workpiece during the welding process to detect the center of the weld groove [4]. It is simple in structure, easy to operate, and has strong resistance to arc light, high temperature and magnetic field. It is widely used in welding seam tracking, and is currently used most in arc welding robots [5]. Arc sensors can be divided into mechanical arc sensors and magnetron arc sensors. Li W. et al. [6] designed a rotating arc sensing system. The motor drives the optical code disc to make the welding wire rotate at the same speed, and obtains information such as frequency and electrical

signals to reflect the state of the weld groove. Hong Bo et al. [7] proposed a magnetron arc sensor, which applies an external magnetic field near the nozzle to control the arc to scan the weld seam in a swinging or rotating manner to obtain weld position information. In the current pipeline welding engineering, arc sensing is usually still used as a means of characterizing the position of the weld, rather than extracting the features of the weld.

Laser vision sensing is based on the principle of laser triangulation measurement to obtain three-dimensional information of welds. Its image processing process is simple and the feature recognition is accurate. It has been widely used in weld identification [8][9]. In recent years, the research on welding seam tracking algorithm has progressed rapidly. Borer et al. [10] proposed a welding seam tracking algorithm based on laser displacement sensor, which effectively solved the problem of butt weld detection in irregular workspace. Zhao et al. [11] proposed a laser streak detection neural network based on structured light visual sensing, and designed a single-line structured light visual navigation sensor to improve the robustness of feature extraction. In the engineering practice of pipeline weld seam identification, high surface reflection and severe welding arc are still the key factors restricting the application of laser weld seam tracking. The current pipeline welding engineering field urgently needs a laser welding seam tracking sensor with simple structure, strong anti-interference ability, no complex post-processing algorithm and accurate feature recognition.

In this paper, a line scanning laser welding seam tracking sensor is first developed, and the real-time tracking test of U-shaped, V-shaped and flat-bottom grooves of pipeline welds is carried out. It is considered that the problem of insufficient recognition accuracy of the welding seam width at the bottom of U-shaped grooves is caused by the high reflectivity of the seam groove and the severe arc light. Then, a laser welding seam tracking sensing technology based on a swing mirror is proposed. The strategy of reciprocating the spot light on the weld surface greatly improves the anti-interference ability of the sensor. In the identification test of U-shaped, V-shaped, flat bottom groove, extremely deep and wide weld (40mm depth&30mm width), extremely deep and narrow weld (40mm depth&12mm width), the sensor has good recognition accuracy and stability.

978-1-6654-8699-6/22 $31.00 © 2022 IEEE

II. PRINCIPLE OF LASER WELDING SEAM TRACKING BASED ON SWING MIRROR

A. Line Laser Sensing Technology

As shown in Fig. 1, the line scan laser sensing technology is developed based on the principle of laser triangulation, and has been widely used in weld feature recognition. The process of identifying weld features is as follows:

(1) The laser emits a linear laser beam to the surface of the weld, and the image collector collects grayscale images;

(2) The center position of the laser line is extracted, and the depth and position data of the weld are obtained by calculating the principle of triangulation;

(3) Based on the calculated data, use filtering and feature point recognition algorithms to identify weld feature points

The application of line scanning laser sensor in pipeline weld identification is shown in Fig. 2. It can be seen that the reflection on the surface of the weld and the arc light during the welding process are relatively strong. The feature recognition of the weld are shown in Fig. 3. It can be seen that the V-shaped groove and filled groove (irregular flat bottom) have clear contours and accurate feature recognition, while the U-shaped groove has outlier interference points on the left and right sides of the bottom, resulting in the wrong feature recognition of the left and right sides of the bottom. Taking the bottom right end point as an example, as shown in Fig. 4, for the V-shaped groove, the right end point of the bottom is an obvious intersection of two polylines, and the boundary is clear and easy to identify; while for the U-shaped groove, the bottom right end point is a circle. The arc segment has no obvious intersection point of broken lines, and the bottom is gentler than the V-shaped groove, and the right feature point recognition of the bottom is easily affected by its surrounding area. When facing a highly reflective surface and severe welding arc, the laser grayscale data is easy to be saturated. The influence will be stronger, resulting in the distortion of the weld profile, which in turn affects the weld feature recognition. Even if complex filtering and feature point recognition algorithms are used in the later stage [12], the recognition accuracy will also drop significantly.

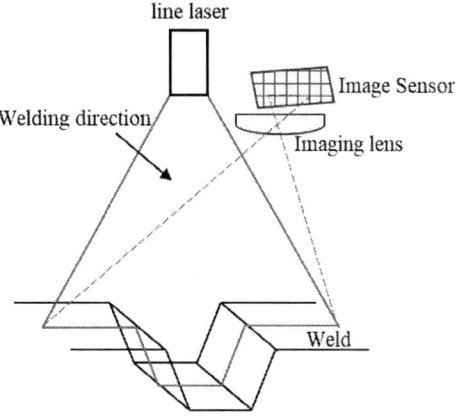

Figure 1. Principle diagram of line laser weld feature recognition (laser triangulation measurement principle)

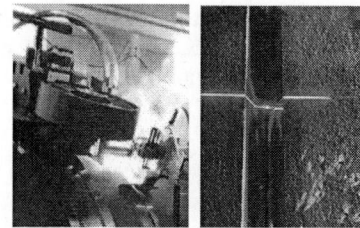

Figure 2. Application of line laser sensor in pipeline weld identification

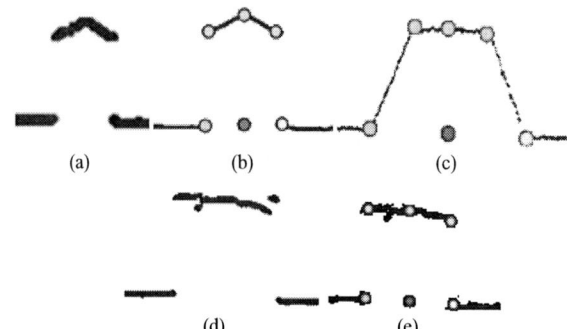

Figure 3. Line laser sensor to identify pipeline weld results (a) V-shaped groove profile (b) V-shaped groove feature points (c) Filled groove (irregular flat bottom) feature points (d) U-shaped groove profile (e) U-shaped groove feature points

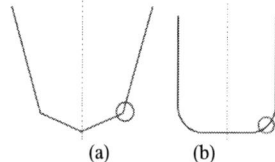

Figure 4. Schematic diagram of the right end point at the bottom of the weld (a) V-shaped groove (b) U-shaped groove

B. Weld Seam Tracking Sensing Technology Based on Swing Mirror

Aiming at the limitation that the line scanning laser cannot be applied to the surface feature recognition of pipeline welds under the conditions of high reflection and strong arc light, a welding seam recognition technology based on swing mirror is proposed, and its principle is shown in Fig. 5. A beam of point light source is emitted by a laser and reflected by a swing mirror driven by a motor to form a periodic reciprocating point-shaped spot within the fan-shaped range of the weld surface. The point-like light spot on the surface of the weld is diffusely reflected, and is collected by the image sensor through the imaging lens. The image sensor and the imaging lens are at the front of the welding direction in order to avoid being blocked. The point-like light spot of different depths and positions occupies different positions on the image collector. According to the positional relationship between the bright spot of the image sensor and the actual depth of the weld seam, the depth and position information of the weld seam at the point-shaped spot is obtained. The motor drives the swing mirror to rotate, so that the point-like light spot scans the entire weld interface, and the spatial coordinates of the point-like light spot on the entire section are calculated accordingly. Compared with the swing laser illumination scheme, the swing mirror scheme only needs to rotate the mirror, while the laser is fixed, which improves the system reliability. The whole system is fixed on the welding machine equipment, and follows the welding equipment to track the welding seam appearance in real time.

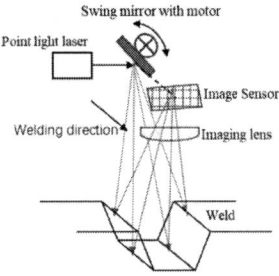

Figure 5. Schematic diagram of laser welding seam tracking based on swing mirror

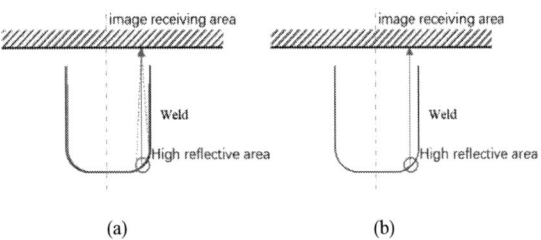

Figure 6. Schematic diagram of the two systems of laser seam tracking sensor for identifying the right end point at the bottom of the U-shaped groove (a) Line laser sensor (b) laser sensor based on swing mirror

Compared with the line laser, the welding seam recognition technology based on the swing mirror uses a point light source, which can achieve higher laser brightness, greatly enhance the relative intensity of the active light source, and make the weld seam characteristic signal have a higher signal-to-noise ratio. It has good anti-interference ability for surface reflection and welding arc light, and does not require complex and time-consuming image processing algorithms. The principle is shown in Fig. 6. For the line scanning laser sensor, under the conditions of strong reflection and arc light, the stray light reflected by the strong reflection area will easily lead to the saturation of the gray level of the identification area of the right end point, which affects the recognition of feature points. For the laser sensor, only the right end recognition area has a light spot, and the stray light is greatly weakened, which makes it have better anti-interference ability.

III. MEASURED RESULTS OF LASER WELDING SEAM TRACKING SENSOR BASED ON SWING MIRROR

A. Test Results of Different Groove Forms

Use a laser weld seam tracking sensor based on a swing mirror to conduct pipeline weld seam identification testing. The test site is shown in Fig. 7. The sensor is installed at the front of the welding equipment in the welding direction, and moves around the pipeline with the welding equipment.

During the test, the sensor faces strong interference such as welding arc light and reflection on the groove surface.

A total of U-shaped, V-shaped grooves of pipeline welds, and flat-bottom grooves after the first pass welding were tested in the testing. The results showed that the welding seam was recognized, stable, and accurate to meet the needs of guiding the movement of the welding torch.

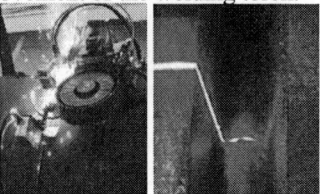

Figure 7. Experiment of laser weld seam tracking sensor based on swing mirror to identify pipeline weld seam

Figure 8. Weld contour identification (a) V-shaped groove (b) U-shaped groove

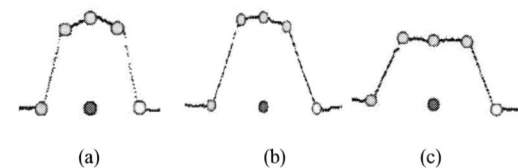

Figure 9. Weld feature point identification (a) V-shaped groove (b) U-shaped groove (c) Filled groove (irregular flat bottom)

The V-shaped and U-shaped (slightly misaligned) weld profiles are shown in Fig. 8. It can be seen that the weld profiles obtained by scanning are clear, and are basically not affected by discrete interference points under the conditions of high reflection and strong arc light. The feature point identification of different weld topography is shown in Fig. 9. The feature identification is carried out for V-shaped, U-shaped (slightly misaligned), and multi-layer multi-pass welding with irregular flat bottom grooves. The results show a variety of weld topography. Feature points can be effectively identified: the midpoint of the weld, the left and right endpoints of the weld, the midpoint of the bottom of the weld, and the left and right endpoints of the bottom of the weld. According to the position of the feature point and the depth coordinates, the weld width and depth features here are obtained. The feature data according to the identification are shown in Table I.

TABLE I. WELD FEATURE IDENTIFICATION DATA

Groove Form	Weld Feature		
	Weld width(mm)	Weld depth(mm)	Weld bottom width(mm)
V-groove	7.789	14.643	4.531
U-groove	13.289	17.733	6.598

Groove Form	Weld Feature		
	Weld width(mm)	*Weld depth(mm)*	*Weld bottom width(mm)*
Fill groove (irregular flat bottom)	7.779	7.066	4.890

B. Limit Width and Depth Weld Test Results

The actual machine test is carried out on the limit width and depth of the weld, and the test results are shown in Fig. 10, for weld with 40mm depth&30mm width, and weld with 40mm depth&12mm width, the two extreme position welds, the laser weld tracking sensor based on the swing mirror can accurately identify the weld outline and extract the seam feature accurately. The limit size weld feature recognition data is shown in Table II. This sensor can accurately recognition the feature data in these limit situation.

TABLE II. LIMIT SIZE WELD FEATURE RECOGNITION DATA

Groove Form	Weld Feature		
	Weld width(mm)	*Weld depth(mm)*	*Weld bottom width(mm)*
40mm weld depth 30mm weld width	30.179	40.398	5.923
40mm weld depth 12mm weld width	11.970	40.673	5.790

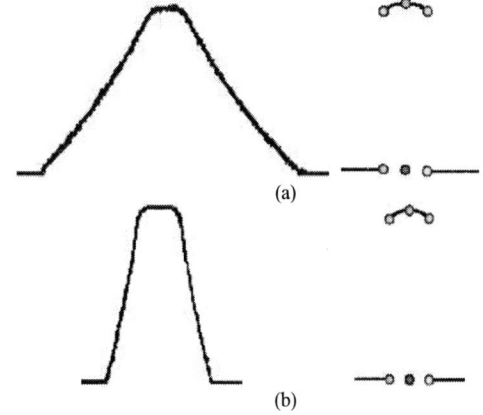

Figure 10. Weld feature identification (a) 40mm weld depth + 30mm weld width (b) 40mm weld depth + 12mm weld width

IV. CONCLUSION

Since the reflection on the surface of the pipeline weld and strong welding arc has greatly interference with the laser weld tracking, this paper proposes a laser weld tracking sensor technology based on a swing mirror, which introduces a point light source, which is reflected by the swing mirror to form a reciprocating motion on the weld surface. For the moving point-shaped light spots, the image collector collects and connects the point-shaped light spots in a cycle to form a cross-sectional image of the weld surface in the cycle. Compared with the traditional line scanning laser welding seam tracking device, due to the use of point-shaped light spots, the relative light intensity of the spot area is higher, and the signal-to-noise ratio of the collected welding seam profile characteristic signals is higher, which is beneficial to the welding reflection and arc light in the pipeline. In more severe cases, the stable and accurate acquisition of weld features can be achieved. The sensor can accurately measure the shape of V-shaped, U-shaped and flat-bottomed grooves, extremely deep and wide weld (40mm depth&30mm width), and extremely deep and narrow weld (40mm depth&12mm width). This system can provide a beneficial industrial application for laser seam tracking. In the future, a special feature extraction algorithm can be developed for the laser seam tracking sensor based on the swing mirror to achieve faster and more robust feature extraction.

REFERENCES

[1] YIN Tie, ZHAO Hong, ZHANG Qian, WU Tingting, ZHOU Lun, WANG Xinsheng. Current situation and development of welding robots for long distance oil and gas pipelines[J]. Petroleum Science Bulletin, 2021, 01:145-157.doi:10.3969/j.issn.2096-1693.2021.01.012.

[2] HU Sheng, LUO Yu, JIAO Xiangdong, et al. Research status of seam tracking methods for orbital pipe welding robots[J]. Journal of Hebei University of Science and Technology, 2014, 36(2):126-133.

[3] Zhang Gong, Tuo Shuaihua, Cao Xuepeng, et al. The state-of-the-art and developing trends of weld seam tracking technology of welding robot[J]. Science Technology and Engineering, 2021, 21(10) : 3868-3876.

[4] Liu W. J., Guan Z. Y., Jiang X., et al. Research on the seam tracking of narrow gap P-GMAW based on arc sound sensing[J]. Sensors and ActuatorsA:Physical,2019,292:205-216, doi: 10. 1016/ j. sna. 2019. 04. 015.

[5] Hu Sheng, Luo Yu, Jiao Xiangdong, Zhou Canfeng, Zhang Zhongliang. Swing Arc Sensor and its Signal Processing[J]. Journal of Shanghai Jiaotong University, 2016, 50(S1):59-61+65. doi:10.16183/j.cnki.jsjtu.2016.S.015.

[6] Li, W., Gao, K., Wu, J. et al. Groove sidewall penetration modeling for rotating arc narrow gap MAG welding[J]. Int J Adv Manuf Technol 78, 573–581 (2015). doi: 10.1007/s00170-014-6678-6.

[7] Hong Bo, Liu Jian, Hong Yuxiang, Wang Qian. Study on deviation prediction of seam tracking using magnetron rotating arc sensor with Kalman filter[J]. Transactions of The China Welding Institution,2015,36(05):55-58+116.

[8] Wang Zuoshan. Research on Seam Tracking Technology Based on Laser Structured Light Vision Guidance[D]. Shandong University, 2019:3-5.

[9] Wang Zhijiang, Xue Kunxi, Wu Dingyong, et al . Robotic weld seam correction control system based on visual sensing Journal of Mechanical Engineering[J], 2019, 55(17) : 48-55.

[10] Borer D, Delbruck T, osgen T. Three-dimensional particle tracking velocimetry using dynamic vision sensors. Experiments in Fluids, 2017, 58(12) : 165.

[11] Zhao C. Y., Yang J. N., Zhou F. Q., et al. A robust laser stripe extraction method for structured-light vision sensing[J]. Sensors, 2020, 20(16) : 4544.

[12] Zhang Shikuan, Wu Qingxiao, Lin Zhiyuan, Detection and Segmentation of Structured Light Stripe in Weld Image[J]. Acta Optica Sinica, ,2021,41(05):88-96.

Author Index

Abdulaziz Al-Amodi	50	Dan Luo	142
Ai Na Gong	110	Daofeng He	239
Alessio Corazza	163	Deng Loulou	191
An Yan	41, 227	Dongpei Shen	70
Aolong Sun	41, 227	Duanming Li	105
Aoyan Zhang	133	Enea Rizzi	163
Baikui Li	183	Evgenii Kuznetsov	1
Baonan Jia	70	F. Lei	265
Baoquan Jin	96, 137	Fang Zhao	115
Beilei Wu	86	Fei Liu	124
Bin Chen	210	Fei Xu	245
Bin Liang	137	Feihong Zhang	183
Bin Wang	239	Fen Zhang	177
Bingkun Zhou	55	Feng Xu	245
Bowen Lu	171	Fuyin Wang	235
Boyao Li	119	G. H. Zhang	265
Caili Gong	30	Gangxiang Shen	63, 210
Caoyang Liu	63	Giovanni Zafarana	163
Changming Xia	22	Gordon Ning Liu	45, 63, 210
Changyuan Yu	115, 149	Guanghui Wang	245
Chao Li	222	Gui Yao Zhou	15
Chao Shen	41	Guiyao Zhou	6, 22, 133
Chao Zuo	270	Guo Zhu	124
Chaodan Chi	158, 167	Guoqiang Li	41
Chaoyu Xu	101	Guoying Liu	248
Chen Ji	158, 167	H. F. Guan	265
Chen Weiting	75	Haibin Su	248
Chen Xueye	25	Haizhi Song	70
Chen Zhao	55	Han Liu	195
Chenkun Ge	200	Hanyu Zhao	124
Chenlong Xue	248	Hao Yang	214
Chi Liu	110	Haonan Zheng	274
Chuan Shen	239	Hengyu Wang	231
Chunmei Ouyang	253, 261	Hong Cheng	177, 274
Chunran Sun	86	Hong Dang	248
Chunyan Cao	235	Hongcan Gu	81
Dan Lu	248	Honglei Wu	183

Hongqi Han	171	Kai Shen	19
Hongyi Li	253, 261	Kan Wu	222
Houling Ji	218	Kangqi Zhu	55
Hua Baocheng	191	Kong Delong	75
Huanhuan Liu	142, 248	Kwai Hei Li	154
HuiMin Lu	195	L. Wang	265
Igor Savelev	1	Lei Kaiyu	191
J. C. Yuan	265	Lei Yue	171
Jia ao Lu	22	Li Xiaoyu	25
Jiahao Yin	154	Li Yahong	25
Jiahui Zhang	245	Libo Yuan	91
Jiajun Ma	253, 261	Lili Kong	145
Jialiang Chen	30	Lin Sun	45, 63, 210
Jialong Li	55, 133	Liu jing	191
Jian Tang	36	Liu Qihai	191
Jianbo Tang	205	Liu Wang	239
Jiangfei Hu	105	Liu Weiren	75
Jianhua Mo	200	Liu Xingtan	191
Jianping Chen	222	Liyang Shao	115, 142
JianPing Wang	195	Longqing Cong	248
Jiantao Liu	22	Lu Xiangxiang	75
Jianyang Shi	41	Luca Mauri	163
Jiaqi Hu	248	Luxiao Zhang	210
Jiawang Xiao	63	M. Z. M. Khan	50
Jiaxuan Long	222	Marco Moraja	163
Jiayu Lu	145	Meng Zou	19
Jiayuan Min	142	Ming Chen	91
Jie Hu	115, 142	Minghui Niu	142
Jie Liang	257	Mingshuai Li	200
Jindong Wang	128	Minxue Gu	105
Jing Li	91	Mudassir Masood	50
Jingwen Li	101	Nan Chi	41, 227
Jinhui Yuan	124	Nan Hua	55
Jinna Chen	248	Ning Cui	96
Junbin Huang	81	Pei Dongliang	75
Junhong Su	10	Pengfei Lu	70
Junhong Wang	137	Pengfei Zhu	70
Junqi Xu	10	Penghao Luo	41, 227
Junwen Zhang	41, 227	Perry Ping Shum	115, 133, 142, 248
Junxiong Zhou	205	Q. Y. Liu	265

Qiancheng Zhao	218	W. Chen	265
Qianqian Wang	257	W. Sun	265
Qing Bai	96, 137	Wang Jieying	75
Qiong Yao	235	Wanshu Xiong	158, 167
Qixiang Gao	187	Wei Junxin	75
Qizhen Sun	19	Wei Li	177
Runnan Zhang	270	Wei Zhang	105
Runze Li	257	Weichao Ma	6, 22
Ruoyun Yao	158, 167	Weihao Lin	115
Sergii Golovynskyi	183	Weihao Yuan	149
Shaonan Zheng	231	Weiren Cheng	218
Shilei Liu	253, 261	Wu Xiuyu	191
Shiying Xiao	86	Xia Ji	235
Shizhuo Xi	15	Xian Zhou	124
Shuidong Xiong	235	Xianyi Cao	222
Siming Sun	115	Xiaojun Liu	124
Siwei Sun	274	Xiaojun Yu	200
Sizhe Xing	41	Xiaoning Guan	70
Song Xuhao	278	Xiaoping Zheng	55
Song Zhou	245	Xiaosheng Xiao	145
Songnian Fu	101	Xiaotong Li	257
Su Wu	81	Xiawei Feng	171
Sui Wei	239	Ximin Wang	30
Sumei Huang	91	Xing Zhong	187
Suotao Dong	66	Xingyan Zhao	231
Tao Liqing	191	Xiuhua Fu	66
Tao Shen	110	Xunting Yang	177
Tao Yang	60	Y. L. Wang	265
Tao Zhu	128	Y. T. Li	265
Tatiana Soloveva	1	Yan Li	137
Tian Changyong	278	Yandong Pang	81
Tianrong Huang	119	Yang Li	10
Tianye Huang	101	Yanhe Li	55
Tianyi Li	222	Yanqun Xiang	248
Tianyu Yang	110	Yi Cai	63, 210
Tielin Lu	171	Yi Li	218
Ting Hu	231	Yi Liu	253, 261
Tong Sun	70	Yifan Zhang	6, 15
Tong Wu	22	Yifei Zhao	6, 15
Tongle Yuan	91	Yin Tie	278

Yingxi Miao	30
Yiti Xiong	158, 167
Yongfeng Wei	30
Youchao Jiang	86
Yu Chen	177
Yu Cheng	91
Yu Li	187
Yu Wang	96, 137
Yuan Dong	231
Yuanhang Wang	187
Yue Yuan	110
Yumeng Luo	154
Yuntao Li	205
Yuqing Yang	30
Yury Golyaev	1
Yury Kolbas	1
Yuting Liu	96
Yuzhong Ma	45
Z. Jiang	265
Zeng Xiangwei	25
Zewei Cai	270
Zhang Quanzhong	25
Zhang Yi	278
Zhangwan Peng	158, 167
Zhaoting Geng	218

Zhe Yu	101
Zhen Gao	248
Zheng Liu	10
Zhengguang Tian	177
Zhengji Xu	231
Zhenhua Sun	183
Zhenming Liang	36
Zhichao Wu	101
Zhijie Sun	96
Zhijun Yan	19
Zhipeng Deng	133
Zhiqiang Zhang	81
Zhiyuan Wang	128, 183
Zhiyun Hou	22
Zhuang Wang	239
Zhuoyu Yu	218
Zhuoyuan Huang	22
Zijing Huang	45
Ziwei Li	41
Zixiao Wang	86
Ziyan Zhao	142
Ziyao Yang	36
Zongru Yang	149
Zuqing Zhu	214

IEEE
445 Hoes Lane
Piscataway, NJ 08854-4141

ISBN 978-1-6654-8699-6